Communications and Control Engineering

For further volumes:
www.springer.com/series/61

Huaguang Zhang · Derong Liu · Yanhong Luo ·
Ding Wang

Adaptive Dynamic Programming for Control

Algorithms and Stability

Springer

Huaguang Zhang
College of Information Science Engin.
Northeastern University
Shenyang
People's Republic of China

Yanhong Luo
College of Information Science Engin.
Northeastern University
Shcnyang
People's Republic of China

Derong Liu
Institute of Automation, Laboratory of Complex Systems
Chinese Academy of Sciences
Beijing
People's Republic of China

Ding Wang
Institute of Automation, Laboratory of Complex Systems
Chinese Academy of Sciences
Beijing
People's Republic of China

ISSN 0178-5354 Communications and Control Engineering
ISBN 978-1-4471-4756-5 ISBN 978-1-4471-4757-2 (eBook)
DOI 10.1007/978-1-4471-4757-2
Springer London Heidelberg New York Dordrecht

Library of Congress Control Number: 2012955288

Printed on acid-free paper

Springer is part of Springer Science+Business Media (www.springer.com)

Preface

Background of This Book

Optimal control, once thought of as one of the principal and complex domains in the control field, has been studied extensively in both science and engineering for several decades. As is known, dynamical systems are ubiquitous in nature and there exist many methods to design stable controllers for dynamical systems. However, stability is only a bare minimum requirement in the design of a system. Ensuring optimality guarantees the stability of nonlinear systems. As an extension of the calculus of variations, optimal control theory is a mathematical optimization method for deriving control policies. Dynamic programming is a very useful tool in solving optimization and optimal control problems by employing the principle of optimality. However, it is often computationally untenable to run true dynamic programming due to the well-known "curse of dimensionality". Hence, the adaptive dynamic programming (ADP) method was first proposed by Werbos in 1977. By building a system, called "critic", to approximate the cost function in dynamic programming, one can obtain the approximate optimal control solution to dynamic programming. In recent years, ADP algorithms have gained much attention from researchers in control fields. However, with the development of ADP algorithms, more and more people want to know the answers to the following questions:

(1) Are ADP algorithms convergent?
(2) Can the algorithm stabilize a nonlinear plant?
(3) Can the algorithm be run on-line?
(4) Can the algorithm be implemented in a finite time horizon?
(5) If the answer to the first question is positive, the subsequent questions are where the algorithm converges to, and how large the error is.

Before ADP algorithms can be applied to real plants, these questions need to be answered first. Throughout this book, we will study all these questions and give specific answers to each question.

Why This Book?

Although lots of monographs on ADP have appeared, the present book has unique features, which distinguish it from others.

First, the types of system involved in this monograph are rather extensive. From the point of view of models, one can find affine nonlinear systems, non-affine nonlinear systems, switched nonlinear systems, singularly perturbed systems and time-delay nonlinear systems in this book; these are the main mathematical models in the control fields.

Second, since the monograph is a summary of recent research works of the authors, the methods presented here for stabilizing, tracking, and games, which to a great degree benefit from optimal control theory, are more advanced than those appearing in introductory books. For example, the dual heuristic programming method is used to stabilize a constrained nonlinear system, with convergence proof; a data-based robust approximate optimal controller is designed based on simultaneous weight updating of two networks; and a single network scheme is proposed to solve the non-zero-sum game for a class of continuous-time systems.

Last but not least, some rather unique contributions are included in this monograph. One notable feature is the implementation of finite horizon optimal control for discrete-time nonlinear systems, which can obtain suboptimal control solutions within a fixed finite number of control steps. Most existing results in other books discuss only the infinite horizon control, which is not preferred in real-world applications. Besides this feature, another notable feature is that a pair of mixed optimal policies is developed to solve nonlinear games for the first time when the saddle point does not exist. Meanwhile, for the situation that the saddle point exists, existence conditions of the saddle point are avoided.

The Content of This Book

The book involves ten chapters. As implied by the book title, the main content of the book is composed of three parts; that is, optimal feedback control, nonlinear games, and related applications of ADP. In the part on optimal feedback control, the edge-cutting results on ADP-based infinite horizon and finite horizon feedback control, including stabilization control, and tracking control are presented in a systematic manner. In the part on nonlinear games, both zero-sum game and non-zero-sum games are studied. For the zero-sum game, it is proved for the first time that the iterative policies converge to the mixed optimal solutions when the saddle point does not exist. For the non-zero-sum game, a single network is proposed to seek the Nash equilibrium for the first time. In the part of applications, a self-learning call admission control scheme is proposed for CDMA cellular networks, and meanwhile an engine torque and air-fuel ratio control scheme is studied in detail, based on ADP.

In Chap. 1, a brief introduction to the background and development of ADP is provided. The review begins with the origin of ADP, and the basic structures and algorithm development are narrated in chronological order. After that, we turn attention to control problems based on ADP. We present this subject regarding two aspects: feedback control based on ADP and nonlinear games based on ADP. We mention a few iterative algorithms from recent literature and point out some open problems in each case.

In Chap. 2, the optimal state feedback control problem is studied based on ADP for both infinite horizon and finite horizon. Three different structures of ADP are utilized to solve the optimal state feedback control strategies, respectively. First, considering a class of affine constrained systems, a new DHP method is developed to stabilize the system, with convergence proof. Then, due to the special advantages of GDHP structure, a new optimal control scheme is developed with discounted cost functional. Moreover, based on a least-square successive approximation method, a series of GHJB equations are solved to obtain the optimal control solutions. Finally, a novel finite-horizon optimal control scheme is developed to obtain the suboptimal control solutions within a fixed finite number of control steps. Compared with the existing results in the infinite-horizon case, the present finite-horizon optimal controller is preferred in real-world applications.

Chapter 3 presents some direct methods for solving the closed-loop optimal tracking control problem for discrete-time systems. Considering the fact that the performance index functions of optimal tracking control problems are quite different from those of optimal state feedback control problems, a new type of performance index function is defined. The methods are mainly based on iterative HDP and GDHP algorithms. We first study the optimal tracking control problem of affine nonlinear systems, and after that we study the optimal tracking control problem of non-affine nonlinear systems. It is noticed that most real-world systems need to be effectively controlled within a finite time horizon. Hence, based on the above results, we further study the finite-horizon optimal tracking control problem, using the ADP approach in the last part of Chap. 3.

In Chap. 4, the optimal state feedback control problems of nonlinear systems with time delays are studied. In general, the optimal control for time-delay systems is an infinite-dimensional control problem, which is very difficult to solve; there are presently no good methods for dealing with this problem. In this chapter, the optimal state feedback control problems of nonlinear systems with time delays both in states and controls are investigated. By introducing a delay matrix function, the explicit expression of the optimal control function can be obtained. Next, for nonlinear time-delay systems with saturating actuators, we further study the optimal control problem using a non-quadratic functional, where two optimization processes are developed for searching the optimal solutions. The above two results are for the infinite-horizon optimal control problem. To the best of our knowledge, there are no results on the finite-horizon optimal control of nonlinear time-delay systems. Hence, in the last part of this chapter, a novel optimal control strategy is developed to solve the finite-horizon optimal control problem for a class of time-delay systems.

In Chap. 5, the optimal tracking control problems of nonlinear systems with time delays are studied using the HDP algorithm. First, the HJB equation for discrete time-delay systems is derived based on state error and control error. Then, a novel iterative HDP algorithm containing the iterations of state, control law, and cost functional is developed. We also give the convergence proof for the present iterative HDP algorithm. Finally, two neural networks, i.e., the critic neural network and the action neural network, are used to approximate the value function and the corresponding control law, respectively. It is the first time that the optimal tracking control problem of nonlinear systems with time delays is solved using the HDP algorithm.

In Chap. 6, we focus on the design of controllers for continuous-time systems via the ADP approach. Although many ADP methods have been proposed for continuous-time systems, a suitable framework in which the optimal controller can be designed for a class of general unknown continuous-time systems still has not been developed. In the first part of this chapter, we develop a new scheme to design optimal robust tracking controllers for unknown general continuous-time nonlinear systems. The merit of the present method is that we require only the availability of input/output data, instead of an exact system model. The obtained control input can be guaranteed to be close to the optimal control input within a small bound. In the second part of the chapter, a novel ADP-based robust neural network controller is developed for a class of continuous-time non-affine nonlinear systems, which is the first attempt to extend the ADP approach to continuous-time non-affine nonlinear systems.

In Chap. 7, several special optimal feedback control schemes are investigated. In the first part, the optimal feedback control problem of affine nonlinear switched systems is studied. To seek optimal solutions, a novel two-stage ADP method is developed. The algorithm can be divided into two stages: first, for each possible mode, calculate the associated value function, and then select the optimal mode for each state. In the second and third parts, the near-optimal controllers for nonlinear descriptor systems and singularly perturbed systems are solved by iterative DHP and HDP algorithms, respectively. In the fourth part, the near-optimal state-feedback control problem of nonlinear constrained discrete-time systems is solved via a single network ADP algorithm. At each step of the iterative algorithm, a neural network is utilized to approximate the costate function, and then the optimal control policy of the system can be computed directly according to the costate function, which removes the action network appearing in the ordinary ADP structure.

Game theory is concerned with the study of decision making in a situation where two or more rational opponents are involved under conditions of conflicting interests. In Chap. 8, zero-sum games are investigated for discrete-time systems based on the model-free ADP method. First, an effective data-based optimal control scheme is developed via the iterative ADP algorithm to find the optimal controller of a class of discrete-time zero-sum games for Roesser type 2-D systems. Since the exact models of many 2-D systems cannot be obtained inherently, the iterative ADP method is expected to avoid the requirement of exact system models. Second, a data-based optimal output feedback controller is developed for solving the zero-sum games of a class of discrete-time systems, whose merit is that knowledge of the model of the system is not required, nor the information of system states.

In Chap. 9, nonlinear game problems are investigated for continuous-time systems, including infinite horizon zero-sum games, finite horizon zero-sum games and non-zero-sum games. First, for the situations that the saddle point exists, the ADP technique is used to obtain the optimal control pair iteratively. The present approach makes the performance index function reach the saddle point of the zero-sum differential games, while complex existence conditions of the saddle point are avoided. For the situations that the saddle point does not exist, the mixed optimal control pair is obtained to make the performance index function reach the mixed optimum. Then, finite horizon zero-sum games for a class of nonaffine nonlinear systems are studied. Moreover, besides the zero-sum games, the non-zero-sum differential games are studied based on single network ADP algorithm. For zero-sum differential games, two players work on a cost functional together and minimax it. However, for non-zero-sum games, the control objective is to find a set of policies that guarantee the stability of the system and minimize the individual performance function to yield a Nash equilibrium.

In Chap. 10, the optimal control problems of modern wireless networks and automotive engines are studied by using ADP methods. In the first part, a novel learning control architecture is proposed based on adaptive critic designs/ADP, with only a single module instead of two or three modules. The choice of utility function for the present self-learning control scheme makes the present learning process much more efficient than existing learning control methods. The call admission controller can perform learning in real time as well as in off-line environments, and the controller improves its performance as it gains more experience. In the second part, an ADP-based learning algorithm is designed according to certain criteria and calibrated for vehicle operation over the entire operating regime. The algorithm is optimized for the engine in terms of performance, fuel economy, and tailpipe emissions through a significant effort in research and development and calibration processes. After the controller has learned to provide optimal control signals under various operating conditions off-line or on-line, it is applied to perform the task of engine control in real time. The performance of the controller can be further refined and improved through continuous learning in real-time vehicle operations.

Acknowledgments

The authors would like to acknowledge the help and encouragement they received during the course of writing this book. A great deal of the materials presented in this book is based on the research that we conducted with several colleagues and former students, including Q.L. Wei, Y. Zhang, T. Huang, O. Kovalenko, L.L. Cui, X. Zhang, R.Z. Song and N. Cao. We wish to acknowledge especially Dr. J.L. Zhang and Dr. C.B. Qin for their hard work on this book. The authors also wish to thank Prof. R.E. Bellman, Prof. D.P. Bertsekas, Prof. F.L. Lewis, Prof. J. Si and Prof. S. Jagannathan for their excellent books on the theory of optimal control and adaptive dynamic programming. We are very grateful to the National Natural Science Foundation of China (50977008, 60904037, 61034005, 61034002, 61104010), the Science and Technology Research Program of The Education Department of Liaoning Province (LT2010040), which provided necessary financial support for writing this book.

Shenyang, China
Beijing, China
Chicago, USA

Huaguang Zhang
Derong Liu
Yanhong Luo
Ding Wang

Contents

Chapter 1
Overview

1.1 Challenges of Dynamic Programming

As is known, there are many methods to design stable controllers for non-linear systems. However, stability is only a bare minimum requirement in system design. Ensuring optimality guarantees the stability of the non-linear system. However, optimal control of non-linear systems is a difficult and challenging topic [8]. *Dynamic programming* is a very useful tool in solving optimization and optimal control problems by employing the principle of optimality. In particular, it can easily be applied to non-linear systems with or without constraints on the control and state variables. In [13], the principle of optimality is expressed as: "An optimal policy has the property that, whatever the initial state and initial decision are, the remaining decisions must constitute an optimal policy with regard to the state resulting from the first decision." There are several options for dynamic programming. One can consider discrete-time systems or continuous-time systems, linear systems or non-linear systems, time-invariant systems or time-varying systems, deterministic systems or stochastic systems, etc.

We first take a look at discrete-time non-linear (time-varying) dynamical (deterministic) systems. Time-varying non-linear systems cover most of the application areas and a discrete time is the basic consideration for digital computation. Suppose that one is given a discrete-time non-linear (time-varying) dynamical system

$$x(k+1) = F\left[x(k), u(k), k\right], k = 0, 1, \ldots, \tag{1.1}$$

where $x(k) \in \mathbb{R}^n$ represents the state vector of the system and $u(k) \in \mathbb{R}^m$ denotes the control action. Suppose that the cost functional that is associated with this system is

$$J\left[x(i), i\right] = \sum_{k=i}^{\infty} \gamma^{k-i} l\left[x(k), u(k), k\right], \tag{1.2}$$

where l is called the *utility function* and γ is the discount factor with $0 < \gamma \le 1$. Note that the functional J is dependent on the initial time i and the initial state

H. Zhang et al., *Adaptive Dynamic Programming for Control*,
Communications and Control Engineering, DOI 10.1007/978-1-4471-4757-2_1,

$x(i)$, and it is referred to as the cost-to-go of state $x(i)$. The objective of dynamic programming problem is to choose a control sequence $u(k), k = i, i+1, \ldots$, so that the functional J (i.e., the cost) in (1.2) is minimized. According to Bellman, the optimal value function is equal to

$$J^*(x(k)) = \min_{u(k)} \left\{ l(x(k), u(k)) + \gamma J^*(x(k+1)) \right\}. \tag{1.3}$$

The optimal control $u^*(k)$ at time k is the $u(k)$ which achieves this minimum, i.e.,

$$u^*(k) = \arg\min_{u(k)} \left\{ l(x(k), u(k)) + \gamma J^*(x(k+1)) \right\}. \tag{1.4}$$

Equation (1.3) is the principle of optimality for discrete-time systems. Its importance lies in the fact that it allows one to optimize over only one control vector at a time by working backward in time.

In the non-linear continuous-time case, the system can be described by

$$\dot{x}(t) = F[x(t), u(t), t], t \geq t_0. \tag{1.5}$$

The cost functional in this case is defined as

$$J(x(t), u) = \int_t^{\infty} l(x(\tau), u(\tau)) d\tau. \tag{1.6}$$

For continuous-time systems, Bellman's principle of optimality can be applied, too. The optimal value function $J^*(x_0) = \min J(x_0, u(t))$ will satisfy the Hamilton–Jacobi–Bellman equation,

$$\begin{aligned} -\frac{\partial J^*(x(t))}{\partial t} &= \min_{u \in U} \left\{ l(x(t), u(t), t) + \left(\frac{\partial J^*(x(t))}{\partial x(t)} \right)^{\mathrm{T}} \right. \\ &\qquad \left. \times F(x(t), u(t), t) \right\} \\ &= l\left(x(t), u^*(t), t\right) + \left(\frac{\partial J^*(x(t))}{\partial x(t)} \right)^{\mathrm{T}} \\ &\qquad \times F\left(x(t), u^*(t), t\right). \end{aligned} \tag{1.7}$$

Equations (1.3) and (1.7) are called the optimality equation of dynamic programming which are the basis for computer implementation of dynamic programming. In the above, if the function F in (1.1), (1.5) and the cost functional J in (1.2), (1.6) are known, obtaining the solution of $u(t)$ becomes a simple optimization problem. If the system is modeled by linear dynamics and the cost functional to be minimized is quadratic in the state and control, then the optimal control is a linear feedback of the states, where the gains are obtained by solving a standard Riccati equation [56]. On the other hand, if the system is modeled by the non-linear dynamics or the cost functional is non-quadratic, the optimal state feedback control will depend upon obtaining the solution to the *Hamilton–Jacobi–Bellman (HJB) equation*, which is generally a non-linear partial differential equation or difference equation [58]. However, it is often computationally untenable to run true dynamic programming due to the backward numerical process required for its solutions, i.e., as a result of the well-known "curse of dimensionality" [13, 30]. In [75], three curses are displayed in resource management and control problems to show that the optimal value function J^*, i.e., the theoretical solution of the HJB equation is very difficult to obtain, except for systems satisfying some very good conditions.

1.2 Background and Development of Adaptive Dynamic Programming

Over the last 30 years, progress has been made to circumvent the "*curse of dimensionality*" by building a system, called "critic," to approximate the cost function in dynamic programming (cf. [68, 76, 81, 97, 99, 100]). The idea is to approximate dynamic programming solutions by using a function approximation structure such as neural networks to approximate the cost function. The earliest research refers to reference [96] in 1977, where Werbos introduced an approach for ADP that was later called *adaptive critic designs (ACDs)*. Then, *adaptive dynamic programming (ADP)* algorithms gained much attention from a lot of researchers, cf. [1, 3, 4, 7, 9, 15, 24, 26, 33, 34, 39, 54, 60–63, 68, 76, 80, 83, 85, 95, 99–102, 104, 105]. In the literature, there are several synonyms used for "Adaptive Critic Designs" [29, 46, 50, 62, 76, 92], including "Approximate Dynamic Programming" [86, 100], "Asymptotic Dynamic Programming" [79], "Adaptive Dynamic Programming" [68, 69], "Heuristic Dynamic Programming" [54, 98], "Neuro-Dynamic Programming" [15], "Neural Dynamic Programming" [86, 106], and "Reinforcement Learning" [87].

In [15], Bertsekas and Tsitsiklis gave an overview of neuro-dynamic programming. They provided the background, gave a detailed introduction to dynamic programming, discussed the neural-network architectures and methods for training them, and developed general convergence theorems for stochastic approximation methods as the foundation for the analysis of various neuro-dynamic programming algorithms. They provided the core neuro-dynamic programming methodology, including many mathematical results and methodological insights. They suggested many useful methodologies to apply in neuro-dynamic programming, like Monte Carlo simulation, on-line and off-line temporal difference methods, Q-learning algorithm, optimistic policy iteration methods, Bellman error methods, approximate linear programming, approximate dynamic programming with cost-to-go function, etc. Particularly impressive successful, greatly motivating subsequent research, was the development of a backgammon playing program by Tesauro [88]. Here a neural network was trained to approximate the optimal cost-to-go function of the game of backgammon by using simulation, that is, by letting the program play against itself.

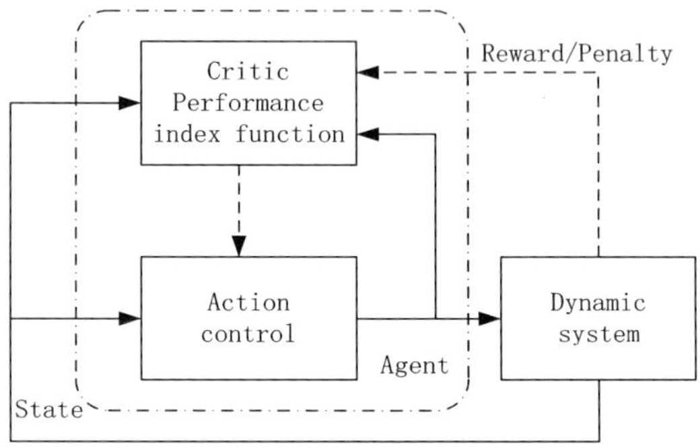

Fig. 1.1 Learning from the environment

Unlike chess programs, this program did not use look-ahead of many steps, so its success can be attributed primarily to the use of a properly trained approximation of the optimal cost-to-go function.

1.2.1 Basic Structures of ADP

To implement the ADP algorithm, Werbos [100] proposed a means to get around this numerical complexity by using "approximate dynamic programming" formulations. His methods approximate the original problem with a discrete formulation. A solution to the ADP formulation is obtained through a neural-network-based adaptive critic approach. The main idea of ADP is shown in Fig. 1.1.

Specifically, Werbos proposed two basic structure of ADP, which are heuristic dynamic programming (HDP) and dual heuristic programming (DHP).

1.2.1.1 Heuristic Dynamic Programming (HDP)

HDP is the most basic and widely applied structure of ADP [10, 42, 82, 98, 121]. The structure of HDP is shown in Fig. 1.2. In HDP, the critic network will give an estimation of the cost function J, which is guaranteed to be a Lyapunov function, at least for deterministic systems. Lyapunov stability theory in general has hugely influenced control theory, physics, and many other disciplines. Within the disciplines of control theory and robotics, many researchers have tried to stabilize complex systems by first deriving Lyapunov functions for those systems. In some cases, the Lyapunov functions have been derived analytically by solving the multiperiod optimization problem in an analytic fashion.

In the presented HDP structure, there are two critic networks. During the ADP algorithm based on HDP, there are two iteration loops, i.e., an outer iteration loop and an inner iteration loop. The weights of critic network 1 are updated at each outer loop iteration step, and the weights of critic network 2 are updated at each inner loop iteration step. During the inner loop iteration, the weights of critic network 1 are kept

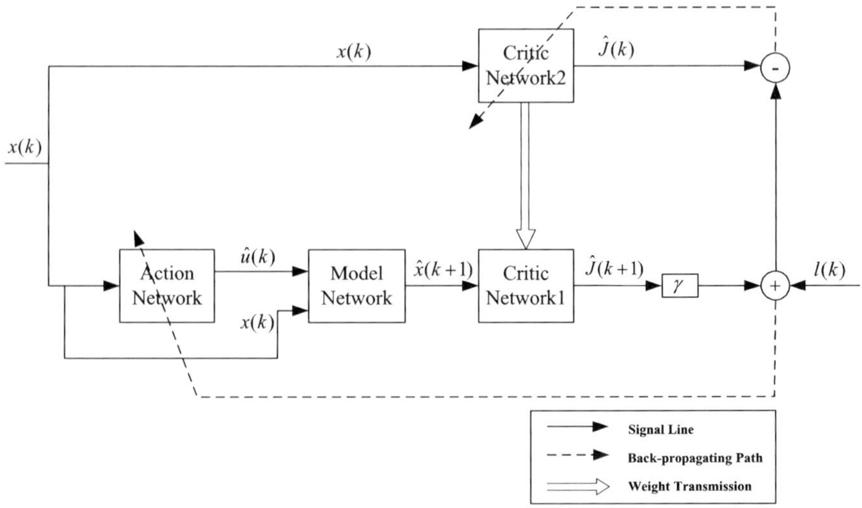

Fig. 1.2 The HDP structure diagram

unchanged. Once the whole inner loop iteration process is finished, the weights of critic network 2 are transferred to critic network 1. The output of critic network 2 is $\hat{J}$, which is the estimate of J in (1.2). This is done by minimizing the following square tracking error measure over time:

$$\|E_h\| = \sum_k E_h(k) = \frac{1}{2}\sum_k \left[\hat{J}(k) - l(k) - \gamma \hat{J}(k+1)\right]^2, \tag{1.8}$$

where $\hat{J}(k) = \hat{J}\left[x(k), u(k), k, W_C, \right]$ and W_C represents the parameters of the critic network. When $E_h(k) = 0$ holds for all k, (1.8) implies that

$$\hat{J}(k) = l(k) + \gamma \hat{J}(k+1) \tag{1.9}$$

and $\hat{J}(k) = \sum_{i=k}^{\infty} \gamma^{i-k} l(i)$, which is the same as (1.2).

1.2.1.2 Dual Heuristic Programming (DHP)

DHP is a structure for estimating the gradient of the value function, rather than J itself. To do this, a function is needed to describe the gradient of the instantaneous reward function with respect to the state of the model. In the DHP structure, the action network remains the same, but for the critic network, the costate vector is the output and the state variables are its input. The structure of DHP is shown in Fig. 1.3, where

$$\mathrm{DER} = \left(\frac{\partial \hat{x}(k+1)}{\partial x(k)} + \frac{\partial \hat{x}(k+1)}{\partial u(k)} \frac{\partial u(k)}{\partial x(k)}\right)^{\mathrm{T}}.$$

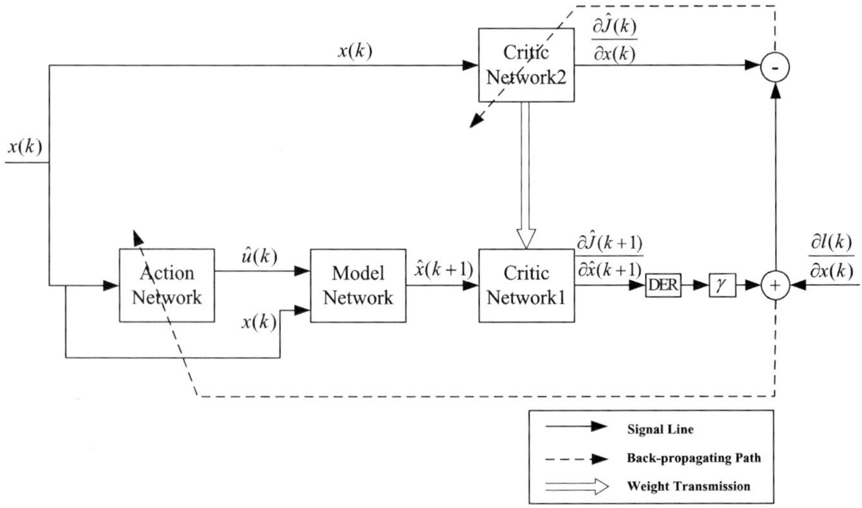

Fig. 1.3 The DHP structure diagram

The critic network's training is more complicated than that in HDP since we need to take into account all relevant pathways of back-propagation. Specifically, this training is done by minimizing the following square tracking error measure over time:

$$\|E_D\| = \sum_k E_D(k) = \frac{1}{2}\sum_k \left[\frac{\partial \hat{J}(k)}{\partial x(k)} - \frac{\partial l(k)}{\partial x(k)} - \gamma \frac{\partial \hat{J}(k+1)}{\partial x(k)}\right]^2, \tag{1.10}$$

where $\frac{\partial \hat{J}(k)}{\partial x(k)} = \frac{\partial \hat{J}[x(k),u(k),k,W_C]}{\partial x(k)}$, and W_C represents the parameters of the critic network. When $E_D(k) = 0$ holds for all k, (1.10) implies that

$$\frac{\partial \hat{J}(k)}{\partial x(k)} = \frac{\partial l(k)}{\partial x(k)} + \gamma \frac{\partial \hat{J}(k+1)}{\partial x(k)}. \tag{1.11}$$

1.2.2 Recent Developments of ADP

1.2.2.1 Development of ADP Structures

In [100], Werbos further presented two other versions called "action-dependent critics," namely, ADHDP and ADDHP. In the two ADP structures, the control is also the input of the critic networks. The two ADP structures are also summarized in [76], where a detailed summary of the major developments of adaptive critic designs up to 1997 is presented and another two ADP structures known as GDHP and ADGDHP are proposed. The GDHP or ADGDHP structure minimizes the error

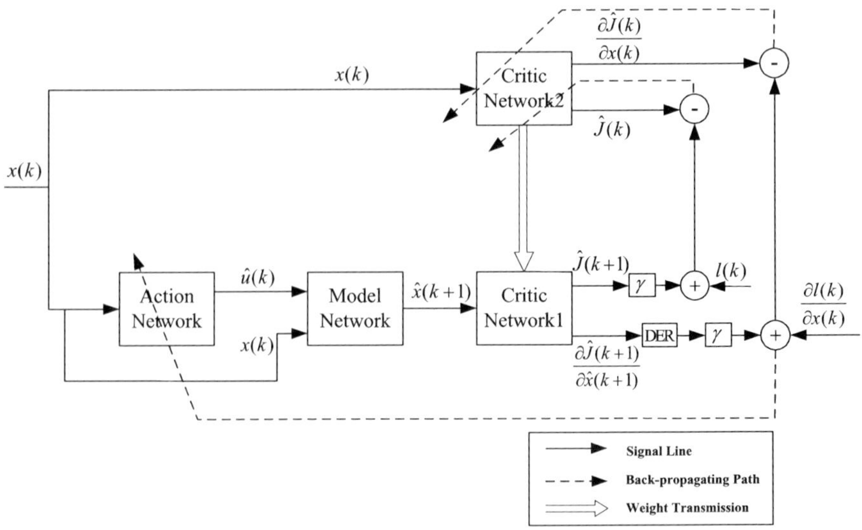

Fig. 1.4 The GDHP structure diagram

with respect to both the cost and its derivatives. While it is more complex to do this simultaneously, the resulting behavior is expected to be superior. The diagram of GDHP structure is shown in Fig. 1.4.

In [108], GDHP serves as a reconfigurable controller to deal with both abrupt and incipient changes in the plant dynamics due to faults. A novel Fault Tolerant Control (FTC) supervisor is combined with GDHP for the purpose of improving the performance of GDHP for fault tolerant control. When the plant is affected by a known abrupt fault, the new initial conditions of GDHP are loaded from a dynamic model bank (DMB). On the other hand, if the fault is incipient, the reconfigurable controller maintains normal performance by continuously modifying itself without supervisor intervention. It is noted that the training of three networks used to implement the GDHP is in an on-line fashion by utilizing two distinct networks to implement the critic unit. The first critic network is trained at every iteration, but the second one is updated at a given period of iterations. During each period of iterations, the weight parameters of the second critic network keep unchanged, which are a copy of the first one.

It should be mentioned that all the ADP structures can realize the same function, that is, to obtain the optimal control while the computation precision and speed are different. Generally, the computation burden of HDP is lowest but the computation precision is low; while GDHP possesses the most excellent precision but the computation process will take the longest time. A detailed comparison can be seen in [76].

In [33, 85], the schematic of direct heuristic dynamic programming is developed. Using the approach of [85], the model network in Fig. 1.2 is not needed anymore. Reference [106] makes significant contributions to model-free adaptive critic designs. Several practical examples are included in [106] for demonstration,

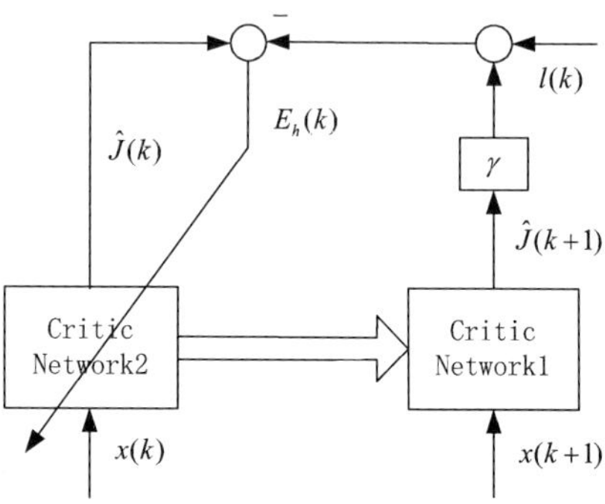

Fig. 1.5 Forward-in-time approach

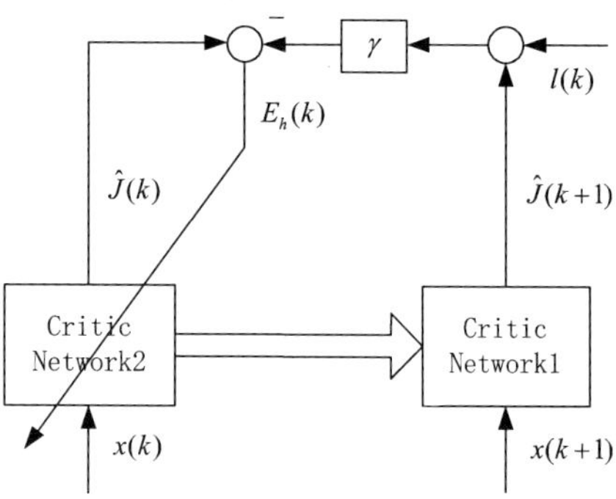

Fig. 1.6 Backward-in-time approach

which include a single inverted pendulum and a triple inverted pendulum. The reinforcement learning-based controller design for non-linear discrete-time systems with input constraints is presented by [40], where the non-linear tracking control is implemented with filtered tracking error using direct HDP designs. For similar work also see [41]. Reference [62] is also about model-free adaptive critic designs. Two approaches for the training of a critic network are provided in [62]: A forward-in-time approach and a backward-in-time approach. Figure 1.5 shows the diagram of the forward-in-time approach. In this approach, we view $\hat{J}(k)$ in (1.9) as the output of the critic network to be trained and choose $l(k) + \hat{J}(k+1)$ as the training target. Note that $\hat{J}(t)$ and $\hat{J}(k+1)$ are obtained using state variables at different time instances. Figure 1.6 shows the diagram of backward-in-time approach. In this approach, we view $\hat{J}(k+1)$ in (1.9) as the output of the critic network to be trained and choose $(\hat{J} - l)/\gamma$ as the training target. The training approach of [106] can be considered as a backward-in-time approach. In Figs. 1.5 and 1.6, $x(k+1)$ is the output of the model network.

Further, an improvement and modification to the action-critic network architecture, which is called the "single network adaptive critic (SNAC)," has been developed in [72]. This approach eliminates the action network. As a consequence, the SNAC architecture offers three potential advantages: a simpler architecture, less computational load (about half of the dual network algorithm), and no approximate error due to the elimination of the action network. The SNAC approach is applicable to a wide class of non-linear systems where the optimal control (stationary) equation can be explicitly expressed in terms of the state and the costate variables. Most of the problems in aerospace, automobile, robotics, and other engineering disciplines can be characterized by the non-linear control-affine equations that yield such a relation. SNAC-based controllers yield excellent tracking performances in applications to microelectronic mechanical systems, chemical reactors, and high-speed reentry problems. Padhi et al. [72] have proved that for linear systems (where the mapping between the costate at stage $k + 1$ and the state at stage k is linear), the solution obtained by the algorithm based on the SNAC structure converges to the solution of discrete Riccati equation.

1.2.2.2 Development of Algorithms and Convergence Analysis

The exact solution of the HJB equation is generally impossible to obtain for non-linear systems. To overcome the difficulty in solving the HJB equation, recursive methods are employed to obtain the solution of the HJB equation indirectly. In 1983, Barto et al. [9] developed a neural computation-based adaptive critic learning method. They divide the state space into boxes and store the learned information for each box. The algorithm works well but the number of boxes may be very large for a complicated system. In 1991, Lin and Kim [59] integrated the cerebellar model articulation controller technique with the box-based scheme. A large state space is mapped into a smaller physical memory space. With the distributed information storage, there is no need to reserve memory for useless boxes; this makes the structure applicable to problems of larger size. Kleinman [49] pointed out that the solution of the Riccati equation can be obtained by successively solving a sequence of Lyapunov equations, which is linear with respect to the cost function of the system, and, thus, it is easier to solve than a Riccati equation, which is non-linear with respect to the cost function. Saridis [80] extended this idea to the case of non-linear continuous-time systems where a recursive method is used to obtain the optimal control of continuous system by successively solving the *generalized Hamilton–Jacobi–Bellman (GHJB) equation*, and then updating the control action if an admissible initial control is given.

Although the GHJB equation is linear and easier to solve than a HJB equation, no general solution for GHJB is supplied. Therefore, successful application of the successive approximation method was limited until the novel work of Beard et al. in [12], where they used a Galerkin spectral approximation method at each iteration to find approximate solutions to the GHJB equations. Then Beard [11] employed a series of polynomial functions as basic functions to solve the approximate GHJB equation in continuous time, but this method requires the computation of a large number of integrals and it is not obvious how to handle explicit constraints on the controls. However, most of the above papers discussed the GHJB method for continuous-time systems, and there are few results available on the GHJB method for discrete-time non-linear systems. The discrete-time version of the approximate

GHJB-equation-based control is important since all the controllers are typically implemented by using embedded digital hardware. In [24], a successive approximation method using the GHJB equation was proposed to solve the near-optimal control problem for affine non-linear discrete-time systems, which requires the small perturbation assumption and an initially stable policy. The theory of GHJB in discrete time has also been applied to the linear discrete-time case, which indicates that the optimal control is nothing but the solution of the standard Riccati equation. On the other hand, in [19], Bradtke et al. implemented a Q-learning policy iteration method for the discrete-time linear-quadratic optimal control problem which required an *initially stable policy*. Furthermore, Landelius [51] applied HDP, DHP, ADHDP and ADDHP techniques to the discrete-time linear-quadratic optimal control problem without the initially stable conditions and discussed their convergence.

On the other hand, based on the work of Lyshevski [66], Lewis and Abu-Khalaf employed a non-quadratic performance functional to solve constrained control problems for general affine non-linear continuous-time systems using neural networks (NNs) in [1]. In addition, one showed how to formulate the associated *Hamilton–Jacobi–Isaac (HJI) equation* using special non-quadratic supply rates to obtain the non-linear state feedback control in [2]. Next, the fixed-final-time-constrained optimal control of non-linear systems was studied in [26, 27] based on the neural-network solution of the GHJB equation. In order to enhance learning speed and improve the performance, Wiering and Hasselt combined multiple different reinforcement learning algorithms to design and implement four different ensemble methods in [103]. In [35], another novel approach for designing the ADP neural-network controllers was presented. The control performance and the closed-loop stability in the linear parameter-varying (LPV) regime are formulated as a set of design equations that are linear with respect to matrix functions of NN parameters. Moreover, in [48], a new algorithm for the closed-loop parallel optimal control of weakly coupled non-linear systems was developed using the successive Galerkin approximation. In [53], the author inspired researchers to develop an experience-based approach, selecting a controller that is appropriate to the current situation from a repository of existing controller solutions. Moreover, in [82], the HJB equations were derived and proven on various time scales. The authors connected the calculus of time scales and stochastic control via an ADP algorithm and further pointed out three significant directions for the investigation of ADP on the time scales. In the past two years, there have also been published some results on ADP and reinforcement learning algorithms, such as [17, 21, 57, 94] and so on.

1.2.2.3 Applications of ADP Algorithms

As for the industrial application of ADP algorithm, it most focuses on missile systems [16], autopilot systems [34], generators [74], communication systems [63] and so on. In [109], an improved reinforcement learning method was proposed to perform navigation in dynamic environments. The difficulties of the traditional reinforcement learning were presented in autonomous navigating and three effective solutions were proposed to overcome these difficulties which were forgetting Q-learning, feature based Q-learning, and hierarchical Q-learning, respectively. Forgetting Q-learning was proposed to improve performance in a dynamic environment by maintaining possible navigation paths, which would be considered unacceptable by traditional Q-learning. Hierarchical Q-learning was proposed as a method of subdividing the problem domain into a set of more manageable ones. Feature-based Q-learning was proposed as a method of enhancing hierarchical Q-learning.

Applications of adaptive critics in the continuous-time domain were mainly done by using the discretization and the well-established discrete-time results (e.g., [89]). Various types of continuous-time nondynamic reinforcement learning were discussed by Campos and Lewis [22] and Rovithakis [78], who approximated a Lyapunov function derivative. Liu [61] proposed an improved ADHDP for on-line control and Abu-Khalaf [1] gave the optimal control scheme under constraint conditions in the actuators. Lu, Si and Xie [65] applied direct heuristic dynamic programming (direct HDP) to a large power system stability control problem. A direct HDP controller learned to cope with model deficiencies for non-linearities and uncertainties on the basis of real system responses instead of a system model. Ray et al. [77] reported a comparison of adaptive critic-based and classical wide-area controllers for power systems. Liu et al. [64] demonstrated a good engine torque and exhaust air-fuel ratio (AFR) control with adaptive critic techniques for an engine application. The design based on the neural network to automatically learn the inherent dynamics and advanced the development of a virtual powertrain to improve their performance during the actual vehicle operations. In [3] a greedy iterative HDP algorithm to solve the discrete-time Hamilton–Jacobi–Bellman (DTHJB) equation of the optimal control problem for general discrete-time non-linear systems was proposed. In [68] a convergent ADP method was developed for stabilizing the continuous-time non-linear systems and one succeeded to improve the autolanding control of aircraft.

Enns and Si [32] presented a lucid article on model-free approach to helicopter control. Recent work by Lewis et al. and Jagannathan et al. has been quite rigorous in theory and useful in practical applications. Jagannathan [84] has extended stability proofs for systems with observers in the feedback loop and applied to spark engine EGR operation on the basis of reinforcement learning dual control [41]. In order to enhance learning speed and final performance, Wiering and Hasselt combined multiple different reinforcement learning algorithms to design and implement four different ensemble methods in [103].

1.3 Feedback Control Based on Adaptive Dynamic Programming

In the most recent years, research on the ADP algorithm has made significant progress. On the one hand, for discrete-time systems, a greedy iterative HDP scheme with convergence proof was proposed in [3] for solving the optimal control problem of non-linear discrete-time systems with a known mathematical model, which did not require an initially stable policy. The basic iterative ADP algorithm for discrete-time non-linear systems, which is proposed based on Bellman's principle of optimality and the greedy iteration principle, is given as follows.

First, one start with the initial value function $V_0(\cdot) = 0$, which is not necessarily the optimal value function. Then, the law of a single control vector $v_0(x)$ can be obtained as follows:

$$v_0(x(k)) = \arg \min_{u(k)} \left\{ x^{\mathrm{T}}(k) Q x(k) + u(k)^{\mathrm{T}} R u(k) + V_0(x(k+1)) \right\}, \tag{1.12}$$

and the value function can be updated as

$$V_1(x(k)) = x^{\mathrm{T}}(k) Q x(k) + v_0(x(k))^{\mathrm{T}} R v_0(x(k)). \tag{1.13}$$

Therefore, for $i = 1, 2, \ldots$, the iterative ADP algorithm then iterates between

$$v_i(x(k)) = \arg \min_{u(k)} \left\{ x^{\mathrm{T}}(k) Q x(k) + u(k)^{\mathrm{T}} R u(k) + V_i(x(k+1)) \right\} \tag{1.14}$$

and

$$V_{i+1}(x(k)) = x^{\mathrm{T}}(k) Q x(k) + v_i(x(k))^{\mathrm{T}} R v_i(x(k)) + V_i\left(x(k+1)\right). \tag{1.15}$$

In summary, in this iterative algorithm, the value function sequence $\{V_i\}$ and control law sequence $\{v_i\}$ are updated by implementing the recurrent iteration between (1.14) and (1.15) with the iteration number i increasing from 0 to ∞.

On the other hand, there are also corresponding developments in the ADP techniques for non-linear continuous-time systems. Murray et al. proposed an iterative ADP scheme in [68] for a class of continuous-time non-linear systems with respect to the quadratic cost function and succeeded to improve the autolanding control of aircraft. The iteration was required to begin with an initially stable policy, and after each iteration the cost function was updated. So the iterative policy is also called "cost iteration." The specific algorithm is given as follows.

Consider the following continuous-time systems:

$$\dot{x} = F(x) + B(x)u, \quad x(t_0) = x_0, \tag{1.16}$$

with the cost functional given by

$$J(x) = \int_{t_0}^{\infty} l\left(x(\tau), u(\tau)\right) d\tau, \tag{1.17}$$

where $l(x, u) = q(x) + u^{\mathrm{T}} r(x) u$ is a nonnegative function and $r(x) > 0$. Similar to [81], an iterative process is proposed to obtain the control law. In this case, the optimal control can be simplified to

$$u^*(x) = -\frac{1}{2} r^{-1}(x) B^{\mathrm{T}}(x) \left[\frac{\mathrm{d}J^*(x)}{\mathrm{d}x} \right]^{\mathrm{T}}. \tag{1.18}$$

Starting from any stable Lyapunov function J_0 (or alternatively, starting from an arbitrary stable controller u_0) and replacing J^* by J_i, (1.18) becomes

$$u_i(x) = -\frac{1}{2}r^{-1}(x)B^{\mathrm{T}}(x)\left[\frac{\mathrm{d}J_i(x)}{\mathrm{d}x}\right]^{\mathrm{T}}, \tag{1.19}$$

where $J_i = \int_{t_0}^{+\infty} l\,(x_{i-1}, u_{i-1})\,d\tau$ is the cost of the trajectory $x_{i-1}(t)$ of plant (1.16) under the input $u(t) = u_{i-1}(t)$. Furthermore, Murray et al. gave a convergence analysis of the iterative ADP scheme and a stability proof of the system. Before that, most of the ADP analysis was based on the Riccati equation for linear systems. In [1], based on the work of Lyshevski [66], an iterative ADP method was used to obtain an approximate solution of the optimal value function of the HJB equation using NNs. Different from the iterative ADP scheme in [68], the iterative scheme in [1] adopted policy iteration, which meant that after each iteration the policy (or control) function was updated. The convergence and stability analysis can also be found in [1].

Moreover, Vrabie et al. [93] proposed a new policy iteration technique to solve on-line the continuous-time LQR problem for a partially model-free system (internal dynamics unknown). They presented an on-line adaptive critic algorithm in which the actor performed continuous-time control, whereas the critic's correction of the actor's behavior was discrete in time, until best performance was obtained. The critic evaluated the actor's performance over a period of time and formulated it in a parameterized form. Policy update was implemented based on the critic's evaluation on the actor. Convergence of the proposed algorithm was established by proving equivalence with an established algorithm [49]. In [35], a novel linear parameter-varying (LPV) approach for designing the ADP neural-network controllers was presented. The control performance and the closed-loop stability of the LPV regime were formulated as a set of design equations that were linear with respect to matrix functions of NN parameters.

It can be seen that most existing results, including the optimal control scheme proposed by Murray et al., require one to implement the algorithm by recurrent iteration between the value function and control law, which is not expected in real-time industrial applications. Therefore, in [91] and [119], new ADP algorithms were proposed to solve the optimal control in an on-line fashion, where the value functions and control laws were updated simultaneously. The optimal control scheme is reviewed in the following.

Consider the non-linear system (1.16), and define the infinite-horizon cost functional as follows:

$$J(x,u) = \int_t^{\infty} l(x(\tau), u(\tau))d\tau, \tag{1.20}$$

where $l(x,u) = x^{\mathrm{T}}Qx + u^{\mathrm{T}}Ru$ is the utility function, and Q and R are symmetric positive definite matrices with appropriate dimensions.

Define the *Hamilton function* as

$$H(x,u,J_x) = J_x^{\mathrm{T}}(F(x) + B(x)u) + x^{\mathrm{T}}Qx + u^{\mathrm{T}}Ru, \tag{1.21}$$

where $J_x = \partial J(x)/\partial x$.

The optimal value function $J^*(x)$ is defined as

$$J^*(x) = \min_{u \in \psi(\Omega)} \int_t^{\infty} l(x(\tau), u(x(\tau)))d\tau \tag{1.22}$$

and satisfies

$$0 = \min_{u \in \psi(\Omega)} (H(x, u, J_x^*)). \tag{1.23}$$

Therefore, we obtain the optimal control u^* by solving $\partial H(x, u, J_x^*)/\partial u = 0$, thus:

$$u^* = -\frac{1}{2} R^{-1} B^{\mathrm{T}}(x) J_x^*, \tag{1.24}$$

where $J_x^* = \partial J^*(x)/\partial x$.

In the following, by employing the critic NN and the action NN, the ADP algorithm is implemented to seek for an optimal feedback control law.

First, a neural network is utilized to approximate $J(x)$ as follows:

$$J(x) = W_c^{\mathrm{T}} \phi_c(x) + \varepsilon_c, \tag{1.25}$$

where W_c is for the unknown ideal constant weights and $\phi_c(x) : \mathbb{R}^n \to \mathbb{R}^{N_1}$ is called the critic NN activation function vector; N_1 is the number of neurons in the hidden layer, and ε_c is the critic NN approximation error.

The derivative of the cost function $J(x)$ with respect to x is

$$J_x = \nabla \phi_c^{\mathrm{T}} W_c + \nabla \varepsilon_c, \tag{1.26}$$

where $\nabla \phi_c \triangleq \partial \phi_c(x)/\partial x$ and $\nabla \varepsilon_c \triangleq \partial \varepsilon_c/\partial x$.

Let $\hat{W}_c$ be an estimate of W_c; then we have the estimate of $J(x)$ as follows:

$$\hat{J}(x) = \hat{W}_c^{\mathrm{T}} \phi_c(x). \tag{1.27}$$

Then the approximate Hamilton function can be derived as follows:

$$\begin{aligned} H(x, u, \hat{W}_a) &= \hat{W}_c^{\mathrm{T}} \nabla \phi_c (F(x) + B(x)u) + x^{\mathrm{T}} Q x + u^{\mathrm{T}} R u \\ &= e_c. \end{aligned} \tag{1.28}$$

Given any admissible control law u, it is desired to select $\hat{W}_c$ to minimize the squared residual error $E_c(\hat{W}_c)$ as follows:

$$E_c(\hat{W}_c) = \frac{1}{2} e_c^{\mathrm{T}} e_c. \tag{1.29}$$

The weight update law for the critic NN is presented based on a gradient descent algorithm, which is given by

$$\dot{\hat{W}}_c = -\alpha_c \sigma_c (\phi_c^{\mathrm{T}} \hat{W}_c + x^{\mathrm{T}} Q x + u^{\mathrm{T}} R u), \tag{1.30}$$

where $\alpha_c > 0$ is the adaptive gain of the critic NN, $\sigma_c = \sigma/(\sigma^{\mathrm{T}}\sigma + 1)$, $\sigma = \nabla\phi_c(F(x) + B(x)u)$.

On the other hand, the feedback control u is approximated by the action NN as

$$u = W_a^{\mathrm{T}}\phi_a(x) + \varepsilon_a, \tag{1.31}$$

where W_a is the matrix of unknown ideal constant weights and $\phi_a(x) : \mathbb{R}^n \to \mathbb{R}^{N_2}$ is called the action NN activation function vector, N_2 is the number of neurons in the hidden layer, and ε_a is the action NN approximation error.

Let $\hat{W}_a$ be an estimate of W_a; then the actual output can be expressed as

$$\hat{u} = \hat{W}_a^{\mathrm{T}}\phi_a(x). \tag{1.32}$$

The feedback error signal used for tuning action NN is defined as

$$e_a = \hat{W}_a^{\mathrm{T}}\phi_a + \frac{1}{2}R^{-1}C_u\nabla\phi_c^{\mathrm{T}}\hat{W}_c. \tag{1.33}$$

The objective function to be minimized by the action NN is defined as

$$E_a(\hat{W}_a) = \frac{1}{2}e_a^{\mathrm{T}}e_a. \tag{1.34}$$

The weight update law for the action NN is designed based on the gradient descent algorithm, which is given by

$$\dot{\hat{W}}_a = -\alpha_a\phi_a\left(\hat{W}_a^{\mathrm{T}}\phi_a + \frac{1}{2}R^{-1}C_u\nabla\phi_c^{\mathrm{T}}\hat{W}_c\right)^{\mathrm{T}}, \tag{1.35}$$

where $\alpha_a > 0$ is the adaptive gain of the action NN.

After the presentation of the weight update rule, a stability analysis of the closed-loop system can be performed based on the Lyapunov approach to guarantee the boundness of the weight parameters [119].

It should be mentioned that most of the above results require the models of the controlled plants to be known or at least partially known. However, in practical applications, most models cannot be obtained. Therefore, it is necessary to reconstruct the non-linear systems with function approximators. Recurrent neural networks (RNNs) are one kind of NN models, which are widely used in the dynamical analysis of non-linear systems, such as [115, 118, 123]. In this book, we will present the specific method for modeling the non-linear systems with RNN. Based on the RNN model, the ADP algorithm can be properly introduced to deal with the optimal control problems of unknown non-linear systems.

Meanwhile, saturation, dead-zones, backlash, and hysteresis are the most common actuator non-linearities in practical control system applications. Saturation non-linearity is unavoidable in most actuators. Due to the nonanalytic nature of the actuator's non-linear dynamics and the fact that the exact actuator's non-linear functions are unknown, such systems present a challenge to control engineers. As far as we know, most of the existing results of dealing with the control of systems with saturating actuators do not refer to the optimal control laws. Therefore, this problem is worthy of study in the framework of the HJB equation. To the best of our knowledge, though ADP algorithms have made large progress in the optimal control field, it is still an open problem how to solve the optimal control problem for discrete-time systems with control constraints based on ADP algorithms. If the actuator has saturating characteristic, how do we find a constrained optimal control? In this book, we shall give positive answers to these questions.

Moreover, traditional optimal control approaches are mostly implemented in an infinite time horizon. However, most real-world systems need to be effectively controlled within a finite time horizon (finite horizon for brief), such as stabilized ones or ones tracked to a desired trajectory in a finite duration of time. The design of finite-horizon optimal controllers faces a huge obstacle in comparison to the infinite-horizon one. An infinite-horizon optimal controller generally obtains an asymptotic result for the controlled systems [73]. That is, the system will not be stabilized or tracked until the time reaches infinity, while for finite-horizon optimal control problems, the system must be stabilized or tracked to a desired trajectory in a finite duration of time [20, 70, 90, 107, 111]. Furthermore, in the case of discrete-time systems, a determination of the number of optimal control steps is necessary for finite-horizon optimal control problems, while for the infinite-horizon optimal control problems, the number of optimal control steps is infinite in general. The finite-horizon control problem has been addressed by many researchers [18, 28, 37, 110, 116]. But most of the existing methods consider only stability problems of systems under finite-horizon controllers [18, 37, 110, 116]. Due to the lack of methodology and the fact that the number of control steps is difficult to determine, the optimal controller design of finite-horizon problems still presents a major challenge to control engineers.

In this book, we will develop a new ADP scheme for finite-horizon optimal control problems. We will study the optimal control problems with an ε-error bound using ADP algorithms. First, the HJB equation for finite-horizon optimal control of discrete-time systems is derived. In order to solve this HJB equation, a new iterative ADP algorithm is developed with convergence and optimality proofs. Second, the difficulties of obtaining the optimal solution using the iterative ADP algorithm is presented and then the ε-optimal control algorithm is derived based on the iterative ADP algorithms. Next, it will be shown that the ε-optimal control algorithm can obtain suboptimal control solutions within a fixed finite number of control steps that make the value function converge to its optimal value with an ε-error.

It should be mentioned that all the above results based on ADP do not refer to the systems with time delays. Actually, time delay often occurs in the transmission between different parts of systems. Transportation systems, communication systems, chemical processing systems, metallurgical processing systems and power systems are examples of time-delay systems. Therefore, the investigation of time-delay systems is significant. In recent years, much researches has been performed on decentralized control, synchronization control and stability analysis [112, 114, 117, 122]. However, the optimal control problem is often encountered in industrial production. In general, optimal control for time-delay systems is an infinite-dimensional control problem [67], which is very difficult to solve. The analysis of systems with time delays is much more difficult than that of systems without delays, and there is no method strictly facing this problem for non-linear time-delay systems. So in this book, optimal state feedback control problems of non-linear systems with time delays will also be discussed.

1.4 Non-linear Games Based on Adaptive Dynamic Programming

All of the above results discuss the situation that there is only one controller to be designed. However, as is known, a large class of real systems are controlled by more than one controller or decision maker with each using an individual strategy. These controllers often operate in a group with a general quadratic cost functional as a game [45]. Zero-sum differential game theory has been widely applied to decision making problems [23, 25, 38, 44, 52, 55], stimulated by a vast number of applications, including those in economy, management, communication networks, power networks, and in the design of complex engineering systems.

In recent years, based on the work of [51], approximate dynamic programming (ADP) techniques have further been extended to the zero-sum games of linear and non-linear systems. In [4, 5], HDP and DHP structures were used to solve the discrete-time linear-quadratic zero-sum games appearing in the H_∞ optimal control problem. The optimal strategies for discrete-time quadratic zero-sum games related to the H_∞ optimal control problem were solved forward in time. The idea is to solve for an action-dependent cost function $Q(x, u, w)$ of the zero-sum games, instead of solving for the state-dependent cost function $J(x)$ which satisfies a corresponding game algebraic Riccati equation (GARE). Using the Kronecker method, two action networks and one critic network were adaptively tuned forward in time using adaptive critic methods without the information of a model of the system. The algorithm was proved to converge to the Nash equilibrium of the corresponding zero-sum games. The performance comparisons were carried out on an F-16 autopilot. Then, in [6] these results were extended to a model-free environment for the control of a power generator system. In the paper, the on-line model-free adaptive critic schemes based on ADP were presented by the authors to solve optimal control problems in both discrete-time and continuous-time domains for linear systems with unknown dynamics. In the discrete-time case, the solution process leads to solving the underlying game GARE of the corresponding optimal control problem or zero-sum games. In the continuous-time domain, their ADP scheme solves the underlying algebraic Riccati equation (ARE) of the optimal control problem. They show that their continuous-time ADP scheme is nothing but a quasi-Newton method to solve the ARE. Either in continuous-time domain or discrete-time domain, the adaptive critic algorithms are easy to initialize considering that initial policies are not required to be stabilizing.

In the following, we present some basic knowledge regarding non-linear zero-sum differential games first [120].

Consider the following two-person zero-sum differential games. The state trajectory of the game is described by the following continuous-time affine non-linear function:

$$\dot{x} = f(x, u, w) = a(x) + b(x)u + c(x)w, \tag{1.36}$$

where $x \in \mathbb{R}^n$, $u \in \mathbb{R}^k$, $w \in \mathbb{R}^m$ and the initial condition $x(0) = x_0$ is given.

The two control variables u and w are functions chosen on $[0, \infty)$ by player I and player II from some control sets $U[0, \infty)$ and $W[0, \infty)$, respectively, subject to the constraints $u \in U(t)$, and $w \in W(t)$ for $t \in [0, \infty)$, for given convex and compact sets $U(t) \subset \mathbb{R}^k$, $W(t) \subset \mathbb{R}^m$. The cost functional is a generalized quadratic form given by

$$\begin{aligned} J(x(0), u, w) = \int_0^{\infty} & (x^{\mathrm{T}} Ax + u^{\mathrm{T}} Bu + w^{\mathrm{T}} Cw \\ & + 2u^{\mathrm{T}} Dw + 2x^{\mathrm{T}} Eu + 2x^{\mathrm{T}} Fw)dt, \end{aligned} \tag{1.37}$$

where matrices A, B, C, D, E, F have suitable dimension and $A \geq 0$, $B > 0$, $C < 0$. So we see, for $\forall t \in [0, \infty)$, that the cost functional $J(x(t), u, w)$ (denoted by $J(x)$ for brevity in the sequel) is convex in u and concave in w. $l(x, u, w) = x^{\mathrm{T}} Ax + u^{\mathrm{T}} Bu + w^{\mathrm{T}} Cw + 2u^{\mathrm{T}} Dw + 2x^{\mathrm{T}} Eu + 2x^{\mathrm{T}} Fw$ is the general quadratic utility function. For the above zero-sum differential games, there are two controllers or players where player I tries to minimize the cost functional $J(x)$, while player II attempts to maximize it. According to the situation of the two players, the following definitions are presented first.

Let

$$\overline{J}(x) := \inf_{u \in U[t,\infty)} \sup_{w \in W[t,\infty)} J(x, u, w) \tag{1.38}$$

be the upper value function and

$$\underline{J}(x) := \sup_{w \in W[t,\infty)} \inf_{u \in U[t,\infty)} J(x, u, w) \tag{1.39}$$

be the lower value function with the obvious inequality $\overline{J}(x) \geq \underline{J}(x)$. Define the optimal control pairs be $(\overline{u}, \overline{w})$ and $(\underline{u}, \underline{w})$ for upper and lower value function, respectively. Then, we have

$$\overline{J}(x) = J(x, \overline{u}, \overline{w}) \tag{1.40}$$

and

$$\underline{J}(x) = J(x, \underline{u}, \underline{w}). \tag{1.41}$$

If both $\overline{J}(x)$ and $\underline{J}(x)$ exist and

$$\overline{J}(x) = \underline{J}(x) = J^*(x), \tag{1.42}$$

we say that the optimal value function of the zero-sum differential games or the saddle point exists and the corresponding optimal control pair is denoted by (u^*, w^*).

As far as we know, traditional approaches in dealing with zero-sum differential games are to find the optimal solution or the saddle point of the games. So many results are developed to discuss the existence conditions of the differential zero-sum games [36, 113].

In the real world, however, the existence conditions of the saddle point for zero-sum differential games are so difficult to satisfy that many applications of the zero-sum differential games are limited to linear systems [31, 43, 47]. On the other hand, for many zero-sum differential games, especially in the non-linear case, the optimal solution of the game (or saddle point) does not exist inherently. Therefore, it is necessary to study the optimal control approach for the zero-sum differential games where the saddle point is invalid. The earlier optimal control scheme is to adopt the mixed trajectory method [14, 71], in which one player selects an optimal probability distribution over his control set and the other player selects an optimal probability distribution over his own control set, and then the expected solution of the game can be obtained in the sense of probability. The expected solution of the game is called a *mixed optimal solution* and the corresponding value function is the *mixed optimal value function*. The main difficulty of the mixed trajectory for the zero-sum differential games is that the optimal probability distribution is too hard, if not impossible, to obtain for the whole real space. Furthermore, the mixed optimal solution is hardly reached once the control schemes are determined. In most cases (i.e. in engineering cases), however, the optimal solution or mixed optimal solution of the zero-sum differential games has to be achieved by a determined optimal or mixed optimal control scheme. In order to overcome these difficulties, a new iterative approach is developed in this book to solve zero-sum differential games for a non-linear system.

1.5 Summary

In this chapter, we briefly introduced the variations on the structure of ADP schemes and stated the development of the iterative ADP algorithms, and, finally, we recall the industrial application of ADP schemes. Due to the focus of the book, we do not list all the methods developed in ADP. Our attention is to give an introduction to the development of theory so as to provide a rough description of ADP for a new-comer in research in this area.

References

1. Abu-Khalaf M, Lewis FL (2005) Nearly optimal control laws for nonlinear systems with saturating actuators using a neural network HJB approach. Automatica 41(5):779–791
2. Abu-Khalaf M, Lewis FL, Huang J (2006) Policy iterations on the Hamilton–Jacobi–Isaacs equation for state feedback control with input saturation. IEEE Trans Autom Control 51(12):1989–1995
3. Al-Tamimi A, Lewis FL (2007) Discrete-time nonlinear HJB solution using approximate dynamic programming: convergence proof. In: Proceedings of IEEE international symposium on approximate dynamic programming and reinforcement learning, Honolulu, HI, pp 38–43

4. Al-Tamimi A, Abu-Khalaf M, Lewis FL (2007) Adaptive critic designs for discrete-time zero-sum games with application to H_∞ control. IEEE Trans Syst Man Cybern, Part B, Cybern 37(1):240–247
5. Al-Tamimi A, Lewis FL, Abu-Khalaf M (2007) Model-free Q-learning designs for linear discrete-time zero-sum games with application to H-infinity control. Automatica 43(3):473–481
6. Al-Tamimi A, Lewis FL, Wang Y (2007) Model-free H-infinity load-frequency controller design for power systems. In: Proceedings of IEEE international symposium on intelligent control, pp 118–125
7. Al-Tamimi A, Lewis FL, Abu-Khalaf M (2008) Discrete-time nonlinear HJB solution using approximate dynamic programming: convergence proof. IEEE Trans Syst Man Cybern, Part B, Cybern 38(4):943–949
8. Bardi M, Capuzzo-Dolcetta I (1997) Optimal control and viscosity solutions of Hamilton–Jacobi–Bellman equations. Birkhauser, Boston
9. Barto AG, Sutton RS, Anderson CW (1983) Neuronlike adaptive elements that can solve difficult learning control problems. IEEE Trans Syst Man Cybern 13(5):835–846
10. Beard RW (1995) Improving the closed-loop performance of nonlinear systems. PhD dissertation, Rensselaer Polytech Institute, Troy, NY
11. Beard RW, Saridis GN (1998) Approximate solutions to the timeinvariant Hamilton–Jacobi–Bellman equation. J Optim Theory Appl 96(3):589–626
12. Beard RW, Saridis GN, Wen JT (1997) Galerkin approximations of the generalized Hamilton–Jacobi–Bellman equation. Automatica 33(12):2159–2177
13. Bellman RE (1957) Dynamic programming. Princeton University Press, Princeton
14. Bertsekas DP (2003) Convex analysis and optimization. Athena Scientific, Belmont
15. Bertsekas DP, Tsitsiklis JN (1996) Neuro-dynamic programming. Athena Scientific, Belmont
16. Bertsekas DP, Homer ML, Logan DA, Patek SD, Sandell NR (2000) Missile defense and interceptor allocation by neuro-dynamic programming. IEEE Trans Syst Man Cybern, Part A, Syst Hum 30(1):42–51
17. Bhasin S, Sharma N, Patre P, Dixon WE (2011) Asymptotic tracking by a reinforcement learning-based adaptive critic controller. J Control Theory Appl 9(3):400–409
18. Blackmore L, Rajamanoharan S, Williams BC (2008) Active estimation for jump Markov linear systems. IEEE Trans Autom Control 53(10):2223–2236
19. Bradtke SJ, Ydestie BE, Barto AG (1994) Adaptive linear quadratic control using policy iteration. In: Proceedings of the American control conference, Baltimore, Maryland, pp 3475–3476
20. Bryson AE, Ho YC (1975) Applied optimal control: optimization, estimation, and control. Hemisphere–Wiley, New York
21. Busoniu L, Babuska R, Schutter BD, Ernst D (2010) Reinforcement learning and dynamic programming using function approximators. CRC Press, Boca Raton
22. Campos J, Lewis FL (1999) Adaptive critic neural network for feedforward compensation. In: Proceedings of American control conference, San Diego, CA, pp 2813–2818
23. Chang HS, Marcus SI (2003) Two-person zero-sum Markov games: receding horizon approach. IEEE Trans Autom Control 48(11):1951–1961
24. Chen Z, Jagannathan S (2008) Generalized Hamilton–Jacobi–Bellman formulation-based neural network control of affine nonlinear discrete-time systems. IEEE Trans Neural Netw 19(1):90–106
25. Chen BS, Tseng CS, Uang HJ (2002) Fuzzy differential games for nonlinear stochastic systems: suboptimal approach. IEEE Trans Fuzzy Syst 10(2):222–233
26. Cheng T, Lewis FL, Abu-Khalaf M (2007) Fixed-final-time-constrained optimal control of nonlinear systems using neural network HJB approach. IEEE Trans Neural Netw 18(6):1725–1736
27. Cheng T, Lewis FL, Abu-Khalaf M (2007) A neural network solution for fixed-final time optimal control of nonlinear systems. Automatica 43(3):482–490

28. Costa OLV, Tuesta EF (2003) Finite horizon quadratic optimal control and a separation principle for Markovian jump linear systems. IEEE Trans Autom Control 48:1836–1842
29. Dalton J, Balakrishnan SN (1996) A neighboring optimal adaptive critic for missile guidance. Math Comput Model 23:175–188
30. Dreyfus SE, Law AM (1977) The art and theory of dynamic programming. Academic Press, New York
31. Engwerda J (2008) Uniqueness conditions for the affine open-loop linear quadratic differential game. Automatica 44(2):504–511
32. Enns R, Si J (2002) Apache helicopter stabilization using neural dynamic programming. J Guid Control Dyn 25(1):19–25
33. Enns R, Si J (2003) Helicopter trimming and tracking control using direct neural dynamic programming. IEEE Trans Neural Netw 14(4):929–939
34. Ferrari S, Stengel RF (2004) Online adaptive critic flight control. J Guid Control Dyn 27(5):777–786
35. Ferrari S, Steck JE, Chandramohan R (2008) Adaptive feedback control by constrained approximate dynamic programming. IEEE Trans Syst Man Cybern, Part B, Cybern 38(4):982–987
36. Goebel R (2002) Convexity in zero-sum differential games. In: Proceedings of the 41th IEEE conference on decision and control, Las Vegas, Nevada, pp 3964–3969
37. Goulart PJ, Kerrigan EC, Alamo T (2009) Control of constrained discrete-time systems with bounded L_2 gain. IEEE Trans Autom Control 54(5):1105–1111
38. Gu D (2008) A differential game approach to formation control. IEEE Trans Control Syst Technol 16(1):85–93
39. Hanselmann T, Noakes L, Zaknich A (2007) Continuous-time adaptive critics. IEEE Trans Neural Netw 18(3):631–647
40. He P, Jagannathan S (2005) Reinforcement learning-based output feedback control of nonlinear systems with input constraints. IEEE Trans Syst Man Cybern, Part B, Cybern 35(1):150–154
41. He P, Jagannathan S (2007) Reinforcement learning neural-network-based controller for nonlinear discrete-time systems with input constraints. IEEE Trans Syst Man Cybern, Part B, Cybern 37(2):425–436
42. Hou ZG, Wu CP (2005) A dynamic programming neural network for large-scale optimization problems. Acta Autom Sin 25(1):46–51
43. Hua X, Mizukami K (1994) Linear-quadratic zero-sum differential games for generalized state space systems. IEEE Trans Autom Control 39(1):143–147
44. Hwnag KS, Chiou JY, Chen TY (2004) Reinforcement learning in zero-sum Markov games for robot soccer systems. In: Proceedings of the 2004 IEEE international conference on networking, sensing and control, Taipei, Taiwan, pp 1110–1114
45. Jamshidi M (1982) Large-scale systems-modeling and control. North-Holland, Amsterdam
46. Javaherian H, Liu D, Zhang Y, Kovalenko O (2004) Adaptive critic learning techniques for automotive engine control. In: Proceedings of American control conference, Boston, MA, pp 4066–4071
47. Jimenez M, Poznyak A (2006) Robust and adaptive strategies with pre-identification via sliding mode technique in LQ differential games. In: Proceedings of American control conference Minneapolis, Minnesota, USA, pp 14–16
48. Kim YJ, Lim MT (2008) Parallel optimal control for weakly coupled nonlinear systems using successive Galerkin approximation. IEEE Trans Autom Control 53(6):1542–1547
49. Kleinman D (1968) On an iterative technique for Riccati equation computations. IEEE Trans Autom Control 13(1):114–115
50. Kulkarni NV, KrishnaKumar K (2003) Intelligent engine control using an adaptive critic. IEEE Trans Control Syst Technol 11:164–173
51. Landelius T (1997) Reinforcement learning and distributed local model synthesis. PhD dissertation, Linkoping University, Sweden

52. Laraki R, Solan E (2005) The value of zero-sum stopping games in continuous time. SIAM J Control Optim 43(5):1913–1922
53. Lendaris GG (2008) Higher level application of ADP: a nextphase for the control field. IEEE Trans Syst Man Cybern, Part B, Cybern 38(4):901–912
54. Lendaris GG, Paintz C (1997) Training strategies for critic and action neural networks in dual heuristic programming method. In: Proceedings of the 1997 IEEE international conference on neural networks, Houston, TX, pp 712–717
55. Leslie DS, Collins EJ (2005) Individual Q-learning in normal form games. SIAM J Control Optim 44(2):495–514
56. Lewis FL (1992) Applied optimal control and estimation. Texas instruments. Prentice Hall, Englewood Cliffs
57. Lewis FL, Liu D (2012) Reinforcement learning and approximate dynamic programming for feedback control. IEEE press series on computational intelligence. Wiley, New York
58. Lewis FL, Syrmos VL (1992) Optimal control. Wiley, New York
59. Lin CS, Kim H (1991) CMAC-based adaptive critic self-learning control. IEEE Trans Neural Netw 2(5):530–533
60. Liu X, Balakrishnan SN (2000) Convergence analysis of adaptive critic based optimal control. In: Proceedings of American control conference, Chicago, Illinois, pp 1929–1933
61. Liu D, Zhang HG (2005) A neural dynamic programming approach for learning control of failure avoidance problems. Int J Intell Control Syst 10(1):21–32
62. Liu D, Xiong X, Zhang Y (2001) Action-dependent adaptive critic designs. In: Proceeding of international joint conference on neural networks, Washington, DC, pp 990–995
63. Liu D, Zhang Y, Zhang HG (2005) A self-learning call admission control scheme for CDMA cellular networks. IEEE Trans Neural Netw 16(5):1219–1228
64. Liu D, Javaherian H, Kovalenko O, Huang T (2008) Adaptive critic learning techniques for engine torque and air–fuel ratio control. IEEE Trans Syst Man Cybern, Part B, Cybern 38(4):988–993
65. Lu C, Si J, Xie X (2008) Direct heuristic dynamic programming for damping oscillations in a large power system. IEEE Trans Syst Man Cybern, Part B, Cybern 38(4):1008–1013
66. Lyshevski SE (2002) Optimization of dynamic systems using novel performance functionals. In: Proceedings of 41st IEEE conference on decision and control, Las Vegas, Nevada, pp 753–758
67. Malek-Zavarei M, Jashmidi M (1987) Time-delay systems: analysis, optimization and applications North-Holland, Amsterdam, pp 80–96
68. Murray JJ, Cox CJ, Lendaris GG, Saeks R (2002) Adaptive dynamic programming. IEEE Trans Syst Man Cybern, Part C, Appl Rev 32(2):140–153
69. Murray JJ, Cox CJ, Saeks R (2003) The adaptive dynamic programming theorem. In: Liu D, Antsaklis PJ (eds) Stability and control of dynamical systems with applications. Birkhäser, Boston, pp 379–394
70. Necoara I, Kerrigan EC, Schutter BD, Boom T (2007) Finite-horizon min–max control of max-plus-linear systems. IEEE Trans Autom Control 52(6):1088–1093
71. Owen G (1982) Game theory. Academic Press, New York
72. Padhi R, Unnikrishnan N, Wang X, Balakrishnan SN (2006) A single network adaptive critic (SNAC) architecture for optimal control synthesis for a class of nonlinear systems. Neural Netw 19(10):1648–1660
73. Parisini T, Zoppoli R (1998) Neural approximations for infinite-horizon optimal control of nonlinear stochastic systems. IEEE Trans Neural Netw 9(6):1388–1408
74. Park JW, Harley RG, Venayagamoorthy GK (2003) Adaptive-critic-based optimal neurocontrol for synchronous generators in a power system using MLP/RBF neural networks. IEEE Trans Ind Appl 39:1529–1540
75. Powell WB (2011) Approximate dynamic programming: solving the curses of dimensionality, 2nd edn. Wiley, Princeton
76. Prokhorov DV, Wunsch DC (1997) Adaptive critic designs. IEEE Trans Neural Netw 8(5):997–1007

77. Ray S, Venayagamoorthy GK, Chaudhuri B, Majumder R (2008) Comparison of adaptive critic-based and classical wide-area controllers for power systems. IEEE Trans Syst Man Cybern, Part B, Cybern 38(4):1002–1007
78. Rovithakis GA (2001) Stable adaptive neuro-control design via Lyapunov function derivative estimation. Automatica 37(8):1213–1221
79. Saeks RE, Cox CJ, Mathia K, Maren AJ (1997) Asymptotic dynamic programming: preliminary concepts and results. In: Proceedings of the 1997 IEEE international conference on neural networks, Houston, TX, pp 2273–2278
80. Saridis GN, Lee CS (1979) An approximation theory of optimal control for trainable manipulators. IEEE Trans Syst Man Cybern 9(3):152–159
81. Saridis GN, Wang FY (1994) Suboptimal control of nonlinear stochastic systems. Control Theory Adv Technol 10(4):847–871
82. Seiffertt J, Sanyal S, Wunsch DC (2008) Hamilton–Jacobi–Bellman equations and approximate dynamic programming on time scales. IEEE Trans Syst Man Cybern, Part B, Cybern 38(4):918–923
83. Shervais S, Shannon TT, Lendaris GG (2003) Intelligent supply chain management using adaptive critic learning. IEEE Trans Syst Man Cybern, Part A, Syst Hum 33(2):235–244
84. Shih P, Kaul B, Jagannathan S, Drallmeier J (2007) Near optimal output-feedback control of nonlinear discrete-time systems in nonstrict feedback form with application to engines. In: Proceedings of international joint conference on neural networks, Orlando, Florida, pp 396–401
85. Si J, Wang YT (2001) On-line learning control by association and reinforcement. IEEE Trans Neural Netw 12(2):264–276
86. Si J, Barto A, Powell W, Wunsch D (2004) Handbook of learning dynamic programming. Wiley, New Jersey
87. Sutton RS, Barto AG (1998) Reinforcement learning: an introduction. MIT Press, Cambridge
88. Tesauro GJ (2000) Practical issues in temporal difference learning. Mach Learn 8:257–277
89. Tsitsiklis JN (1995) Efficient algorithms for globally optimal trajectories. IEEE Trans Autom Control 40(9):1528–1538
90. Uchida K, Fujita M (1992) Finite horizon H_∞ control problems with terminal penalties. IEEE Trans Autom Control 37(11):1762–1767
91. Vamvoudakis KG, Lewis FL (2010) Online actor-critic algorithm to solve the continuous-time infinite horizon optimal control problem. Automatica 46:878–888
92. Venayagamoorthy GK, Harley RG, Wunsch DG (2002) Comparison of heuristic dynamic programming and dual heuristic programming adaptive critics for neurocontrol of a turbogenerator. IEEE Trans Neural Netw 13:764–773
93. Vrabie D, Abu-Khalaf M, Lewis FL, Wang Y (2007) Continuous-time ADP for linear systems with partially unknown dynamics. In: Proceedings of the 2007 IEEE symposium on approximate dynamic programming and reinforcement learning, Honolulu, USA, pp 247–253
94. Vrabie D, Vamvoudakis KG, Lewis FL (2012) Optimal adaptive control and differential games by reinforcement learning principles. IET Press, London
95. Watkins C (1989) Learning from delayed rewards. PhD dissertation, Cambridge University, Cambridge, England
96. Werbos PJ (1977) Advanced forecasting methods for global crisis warning and models of intelligence. Gen Syst Yearbook 22:25–38
97. Werbos PJ (1987) Building and understanding adaptive systems: a statistical/numerical approach to factory automation and brain research. IEEE Trans Syst Man Cybern 17(1):7–20
98. Werbos PJ (1990) Consistency of HDP applied to a simple reinforcement learning problem. Neural Netw 3(2):179–189
99. Werbos PJ (1990) A menu of designs for reinforcement learning over time. In: Miller WT, Sutton RS, Werbos PJ (eds) Neural networks for control. MIT Press, Cambridge, pp 67–95

100. Werbos PJ (1992) Approximate dynamic programming for real-time control and neural modeling. In: White DA, Sofge DA (eds) Handbook of intelligent control: neural, fuzzy and adaptive approaches. Van Nostrand, New York, chap 13
101. Werbos PJ (2007) Using ADP to understand and replicate brain intelligence: the next level design. In: Proceedings of IEEE symposium on approximate dynamic programming and reinforcement learning, Honolulu, HI, pp 209–216
102. Widrow B, Gupta N, Maitra S (1973) Punish/reward: learning with a critic in adaptive threshold systems. IEEE Trans Syst Man Cybern 3(5):455–465
103. Wiering MA, Hasselt HV (2008) Ensemble algorithms in reinforcement learning. IEEE Trans Syst Man Cybern, Part B, Cybern 38(4):930–936
104. Yadav V, Padhi R, Balakrishnan SN (2007) Robust/optimal temperature profile control of a high-speed aerospace vehicle using neural networks. IEEE Trans Neural Netw 18(4):1115–1128
105. Yang Q, Jagannathan S (2007) Online reinforcement learning neural network controller design for nanomanipulation. In: Proceedings of IEEE symposium on approximate dynamic programming and reinforcement learning, Honolulu, HI, pp 225–232
106. Yang L, Enns R, Wang YT, Si J (2003) Direct neural dynamic programming. In: Liu D, Antsaklis PJ (eds) Stability and control of dynamical systems with applications. Birkhauser, Boston
107. Yang F, Wang Z, Feng G, Liu X (2009) Robust filtering with randomly varying sensor delay: the finite-horizon case. IEEE Trans Circuits Syst I, Regul Pap 56(3):664–672
108. Yen GG, DeLima PG (2005) Improving the performance of globalized dual heuristic programming for fault tolerant control through an online learning supervisor. IEEE Trans Autom Sci Eng 2(2):121–131
109. Yen GG, Hickey TW (2004) Reinforcement learning algorithms for robotic navigation in dynamic environments. ISA Trans 43:217–230
110. Zadorojniy A, Shwartz A (2006) Robustness of policies in constrained Markov decision processes. IEEE Trans Autom Control 51(4):635–638
111. Zattoni E (2008) Structural invariant subspaces of singular Hamiltonian systems and nonrecursive solutions of finite-horizon optimal control problems. IEEE Trans Autom Control 53(5):1279–1284
112. Zhang HG, Wang ZS (2007) Global asymptotic stability of delayed cellular neural networks. IEEE Trans Neural Netw 18(3):947–950
113. Zhang P, Deng H, Xi J (2005) On the value of two-person zero-sum linear quadratic differential games. In: Proceedings of the 44th IEEE conference on decision and control, and the European control conference. Seville, Spain, pp 12–15
114. Zhang HG, Lun SX, Liu D (2007) Fuzzy H(infinity) filter design for a class of nonlinear discrete-time systems with multiple time delays. IEEE Trans Fuzzy Syst 15(3):453–469
115. Zhang HG, Wang ZS, Liu D (2007) Robust exponential stability of recurrent neural networks with multiple time-varying delays. IEEE Trans Circuits Syst II, Express Briefs 54(8):730–734
116. Zhang HS, Xie L, Duan G (2007) H_∞ control of discrete-time systems with multiple input delays. IEEE Trans Autom Control 52(2):271–283
117. Zhang HG, Yang DD, Chai TY (2007) Guaranteed cost networked control for T-S fuzzy systems with time delays. IEEE Trans Syst Man Cybern, Part C, Appl Rev 37(2):160–172
118. Zhang HG, Ma TD, Huang GB (2010) Robust global exponential synchronization of uncertain chaotic delayed neural networks via dual-stage impulsive control. IEEE Trans Syst Man Cybern, Part B, Cybern 40(3):831–844
119. Zhang HG, Cui LL, Zhang X, Luo YH (2011) Data-driven robust approximate optimal tracking control for unknown general nonlinear systems using adaptive dynamic programming method. IEEE Trans Neural Netw 22(12):2226–2236
120. Zhang HG, Wei QL, Liu D (2011) An iterative approximate dynamic programming method to solve for a class of nonlinear zero-sum differential games. Automatica 47(1):207–214

121. Zhao Y, Patek SD, Beling PA (2008) Decentralized Bayesian search using approximate dynamic programming methods. IEEE Trans Syst Man Cybern, Part B, Cybern 38(4):970–975
122. Zheng CD, Zhang HG, Wang ZS (2010) An augmented LKF approach involving derivative information of both state and delay. IEEE Trans Neural Netw 21(7):1100–1109
123. Zheng CD, Zhang HG, Wang ZS (2011) Novel exponential stability criteria of high-order neural networks with time-varying delays. IEEE Trans Syst Man Cybern, Part B, Cybern 41(2):486–496

Chapter 2
Optimal State Feedback Control for Discrete-Time Systems

2.1 Introduction

The optimal control problem of nonlinear systems has always been the key focus of control fields in the past several decades. Traditional optimal control approaches are mostly based on linearization methods or numerical computation methods. However, closed-loop optimal feedback control is desired for most researchers in practice. Therefore, in this chapter, several near-optimal control scheme will be developed for different nonlinear discrete-time systems by introducing the different iterative ADP algorithms.

First, an infinite-horizon optimal state feedback controller is developed for a class of discrete-time systems based on DHP. Then, due to the special advantages of GDHP algorithm, a new optimal control scheme is developed with discounted cost functional. Moreover, based on GHJB algorithm, an infinite-horizon optimal state feedback stabilizing controller is designed. Further, most existing controllers are implemented in infinite time horizon. However, many real-world systems need to be effectively controlled within a finite time horizon. Therefore, we further propose a finite-horizon optimal controllers with ε-error bound, where the number of optimal control steps can be determined definitely.

2.2 Infinite-Horizon Optimal State Feedback Control Based on DHP

Saturation, dead-zone, backlash, and hysteresis are the most common actuator nonlinearities in practical control system applications. Due to the nonanalytic nature of the actuator nonlinear dynamics and the fact that the exact actuator nonlinear functions are unknown, the systems with saturation present a challenge to control engineers. In this section, we study this problem in the framework of the HJB equation from optimal control theory. First, based on nonquadratic functionals, the HJB equation is formulated, whose solution results in a smooth saturated controller. Then

H. Zhang et al., *Adaptive Dynamic Programming for Control*,
Communications and Control Engineering, DOI 10.1007/978-1-4471-4757-2_2,

a new iterative ADP algorithm is presented with convergence proof to solve the HJB equation derived.

2.2.1 Problem Formulation

Consider a class of discrete-time affine nonlinear systems as follows:

$$x(k+1) = f(x(k)) + g(x(k))u(k), \tag{2.1}$$

where $x(k) \in \mathbb{R}^n$ is the state vector, and $f\colon \mathbb{R}^n \to \mathbb{R}^n$ and $g\colon \mathbb{R}^n \to \mathbb{R}^{n\times m}$ are differentiable in their arguments with $f(0) = 0$. Assume that $f + gu$ is Lipschitz continuous on a set Ω in $\mathbb{R}^n$ containing the origin, and that the system (2.1) is controllable in the sense that there exists at least a continuous control law on Ω that asymptotically stabilizes the system. We denote $\Omega_u = \{u(k) \mid u(k) = [u_1(k), u_2(k), \dots, u_m(k)]^{\mathrm{T}} \in \mathbb{R}^m,\ |u_i(k)| \le \bar{u}_i,\ i = 1, \dots, m\}$, where $\bar{u}_i$ is the saturating bound for the ith actuator. Let $\bar{U} \in \mathbb{R}^{m\times m}$ be the constant diagonal matrix given by $\bar{U} = \operatorname{diag}\{\bar{u}_1, \bar{u}_2, \dots, \bar{u}_m\}$.

In this subsection, we mainly discuss how to design an optimal state feedback controller for this class of constrained discrete-time systems. It is desired to find the optimal control law $v(x)$ so that the control sequence $u(\cdot) = (u(i), u(i+1), \dots)$ with each $u(i) \in \Omega_u$ minimizes the generalized *cost functional* as follows:

$$J(x(k), u(\cdot)) = \sum_{i=k}^{\infty} \left\{x^{\mathrm{T}}(i)Qx(i) + W(u(i))\right\}, \tag{2.2}$$

where $u(i) = v(x(i))$, $W(u(i)) \in \mathbb{R}$ is positive definite, and the weight matrix Q is also positive definite.

For optimal control problems, the state feedback control law $v(x)$ must not only stabilize the system on Ω but also guarantee that (2.2) is finite. Such a control law is said to be admissible.

Definition 2.1 A control law $v(x)$ is said to be admissible with respect to (2.2) on Ω if $v(x)$ is continuous with $v(x(k)) \in \Omega_u$ for $\forall x(k) \in \Omega$ and stabilizes (2.1) on Ω, $v(0) = 0$, and for $\forall x(0) \in \Omega$, $J(x(0), u(\cdot))$ is finite, where $u(\cdot) = (u(0), u(1), \dots)$ and $u(k) = v(x(k))$, $k = 0, 1, \dots$.

Based on the above definition, we are ready to explain the *admissible control law* sequence. A control law sequence $\{\eta_i\} = (\eta_0, \eta_1, \dots, \eta_\infty)$ is called admissible if the resultant control sequence $(u(0), u(1), \dots, u(\infty))$ stabilizes the system (2.1) with any initial state $x(0)$ and guarantees that $J(x(0), u(\cdot))$ is finite. It should be mentioned that, in this case, each control action obeys a different control law, i.e., $u(i)$ is produced by a control law η_i for $i = 0, 1, \dots$. The control law sequence $\{\eta_i\} = (\eta_0, \eta_1, \dots, \eta_\infty)$ is also called a nonstationary policy in the literature [2].

For convenience, in the sequel $J^*(x(k))$ is used to denote the optimal value function which is defined as $J^*(x(k)) = \min_{u(\cdot)} J(x(k), u(\cdot))$, and $u^*(x)$ is used to denote the corresponding optimal control law.

For the unconstrained control problem, $W(u(i))$ in the performance functional (2.2) is commonly chosen as the quadratic form of the control input $u(i)$. However, in this subsection, to confront the bounded control problem, we employ a *non-quadratic functional* as follows:

$$W(u(i)) = 2\int_0^{u(i)} \varphi^{-\mathrm{T}}(\bar{U}^{-1}s)\bar{U}R\,ds, \tag{2.3}$$

$$\varphi^{-1}(u(i)) = [\varphi^{-1}(u_1(i)), \varphi^{-1}(u_2(i)), \dots, \varphi^{-1}(u_m(i))]^{\mathrm{T}},$$

where R is positive definite and assumed to be diagonal for simplicity of analysis, $s \in \mathbb{R}^m$, $\varphi \in \mathbb{R}^m$, $\varphi(\cdot)$ is a bounded one-to-one function satisfying $|\varphi(\cdot)| \le 1$ and belonging to C^p $(p \ge 1)$ and $L_2(\Omega)$. Moreover, it is a monotonic increasing odd function with its first derivative bounded by a constant M. Such a function is easy to find, and one example is the hyperbolic tangent function $\varphi(\cdot) = \tanh(\cdot)$. It should be noticed that, by the definition above, $W(u(i))$ is ensured to be positive definite because $\varphi^{-1}(\cdot)$ is a monotonic odd function and R is positive definite.

According to Bellman's principle of optimality, the *optimal value function* $J^*(x)$ should satisfy the following HJB equation:

$$\begin{aligned} J^*(x(k)) &= \min_{u(\cdot)} \sum_{i=k}^{\infty} \left\{ x^{\mathrm{T}}(i)Qx(i) + 2\int_0^{u(i)} \varphi^{-\mathrm{T}}(\bar{U}^{-1}s)\bar{U}R\,ds \right\} \\ &= \min_{u(k)} \left\{ x^{\mathrm{T}}(k)Qx(k) + 2\int_0^{u(k)} \varphi^{-\mathrm{T}}(\bar{U}^{-1}s)\bar{U}R\,ds \right. \\ &\qquad \left. + J^*(x(k+1)) \right\}. \end{aligned} \tag{2.4}$$

The optimal control law $u^*(x)$ should satisfy

$$\begin{aligned} u^*(x(k)) = \arg\min_{u(k)} &\left\{ x^{\mathrm{T}}(k)Qx(k) + 2\int_0^{u(k)} \varphi^{-\mathrm{T}}(\bar{U}^{-1}s)\bar{U}R\,ds \right. \\ &\left. + J^*(x(k+1)) \right\}. \end{aligned} \tag{2.5}$$

The optimal control problem can be solved if the optimal value function $J^*(x)$ can be obtained from (2.4). However, there is currently no method for solving this value function of the constrained optimal control problem. Therefore, in the next subsection we will discuss how to utilize the iterative ADP algorithm to seek the near-optimal control solution.

2.2.2 Infinite-Horizon Optimal State Feedback Control via DHP

Since direct solution of the HJB equation is computationally intensive, we develop in this subsection an iterative ADP algorithm, based on Bellman's principle of optimality and the greedy iteration principle.

First, we start with initial value function $V_0(\cdot) = 0$ which is not necessarily the optimal value function. Then, we find the law of single control vector $v_0(x)$ as follows:

$$v_0(x(k)) = \arg\min_{u(k)} \Big\{ x^{\mathrm{T}}(k)Qx(k) + 2\int_0^{u(k)} \varphi^{-\mathrm{T}}(\bar{U}^{-1}s)\bar{U}R\,ds + V_0(x(k+1)) \Big\}, \tag{2.6}$$

and we update the value function by

$$V_1(x(k)) = x^{\mathrm{T}}(k)Qx(k) + 2\int_0^{v_0(x(k))} \varphi^{-\mathrm{T}}(\bar{U}^{-1}s)\bar{U}R\,ds. \tag{2.7}$$

Therefore, for $i = 1, 2, \ldots$, the iterative ADP algorithm iterates between

$$v_i(x(k)) = \arg\min_{u(k)} \Big\{ x^{\mathrm{T}}(k)Qx(k) + 2\int_0^{u(k)} \varphi^{-\mathrm{T}}(\bar{U}^{-1}s)\bar{U}R\,ds + V_i(x(k+1)) \Big\} \tag{2.8}$$

and

$$V_{i+1}(x(k)) = \min_{u(k)} \Big\{ x^{\mathrm{T}}(k)Qx(k) + 2\int_0^{u(k)} \varphi^{-\mathrm{T}}(\bar{U}^{-1}s)\bar{U}R\,ds + V_i(x(k+1)) \Big\}. \tag{2.9}$$

It can be seen that, based on (2.8), (2.9) can further be written as

$$V_{i+1}(x(k)) = x^{\mathrm{T}}(k)Qx(k) + 2\int_0^{v_i(x(k))} \varphi^{-\mathrm{T}}(\bar{U}^{-1}s)\bar{U}R\,ds + V_i\left(x(k+1)\right), \tag{2.10}$$

where $x(k+1) = f(x(k)) + g(x(k))v_i(x(k))$.

In summary, in this iterative algorithm, the value function sequence $\{V_i\}$ and control law sequence $\{v_i\}$ are updated by implementing the recurrent iteration between (2.8) and (2.10) with the iteration number i increasing from 0 to ∞.

To further explain the iteration process, next we are ready to analyze this iterative algorithm. First, based on (2.10) we obtain

$$V_i(x(k+1)) = x^{\mathrm{T}}(k+1)Qx(k+1) + 2\int_0^{v_{i-1}(x(k+1))} \varphi^{-\mathrm{T}}(\bar{U}^{-1}s)\bar{U}R\,ds + V_{i-1}(x(k+2)), \tag{2.11}$$

where $x(k+2) = f(x(k+1)) + g(x(k+1))v_{i-1}(x(k+1))$. Then, by further expanding (2.10), we have

$$\begin{aligned} V_{i+1}(x(k)) = {} & x^{\mathrm{T}}(k)Qx(k) + 2\int_0^{v_i(x(k))} \varphi^{-\mathrm{T}}(\bar{U}^{-1}s)\bar{U}R\,ds \\ & + x^{\mathrm{T}}(k+1)Qx(k+1) + 2\int_0^{v_{i-1}(x(k+1))} \varphi^{-\mathrm{T}}(\bar{U}^{-1}s)\bar{U}R\,ds \\ & + \cdots + x^{\mathrm{T}}(k+i)Qx(k+i) \\ & + 2\int_0^{v_0(x(k+i))} \varphi^{-\mathrm{T}}(\bar{U}^{-1}s)\bar{U}R\,ds + V_0(x(k+i+1)), \end{aligned} \tag{2.12}$$

where $V_0(x(k+i+1)) = 0$.

From (2.12), it can be seen that during the iteration process, the control actions for different control steps obey different control laws. After the iteration number $i+1$, the obtained control law sequence is $(v_i, v_{i-1}, \ldots, v_0)$. With the iteration number i increasing to ∞, the obtained control law sequence has a length of ∞. For the infinite-horizon problem, both the optimal value function and the optimal control law are unique. Therefore, it is desired that the control law sequence will converge when the iteration number $i \to \infty$. In the following, we will prove that both the value function sequence $\{V_i\}$ and the control law sequence $\{v_i\}$ are convergent.

In this subsection, in order to prove the convergence characteristics of the iterative ADP algorithm for the constrained nonlinear system, we first present two lemmas before presenting our theorems. For convenience, the nonquadratic functional $2\int_0^{u(k)} \varphi^{-\mathrm{T}}(\bar{U}^{-1}s)\bar{U}R\,ds$ will be written as $W(u(k))$ in the sequel.

Lemma 2.2 *Let $\{\mu_i\}$ be an arbitrary sequence of control laws, and $\{v_i\}$ be the control law sequence as in* (2.8). *Let V_i be as in* (2.9) *and Λ_i be*

$$\Lambda_{i+1}(x(k)) = x^{\mathrm{T}}(k)Qx(k) + W(\mu_i(x(k))) + \Lambda_i(x(k+1)). \tag{2.13}$$

If $V_0(\cdot) = \Lambda_0(\cdot) = 0$, then $V_i(x) \le \Lambda_i(x)$, $\forall i$.

Proof It is clear from the fact that V_{i+1} is the result of minimizing the right hand side of (2.9) with respect to the control input $u(k)$, while Λ_{i+1} is a result of arbitrary control input. □

Lemma 2.3 *Let the sequence $\{V_i\}$ be defined as in* (2.9). *If the system is controllable, then there is an upper bound Y such that $0 \le V_i(x(k)) \le Y$, $\forall i$.*

Proof Let $\{\eta_i(x)\}$ be a sequence of stabilizing and admissible control laws, and let $V_0(\cdot) = P_0(\cdot) = 0$, where V_i is updated by (2.9) and P_i is updated by

$$P_{i+1}(x(k)) = x^{\mathrm{T}}(k)Qx(k) + W(\eta_i(x(k))) + P_i(x(k+1)). \tag{2.14}$$

From (2.14), we further obtain

$$\begin{aligned} P_i(x(k+1)) = {} & x^{\mathrm{T}}(k+1)Qx(k+1) + W(\eta_{i-1}(x(k+1))) \\ & + P_{i-1}(x(k+2)). \end{aligned} \tag{2.15}$$

Thus, the following relation can be obtained:

$$\begin{aligned} P_{i+1}(x(k)) &= x^{\mathrm{T}}(k)Qx(k) + W(\eta_i(x(k))) \\ &\quad + x^{\mathrm{T}}(k+1)Qx(k+1) + W(\eta_{i-1}(x(k+1))) \\ &\quad + P_{i-1}(x(k+2)) \\ &= x^{\mathrm{T}}(k)Qx(k) + W(\eta_i(x(k))) \\ &\quad + x^{\mathrm{T}}(k+1)Qx(k+1) + W(\eta_{i-1}(x(k+1))) \\ &\quad + x^{\mathrm{T}}(k+2)Qx(k+2) + W(\eta_{i-2}(x(k+2))) \\ &\quad + P_{i-2}(x(k+3)) \\ &\quad \vdots \\ &= x^{\mathrm{T}}(k)Qx(k) + W(\eta_i(x(k))) \\ &\quad + x^{\mathrm{T}}(k+1)Qx(k+1) + W(\eta_{i-1}(x(k+1))) \\ &\quad + x^{\mathrm{T}}(k+2)Qx(k+2) + W(\eta_{i-2}(x(k+2))) \\ &\quad + \dots \\ &\quad + x^{\mathrm{T}}(k+i)Qx(k+i) + W(\eta_0(x(k+i))) \\ &\quad + P_0(x(k+i+1)), \end{aligned} \tag{2.16}$$

where $P_0(x(k+i+1)) = 0$.

Let $l_i(x(k)) = x^{\mathrm{T}}(k)Qx(k) + W(\eta_i(x(k)))$, and then (2.16) can further be written as

$$\begin{aligned} P_{i+1}(x(k)) &= \sum_{j=0}^{i} l_{i-j}(x(k+j)) \\ &= \sum_{j=0}^{i} \left\{ x^{\mathrm{T}}(k+j)Qx(k+j) + W(\eta_{i-j}(x(k+j))) \right\} \end{aligned}$$

$$\leq \lim_{i\to\infty} \sum_{j=0}^{i} \Big\{ x^{\mathrm{T}}(k+j)Qx(k+j) + W\big(\eta_{i-j}(x(k+j))\big)\Big\}. \tag{2.17}$$

Note that $\{\eta_i(x)\}$ is an admissible control law sequence, i.e., $x(k) \to 0$ as $k \to \infty$. Therefore there exists an upper bound Y such that

$$\forall i: \quad P_{i+1}(x(k)) \leq \lim_{i\to\infty} \sum_{j=0}^{i} l_{i-j}(x(k+j)) \leq Y. \tag{2.18}$$

Combining with Lemma 2.2, we obtain

$$\forall i: \quad V_{i+1}(x(k)) \leq P_{i+1}(x(k)) \leq Y. \tag{2.19}$$

This completes the proof. □

Next, Lemmas 2.2 and 2.3 will be used in the proof of our main theorems.

Theorem 2.4 (cf. [17]) *Define the value function sequence $\{V_i\}$ as in* (2.10) *with $V_0(\cdot) = 0$, and the control law sequence $\{v_i\}$ as in* (2.8). *Then, we can conclude that $\{V_i\}$ is a nondecreasing sequence satisfying $V_{i+1}(x(k)) \geq V_i(x(k))$, $\forall i$.*

Proof For convenience of analysis, define a new sequence $\{\Phi_i\}$ as follows:

$$\Phi_{i+1}(x(k)) = x^{\mathrm{T}}(k)Qx(k) + W(v_{i+1}(x(k))) + \Phi_i(x(k+1)), \tag{2.20}$$

where $\Phi_0(\cdot) = V_0(\cdot) = 0$. The control law sequence $\{v_i\}$ is updated by (2.8) and the value function sequence $\{V_i\}$ is updated by (2.10).

In the following, we prove that $\Phi_i(x(k)) \leq V_{i+1}(x(k))$ by mathematical induction.

First, we prove that it holds for $i = 0$. Noticing that

$$V_1(x(k)) - \Phi_0(x(k)) = x^{\mathrm{T}}(k)Qx(k) + W(v_0(x(k))) \geq 0, \tag{2.21}$$

thus for $i = 0$, we have

$$V_1(x(k)) \geq \Phi_0(x(k)). \tag{2.22}$$

Second, we assume that it holds for $i-1$. That is to say, for any $x(k)$, we have $V_i(x(k)) \geq \Phi_{i-1}(x(k))$. Then, for i, since

$$\Phi_i(x(k)) = x^{\mathrm{T}}(k)Qx(k) + W(v_i(x(k))) + \Phi_{i-1}(x(k+1)) \tag{2.23}$$

and

$$V_{i+1}(x(k)) = x^{\mathrm{T}}(k)Qx(k) + W(v_i(x(k))) + V_i(x(k+1)) \tag{2.24}$$

hold, we obtain

$$V_{i+1}(x(k)) - \Phi_i(x(k)) = V_i(x(k+1)) - \Phi_{i-1}(x(k+1)) \geq 0, \tag{2.25}$$

i.e., the following equation holds:

$$\Phi_i(x(k)) \le V_{i+1}(x(k)). \tag{2.26}$$

Therefore, (2.26) is proved for any i by mathematical induction.

Furthermore, from Lemma 2.2 we know that $V_i(x(k)) \le \Phi_i(x(k))$. Therefore we have

$$V_i(x(k)) \le \Phi_i(x(k)) \le V_{i+1}(x(k)). \tag{2.27}$$

The proof is completed. □

Next, we are ready to exploit the limit of the value function sequence $\{V_i\}$ when $i \to \infty$.

Let $\{\eta_i^{(l)}\}$ be the lth admissible control law sequence, similar to the proof of Lemma 2.3, we can construct the associated sequence $P_i^{(l)}(x)$ as follows:

$$P_{i+1}^{(l)}(x(k)) = x^{\mathrm{T}}(k)Qx(k) + W(\eta_i^{(l)}(x(k))) + P_i^{(l)}(x(k+1)), \tag{2.28}$$

with $P_0^{(l)}(\cdot) = 0$.

Let $l_i^{(l)}(x(k)) = x^{\mathrm{T}}(k)Qx(k) + W(\eta_i^{(l)}(x(k)))$. Then, the following relation can be obtained similarly:

$$P_{i+1}^{(l)}(x(k)) = \sum_{j=0}^{i} l_{i-j}^{(l)}(x(k+j)). \tag{2.29}$$

Let $i \to \infty$; we have

$$P_{\infty}^{(l)}(x(k)) = \lim_{i\to\infty} \sum_{j=0}^{i} l_{i-j}^{(l)}(x(k+j)). \tag{2.30}$$

Combining (2.29) with (2.30), we obtain

$$P_{i+1}^{(l)}(x(k)) \le P_{\infty}^{(l)}(x(k)). \tag{2.31}$$

Theorem 2.5 (cf. [17]) *Define $P_{\infty}^{(l)}(x(k))$ as in (2.30), and the value function sequence $\{V_i\}$ as in (2.10) with $V_0(\cdot) = 0$. For any state vector $x(k)$, define $J^*(x(k)) = \inf_l\{P_{\infty}^{(l)}(x(k))\}$, which can be considered as the "optimal" value function starting from $x(k)$ under all admissible control law sequences with length of ∞. Then, we can conclude that J^* is the limit of the value function sequence $\{V_i\}$.*

Proof According to the definition of $P_{\infty}^{(l)}(x(k))$, the associated control law sequence $\{\eta_i^{(l)}(x)\}$ is admissible. Thus, it is guaranteed that $\lim_{i\to\infty}\sum_{j=0}^{i} l_{i-j}^{(l)}(x(k+j))$ is finite, i.e., $P_\infty^{(l)}(x(k))$ is finite. Hence for any l, there exists an upper bound Y_l such that

$$P_{i+1}^{(l)}(x(k)) \le P_\infty^{(l)}(x(k)) \le Y_l. \tag{2.32}$$

Combining with Lemma 2.2, we further obtain

$$\forall l, i: \ V_{i+1}(x(k)) \le P_{i+1}^{(l)}(x(k)) \le Y_l. \tag{2.33}$$

Since $J^*(x(k)) = \inf_l\{P_\infty^{(l)}(x(k))\}$, for any $\epsilon > 0$, there exists a sequence of admissible control laws $\{\eta_i^{(K)}\}$ such that the associated value function satisfies $P_\infty^{(K)}(x(k)) \le J^*(x(k)) + \epsilon$. According to (2.33), we have $V_i(x(k)) \le P_i^{(l)}(x(k))$ for any l and i. Thus, we obtain $\lim_{i\to\infty} V_i(x(k)) \le P_\infty^{(K)}(x(k)) \le J^*(x(k)) + \epsilon$. Noting that ϵ is chosen arbitrarily, we have

$$\lim_{i\to\infty} V_i(x(k)) \le J^*(x(k)). \tag{2.34}$$

On the other hand, since $V_{i+1}(x(k)) \le P_{i+1}^{(l)}(x(k)) \le Y_l, \forall l, i$, we have $\lim_{i\to\infty} V_i(x(k)) \le \inf_l \{Y_l\}$. According to the definition of admissible control law sequence, the control law sequence associated with the value function $\lim_{i\to\infty} V_i(x(k))$ must be an admissible control law sequence, i.e., there exists a sequence of admissible control laws $\{\eta_i^{(N)}\}$ such that $\lim_{i\to\infty} V_i(x(k)) = P_\infty^{(N)}(x(k))$. Combining with the definition $J^*(x(k)) = \inf_l\{P_\infty^{(l)}(x(k))\}$, we can obtain

$$\lim_{i\to\infty} V_i(x(k)) \ge J^*(x(k)). \tag{2.35}$$

Therefore, combining (2.34) with (2.35), we can conclude that $\lim_{i\to\infty} V_i(x(k)) = J^*(x(k))$, i.e., J^* is the limit of the value function sequence $\{V_i\}$.

The proof is completed. □

Next, let us consider what will happen when we make $i \to \infty$ in (2.9). The left hand side is simply $V_\infty(x)$. But for the right hand side, it is not obvious to see since the minimum will reach at different $u(k)$ for different i. However, the following result can be proved.

Theorem 2.6 *For any state vector $x(k)$, the "optimal" value function $J^*(x)$ satisfies the HJB equation*

$$J^*(x(k)) = \inf_{u(k)} \left\{x^{\mathrm{T}}(k)Qx(k) + W(u(k)) + J^*(x(k+1))\right\}.$$

Proof For any $u(k)$ and i, according to (2.9), we have

$$V_i(x(k)) \le x^{\mathrm{T}}(k)Qx(k) + W(u(k)) + V_{i-1}(x(k+1)). \tag{2.36}$$

According to Theorems 2.4 and 2.5, the value function sequence $\{V_i\}$ is a nondecreasing sequence satisfying $\lim_{i\to\infty} V_i(x(k)) = J^*(x(k))$, hence the relation $V_{i-1}(x(k+1)) \le J^*(x(k+1))$ holds for any i. Thus, we obtain

$$V_i(x(k)) \le x^{\mathrm{T}}(k)Qx(k) + W(u(k)) + J^*(x(k+1)). \tag{2.37}$$

Let $i \to \infty$; we have

$$J^*(x(k)) \le x^{\mathrm{T}}(k)Qx(k) + W(u(k)) + J^*(x(k+1)). \tag{2.38}$$

Since $u(k)$ in the above equation is chosen arbitrarily, the following equation holds:

$$J^*(x(k)) \le \inf_{u(k)} \left\{x^{\mathrm{T}}(k)Qx(k) + W(u(k)) + J^*(x(k+1))\right\}. \tag{2.39}$$

On the other hand, for any i the value function sequence satisfies

$$V_i(x(k)) = \min_{u(k)} \left\{x^{\mathrm{T}}(k)Qx(k) + W(u(k)) + V_{i-1}(x(k+1))\right\}. \tag{2.40}$$

Combining with $V_i(x(k)) \le J^*(x(k)), \forall i$, we have

$$J^*(x(k)) \ge \inf_{u(k)} \left\{x^{\mathrm{T}}(k)Qx(k) + W(u(k)) + V_{i-1}(x(k+1))\right\}. \tag{2.41}$$

Let $i \to \infty$; then we obtain

$$J^*(x(k)) \ge \inf_{u(k)} \left\{x^{\mathrm{T}}(k)Qx(k) + W(u(k)) + J^*(x(k+1))\right\}. \tag{2.42}$$

Combining (2.39) and (2.42), we have

$$J^*(x(k)) = \inf_{u(k)} \left\{x^{\mathrm{T}}(k)Qx(k) + W(u(k)) + J^*(x(k+1))\right\}. \tag{2.43}$$

The proof is completed. □

According to Theorems 2.4 and 2.5, we can conclude that $V_i(x(k)) \le V_{i+1}(x(k))$, $\forall i$ and $\lim_{i\to\infty} V_i(x(k)) = J^*(x(k))$. Furthermore, according to Theorem 2.6, we have $J^*(x(k)) = \inf_{u(k)}\{x^{\mathrm{T}}(k)Qx(k) + W(u(k)) + J^*(x(k+1))\}$. Therefore, we can conclude that the value function sequence $\{V_i\}$ converges to the optimal value function of the discrete-time HJB equation, i.e., $V_i \to J^*$ as $i \to \infty$. Since the value function sequence is convergent, according to (2.5) and (2.8), we can conclude that the corresponding control law sequence $\{v_i\}$ converges to the optimal control law u^* as $i \to \infty$.

It should be mentioned that the value function $V_i(x)$ we constructed is a new function that is different from ordinary cost function. Via Lemma 2.3 and Theorem 2.4, we have showed that for any $x(k) \in \Omega$, the function sequence $\{V_i(x(k))\}$ is a nondecreasing sequence, which will increase its value with an upper bound. This is in contrast to other work in the literature, e.g., [5], where the value functions are constructed as a nonincreasing sequence with lower bound. Moreover, it should be noted that we do not require every control law in the sequence $\{v_i\}$ to be admissible. What we need is a control law sequence to be admissible, i.e., the resultant sequence of control vectors can stabilize the system.

Next, we are ready to discuss the implementation of the iterative ADP algorithm.

(1) *Derivation of the iterative DHP algorithm.* First, we assume that the value function $V_i(x)$ is smooth. In order to implement the iteration between (2.8) and (2.10), for $i = 0, 1, \dots$, we further assume that the minimum of the right hand side of (2.8) can be exactly solved by letting the gradient of the right hand side of (2.8) with respect to $u(k)$ equal to zero, i.e.,

$$\frac{\partial\left(x^{\mathrm{T}}(k)Qx(k)+W(u(k))\right)}{\partial u(k)}+\left(\frac{\partial x(k+1)}{\partial u(k)}\right)^{\mathrm{T}}\frac{\partial V_i(x(k+1))}{\partial x(k+1)}=0. \tag{2.44}$$

Therefore, for $i = 0, 1, \dots$, the corresponding control law $v_i(x)$ can be obtained by solving the above equation, i.e.,

$$v_i(x(k))=\bar{U}\varphi\left(-\frac{1}{2}(\bar{U}R)^{-1}g^{\mathrm{T}}(x(k))\frac{\partial V_i(x(k+1))}{\partial x(k+1)}\right). \tag{2.45}$$

From (2.45), we find that the control law $v_i(x)$ at each step of iteration has to be computed by $\partial V_i(x(k+1))/\partial x(k+1)$, which is not an easy task. Furthermore, at each iteration step of value function $V_{i+1}(x(k))$ in (2.10), there exists an integral term $2\int_0^{v_i(x(k))}\varphi^{-\mathrm{T}}(\bar{U}^{-1}s)\bar{U}R\,ds$ to compute, which is a large computing burden. Therefore, in the following we will present another method called iterative DHP algorithm to implement the iterative ADP algorithm.

Define the costate function $\lambda(x)=\partial V(x)/\partial x$. Here, we assume that the value function $V(x)$ is smooth so that $\lambda(x)$ exists. Then, the recurrent iteration between (2.8) and (2.10) can be implemented as follows.

First, we start with an initial costate function $\lambda_0(\cdot)=0$. Then, for $i = 0, 1, \dots$, by substituting $\lambda_i(x)=\partial V_i(x)/\partial x$ into (2.45), we obtain the corresponding control law $v_i(x)$ as

$$v_i(x(k))=\bar{U}\varphi\Big(-\frac{1}{2}(\bar{U}R)^{-1}g^{\mathrm{T}}(x(k))\lambda_i(x(k+1))\Big). \tag{2.46}$$

For $\lambda_{i+1}(x(k))=\dfrac{\partial V_{i+1}(x(k))}{\partial x(k)}$, according to (2.10) we can obtain

$$\begin{aligned}\lambda_{i+1}(x(k))={}&\frac{\partial\left(x^{\mathrm{T}}(k)Qx(k)+W(v_i(x(k)))\right)}{\partial x(k)}\\&+\left(\frac{\partial v_i(x(k))}{\partial x(k)}\right)^{\mathrm{T}}\frac{\partial\left(x^{\mathrm{T}}(k)Qx(k)+W(v_i(x(k)))\right)}{\partial v_i(x(k))}\\&+\left(\frac{\partial x(k+1)}{\partial x(k)}\right)^{\mathrm{T}}\frac{\partial V_i(x(k+1))}{\partial x(k+1)}\end{aligned}$$

$$
\begin{aligned}
&+\left(\frac{\partial v_i(x(k))}{\partial x(k)}\right)^{\mathrm{T}}\left(\frac{\partial x(k+1)}{\partial v_i(x(k))}\right)^{\mathrm{T}}\frac{\partial V_i(x(k+1))}{\partial x(k+1)}\\
&=\frac{\partial\left(x^{\mathrm{T}}(k)Qx(k)+W(v_i(x(k)))\right)}{\partial x(k)}\\
&+\left(\frac{\partial v_i(x(k))}{\partial x(k)}\right)^{\mathrm{T}}\left[\frac{\partial\left(x^{\mathrm{T}}(k)Qx(k)+W(v_i(x(k)))\right)}{\partial v_i(x(k))}\right.\\
&\left.\quad+\left(\frac{\partial x(k+1)}{\partial v_i(x(k))}\right)^{\mathrm{T}}\frac{\partial V_i(x(k+1))}{\partial x(k+1)}\right]\\
&+\left(\frac{\partial x(k+1)}{\partial x(k)}\right)^{\mathrm{T}}\frac{\partial V_i(x(k+1))}{\partial x(k+1)}.
\end{aligned}
\tag{2.47}
$$

According to (2.44) and (2.45), we have

$$
\frac{\partial\left(x^{\mathrm{T}}(k)Qx(k)+W(v_i(x(k)))\right)}{\partial v_i(x(k))}+\left(\frac{\partial x(k+1)}{\partial v_i(x(k))}\right)^{\mathrm{T}}\frac{\partial V_i(x(k+1))}{\partial x(k+1)}=0. \tag{2.48}
$$

Therefore (2.47) can further be written as

$$
\begin{aligned}
\lambda_{i+1}(x(k))=&\frac{\partial\left(x^{\mathrm{T}}(k)Qx(k)+W(v_i(x(k)))\right)}{\partial x(k)}\\
&+\left(\frac{\partial x(k+1)}{\partial x(k)}\right)^{\mathrm{T}}\frac{\partial V_i(x(k+1))}{\partial x(k+1)},
\end{aligned}
\tag{2.49}
$$

i.e.,

$$
\lambda_{i+1}(x(k))=2Qx(k)+\left(\frac{\partial x(k+1)}{\partial x(k)}\right)^{\mathrm{T}}\lambda_i(x(k+1)). \tag{2.50}
$$

Therefore, the iteration between (2.46) and (2.50) is an implementation of the iteration between (2.8) and (2.10). From (2.46) the control law v_i can directly be obtained by the costate function. Hence the iteration of value function in (2.10) can be omitted in the implementation of this iterative algorithm. Considering the principle of DHP algorithm in Chap. 1, we call such iterative algorithm as iterative DHP algorithm.

Next, we present a *convergence analysis* of the iteration between (2.46) and (2.50).

Theorem 2.7 *Define the control law sequence* $\{v_i\}$ *as in* (2.8), *and update the value function sequence* $\{V_i\}$ *by* (2.10) *with* $V_0(\cdot)=0$. *Define the costate function sequence* $\{\lambda_i\}$ *as in* (2.50) *with* $\lambda_0(\cdot)=0$. *Then, the costate function sequence* $\{\lambda_i\}$ *and the control law sequence* $\{v_i\}$ *are convergent as* $i\to\infty$. *The optimal value* λ^* *is defined as the limit of the costate function* λ_i *when* v_i *approaches the optimal value* u^*.

Proof According to Theorems 2.4–2.6, we have proved that $\lim_{i\to\infty} V_i(x(k)) = J^*(x(k))$, and $J^*(x(k))$ satisfies the corresponding HJB equation, i.e.,

$$J^*(x(k)) = \inf_{u(k)}\{x^{\mathrm{T}}(k)Qx(k) + W(u(k)) + J^*(x(k+1))\}.$$

Therefore, we conclude that the value function sequence $\{V_i\}$ converges to the optimal value function of the DTHJB equation, i.e., $V_i \to J^*$ as $i \to \infty$. With $\lambda_i(x(k)) = \partial V_i(x(k))/\partial x(k)$, we conclude that the corresponding costate function sequence $\{\lambda_i\}$ is also convergent with $\lambda_i \to \lambda^*$ as $i \to \infty$. Since the costate function is convergent, we can conclude that the corresponding control law sequence $\{v_i\}$ converges to the optimal control law u^* as $i \to \infty$. □

Remark 2.8 In the iterative DHP algorithm, via the costate sequence (2.50), the corresponding control law sequence can be directly obtained by (2.46), which does not require the computation of $\partial V_i(x(k+1))/\partial x(k+1)$. Furthermore, in (2.10) there is an integral term $2\int_0^{v_i(x(k))} \varphi^{-\mathrm{T}}(\bar{U}^{-1}s)\bar{U}R\,ds$ to compute at each iteration step, which is not an easy task. However, in (2.50) the integral term has been removed, which greatly reduces the computational burden. On the other hand, in order to compute the costate function by (2.50), the internal dynamics $f(x(k))$ and $g(x(k))$ of the system are needed. In the implementation part of the algorithm, a model network is constructed to approximate the nonlinear dynamics of the system, which avoids the requirement of known $f(x(k))$ and $g(x(k))$.

(2) *RBFNN implementation of the iterative DHP algorithm.* In the iterative DHP algorithm, the optimal control is difficult to solve analytically. For example, in (2.46), the control at step k is a function of costate at step $k+1$. A closed-form explicit solution is difficult to solve, if not impossible. Therefore we need to use parametric structures, such as fuzzy models [15] or neural networks, to approximate the costate function and the corresponding control law in the iterative DHP algorithm. In this subsection, we choose radial basis function (RBF) NNs to approximate the nonlinear functions.

An RBFNN consists of three-layers (input, hidden and output). Each input value is assigned to a node in the input layer and passed directly to the hidden layer without weights. Nodes at the hidden layer are called RBF units, determined by a vector called center and a scalar called width. The Gaussian density function is used as an activation function for the hidden neurons. Then, linear output weights connect the hidden and output layers. The overall input–output equation of the RBFNN is given as

$$y_i = b_i + \sum_{j=1}^{h} w_{ji}\phi_j(X), \tag{2.51}$$

where X is the input vector, $\phi_j(X) = \exp(-\|X - C_j\|^2/\sigma_j{}^2)$ is the activation function of the jth RBF unit in the hidden layer, $C_j \in \mathbb{R}^n$ is the center of the jth RBF unit, h is the number of RBF units, b_i and w_{ji} are the bias term and the weight between hidden and output layer, and y_i is the ith output in the m-dimensional space. Once the optimal RBF centers are established over a wide range of operating points of the plant, the width of the ith center in the hidden layer is calculated by the following formula:

$$\sigma_i = \sqrt{\frac{1}{h}\sum_{j=1}^{h}\sum_{k=1}^{n}(\|c_{ki} - c_{kj}\|)}, \tag{2.52}$$

where c_{ki} and c_{kj} are the kth value of the center of the ith and jth RBF units, respectively. In (2.51) and (2.52), $\|\cdot\|$ represents the Euclidean norm. To avoid the extensive computational complexity during training, the batch mode k-means clustering algorithm is used to calculate the centers of the RBF units.

In order to implement the iterative ADP algorithm, i.e., implement the iteration between (2.46) and (2.50), we employ RBFNNs to approximate the costate function $\lambda_i(x)$ and the corresponding control law $v_i(x)$ at each iteration step i. In the implementation of the iterative DHP algorithm, there are three networks, which are model network, critic network and action network, respectively. All the neural networks are chosen as RBF networks. The inputs of the model network are $x(k)$ and $v_i(x(k))$ and the inputs of the critic network and action network are $x(k+1)$ and $x(k)$, respectively. The diagram of the whole structure is shown in Fig. 2.1.

For unknown plants, before carrying out the iterative DHP algorithm, we first train the model network. For any given $x(k)$ and $\hat{v}_i(x(k))$, we obtain $\hat{x}(k+1)$, and the output of the model network is denoted

$$\hat{x}(k+1) = w_m^{\mathrm{T}}\phi(I_m(k)), \tag{2.53}$$

where $I_m(k) = [x^{\mathrm{T}}(k)\hat{v}_i^{\mathrm{T}}(x(k))]^{\mathrm{T}}$ is the input vector of the model network.

We define the error function of the model network as

$$e_m(k) = \hat{x}(k+1) - x(k+1). \tag{2.54}$$

The weights in the model network are updated to minimize the following performance measure:

$$E_m(k) = \frac{1}{2}e_m^{\mathrm{T}}(k)e_m(k). \tag{2.55}$$

The weight updating rule for model network is chosen as a gradient-based adaptation rule

$$w_m(k+1) = w_m(k) - \alpha_m\left[\frac{\partial E_m(k)}{\partial w_m(k)}\right], \tag{2.56}$$

where α_m is the learning rate of the model network.

After the model network is trained, its weights are kept unchanged.

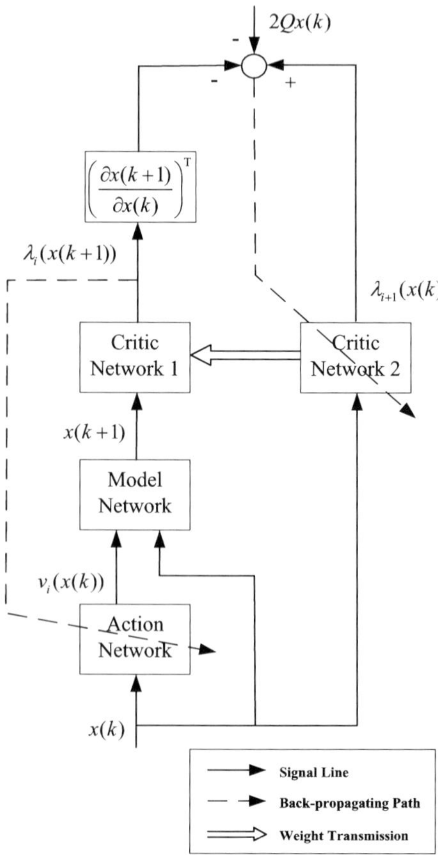

Fig. 2.1 The structure diagram of the iterative DHP algorithm

The critic network is used to approximate the costate function $\lambda_{i+1}(x)$. The output of the critic network is denoted

$$\hat{\lambda}_{i+1}(x(k)) = w_{c(i+1)}^{\mathrm{T}}\phi(x(k)). \tag{2.57}$$

The target costate function is given as in (2.50). Define the error function for the critic network as

$$e_{c(i+1)}(k) = \hat{\lambda}_{i+1}(x(k)) - \lambda_{i+1}(x(k)). \tag{2.58}$$

The objective function to be minimized for the critic network is

$$E_{c(i+1)}(k) = \frac{1}{2}e_{c(i+1)}^{\mathrm{T}}(k)e_{c(i+1)}(k). \tag{2.59}$$

The weight updating rule for the critic network is a gradient-based adaptation given by

$$w_{c(i+1)}(j+1) = w_{c(i+1)}(j) - \alpha_c\left[\frac{\partial E_{c(i+1)}(k)}{\partial w_{c(i+1)}(j)}\right], \tag{2.60}$$

where $\alpha_c > 0$ is the learning rate of the critic network, and j is the inner-loop iteration step for updating the weight parameters.

In the action network, the state $x(k)$ is used as the input of the network and the output can be formulated as

$$\hat{v}_i(x(k)) = w_{ai}^{\mathrm{T}} \phi(x(k)). \tag{2.61}$$

The target value of the control $v_i(x(k))$ is obtained by (2.46). So we can define the error function of the action network as

$$e_{ai}(k) = \hat{v}_i(x(k)) - v_i(x(k)). \tag{2.62}$$

The weights of the action network are updated to minimize the following performance error measure:

$$E_{ai}(k) = \frac{1}{2} e_{ai}^{\mathrm{T}}(k) e_{ai}(k). \tag{2.63}$$

The updating algorithm is then similar to the one for the critic network. By the gradient descent rule, we obtain

$$w_{ai}(j+1) = w_{ai}(j) - \alpha_a \left[\frac{\partial E_{ai}(k)}{\partial w_{ai}(j)} \right], \tag{2.64}$$

where $\alpha_a > 0$ is the learning rate of the action network, and j is the inner-loop iteration step for updating the weight parameters.

From the neural-network implementation, we can find that in this iterative DHP algorithm, $\partial V_i(x(k+1))/\partial x(k+1)$ is replaced by $\hat{\lambda}_i(x(k+1))$, which is just the output of the critic network. Therefore, it is more accurate than computing by backpropagation through the critic network as in [1].

(3) *Design procedure of the approximate optimal controller*. Based on the iterative DHP algorithm, the design procedure of the optimal control scheme is summarized as follows:

1. Choose $i_{\max}$, $j_{\max}^a$, $j_{\max}^c$, ε_m, ε_0, $\bar{U}$, α_m, α_c, α_a and the weight matrices Q and R.
2. Construct the model network $\hat{x}(k+1) = w_m^{\mathrm{T}} \phi(I_m(k))$ with the initial weight parameters w_{m0} chosen randomly from $[-0.1, 0.1]$ and train the model network with a random input vector uniformly distributed in the interval $[-1, 1]$ and arbitrary initial state vector in $[-1, 1]$ till the given accuracy ε_m is reached.
3. Set the iteration step $i = 0$. Set the initial weight parameters of critic network w_{c0} as zero so that the initial value of the costate function $\lambda_0(\cdot) = 0$, and initialize the action network with the weight parameters w_{a0} chosen randomly in $[-0.1, 0.1]$.
4. Choose an array of state vector $x(k) = (x^{(1)}(k), x^{(2)}(k), \ldots, x^{(p)}(k))$ randomly from the operation region and compute the corresponding output target $v_i(x(k)) = (v_i(x^{(1)}(k)), v_i(x^{(2)}(k)), \ldots, v_i(x^{(p)}(k)))$ by (2.46), where the state

vector at the next time instant

$$x(k+1)=\left(x^{(1)}(k+1),x^{(2)}(k+1),\dots,x^{(p)}(k+1)\right)$$

is computed by the model network (2.53). With the same state vector $x(k)=(x^{(1)}(k),x^{(2)}(k),\dots,x^{(p)}(k))$ and

$$x(k+1)=\left(x^{(1)}(k+1),x^{(2)}(k+1),\dots,x^{(p)}(k+1)\right),$$

compute the resultant output target

$$\lambda_{i+1}(x(k))=\left(\lambda_{i+1}(x^{(1)}(k)),\lambda_{i+1}(x^{(2)}(k)),\dots,\lambda_{i+1}(x^{(p)}(k))\right)$$

by (2.50).

5. Set $w_{c(i+1)}=w_{ci}$. With the data set $(x^{(j)}(k),\lambda_{i+1}(x^{(j)}(k)))$, $j=1,2,\dots,p$, update the weight parameters of the critic network $w_{c(i+1)}$ by (2.60) for $j^c_{\max}$ steps to get the approximate costate function $\hat{\lambda}_{i+1}$.
6. With the data set $(x^{(j)}(k),v_i(x^{(j)}(k)))$, $j=1,2,\dots,p$, update the weight parameters of the action network w_{ai} by (2.64) for $j^a_{\max}$ steps to get the approximate control law $\hat{v}_i$.
7. If

$$\|\lambda_{i+1}(x(k))-\lambda_i(x(k))\|^2<\varepsilon_0,$$

go to Step 9; otherwise, go to Step 8.

8. If $i>i_{\max}$, go to Step 9; otherwise, set $i=i+1$ and go to Step 4.
9. Set the final approximate optimal control law $\hat{u}^*(x)=\hat{v}_i(x)$.
10. Stop.

As stated in the last subsection, the iterative algorithm will be convergent with $\lambda_i(x)\to\lambda^*(x)$ and the control sequence $v_i(x)\to u^*(x)$ as $i\to\infty$. However, in practical applications, we cannot implement the iteration till $i\to\infty$. Actually, we iterate the algorithm for a max number $i_{\max}$ or with a pre-specified accuracy ε_0 to test the convergence of the algorithm. In the above procedure, there are two levels of loops. The outer loop starts from Step 3 and ends at Step 8. There are two inner loops in Steps 5 and 6, respectively. The inner loop of Step 5 includes $j^c_{\max}$ iterative steps, and the inner loop of Step 6 includes $j^a_{\max}$ iterative steps. The state vector $x(k)$ is chosen randomly at Step 4. Suppose that the associated random probability density function is nonvanishing everywhere. Then we can assume that all the states will be explored. So we know that the resulting networks tend to satisfy the formulas (2.46) and (2.50) for all state vectors $x(k)$. The limits of $\hat{\lambda}_i$ and $\hat{v}_i$ will approximate the optimal ones λ^* and u^*, respectively. The parameters ε_0 and $i_{\max}$ are chosen by the designer. The smaller the value of ε_0 is set, the more accurate the costate function and the optimal control law will be. If the condition set in Step 7 is satisfied, it implies that the costate function sequence is convergent with the pre-specified accuracy. The larger the value of $i_{\max}$ in Step 8 is set, the more accurate the obtained control law $\hat{v}(x)$ will be at the price of increased computational burden.

2.2.3 Simulations

In this section, two examples are provided to demonstrate the effectiveness of the control scheme developed in this subsection.

Example 2.9 (Nonlinear Discrete-Time System) Consider the following nonlinear system [5]:

$$x(k+1) = f(x(k)) + g(x(k))u(k), \tag{2.65}$$

where $f(x(k)) = \begin{bmatrix} -0.8x_2(k) \\ \sin(0.8x_1(k)-x_2(k))+1.8x_2(k) \end{bmatrix}$, $g(x(k)) = \begin{bmatrix} 0 \\ -x_2(k) \end{bmatrix}$, and assume that the control constraint is set to $|u| \le 0.3$.

Define the cost functional as

$$J(x(k), u(\cdot)) = \sum_{i=k}^{\infty} \left\{ x^{\mathrm{T}}(i)Qx(i) + 2\int_0^{u(i)} \tanh^{-\mathrm{T}}(\bar{U}^{-1}s)\bar{U}R\,ds \right\}, \tag{2.66}$$

where $\bar{U} = 0.3$, and the weight matrices are chosen as $Q = \begin{bmatrix} 1 & 0 \\ 0 & 1 \end{bmatrix}$ and $R = [0.5]$.

First, we perform the simulation of iterative ADP algorithm. In this iterative algorithm, we choose RBFNNs as the critic network, the action network and the model network with the structure 2–9–2, 2–9–1 and 3–9–2, respectively. The training sets are selected as $-1 \le x_1 \le 1$ and $-1 \le x_2 \le 1$, which is the operation region of the system. It should be mentioned that the model network should be trained first. The initial state vectors are chosen randomly from $[-1, 1]$. Under the learning rate of $\alpha_m = 0.1$, the model network is trained until the given accuracy $\varepsilon_m = 10^{-6}$ is reached. After the training of the model network is completed, the weights are kept unchanged. Then, the critic network and the action network are trained with the learning rates $\alpha_a = \alpha_c = 0.1$ and the inner-loop iteration number $j^c_{\max} = j^a_{\max} = 2000$. Meanwhile the pre-specified accuracy ε_0 is set to 10^{-20}. Denote the outer loop iteration number as L. After implementing the outer loop iteration for $L = i_{\max} = 100$, the convergence curves of the costate function are shown in Fig. 2.2. It can be seen that the costate function is basically convergent with the outer loop iteration $L > 15$. In order to compare the different actions of the control laws obtained under different outer loop iteration numbers, for the same initial state vector $x_1(0) = 0.5$ and $x_2(0) = 0.5$, we apply different control laws to the plant for 30 time steps and obtain the simulation results as follows. The state curves are shown in Figs. 2.3 and 2.4, and the corresponding control inputs are shown in Fig. 2.5. It can be seen that the system responses are improved when the outer loop iteration number L is increased. When $L > 80$, the system responses only improve slightly in performance.

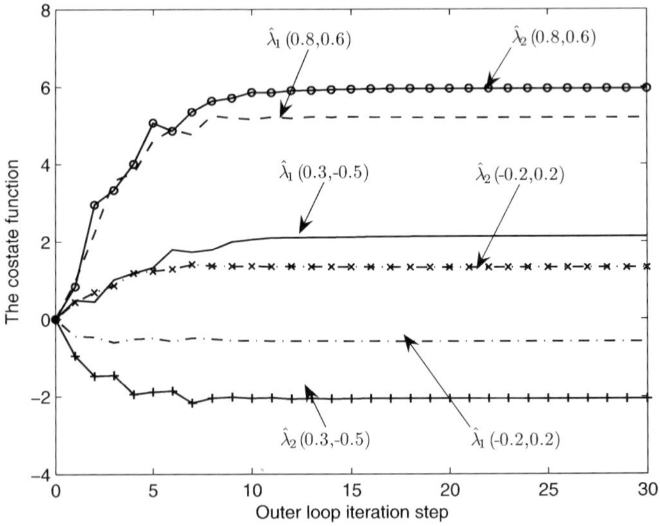

Fig. 2.2 The convergence process of the costate function at $x = (0.3, -0.5)$, $x = (-0.2, 0.2)$, $x = (0.8, 0.6)$

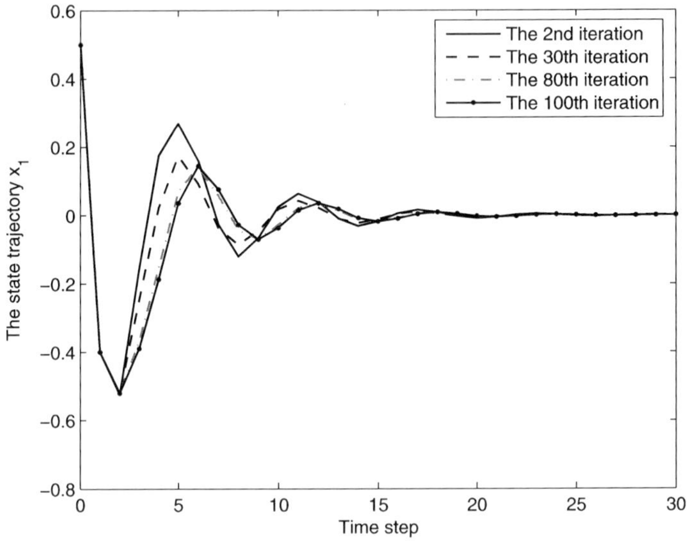

Fig. 2.3 The state trajectory x_1 for $L = 2, 30, 80, 100$

It should be mentioned that in order to show the convergence characteristics of the iterative process more clearly, we set the required accuracy ε_0 to a very small number 10^{-20} and we set the max iteration number to twice of what is needed.

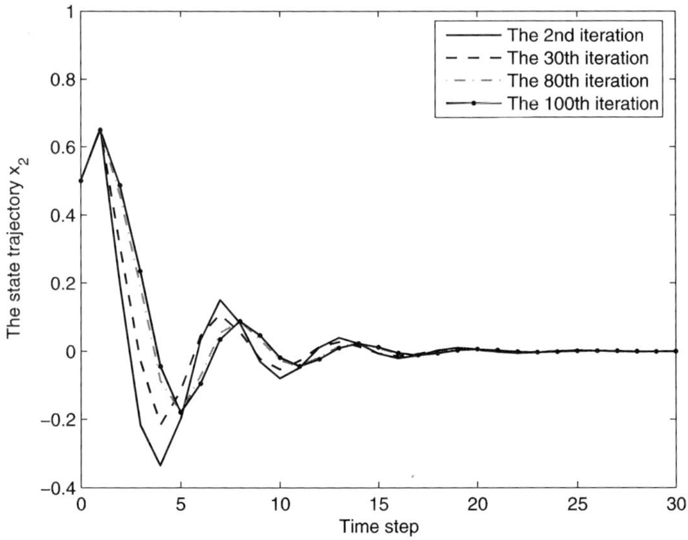

Fig. 2.4 The state trajectory x_2 for $L = 2, 30, 80, 100$

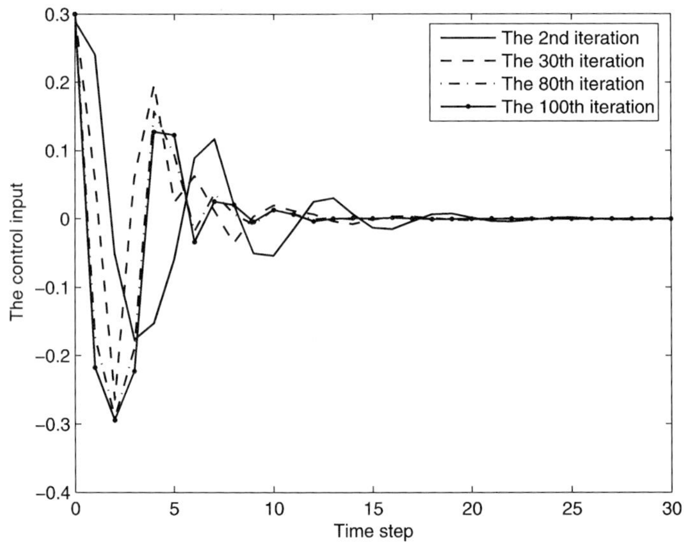

Fig. 2.5 The control input u for $L = 2, 30, 80, 100$

In this way, the given accuracy ε_0 did not take effect even when the max iteration number is reached. Therefore, it seems that the max iteration number $i_{\max}$ becomes the stopping criterion in this case. If the designer wants to save the running time, the

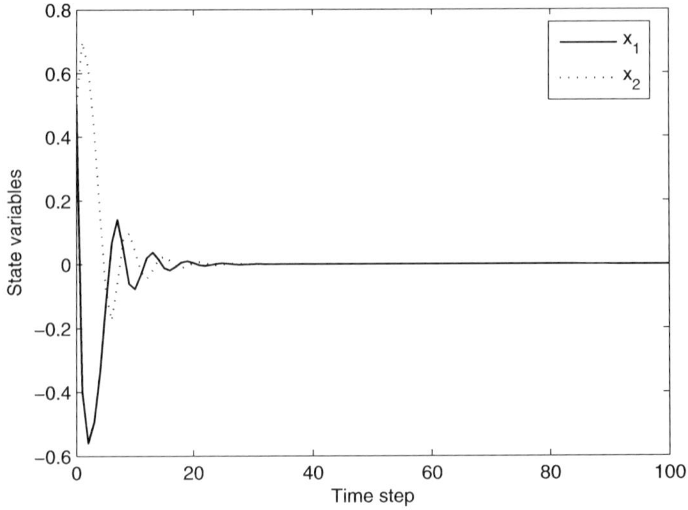

Fig. 2.6 The state variables curves without considering the actuator saturation in the controller design

pre-specified accuracy ε_0 can be set to a normal value so that the iterative process will be stopped once the accuracy ε_0 is reached.

Moreover, in order to make comparison with the controller designed without considering the actuator saturation, we also present the system responses obtained by the controller designed regardless of the actuator saturation. However, the actuator saturation is actually existing, therefore in the simulation if the control input overrun the saturation bound, it is limited to the bound value. After simulation, the state curves are as shown in Fig. 2.6, and the control curve is shown in Fig. 2.7.

From the simulation results, we can see that the iterative costate function sequences do converge to the optimal ones with very fast speed, which also indicates the validity of the iterative ADP algorithm for dealing with constrained nonlinear systems. Comparing Fig. 2.5 with Fig. 2.7, we can see that in Fig. 2.5 the restriction of actuator saturation has been overcome successfully, but in Fig. 2.7 the control input has overrun the saturation bound and therefore be limited to the bound value. From this point, we can conclude that the present iterative ADP algorithm is effective in dealing with the constrained optimal control problem.

Example 2.10 (Mass–Spring System) Consider the following discrete-time nonlinear mass–spring system:

$$\begin{cases} x_1(k+1) = 0.05x_2(k) + x_1(k), \\ x_2(k+1) = -0.0005x_1(k) - 0.0335x_1^3(k) + 0.05u(k) + x_2(k), \end{cases} \tag{2.67}$$

where $x(k)$ is the state vector, and $u(k)$ is the control input.

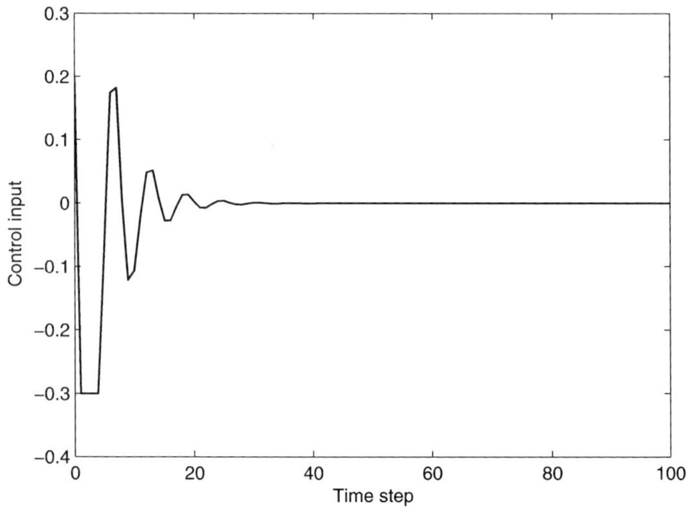

Fig. 2.7 The control input curve without considering the actuator saturation in the controller design

Define the cost functional as

$$J(x(k), u(\cdot)) = \sum_{i=k}^{\infty} \left\{ x^{\mathrm{T}}(i)Qx(i) + 2\int_0^{u(i)} \tanh^{-\mathrm{T}}(\bar{U}^{-1}s)\bar{U}R\,ds \right\}, \tag{2.68}$$

where the control constraint is set to $\bar{U} = 0.6$, and the weight matrices are chosen as $Q = \left[\begin{smallmatrix} 0.5 & 0 \\ 0 & 0.5 \end{smallmatrix}\right]$ and $R = [1]$. The training sets are $-1 \le x_1 \le 1$ and $-1 \le x_2 \le 1$. The critic network, the action network and the model network are chosen as RBF neural networks with the structure of 2–16–2, 2–16–1 and 3–16–2, respectively. In the training process, the learning rates are set to $\alpha_a = \alpha_c = 0.1$. The other parameters are set the same as those in Example 2.9. After implementing the outer loop iteration for $L = i_{\max} = 300$, the convergence curves of the costate function are shown in Fig. 2.8. It can be seen that the costate function is basically convergent with the outer loop iteration $L > 200$. In order to compare the different actions of the control laws obtained under different outer loop iteration numbers, for the same initial state vector $x_1(0) = -1$ and $x_2(0) = 1$, we apply different control laws to the plant for 300 time steps and obtain the simulation results as follows. The state curves are shown in Figs. 2.9, 2.10, and the corresponding control inputs are shown in Fig. 2.11. It can be seen that the closed-loop system is divergent when using the control law obtained by $L = 2$, and the system's responses are improved when the outer loop iteration number L is increased. When $L > 200$, the system

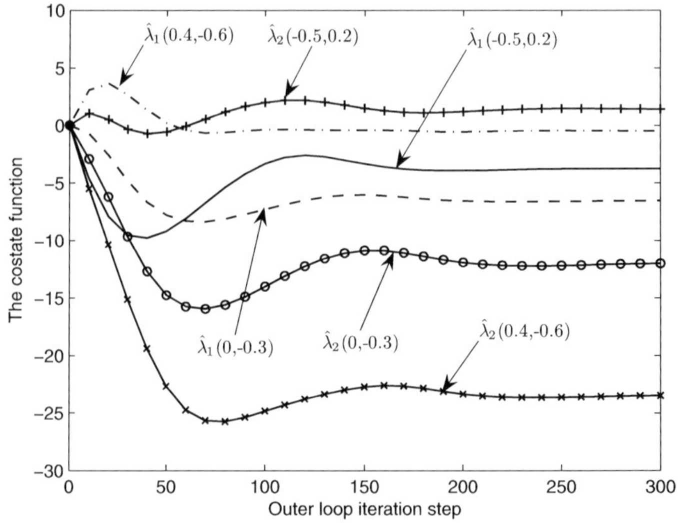

Fig. 2.8 The convergence process of the costate function at $x = (-0.5, 0.2)$, $x = (0.4, -0.6)$, $x = (0, -0.3)$

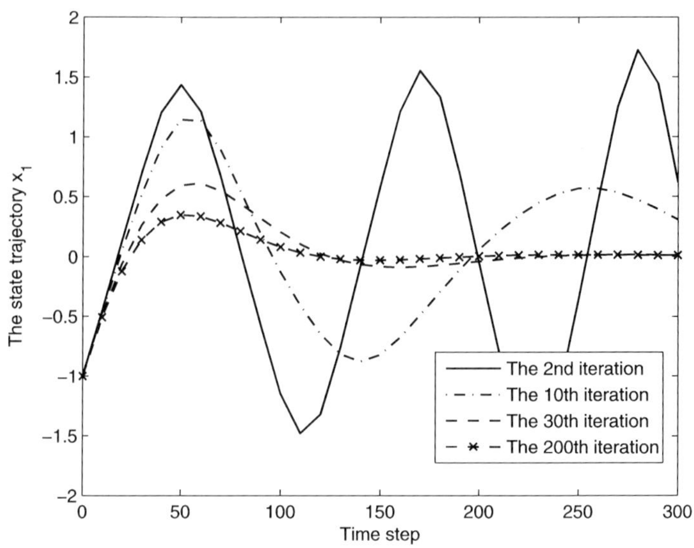

Fig. 2.9 The state trajectory x_1 for $L = 2, 10, 30, 200$

responses basically remain unchanged with no significant improvement in performance.

In order to make comparison with the controller without considering the actuator saturation, we also present the controller designed by iterative ADP algorithm

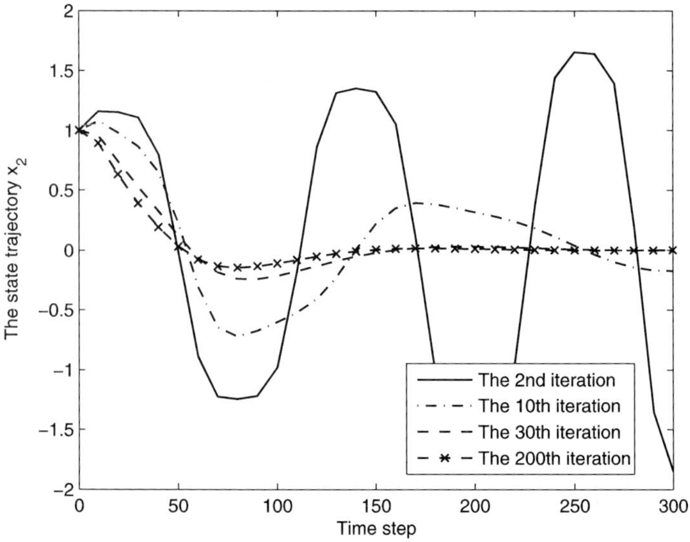

Fig. 2.10 The state trajectory x_2 for $L = 2, 10, 30, 200$

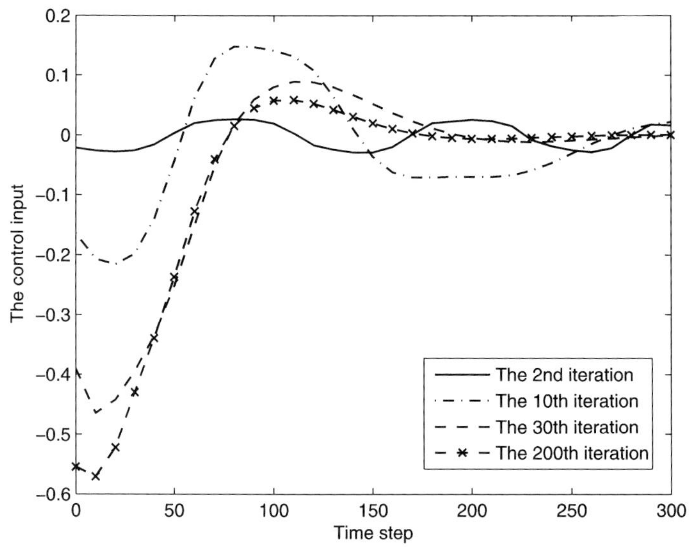

Fig. 2.11 The control input u for $L = 2, 10, 30, 200$

regardless of the actuator saturation. The state curves are shown in Fig. 2.12 and the control curve is shown in Fig. 2.13.

From the simulation results, we can see that the iterative costate function sequence does converge to the optimal one very fast. Comparing Fig. 2.11 with

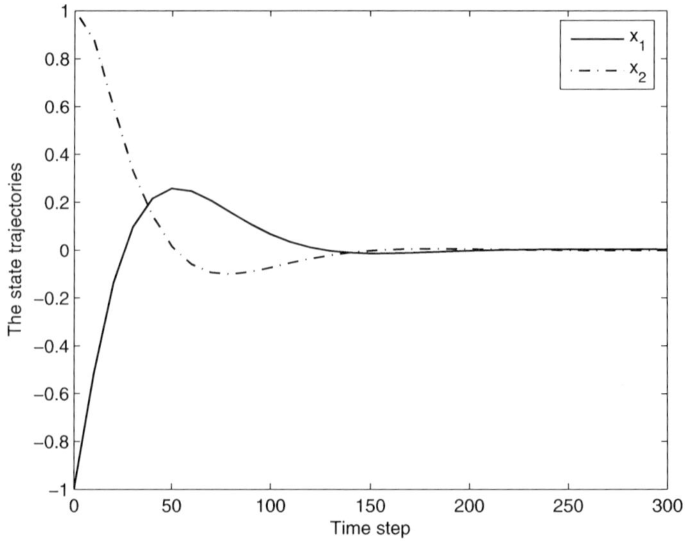

Fig. 2.12 The state curves without considering the actuator saturation in controller design

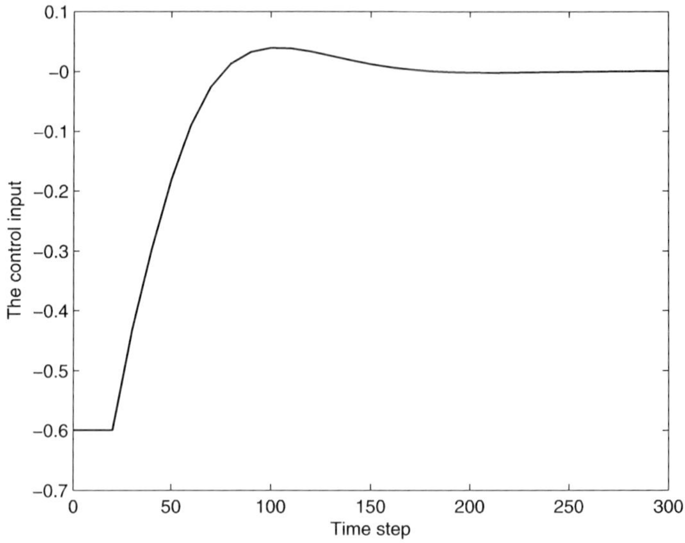

Fig. 2.13 The control curves without considering the actuator saturation in controller design

Fig. 2.13, we can find that in Fig. 2.11 the restriction of actuator saturation has been overcome successfully, which further verifies the effectiveness of the present iterative ADP algorithm.

2.3 Infinite-Horizon Optimal State Feedback Control Based on GDHP

2.3.1 Problem Formulation

In this section, we will study the discrete-time nonlinear systems described by

$$x(k+1) = f(x(k)) + g(x(k))u(k), \tag{2.69}$$

where $x(k) \in \mathbb{R}^n$ is the state vector and $u(k) \in \mathbb{R}^m$ is the control vector, $f(\cdot)$ and $g(\cdot)$ are differentiable in their arguments with $f(0) = 0$. Assume that $f + gu$ is Lipschitz continuous on a set Ω in $\mathbb{R}^n$ containing the origin, and that the system (2.69) is controllable in the sense that there exists a continuous control law on Ω that asymptotically stabilizes the system.

Let $x(0)$ be an initial state and define $\underline{u}_0^{N-1} = (u(0), u(1), u(N-1))$ be a control sequence with which the system (2.69) gives a trajectory starting from $x(0)$: $x(1) = f(x(0)) + g(x(0))u(0)$, $x(2) = f(x(1)) + g(x(1))u(1), \dots, x(N) = f(x(N-1)) + g(x(N-1))u(N-1)$. We call the number of elements in the control sequence $\underline{u}_0^{N-1}$ the length of $\underline{u}_0^{N-1}$ and denote it as $|\underline{u}_0^{N-1}|$. Then, $|\underline{u}_0^{N-1}| = N$. The final state under the control sequence $\underline{u}_0^{N-1}$ can be denoted $x^{(f)}(x(0), \underline{u}_0^{N-1}) = x(N)$. When the control sequence starting from $u(0)$ has infinite length, we denote it as $\underline{u}_0^{\infty} = (u(0), u(1), \dots)$ and then the correspondingly final state can be written as $x^{(f)}(x(0), \underline{u}_0^{\infty}) = \lim_{k\to\infty} x(k)$.

Definition 2.11 A nonlinear dynamical system is said to be stabilizable on a compact set $\Omega \in \mathbb{R}^n$, if for all initial conditions $x(0) \in \Omega$, there exists a control sequence $\underline{u}_0^{\infty} = (u(0), u(1), \dots)$, $u(i) \in \mathbb{R}^m$, $i = 0, 1, \dots$, such that the state $x^{(f)}(x(0), \underline{u}_0^{\infty}) = 0$.

Let $\underline{u}_k^{\infty} = (u(k), u(k+1), \dots)$ be the control sequence starting at k. It is desired to find the control sequence $\underline{u}_k^{\infty}$ which minimizes the infinite-horizon cost functional given by

$$J(x(k), \underline{u}_k^{\infty}) = \sum_{i=k}^{\infty} \gamma^{i-k} l(x(i), u(i)), \tag{2.70}$$

where l is the utility function, $l(0,0) = 0$, $l(x(i), u(i)) \geq 0$ for $\forall x(i), u(i)$, and γ is the discount factor with $0 < \gamma \leq 1$. Generally speaking, the utility function can be chosen as the quadratic form as follows:

$$l(x(i), u(i)) = x^{\mathrm{T}}(i)Qx(i) + u^{\mathrm{T}}(i)Ru(i).$$

For optimal control problems, the designed feedback control must not only stabilize the system on Ω but also guarantee that (2.70) is finite, i.e., the control must be admissible.

It is noted that a control law sequence $\{\eta_i\} = (\eta_N, \dots, \eta_1, \eta_0)$, $N \to \infty$, is called admissible if the resultant control sequence $(u(0), u(1), \dots, u(N))$ stabilizes system (2.69) with any initial state $x(0)$ and guarantees that $J(x(0), \underline{u}_0^N)$ is finite. In this case, it should be mentioned that each control action obeys a different control law, i.e., the control action $u(i)$ is produced by the control law η_{N-i} or $u(i) = \eta_{N-i}(x(i))$, for $i = 0, 1, \dots, N$, $N \to \infty$.

Let

$$\mathfrak{A}_{x(k)} = \left\{\underline{u}_k^\infty : x^{(f)}(x(k), \underline{u}_k^\infty) = 0\right\}$$

be the set of all *infinite-horizon* admissible control sequences of $x(k)$. Define the optimal value function as

$$J^*(x(k)) = \inf_{\underline{u}_k^\infty} \left\{J(x(k), \underline{u}_k^\infty) : \underline{u}_k^\infty \in \mathfrak{A}_{x(k)}\right\}. \tag{2.71}$$

Note that (2.70) can be written as

$$\begin{aligned} J(x(k), \underline{u}_k^\infty) &= x^{\mathrm{T}}(k) Q x(k) + u^{\mathrm{T}}(k) R u(k) + \gamma \sum_{i=k+1}^{\infty} \gamma^{i-k-1} l(x(i), u(i)) \\ &= x^{\mathrm{T}}(k) Q x(k) + u^{\mathrm{T}}(k) R u(k) + \gamma J(x(k+1), \underline{u}_{k+1}^\infty). \end{aligned} \tag{2.72}$$

According to Bellman's optimality principle, it is known that, for the case of infinite-horizon optimization, the optimal value function $J^*(x(k))$ is time invariant and satisfies the DTHJB equation

$$J^*(x(k)) = \min_{u(k)} \left\{x^{\mathrm{T}}(k) Q x(k) + u^{\mathrm{T}}(k) R u(k) + \gamma J^*(x(k+1))\right\}. \tag{2.73}$$

The optimal control u^* satisfies the first-order necessary condition, which is given by the gradient of the right hand side of (2.73) with respect to $u(k)$ as

$$\frac{\partial\big(x^{\mathrm{T}}(k) Q x(k) + u^{\mathrm{T}}(k) R u(k)\big)}{\partial u(k)} + \gamma \left(\frac{\partial x(k+1)}{\partial u(k)}\right)^{\mathrm{T}} \frac{\partial J^*(x(k+1))}{\partial x(k+1)} = 0.$$

Then, we obtain

$$u^*(x(k)) = -\frac{\gamma}{2} R^{-1} g^{\mathrm{T}}(x(k)) \frac{\partial J^*(x(k+1))}{\partial x(k+1)}. \tag{2.74}$$

By substituting (2.74) into (2.73), the DTHJB equation becomes

$$\begin{aligned} J^*(x(k)) = x^{\mathrm{T}}(k) Q x(k) &+ \frac{\gamma^2}{4} \left(\frac{\partial J^*(x(k+1))}{\partial x(k+1)}\right)^{\mathrm{T}} g(x(k)) R^{-1} \\ &\times g^{\mathrm{T}}(x(k)) \frac{\partial J^*(x(k+1))}{\partial x(k+1)} + \gamma J^*(x(k+1)) \end{aligned} \tag{2.75}$$

where $J^*(x(k))$ is the optimal value function corresponding to the optimal control law $u^*(x(k))$. When dealing with the linear quadratic regulator (LQR) optimal control problems, this equation reduces to the Riccati equation which can be efficiently solved. In the general nonlinear case, however, the HJB equation cannot be solved exactly.

2.3.2 *Infinite-Horizon Optimal State Feedback Control Based on GDHP*

Four parts are included in this subsection. In the first part, the unknown nonlinear system is identified via an NN system identification scheme with stability proof. The iterative ADP algorithm is introduced in the second part, while in the third part, the corresponding convergence proof is developed. Then, in the fourth part, the implementation of the iterative ADP algorithm based on NN is described in detail.

2.3.2.1 NN Identification of the Unknown Nonlinear System

For the design of the NN identifier, a three-layer NN is considered as the function approximation structure. Let the number of hidden-layer neurons be denoted by l, the ideal weight matrix between the input layer and hidden layer be denoted by ν_m^*, and the ideal weight matrix between the hidden layer and output layer be denoted by ω_m^*. According to the universal approximation property [8] of NN, the system dynamics (2.69) has a NN representation on a compact set S, which can be written as

$$x(k+1) = \omega_m^{*\mathrm{T}} \sigma\left(\nu_m^{*\mathrm{T}} z(k)\right) + \theta(k). \tag{2.76}$$

In (2.76), $z(k) = [x^{\mathrm{T}}(k)\ u^{\mathrm{T}}(k)]^{\mathrm{T}}$ is the NN input, $\theta(k)$ is the bounded NN functional approximation error according to the universal approximation property, and $[\sigma(\bar{z})]_i = (e^{\bar{z}_i} - e^{-\bar{z}_i})/(e^{\bar{z}_i} + e^{-\bar{z}_i})$, $i = 1, 2, \ldots, l$, are the activation functions selected in this work, where $\bar{z}(k) = \nu_m^{*\mathrm{T}} z(k)$, $\bar{z}(k) \in \mathbb{R}^l$. Additionally, the NN activation functions are bounded such that $\|\sigma(\bar{z}(k))\| \le \sigma_M$ for a constant σ_M.

In the system identification process, we keep the weight matrix between the input layer and the hidden layer as constant while only tune the weight matrix between the hidden layer and the output layer. So, we define the NN system identification scheme as

$$\hat{x}(k+1) = \omega_m^{\mathrm{T}}(k) \sigma(\bar{z}(k)), \tag{2.77}$$

where $\hat{x}(k)$ is the estimated system state vector, and $\omega_m(k)$ is the estimation of the constant ideal weight matrix ω_m^*.

Denote $\tilde{x}(k) = \hat{x}(k) - x(k)$ as the system identification error. Combining (2.76) and (2.77), we can obtain the identification error dynamics as

$$\tilde{x}(k+1) = \tilde{\omega}_m^{\mathrm{T}}(k)\sigma(\bar{z}(k)) - \theta(k), \tag{2.78}$$

where $\tilde{\omega}_m(k) = \omega_m(k) - \omega_m^*$. Let $\psi(k) = \tilde{\omega}_m^{\mathrm{T}}(k)\sigma(\bar{z}(k))$. Then, (2.78) can be rewritten as

$$\tilde{x}(k+1) = \psi(k) - \theta(k). \tag{2.79}$$

The weights in the system identification process are updated to minimize the following performance measure:

$$E(k+1) = \frac{1}{2}\tilde{x}^{\mathrm{T}}(k+1)\tilde{x}(k+1). \tag{2.80}$$

Using the gradient-based adaptation rule, the weights can be updated as

$$\begin{aligned} \omega_m(k+1) &= \omega_m(k) - \alpha_m \left[\frac{\partial E(k+1)}{\partial \omega_m(k)} \right] \\ &= \omega_m(k) - \alpha_m \sigma(\bar{z}(k))\tilde{x}^{\mathrm{T}}(k+1), \end{aligned} \tag{2.81}$$

where $\alpha_m > 0$ is the NN learning rate.

We now give the following assumption before presenting the asymptotic stability proof of the state estimation error $\tilde{x}(k)$.

Assumption 2.12 The NN approximation error term $\theta(k)$ is assumed to be upper bounded by a function of the state estimation error $\tilde{x}(k)$ such that

$$\theta^{\mathrm{T}}(k)\theta(k) \le \theta_{Mk} = \delta\tilde{x}^{\mathrm{T}}(k)\tilde{x}(k), \tag{2.82}$$

where δ is the constant target value with δ_M as its upper bound, i.e., $\|\delta\| \le \delta_M$.

Next, the stability analysis of the present NN-based system identification scheme is presented by using the Lyapunov theory.

Theorem 2.13 (cf. [10]) *Let the identification scheme* (2.77) *be used to identify the nonlinear system* (2.69), *and let the parameter update law given in* (2.81) *be used for tuning the NN weights. Then, the state estimation error dynamics* $\tilde{x}(k)$ *is asymptotically stable while the parameter estimation error* $\tilde{\omega}_m(k)$ *is bounded.*

Proof Consider the following positive definite Lyapunov function candidate:

$$L_k = L_{1k} + L_{2k}, \tag{2.83}$$

where

$$L_{1k} = \tilde{x}^{\mathrm{T}}(k)\tilde{x}(k),$$

$$L_{2k}=\frac{1}{\alpha_m}\mathrm{tr}\big\{\tilde{\omega}_m^{\mathrm{T}}(k)\tilde{\omega}_m(k)\big\}.$$

Taking the first difference of the Lyapunov function (2.83) and substituting the identification error dynamics (2.79) and the NN weight update law (2.81) reveal that

$$\begin{aligned}
\Delta L_{1k}&=\tilde{x}^{\mathrm{T}}(k+1)\tilde{x}(k+1)-\tilde{x}^{\mathrm{T}}(k)\tilde{x}(k)\\
&=\psi^{\mathrm{T}}(k)\psi(k)-2\psi^{\mathrm{T}}(k)\theta(k)+\theta^{\mathrm{T}}(k)\theta(k)-\tilde{x}^{\mathrm{T}}(k)\tilde{x}(k)\\
\Delta L_{2k}&=\frac{1}{\alpha_m}\mathrm{tr}\big\{\tilde{\omega}_m^{\mathrm{T}}(k+1)\tilde{\omega}_m(k+1)-\tilde{\omega}_m^{\mathrm{T}}(k)\tilde{\omega}_m(k)\big\}\\
&=\frac{1}{\alpha_m}\mathrm{tr}\big\{-2\alpha_m\psi(k)\tilde{x}^{\mathrm{T}}(k+1)\\
&\quad+\alpha_m^2\tilde{x}(k+1)\sigma^{\mathrm{T}}(\bar{z}(k))\sigma(\bar{z}(k))\tilde{x}^{\mathrm{T}}(k+1)\big\}\\
&=-2\psi^{\mathrm{T}}(k)\tilde{x}(k+1)+\alpha_m\sigma^{\mathrm{T}}(\bar{z}(k))\sigma(\bar{z}(k))\tilde{x}^{\mathrm{T}}(k+1)\tilde{x}(k+1).
\end{aligned}$$

After applying the Cauchy–Schwarz inequality ($(a_1+a_2+\cdots+a_n)^{\mathrm{T}}(a_1+a_2+\cdots+a_n)\le n(a_1^{\mathrm{T}}a_1+a_2^{\mathrm{T}}a_2+\cdots+a_n^{\mathrm{T}}a_n)$) to ΔL_{2k}, we have

$$\begin{aligned}
\Delta L_{2k}\le{}&-2\psi^{\mathrm{T}}(k)\big(\psi(k)-\theta(k)\big)\\
&+2\alpha_m\sigma^{\mathrm{T}}(\bar{z}(k))\sigma(\bar{z}(k))\big(\psi^{\mathrm{T}}(k)\psi(k)+\theta^{\mathrm{T}}(k)\theta(k)\big).
\end{aligned}$$

Therefore, we can find that

$$\begin{aligned}
\Delta L_k\le{}&-\psi^{\mathrm{T}}(k)\psi(k)+\theta^{\mathrm{T}}(k)\theta(k)-\tilde{x}^{\mathrm{T}}(k)\tilde{x}(k)\\
&+2\alpha_m\sigma^{\mathrm{T}}(\bar{z}(k))\sigma(\bar{z}(k))\big(\psi^{\mathrm{T}}(k)\psi(k)+\theta^{\mathrm{T}}(k)\theta(k)\big).
\end{aligned}$$

Considering $\|\sigma(\bar{z}(k))\|\le\sigma_M$ and (2.82), we obtain

$$\begin{aligned}
\Delta L_k\le{}&-\big(1-2\alpha_m\sigma_M^2\big)\|\psi(k)\|^2\\
&-\big(1-\delta_M-2\alpha_m\delta_M\sigma_M^2\big)\|\tilde{x}(k)\|^2.
\end{aligned}\tag{2.84}$$

Define $\alpha_m\le\rho^2/(2\sigma_M^2)$; then (2.84) becomes

$$\begin{aligned}
\Delta L_k&\le-\big(1-\rho^2\big)\|\psi(k)\|^2-\big(1-\delta_M-\delta_M\rho^2\big)\|\tilde{x}(k)\|^2\\
&=-\big(1-\rho^2\big)\big\|\tilde{\omega}_m^{\mathrm{T}}(k)\sigma(\bar{z}(k))\big\|^2\\
&\quad-\big(1-\delta_M-\delta_M\rho^2\big)\|\tilde{x}(k)\|^2.
\end{aligned}\tag{2.85}$$

From (2.85), we can conclude that $\Delta L_k\le 0$ provided $0<\delta_M<1$ and

$$\max\left\{-\sqrt{\frac{1-\delta_M}{\delta_M}},-1\right\}\le\rho\le\min\left\{\sqrt{\frac{1-\delta_M}{\delta_M}},1\right\},$$

where $\rho \neq 0$. As long as the parameters are selected as discussed above, $\Delta L_k \leq 0$ in (2.85), which shows stability in the sense of Lyapunov. Therefore, $\tilde{x}(k)$ and $\tilde{\omega}_m(k)$ are bounded, provided $\tilde{x}_0$ and $\tilde{\omega}_m(0)$ are bounded in the compact set S. Furthermore, by summing both sides of (2.85) to infinity and taking account of $\Delta L_k \leq 0$, we have

$$\left|\sum_{k=0}^{\infty} \Delta L_k\right| = \left|\lim_{k\to\infty} L_k - L_0\right| < \infty.$$

This implies that

$$\sum_{k=0}^{\infty}\left\{\left(1-\rho^2\right)\left\|\tilde{\omega}_m^{\mathrm{T}}(k)\sigma(\bar{z}(k))\right\|^2 + \left(1-\delta_M-\delta_M\rho^2\right)\|\tilde{x}(k)\|^2\right\} < \infty.$$

Hence, it can be concluded that the estimation error approaches zero, i.e., $\|\tilde{x}(k)\| \to 0$ as $k \to \infty$. □

Remark 2.14 According to Theorem 2.13, after a sufficient learning session, the NN system identification error converges to zero, i.e., we have

$$f(x(k)) + \hat{g}(x(k))u(k) = \omega_m^{\mathrm{T}}(k)\sigma(\bar{z}(k)), \tag{2.86}$$

where $\hat{g}(x(k))$ denotes the estimated value of the control coefficient matrix $g(x(k))$. Taking the partial derivative of both sides of (2.86) with respect to $u(k)$ yields

$$\begin{aligned}\hat{g}(x(k)) &= \frac{\partial\left(\omega_m^{\mathrm{T}}(k)\sigma(\bar{z}(k))\right)}{\partial u(k)}\\ &= \omega_m^{\mathrm{T}}(k)\frac{\partial\sigma(\bar{z}(k))}{\partial\bar{z}(k)}\nu_m^{*\mathrm{T}}\frac{\partial z(k)}{\partial u(k)},\end{aligned} \tag{2.87}$$

where

$$\frac{\partial z(k)}{\partial u(k)} = \begin{bmatrix} 0_{n\times m} \\ \hdashline I_m \end{bmatrix},$$

and I_m is the $m \times m$ identity matrix.

Next, this result will be used in the derivation and implementation of the iterative ADP algorithm for the optimal control of unknown discrete-time nonlinear systems.

2.3.2.2 Derivation of the Iterative ADP Algorithm

In this part, we mainly present the iterative ADP algorithm. First, we start with the initial value function $V_0(\cdot) = 0$, and then solve for the law of single control vector $v_0(x(k))$ as follows:

$$v_0(x(k)) = \arg\min_{u(k)}\left\{x^{\mathrm{T}}(k)Qx(k) + u^{\mathrm{T}}(k)Ru(k) + \gamma V_0(x(k+1))\right\}. \tag{2.88}$$

Once the control law $v_0(x(k))$ is determined, we update the cost function as

$$\begin{aligned} V_1(x(k)) &= \min_{u(k)} \left\{ x^{\mathrm{T}}(k)Qx(k) + u^{\mathrm{T}}(k)Ru(k) + \gamma V_0(x(k+1)) \right\} \\ &= x^{\mathrm{T}}(k)Qx(k) + v_0^{\mathrm{T}}(x(k))Rv_0(x(k)). \end{aligned} \tag{2.89}$$

Therefore, for $i = 1, 2, \ldots$, the iterative ADP algorithm can be used to implement the iteration between the control law

$$\begin{aligned} v_i(x(k)) &= \arg\min_{u(k)} \left\{ x^{\mathrm{T}}(k)Qx(k) + u^{\mathrm{T}}(k)Ru(k) + \gamma V_i(x(k+1)) \right\} \\ &= -\frac{\gamma}{2} R^{-1} \hat{g}^{\mathrm{T}}(x(k)) \frac{\partial V_i(x(k+1))}{\partial x(k+1)} \end{aligned} \tag{2.90}$$

and the value function

$$\begin{aligned} V_{i+1}(x(k)) &= \min_{u(k)} \left\{ x^{\mathrm{T}}(k)Qx(k) + u^{\mathrm{T}}(k)Ru(k) + \gamma V_i(x(k+1)) \right\} \\ &= x^{\mathrm{T}}(k)Qx(k) + v_i^{\mathrm{T}}(x(k))Rv_i(x(k)) + \gamma V_i(x(k+1)). \end{aligned} \tag{2.91}$$

In the above recurrent iteration, i is the iteration index of the control law and value function, while k is the time index of the system's control and state trajectories. The value function and control law are updated until they converge to the optimal ones. In the following part, we will present a proof of convergence of the iteration between (2.90) and (2.91) with the value function $V_i \to J^*$ and the control law $v_i \to u^*$ as $i \to \infty$.

2.3.2.3 Convergence Analysis of the Iterative ADP Algorithm

Lemma 2.15 *Let $\{\mu_i\}$ be an arbitrary sequence of control laws and $\{v_i\}$ be the control law sequence described in* (2.90). *Define V_i as in* (2.91) *and Λ_i as*

$$\Lambda_{i+1}(x(k)) = x^{\mathrm{T}}(k)Qx(k) + \mu_i^{\mathrm{T}}(x(k))R\mu_i(x(k)) + \gamma \Lambda_i(x(k+1)). \tag{2.92}$$

If $V_0(x(k)) = \Lambda_0(x(k)) = 0$, then $V_i(x(k)) \le \Lambda_i(x(k))$, $\forall i$.

Proof It can easily be derived noticing that V_{i+1} is the result of minimizing the right hand side of (2.91) with respect to the control input $u(k)$, while Λ_{i+1} is a result of arbitrary control input. □

Lemma 2.16 *Let the value function sequence $\{V_i\}$ be defined as in* (2.91). *If the system is controllable, then there is an upper bound Y such that $0 \le V_i(x(k)) \le Y$, $\forall i$.*

Proof Let $\{\eta_i(x)\}$ be a sequence of admissible control laws, and let $V_0(\cdot) = Z_0(\cdot) = 0$, where V_i is updated as in (2.91) and Z_i is updated by

$$Z_{i+1}(x(k)) = x^{\mathrm{T}}(k)Qx(k) + \eta_i^{\mathrm{T}}(x(k))R\eta_i(x(k)) + \gamma Z_i(x(k+1)). \tag{2.93}$$

It is clear that

$$\begin{aligned}Z_i(x(k+1)) = {} & x^{\mathrm{T}}(k+1)Qx(k+1) + \eta_{i-1}^{\mathrm{T}}(x(k+1))R\eta_{i-1}(x(k+1)) \\ & + \gamma Z_{i-1}(x(k+2)).\end{aligned} \tag{2.94}$$

Noticing that $l(x(k), \eta_i(x(k))) = x^{\mathrm{T}}(k)Qx(k) + \eta_i^{\mathrm{T}}(x(k))R\eta_i(x(k))$, we can further obtain

$$\begin{aligned}Z_{i+1}(x(k)) = {} & l(x(k), \eta_i(x(k))) + \gamma l(x(k+1), \eta_{i-1}(x(k+1))) \\ & + \gamma^2 Z_{i-1}(x(k+2)) \\ = {} & l(x(k), \eta_i(x(k))) + \gamma l(x(k+1), \eta_{i-1}(x(k+1))) \\ & + \gamma^2 l(x(k+2), \eta_{i-2}(x(k+2))) + \gamma^3 Z_{i-2}(x(k+3)) \\ & \vdots \\ = {} & l(x(k), \eta_i(x(k))) + \gamma l(x(k+1), \eta_{i-1}(x(k+1))) \\ & + \gamma^2 l(x(k+2), \eta_{i-2}(x(k+2))) \\ & + \cdots + \gamma^i l(x(k+i), \eta_0(x(k+i))) \\ & + \gamma^{i+1} Z_0(x(k+i+1)),\end{aligned} \tag{2.95}$$

where $Z_0(x(k+i+1)) = 0$. Then, (2.95) can be written as

$$\begin{aligned}Z_{i+1}(x(k)) &= \sum_{j=0}^{i} \gamma^j l(x(k+j), \eta_{i-j}(x(k+j))) \\ &= \sum_{j=0}^{i} \gamma^j \big(x^{\mathrm{T}}(k+j)Qx(k+j) + \eta_{i-j}^{\mathrm{T}}(x(k+j))R\eta_{i-j}(x(k+j))\big) \\ &\le \lim_{i\to\infty} \sum_{j=0}^{i} \gamma^j \big(x^{\mathrm{T}}(k+j)Qx(k+j) \\ &\quad + \eta_{i-j}^{\mathrm{T}}(x(k+j))R\eta_{i-j}(x(k+j))\big).\end{aligned} \tag{2.96}$$

Since $\{\eta_i(x)\}$ is an admissible control law sequence, we have $x(k) \to 0$ as $k \to \infty$, and there exists an upper bound Y such that

$$Z_{i+1}(x(k)) \le \lim_{i\to\infty} \sum_{j=0}^{i} \gamma^j l(x(k+j), \eta_{i-j}(x(k+j))) \le Y, \ \forall i. \tag{2.97}$$

By using Lemma 2.15, we obtain

$$V_{i+1}(x(k)) \le Z_{i+1}(x(k)) \le Y, \ \forall i. \tag{2.98}$$

□

Based on Lemmas 2.15 and 2.16, we now present our main theorems.

Theorem 2.17 *Define the value function sequence* $\{V_i\}$ *as in* (2.91) *with* $V_0(\cdot) = 0$, *and the control law sequence* $\{v_i\}$ *as in* (2.90). *Then,* $\{V_i\}$ *is a monotonically nondecreasing sequence satisfying* $V_{i+1} \ge V_i, \forall i$.

Proof Define a new sequence

$$\Phi_{i+1}(x(k)) = x^{\mathrm{T}}(k)Qx(k) + v_{i+1}^{\mathrm{T}}(x(k))Rv_{i+1}(x(k)) + \gamma\Phi_i(x(k+1)) \tag{2.99}$$

with $\Phi_0(\cdot) = V_0(\cdot) = 0$. Let the control law sequence $\{v_i\}$ and the value function sequence $\{V_i\}$ be updated as in (2.90) and (2.91), respectively.

In the following part, we prove that $\Phi_i(x(k)) \le V_{i+1}(x(k))$ by mathematical induction.

First, we prove that it holds for $i = 0$. Considering

$$V_1(x(k)) - \Phi_0(x(k)) = x^{\mathrm{T}}(k)Qx(k) + v_0^{\mathrm{T}}(x(k))Rv_0(x(k)) \ge 0$$

then, for $i = 0$, we get

$$V_1(x(k)) \ge \Phi_0(x(k)). \tag{2.100}$$

Second, we assume that it holds for $i - 1$, i.e., $V_i(x(k)) \ge \Phi_{i-1}(x(k)), \forall x(k)$. Then, for i, noticing that

$$V_{i+1}(x(k)) = x^{\mathrm{T}}(k)Qx(k) + v_i^{\mathrm{T}}(x(k))Rv_i(x(k)) + \gamma V_i(x(k+1))$$

and

$$\Phi_i(x(k)) = x^{\mathrm{T}}(k)Qx(k) + v_i^{\mathrm{T}}(x(k))Rv_i(x(k)) + \gamma\Phi_{i-1}(x(k+1)),$$

we get

$$V_{i+1}(x(k)) - \Phi_i(x(k)) = \gamma(V_i(x(k+1)) - \Phi_{i-1}(x(k+1))) \ge 0$$

i.e.,

$$V_{i+1}(x(k)) \geq \Phi_i(x(k)). \tag{2.101}$$

Thus, we complete the proof through mathematical induction.

Furthermore, from Lemma 2.15 we know that $V_i(x(k)) \leq \Phi_i(x(k))$, therefore, we have

$$V_{i+1}(x(k)) \geq \Phi_i(x(k)) \geq V_i(x(k)). \tag{2.102}$$

□

We have reached the conclusion that the value function sequence $\{V_i\}$ is a monotonically nondecreasing sequence with an upper bound, and therefore, its limit exists. Now, we can derive the following theorem.

Theorem 2.18 *For any state vector $x(k)$, define*

$$\lim_{i\to\infty} V_i(x(k)) = V_\infty(x(k))$$

as the limit of the value function sequence $\{V_i\}$. Then, the following equation holds:

$$V_\infty(x(k)) = \min_{u(k)} \left\{x^{\mathrm{T}}(k)Qx(k) + u^{\mathrm{T}}(k)Ru(k) + \gamma V_\infty(x(k+1))\right\}.$$

Proof For any $u(k)$ and i, according to (2.91), we can derive

$$V_i(x(k)) \leq x^{\mathrm{T}}(k)Qx(k) + u^{\mathrm{T}}(k)Ru(k) + \gamma V_{i-1}(x(k+1)).$$

Combining with

$$V_i(x(k)) \leq V_\infty(x(k)), \ \forall i \tag{2.103}$$

which is obtained from Theorem 2.17, we have

$$V_i(x(k)) \leq x^{\mathrm{T}}(k)Qx(k) + u^{\mathrm{T}}(k)Ru(k) + \gamma V_\infty(x(k+1)), \ \forall i.$$

Let $i \to \infty$, we can acquire

$$V_\infty(x(k)) \leq x^{\mathrm{T}}(k)Qx(k) + u^{\mathrm{T}}(k)Ru(k) + \gamma V_\infty(x(k+1)).$$

Note that in the above equation, $u(k)$ is chosen arbitrarily; thus, we obtain

$$V_\infty(x(k)) \leq \min_{u(k)} \left\{x^{\mathrm{T}}(k)Qx(k) + u^{\mathrm{T}}(k)Ru(k) + \gamma V_\infty(x(k+1))\right\}. \tag{2.104}$$

On the other hand, since the value function sequence satisfies

$$V_i(x(k)) = \min_{u(k)} \left\{x^{\mathrm{T}}(k)Qx(k) + u^{\mathrm{T}}(k)Ru(k) + \gamma V_{i-1}(x(k+1))\right\}$$

for any i, considering (2.103), we have

$$V_\infty(x(k)) \geq \min_{u(k)} \left\{x^{\mathrm{T}}(k)Qx(k) + u^{\mathrm{T}}(k)Ru(k) + \gamma V_{i-1}(x(k+1))\right\}, \ \forall i.$$

Let $i \to \infty$; we get

$$V_\infty(x(k)) \geq \min_{u(k)} \left\{x^{\mathrm{T}}(k)Qx(k) + u^{\mathrm{T}}(k)Ru(k) + \gamma V_\infty(x(k+1))\right\}. \tag{2.105}$$

Based on (2.104) and (2.105), we can acquire the conclusion that $V_\infty(x(k)) = \min_{u(k)}\{x^{\mathrm{T}}(k)Qx(k) + u^{\mathrm{T}}(k)Ru(k) + \gamma V_\infty(x(k+1))\}$. □

Next, we will prove that the value function sequence $\{V_i\}$ converges to the optimal value function $J^*(x(k))$ as $i \to \infty$.

Theorem 2.19 (cf. [10]) *Define the value function sequence* $\{V_i\}$ *as in* (2.91) *with* $V_0(\cdot) = 0$. *If the system state* $x(k)$ *is controllable, then* J^* *is the limit of the value function sequence* $\{V_i\}$, *i.e.,*

$$\lim_{i \to \infty} V_i(x(k)) = J^*(x(k)).$$

Proof Let $\{\eta_i^{(l)}\}$ be the lth admissible control law sequence. We construct the associated sequence $\{P_i^{(l)}(x)\}$ as follows:

$$P_{i+1}^{(l)}(x(k)) = x^{\mathrm{T}}(k)Qx(k) + \eta_i^{(l)\mathrm{T}}(x(k))R\eta_i^{(l)}(x(k)) + \gamma P_i^{(l)}(x(k+1)) \tag{2.106}$$

with $P_0^{(l)}(\cdot) = 0$. Similar to the derivation of (2.95), we get

$$P_{i+1}^{(l)}(x(k)) = \sum_{j=0}^{i} \gamma^j l\big(x(k+j), \eta_{i-j}^{(l)}(x(k+j))\big). \tag{2.107}$$

Using Lemmas 2.15 and 2.16, we have

$$V_{i+1}(x(k)) \leq P_{i+1}^{(l)}(x(k)) \leq Y_l, \ \forall l, i \tag{2.108}$$

where Y_l is the upper bound associated with the sequence $\{P_{i+1}^{(l)}(x(k))\}$. Denote

$$\lim_{i \to \infty} P_i^{(l)}(x(k)) = P_\infty^{(l)}(x(k));$$

then, we obtain

$$V_\infty(x(k)) \leq P_\infty^{(l)}(x(k)) \leq Y_l, \ \forall l. \tag{2.109}$$

Let the corresponding control sequence associated with (2.107) be

$$\begin{aligned}{}^{(l)}\underline{\hat{u}}_k^{k+i} &= \left({}^{(l)}\hat{u}(k), {}^{(l)}\hat{u}(k+1), \ldots, {}^{(l)}\hat{u}_{k+i}\right) \\ &= \left(\eta_i^{(l)}(x(k)), \eta_{i-1}^{(l)}(x(k+1)), \ldots, \eta_0^{(l)}(x(k+i))\right);\end{aligned}$$

then we have

$$J\left(x(k), {}^{(l)}\underline{\hat{u}}_k^{k+i}\right) = \sum_{j=0}^{i} \gamma^j l\left(x(k+j), \eta_{i-j}^{(l)}(x(k+j))\right) = P_{i+1}^{(l)}(x(k)). \tag{2.110}$$

Letting $i \to \infty$, and denoting the admissible control sequence related to $P_\infty^{(l)}(x(k))$ with length ∞ as ${}^{(l)}\underline{\hat{u}}_k^{\infty}$, we get

$$J\left(x(k), {}^{(l)}\underline{\hat{u}}_k^{\infty}\right) = \sum_{j=0}^{\infty} \gamma^j l\left(x(k+j), {}^{(l)}\hat{u}(k+j)\right) = P_\infty^{(l)}(x(k)). \tag{2.111}$$

Then, according to the definition of $J^*(x(k))$ in (2.71), for any $\varepsilon > 0$, there exists a sequence of admissible control laws $\{\eta_i^{(M)}\}$ such that the associated cost function

$$J\left(x(k), {}^{(M)}\underline{\hat{u}}_k^{\infty}\right) = \sum_{j=0}^{\infty} \gamma^j l\left(x(k+j), {}^{(M)}\hat{u}(k+j)\right) = P_\infty^{(M)}(x(k)) \tag{2.112}$$

satisfies $J(x(k), {}^{(M)}\underline{\hat{u}}_k^{\infty}) \le J^*(x(k)) + \varepsilon$. Combining with (2.109), we have

$$V_\infty(x(k)) \le P_\infty^{(M)}(x(k)) \le J^*(x(k)) + \varepsilon. \tag{2.113}$$

Since ε is chosen arbitrarily, we get

$$V_\infty(x(k)) \le J^*(x(k)). \tag{2.114}$$

On the other hand, because $V_{i+1}(x(k)) \le P_{i+1}^{(l)}(x(k)) \le Y_l, \forall l, i$, we can get $V_\infty(x(k)) \le \inf_l\{Y_l\}$. According to the definition of admissible control law sequence, the control law sequence associated with the cost function $V_\infty(x(k))$ must be an admissible control law sequence. We can see that there exists a sequence of admissible control laws $\{\eta_i^{(N)}\}$ such that $V_\infty(x(k)) = P_\infty^{(N)}(x(k))$. Combining with (2.111), we get $V_\infty(x(k)) = J(x(k), {}^{(N)}\underline{\hat{u}}_k^{\infty})$. Sine $J^*(x(k))$ is the infimum of all admissible control sequences starting at k with length ∞, we obtain

$$V_\infty(x(k)) \ge J^*(x(k)). \tag{2.115}$$

Based on (2.114) and (2.115), we can conclude that J^* is the limit of the value function sequence $\{V_i\}$, i.e., $V_\infty(x(k)) = J^*(x(k))$. □

From Theorems 2.17 and 2.18, we can derive that the limit of the value function sequence $\{V_i\}$ satisfies the DTHJB equation, i.e., $V_\infty(x(k)) = \min_{u(k)}\{x^{\mathrm{T}}(k)Qx(k)$

$+ u^{\mathrm{T}}(k)Ru(k) + \gamma V_\infty(x(k+1))\}$. Besides, from Theorem 2.19, we can get the result that $V_\infty(x(k)) = J^*(x(k))$. Therefore, we can find that the cost function sequence $\{V_i(x(k))\}$ converges to the optimal value function $J^*(x(k))$ of the DTHJB equation, i.e., $V_i \to J^*$ as $i \to \infty$. Then, according to (2.74) and (2.90), we can conclude the convergence of the corresponding control law sequence. Now, we present the following corollary.

Corollary 2.20 *Define the value function sequence* $\{V_i\}$ *as in* (2.91) *with* $V_0(\cdot) = 0$, *and the control law sequence* $\{v_i\}$ *as in* (2.90). *If the system state* $x(k)$ *is controllable, then the sequence* $\{v_i\}$ *converges to the optimal control law* u^* *as* $i \to \infty$, *i.e.*,

$$\lim_{i\to\infty} v_i(x(k)) = u^*(x(k)).$$

Remark 2.21 Like (2.95), when we further expand (2.91), we obtain a control law sequence $(v_i, v_{i-1}, \ldots, v_0)$ and the resultant control sequence $(v_i(x(0)), v_{i-1}(x(1)), \ldots, v_0(x(i)))$. With the iteration number increasing to ∞, the derived control law sequence has the length of ∞. Then, using the corresponding control sequence, we obtain a state trajectory. However, it is not derived from a single control law. For infinite-horizon optimal control problem, what we should get is a unique optimal control law under which we can obtain the optimal state trajectory. Therefore, we only use the optimal control law u^* obtained in Corollary 2.20 to produce a control sequence when we apply the algorithm to practical systems.

2.3.2.4 NN Implementation of the Iterative ADP Algorithm Using GDHP Technique

When the controlled system is linear and the cost function is quadratic, we can obtain a linear control law. In the nonlinear case, however, this is not necessarily true. Therefore, we need to use function approximation structure, such as NN, to approximate both $v_i(x(k))$ and $V_i(x(k))$.

Now, we implement the iterative GDHP algorithm in (2.90) and (2.91). In the iterative GDHP algorithm, there are three networks, which are model network, critic network and action network. All the networks are chosen as three-layer feedforward NNs. The input of the critic network and action network is $x(k)$, while the input of the model network is $x(k)$ and $\hat{v}_i(x(k))$. The diagram of the whole structure is shown in Fig. 2.14, where

$$\mathrm{DER} = \left(\frac{\partial \hat{x}(k+1)}{\partial x(k)} + \frac{\partial \hat{x}(k+1)}{\partial \hat{v}_i(x(k))}\frac{\partial \hat{v}_i(x(k))}{\partial x(k)}\right)^{\mathrm{T}}.$$

The training of the model network is completed after the system identification process and its weights are kept unchanged. Then, according to Theorem 2.13, when given $x(k)$ and $\hat{v}_i(x(k))$, we can compute $\hat{x}(k+1)$ by (2.77), i.e.,

$$\hat{x}(k+1) = \omega_m^{\mathrm{T}}(k)\sigma\big(\nu_m^{*\mathrm{T}}[x^{\mathrm{T}}(k)\ \hat{v}_i^{\mathrm{T}}(x(k))]^{\mathrm{T}}\big).$$

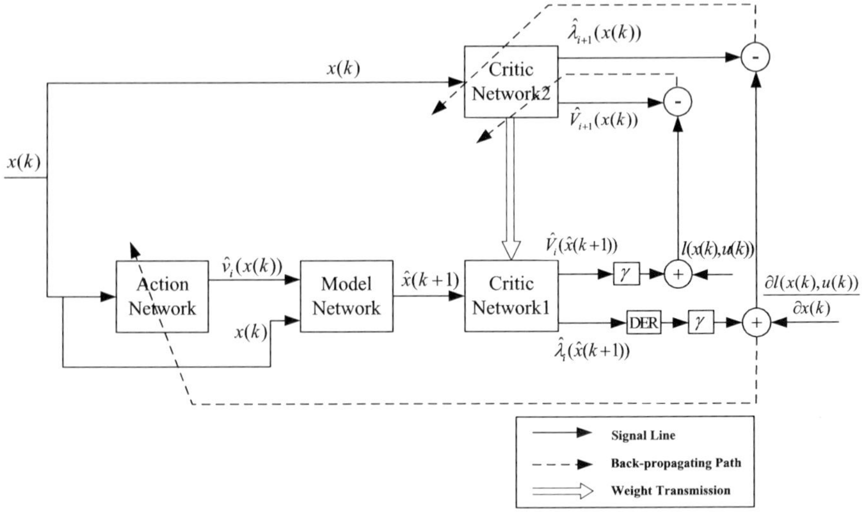

Fig. 2.14 The structure diagram of the iterative GDHP algorithm

As a result, we avoid the requirement of knowing $f(x(k))$ and $g(x(k))$ during the implementation of the iterative GDHP algorithm.

Next, the learned NN system model will be used in the process of training critic network and action network.

The critic network is used to approximate both $V_i(x(k))$ and its derivative $\partial V_i(x(k))/\partial x(k)$, which is named the costate function and denoted $\lambda_i(x(k))$. The output of the critic network is denoted

$$\begin{bmatrix} \hat{V}_i(x(k)) \\ \hat{\lambda}_i(x(k)) \end{bmatrix} = \begin{bmatrix} \omega_{ci}^{1\mathrm{T}} \\ \omega_{ci}^{2\mathrm{T}} \end{bmatrix} \sigma\left(\nu_{ci}^{\mathrm{T}} x(k)\right) = \omega_{ci}^{\mathrm{T}} \sigma\left(\nu_{ci}^{\mathrm{T}} x(k)\right), \tag{2.116}$$

where

$$\omega_{ci} = \left[\omega_{ci}^1 \;\; \omega_{ci}^2\right],$$

i.e.,

$$\hat{V}_i(x(k)) = \omega_{ci}^{1\mathrm{T}} \sigma\left(\nu_{ci}^{\mathrm{T}} x(k)\right) \tag{2.117}$$

and

$$\hat{\lambda}_i(x(k)) = \omega_{ci}^{2\mathrm{T}} \sigma\left(\nu_{ci}^{\mathrm{T}} x(k)\right). \tag{2.118}$$

The target function can be written as

$$V_{i+1}(x(k)) = x^{\mathrm{T}}(k) Q x(k) + v_i^{\mathrm{T}}(x(k)) R v_i(x(k)) + \gamma \hat{V}_i(\hat{x}(k+1)) \tag{2.119}$$

and

$$\begin{aligned}\lambda_{i+1}(x(k)) &= \frac{\partial\big(x^{\mathrm{T}}(k)Qx(k) + v_i^{\mathrm{T}}(x(k))Rv_i(x(k))\big)}{\partial x(k)} + \gamma\frac{\partial \hat{V}_i(\hat{x}(k+1))}{\partial x(k)} \\ &= 2Qx(k) + 2\left(\frac{\partial v_i(x(k))}{\partial x(k)}\right)^{\mathrm{T}} Rv_i(x(k)) \\ &\quad + \gamma\left(\frac{\partial \hat{x}(k+1)}{\partial x(k)} + \frac{\partial \hat{x}(k+1)}{\partial \hat{v}_i(x(k))}\frac{\partial \hat{v}_i(x(k))}{\partial x(k)}\right)^{\mathrm{T}} \hat{\lambda}_i(\hat{x}(k+1)). \end{aligned} \tag{2.120}$$

Then, we define the error function for training the critic network as

$$e_{cik}^1 = \hat{V}_i(x(k)) - V_{i+1}(x(k)) \tag{2.121}$$

and

$$e_{cik}^2 = \hat{\lambda}_i(x(k)) - \lambda_{i+1}(x(k)). \tag{2.122}$$

The objective function to be minimized in the critic network training is

$$E_{cik} = (1-\beta)E_{cik}^1 + \beta E_{cik}^2, \tag{2.123}$$

where

$$E_{cik}^1 = \frac{1}{2}e_{cik}^{1\mathrm{T}}e_{cik}^1 \tag{2.124}$$

and

$$E_{cik}^2 = \frac{1}{2}e_{cik}^{2\mathrm{T}}e_{cik}^2. \tag{2.125}$$

The weight updating rule for training the critic network is also gradient-based adaptation given by

$$\omega_{ci}(j+1) = \omega_{ci}(j) - \alpha_c\left[(1-\beta)\frac{\partial E_{cik}^1}{\partial \omega_{ci}(j)} + \beta\frac{\partial E_{cik}^2}{\partial \omega_{ci}(j)}\right] \tag{2.126}$$

$$\nu_{ci}(j+1) = \nu_{ci}(j) - \alpha_c\left[(1-\beta)\frac{\partial E_{cik}^1}{\partial \nu_{ci}(j)} + \beta\frac{\partial E_{cik}^2}{\partial \nu_{ci}(j)}\right] \tag{2.127}$$

where $\alpha_c > 0$ is the learning rate of the critic network, j is the inner-loop iteration step for updating the weight parameters, and $0 \le \beta \le 1$ is a parameter that adjusts how HDP and DHP are combined in GDHP. For $\beta = 0$, the training of the critic network reduces to a pure HDP, while $\beta = 1$ does the same for DHP.

In the action network, the state $x(k)$ is used as input to obtain the optimal control. The output can be formulated as

$$\hat{v}_i(x(k)) = \omega_{ai}^{\mathrm{T}}\sigma\big(\nu_{ai}^{\mathrm{T}}x(k)\big). \tag{2.128}$$

The target control input is given as

$$v_i(x(k)) = -\frac{\gamma}{2} R^{-1} \hat{g}^{\mathrm{T}}(x(k)) \frac{\partial \hat{V}_i(\hat{x}(k+1))}{\partial \hat{x}(k+1)}. \tag{2.129}$$

The error function of the action network can be defined as

$$e_{aik} = \hat{v}_i(x(k)) - v_i(x(k)). \tag{2.130}$$

The weights of the action network are updated to minimize the following performance error measure:

$$E_{aik} = \frac{1}{2} e_{aik}^{\mathrm{T}} e_{aik}. \tag{2.131}$$

Similarly, the weight updating algorithm is

$$\omega_{ai}(j+1) = \omega_{ai}(j) - \alpha_a \left[\frac{\partial E_{aik}}{\partial \omega_{ai}(j)} \right], \tag{2.132}$$

$$\nu_{ai}(j+1) = \nu_{ai}(j) - \alpha_a \left[\frac{\partial E_{aik}}{\partial \nu_{ai}(j)} \right] \tag{2.133}$$

where $\alpha_a > 0$ is the learning rate of the action network, and j is the inner-loop iteration step for updating the weight parameters.

Remark 2.22 According to Theorem 2.19, $V_i(x(k)) \to J^*(x(k))$ as $i \to \infty$. Since $\lambda_i(x(k)) = \partial V_i(x(k))/\partial x(k)$, we can conclude that the costate function sequence $\{\lambda_i(x(k))\}$ is also convergent with $\lambda_i(x(k)) \to \lambda^*(x(k))$ as $i \to \infty$.

Remark 2.23 From Fig. 2.14, we can see that the outputs of the critic network of the iterative GDHP algorithm contain not only the cost function but also its derivative. This is important because the information associated with the cost function is as useful as the knowledge of its derivative. Besides, as is shown in (2.126) and (2.127), training the critic network of the iterative GDHP algorithm utilizes an error measure which is a combination of the error measures of HDP and DHP. Though it is more complicated to do this, the resulting behavior is expected to be superior to simple ADP methods.

2.3.3 Simulations

An example is provided in this subsection to demonstrate the effectiveness of the control scheme derived by the iterative GDHP algorithm.

Example 2.24 Consider the following nonlinear system:

$$x(k+1) = f(x(k)) + g(x(k))u(k),$$

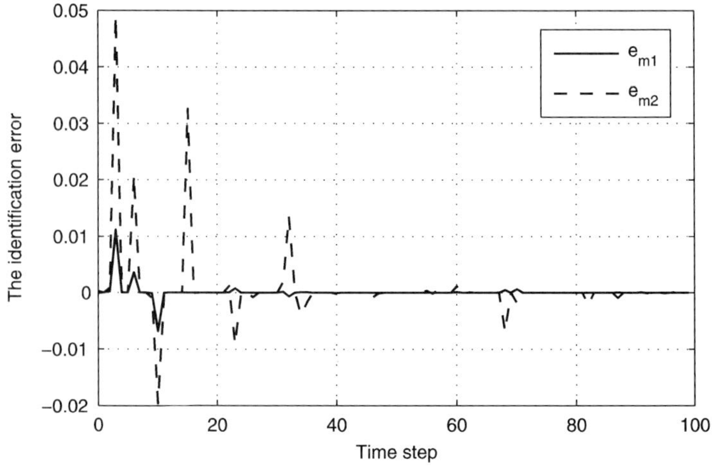

Fig. 2.15 The system identification error ($\tilde{x}_{k1}$ and $\tilde{x}_{k2}$ are denoted e_{m1} and e_{m2}, respectively)

where $x(k) = [x_1(k)\ x_2(k)]^{\mathrm{T}} \in \mathbb{R}^2$ and $u(k) \in \mathbb{R}$ are the state and control variables, respectively. The cost function is chosen as $l(x(k), u(k)) = x^{\mathrm{T}}(k)x(k) + u^{\mathrm{T}}(k)Ru(k)$. The system functions are given as

$$f(x(k)) = \begin{bmatrix} -\sin(0.5x_2(k)) \\ -\cos(1.4x_2(k))\sin(0.9x_1(k)) \end{bmatrix}$$

$$g(x(k)) = \begin{bmatrix} 0 \\ 1 \end{bmatrix}.$$

We choose three-layer feedforward NNs as model network, critic network and action network with the structures 3–8–2, 2–8–3, 2–8–1, respectively. In the system identification process, the initial weights between the input layer and the hidden layer, and the hidden layer and the output layer are chosen randomly in $[-0.5, 0.5]$ and $[-0.1, 0.1]$, respectively. We apply the NN identification scheme for 100 steps under the learning rate $\alpha_m = 0.05$ and obtain the result as shown in Fig. 2.15. It is clearly observed that the NN identifier successfully learns the unknown nonlinear system. Then, we finish the training of the model network and keep its weights unchanged.

The initial weights of the critic network and action network are all set to be random in $[-0.1, 0.1]$. Then, let the discount factor $\gamma = 1$ and the adjusting parameter $\beta = 0.5$, we train the critic network and action network for 10 training cycles with each cycle of 2000 steps. In the training process, the learning rate $\alpha_c = \alpha_a = 0.05$. The convergence process of the value function and its derivative of the iterative GDHP algorithm at time instant $k = 0$ are shown in Fig. 2.16. We can see that the iterative value function sequence does converge to the optimal value function quite rapidly, which also indicates the effectiveness of the iterative GDHP algorithm.

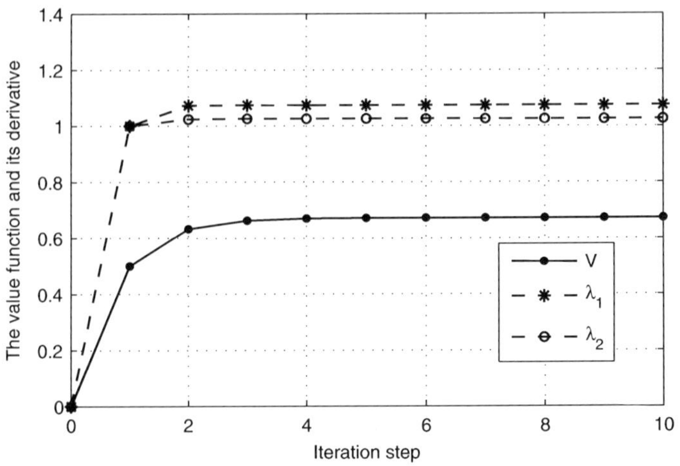

Fig. 2.16 The convergence process of the value function and its derivative of the iterative GDHP algorithm

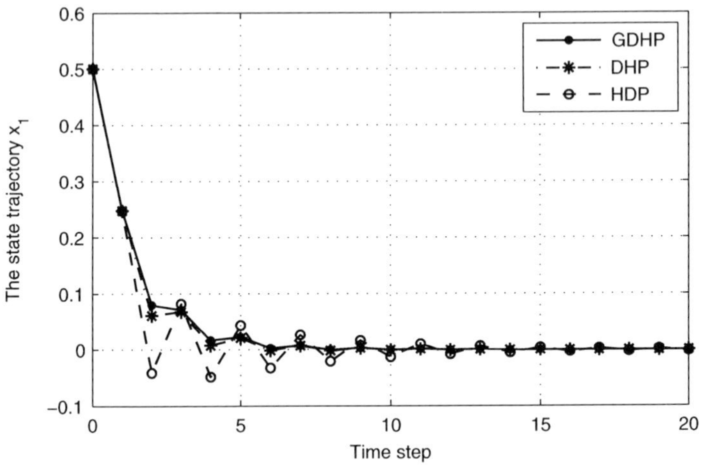

Fig. 2.17 The state trajectory x_1

Moreover, in order to make a comparison with the iterative ADP algorithm using HDP and DHP technique (iterative HDP algorithm and iterative DHP algorithm for brief), we also present the controllers designed by iterative HDP algorithm and iterative DHP algorithm, respectively. Then, for the given initial state $x_1(0) = 0.5$ and $x_2(0) = 0.5$, we apply the optimal control laws designed by iterative GDHP, HDP and DHP algorithm to the controlled system for 20 time steps, respectively, and obtain the state curves as shown in Figs. 2.17 and 2.18. The corresponding control curves are shown in Fig. 2.19. It can be seen from the simulation results that the

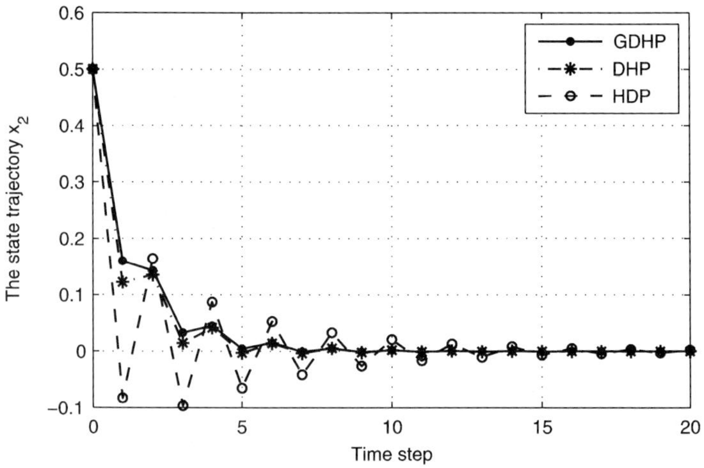

Fig. 2.18 The state trajectory x_2

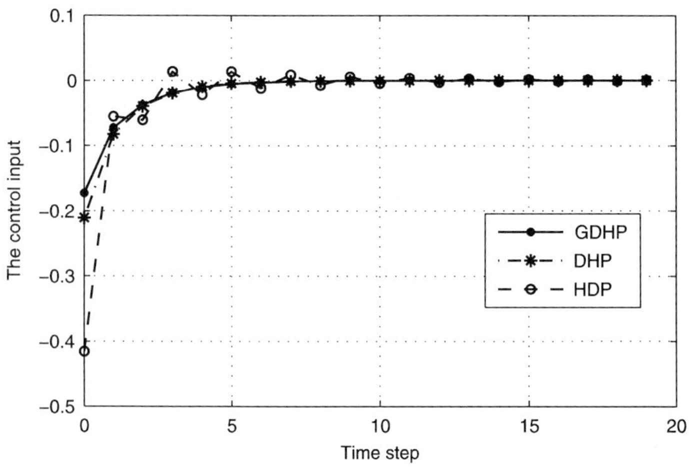

Fig. 2.19 The control input u

controller designed by the iterative GDHP algorithm has better performance than iterative HDP algorithm and iterative DHP algorithm. The most important property that the iterative GDHP algorithm superior to the iterative DHP algorithm is the former can show us the convergence process of the value function sequence. Besides, the time that the iterative GDHP algorithm takes in the entire computation process is much less than that of HDP. For the same problem, the iterative GDHP algorithm takes about 26.6 seconds while the iterative HDP algorithm takes about 61.3 seconds before satisfactory results are obtained.

2.4 Infinite-Horizon Optimal State Feedback Control Based on GHJB Algorithm

2.4.1 Problem Formulation

Consider an affine nonlinear discrete-time system of the form

$$x(k+1) = f(x(k)) + g(x(k))u(k), \tag{2.134}$$

where $x(k) \in \Omega \in \mathbb{R}^n$, $f\colon \mathbb{R}^n \to \mathbb{R}^n$, $g\colon \mathbb{R}^n \to \mathbb{R}^{n\times m}$. The input satisfies $u(k) \in \Omega_u$, $\Omega_u = \{u(k) = [u_1(k), u_2(k), \dots, u_m(k)]^{\mathrm{T}} \in \mathbb{R}^m\colon |u_i(k)| < \bar{u}_i(k), i = 1, 2, \dots, m\}$, where $\bar{u}_i(k)$ is the saturating bound of the ith actuator. Let $\bar{U} = \operatorname{diag}\{\bar{u}_1, \bar{u}_2, \dots, \bar{u}_m\}$. Assuming that $f + gu$ is continuous on a set $\Omega \subseteq \mathbb{R}^n$ containing the origin, and system (2.134) is controllable in the sense that there exists a continuous control on Ω that asymptotically stabilizes the system. In this subsection, the infinite-horizon optimal control problem for nonlinear discrete-time systems with actuator saturation is investigated. It is desired to find the constrained state feedback input $u(x(k))$ which minimizes a generalized cost functional as follows:

$$J(x(0), u) = \sum_{k=0}^{\infty} Q(x(k)) + W(u(x(k))), \tag{2.135}$$

where $Q(x(k))$ and $W(u(x(k)))$ are positive definite on Ω. For optimal control problem, it is worthy to note that $u(x(k))$ must both stabilize the system and make the cost functional finite, i.e., it must be an admissible control.

For system (2.134), the nonlinear discrete-time GHJB equation without considering saturation is given as follows:

$$\nabla V^{\mathrm{T}}(x)(f(x) + g(x)u(x) - x) + Q(x) + W(u(x)) = 0, \tag{2.136}$$

$$V(x)|_{x=0} = 0. \tag{2.137}$$

For a given admissible control u, there exists a positive definite continuously differentiable value function $V(x)$ whose initial value $V(x(0))$ equals $J(x(0), u)$.

For unconstrained control problem, a common choice of function $W(u(x))$ is $u^{\mathrm{T}}(x)Ru(x)$, where $R \in \mathbb{R}^{m\times m}$ is positive definite. On substitution of the optimal control $u^*(x) = -R^{-1}g^{\mathrm{T}}(x)\nabla J^*(x)/2$, where $J^*(x)$ is the optimal value function corresponding to optimal control $u^*(x)$, the GHJB equation (2.136) with the boundary condition (2.137) becomes the well-known HJB equation as follows:

$$\begin{aligned} &\nabla J^{*\mathrm{T}}(x)(f(x) + g(x)u(x) - x) + Q(x) \\ &\quad + \frac{1}{4}\nabla J^{*\mathrm{T}}(x)g(x)R^{-1}g^{\mathrm{T}}(x)\nabla J^*(x) = 0, \end{aligned} \tag{2.138}$$

$$J^*(x)|_{x=0} = 0. \tag{2.139}$$

However, the HJB equation above is not suitable for constrained optimal control problem. To guarantee bounded controls, we introduce a generalized nonquadratic functional as follows:

$$W(u(x)) = 2\int_0^{u(x)} \Phi^{-\mathrm{T}}(\bar{U}^{-1}s)\bar{U}R\,ds, \tag{2.140}$$

where $W(u(x))$ is a scalar, $\Phi(v) = [\varphi(v_1), \dots, \varphi(v_m)]^{\mathrm{T}}$ and

$$\Phi^{-1}(u) = [\varphi(u_1)^{-1}, \dots, \varphi(u_m)^{-1}]^{\mathrm{T}}$$

are bounded one-to-one functions that belong to $C^p(p) \geq 1$ and $L_2(\Omega)$, satisfying $|\varphi(\cdot)| \leq 1$. Moreover, $\varphi(\cdot)$ is a monotonic odd function with its first derivative bounded by a constant M. It is not difficult to find such functions, such as the hyperbolic tangent function $\varphi(\cdot) = \tanh(\cdot)$. R is positive definite and assumed to be symmetric for simplicity of analysis. Substituting (2.140) into (2.136), the constrained discrete-time GHJB equation with boundary condition is derived as follows:

$$\nabla V^{\mathrm{T}}(x)(f(x) + g(x)u(x) - x) + Q(x) + 2\int_0^{u(x)} \Phi^{-\mathrm{T}}(\bar{U}^{-1}s)\bar{U}R\,ds = 0, \tag{2.141}$$

$$V(x)|_{x=0} = 0. \tag{2.142}$$

According to the first-order necessary condition of the optimal control, the constrained optimal state feedback control law can be obtained as follows:

$$u^*(x) = \bar{U}\Phi\left(-\frac{1}{2}(\bar{U}R)^{-1}g^{\mathrm{T}}(x)\nabla J^*(x)\right). \tag{2.143}$$

Substitute (2.143) into (2.141), and the constrained discrete-time HJB equation can be derived as follows:

$$\begin{aligned} &\nabla J^{*\mathrm{T}}(x)\left(f(x) + g\bar{U}\Phi\left(-\frac{1}{2}(\bar{U}R)^{-1}g^{\mathrm{T}}\nabla J^*(x)\right) - x\right) \\ &\quad + Q(x) + 2\int_0^{\bar{U}\Phi(-\frac{1}{2}(\bar{U}R)^{-1}g^{\mathrm{T}}(x)\nabla J^*(x))} \Phi^{-\mathrm{T}}(\bar{U}^{-1}s)\bar{U}R\,ds = 0, \end{aligned}$$

$$J^*(x)|_{x=0} = 0. \tag{2.144}$$

If this HJB equation can be solved for the optimal value function $J^*(x)$, then (2.143) gives the optimal constrained control. However, this equation is generally impossible to solve.

2.4.2 Constrained Optimal Control Based on GHJB Equation

Contrast to HJB equation (2.144) being nonlinear difference equation about $\nabla J^*(x)$, the GHJB equation (2.141) is linear in $\nabla V(x)$. So it is easier to solve the GHJB equation than the HJB equation from a theoretical viewpoint. That is the reason why a successive approximation based on GHJB equation is introduced to solve the HJB equation. Via solving a sequence of $\mathrm{GHJB}(V^{[i]}, u^{[i]}) = 0$ we can obtain a sequence for $V^{[i]}$ and prove $V^{[i]} \to J^*$.

Theorem 2.25 (cf. [6]) *If $u^{[i]}(x) \in \Psi(\Omega)$, $x(0) \in \Omega$, the value function $V^{[i]}$ is positive definite and continuously differentiable on Ω, and satisfies* $\mathrm{GHJB}(V^{[i]}, u^{[i]}) = 0$ *with the boundary condition $V^{[i]}(0) = 0$, then*

$$u^{[i+1]}(x) = \bar{U}\Phi\left(-\frac{1}{2}(\bar{U}R)^{-1}g^{\mathrm{T}}\nabla V^{[i]}(x)\right) \tag{2.145}$$

is an admissible control for (2.134) *on Ω. Moreover, if $\varphi(\cdot)$ is monotone odd, and $V^{[i+1]}$ is the unique positive definite function, which satisfies* $\mathrm{GHJB}(V^{[i+1]}, u^{[i+1]}) = 0$ *with the boundary condition $V^{[i+1]}(0) = 0$, then we can conclude that $V^{(i+1)}(x(0)) \le V^{[i]}(x(0))$.*

Proof First, we should prove that $u^{[i+1]}$ is admissible.

For simplicity, in the following we write $f(x)$ as f, $g(x)$ as g. Taking the difference of the system with the control $u^{[i+1]}$ along the system $(f, g, u^{[i+1]})$, we have

$$\begin{aligned}\Delta V^{[i]}(x(k)) &= V^{[i]}(x(k+1)) - V^{[i]}(x(k)) \\ &\approx \nabla V_k^{[i]\mathrm{T}}\left(f_k + g_k u_k^{[i+1]} - x(k)\right),\end{aligned} \tag{2.146}$$

where

$$\begin{aligned}\nabla V_k &= \left.\frac{\partial V(x)}{\partial x}\right|_{x=x(k)} \\ &= \left.\left[\frac{\partial}{\partial x_1}V(x), \frac{\partial}{\partial x_2}V(x), \ldots, \frac{\partial}{\partial x_n}V(x)\right]^{\mathrm{T}}\right|_{x=x(k)},\end{aligned}$$

$f_k = f(x(k))$, $g_k = g(x(k))$, and $u_k = u(x(k))$. For any $x(k) \in \Omega$, since $\mathrm{GHJB}(V^{[i]}, u^{[i]}) = 0$, we have

$$\nabla V_k^{[i]\mathrm{T}}\left(f_k + g_k u_k^{[i]} - x(k)\right) + Q(x(k)) + 2\int_0^{u_k^{[i]}} \Phi^{-\mathrm{T}}(\bar{U}^{-1}s)\bar{U}R\,ds = 0. \tag{2.147}$$

Substituting (2.147) into (2.146), we have

$$\Delta V^{[i]}(x(k)) = \nabla V_k^{[i]\mathrm{T}} g_k \left(u_k^{[i+1]} - u_k^{[i]}\right) - Q(x(k)) - 2\int_0^{u_k^{[i]}} \Phi^{-\mathrm{T}}(\bar{U}^{-1}s)\bar{U}R ds. \tag{2.148}$$

Since

$$\nabla V_k^{[i]\mathrm{T}}(x) g_k = -2\Phi^{-\mathrm{T}}\left(\bar{U}^{-1}u_k^{[i+1]}\right)\bar{U}R, \tag{2.149}$$

we get

$$\Delta V^{[i]}(x(k)) = -Q(x(k)) + 2\Bigg[\Phi^{-\mathrm{T}}\left(\bar{U}^{-1}u_k^{[i+1]}\right)\bar{U}R\left(u_k^{[i]} - u_k^{[i+1]}\right) - \int_0^{u_k^{[i]}} \Phi^{-\mathrm{T}}(\bar{U}^{-1}s)\bar{U}R ds\Bigg]. \tag{2.150}$$

Because φ and φ^{-1} are monotone odd, the second term of (2.150) is negative. This implies that $\Delta V^{[i]}(x(k)) < 0$ for $x(k) \neq 0$. Thus, $V^{[i]}$ is a Lyapunov function for $u^{[i+1]}$ on Ω and the system (2.134) is asymptotically stable.

Next, we are ready to show that the value function of the system with the updated control $u^{[i+1]}$ is finite.

Since $u^{[i]}$ is admissible, from Definition 2.1, we get

$$V^{[i]}(x(0)) = J(x(0), u^{[i]}) < \infty, x(0) \in \Omega. \tag{2.151}$$

The value function for $u^{[i+1]}$ is

$$V(x(0), u^{[i+1]}) = \sum_{k=0}^{\infty}\left\{Q(x(k) + 2\int_0^{u_k^{[i+1]}} \Phi^{-\mathrm{T}}(\bar{U}^{-1}s)\bar{U}R ds\right\}, \tag{2.152}$$

where $x(k)$ is the state trajectory of system with admissible control $u^{[i+1]}$.

From (2.150) and (2.152), we have

$$\begin{aligned} V^{[i]}(x(\infty)) - V^{[i]}(x(0)) &= \sum_{k=0}^{\infty} \Delta V^{[i]}(k) \\ &= \sum_{k=0}^{\infty}\Bigg\{ - Q(x(k)) + 2[\Phi^{-\mathrm{T}}(\bar{U}^{-1}u_k^{[i+1]})\bar{U}R(u_k^{[i]} - u_k^{[i+1]}) \\ &\quad - \int_0^{u_k^{[i]}} \Phi^{-1}(\bar{U}^{-1}s)\bar{U}R ds]\Bigg\} \end{aligned}$$

$$
\begin{aligned}
&= - J(x(0), u^{[i+1]}) + 2\sum_{k=0}^{\infty}\Bigg\{\Phi^{-\mathrm{T}}(\bar{U}^{-1}u_k^{[i+1]})\bar{U}R(u_k^{[i]} - u_k^{[i+1]}) \\
&\quad + \int_{u_k^{[i]}}^{u_k^{[i+1]}} \Phi^{-\mathrm{T}}(\bar{U}^{-1}s)\bar{U}R ds\Bigg\}.
\end{aligned}
\tag{2.153}
$$

Since $x(\infty) = 0$ and $V^{[i]}(x)|_{x=0} = 0$, we get $V^{[i]}(x(\infty)) = 0$. By rewriting (2.152), we have

$$
\begin{aligned}
J(x(0), u^{[i+1]}) &= V^{[i]}(x(0)) + 2\sum_{k=0}^{\infty}\Bigg\{\Phi^{-\mathrm{T}}(\bar{U}^{-1}u_k^{[i+1]}\bar{U}R(u_k^{[i]} - u_k^{[i+1]}) \\
&\quad + \int_{u_k^{[i]}}^{u_k^{[i+1]}} \Phi^{-\mathrm{T}}(\bar{U}^{-1}s)\bar{U}R ds\Bigg\}.
\end{aligned}
\tag{2.154}
$$

Since φ and φ^{-1} are monotone odd, and the second term of (2.154) is less than 0, we have

$$
J(x(0), u^{[i+1]}) < V^{[i]}(x(0)) = J(x(0), u^{[i]}) < \infty. \tag{2.155}
$$

Because $V^{[i]}$ is continuously differentiable, and $g : R^n \to R^{n\times m}$ is a Lipschitz continuous function, $u^{[i+1]}$ is continuous. Since $V^{[i]}$ is positive definite and attains its minimum at the origin, and $\Delta V^{[i]}$ must approach 0 at the origin, from (2.145) we have $u^{[i+1]}(x)|_{x=0} = 0$.

From Definition 2.1, we know that $u^{[i+1]}$ is an admissible control on Ω. Since $u^{[i+1]}$ is admissible, there exists a $V^{[i+1]}$ satisfying $\mathrm{GHJB}(V^{[i+1]}, u^{[i+1]}) = 0$, and

$$
V^{[i+1]}(x(0)) = J(x(0), u^{[i+1]}). \tag{2.156}
$$

From (2.154) and (2.156), we get

$$
\begin{aligned}
V^{[i+1]}(x(0)) - V^{[i]}(x(0)) &= -2\sum_{k=0}^{\infty}\Bigg\{\Phi^{-\mathrm{T}}(\bar{U}^{-1}u_k^{[i+1]})\bar{U}R(u_k^{[i]} - u_k^{[i+1]}) \\
&\quad + \int_{u_k^{[i]}}^{u_k^{[i+1]}} \Phi^{-\mathrm{T}}(\bar{U}^{-1}s)\bar{U}R ds\Bigg\} \\
&\le 0.
\end{aligned}
\tag{2.157}
$$

The proof is completed. □

Corollary 2.26 *Given $u^{[0]}(x) \in \Psi(\Omega)$, if one iteratively solve the GHJB equation GHJB $(V^{[i]}, u^{[i]}) = 0$ and for $i = 0, 1, 2, \ldots$, update the control as $u^{[i+1]}(x) = \bar{U}\Phi(-\frac{1}{2}(\bar{U}R)^{-1}g^{\mathrm{T}}\nabla V^{[i]}(x))$, then it can be concluded that $V^{[i]}(x) \to J^*(x)$.*

Proof According to Theorem 2.25, $V^{[i]}$ is a decreasing sequence with a lower bound. Since $V^{[i]} > 0$, $V^{[i+1]} - V^{[i]} < 0$, $V^{[i]}$ will converge to a positive definite function $V^{[i+1]} = V^{[i]} = V^d$ when $i \to \infty$. Due to the HJB equation having a unique solution, we just need to prove $V^d = J^*$. When $V^{[i]} = V^{[i+1]} = V^d$, from (2.145) we have

$$u^{[i]}(x) = u^{[i+1]}(x) = \bar{U}\Phi\left(-\frac{1}{2}(\bar{U}R)^{-1}g^{\mathrm{T}}\Delta V^{[i]}(x)\right). \tag{2.158}$$

The GHJB equation with input $u^{[i]}$ can be written as

$$\begin{aligned}\nabla V^{[i]}(x)^{\mathrm{T}}\left(f(x) + g\bar{U}\Phi\left(-\frac{1}{2}(\bar{U}R)^{-1}g^{\mathrm{T}}\nabla V^{[i]}(x)\right) - x\right) + Q(x)\\ + 2\int_0^{\bar{U}\Phi(-\frac{1}{2}(\bar{U}R)^{-1}g^{\mathrm{T}}\nabla V^{[i]}(x))} \Phi^{-\mathrm{T}}(\bar{U}^{-1}s)\bar{U}R\,ds = 0,\end{aligned} \tag{2.159}$$

$$V^{[i]}(x)|_{x=0} = 0. \tag{2.160}$$

From (2.144), the conclusion can be drawn that (2.159) with boundary condition (2.160) is the HJB equation. This implies that $V^{[i]}(x) \to J^*(x)$, $u^{[i]}(x) \to u^*(x)$. □

In the next part, we are ready to discuss how to design the nearly optimal saturated controller using NNs. In general, the closed-form solution of GHJB equation (2.141) cannot be obtained even though solving GHJB equation (2.141) is easier than solving HJB equation (2.144). In this section, a neural network is used to approximate the solution $V(x)$ of constrained nonlinear discrete-time GHJB equation. Finally, the nearly optimal state feedback control is obtained according to (2.143).

$V(x)$ is approximated by a neural network as follows:

$$V_L(x) = \sum_{j=1}^{L} w_j\sigma_j(x) = W_L^{\mathrm{T}}\bar{\sigma}_L(x), \tag{2.161}$$

where w_j are the weights of the neural network, $\sigma_j(x)$ are the activation functions, $\sigma_j(x)$ are continuous and satisfy $\sigma_j(x)|_{x=0} = 0$. L is the number of hidden-layer neurons. $\bar{\sigma}_L(x) \equiv [\sigma_1(x), \sigma_2(x), \ldots, \sigma_L(x)]^{\mathrm{T}}$ is the vector activation function, $W_L(x) \equiv [w_1(x), w_2(x), \ldots, w_L(x)]^{\mathrm{T}}$ is the vector weight. The control objective is to make the residual error minimum in a least-square sense by tuning the weights.

Substituting (2.161) into (2.141), we have

$$\mathrm{GHJB}\left(V_L = \sum_{j=1}^{L} w_j\sigma_j, u\right) = e_L(x). \tag{2.162}$$

The method of weighted residuals is used to find the least-square solution, i.e.,

$$\left\langle \frac{\partial(e_L(x))}{\partial W_L(x)}, e_L(x) \right\rangle = 0, \tag{2.163}$$

where $\langle f, g\rangle = \int_\Omega fg dx$ is a *Lebesgue integral.*

By expanding (2.163), we get

$$\begin{aligned} &\langle \nabla\bar{\sigma}_L(x)\Delta x, \nabla\bar{\sigma}_L(x)\Delta x\rangle \cdot W_L \\ &\quad + \left\langle Q(x) + 2\int_0^{u(x)} \Phi^{-\mathrm{T}}(\bar{U}^{-1}s)\bar{U}Rds, \nabla\bar{\sigma}_L(x)\Delta x \right\rangle = 0. \end{aligned} \tag{2.164}$$

Lemma 2.27 *If the set* $\{\sigma_j(x)\}_1^L$ *is linearly independent and* $u \in \Omega_u$, *then the set* $\{\nabla\sigma_j^{\mathrm{T}}\Delta x\}_1^L$ *is also linearly independent.*

From Lemma 2.27, $\langle \nabla\bar{\sigma}_L(x)\Delta x, \nabla\bar{\sigma}_L(x)\Delta x\rangle$ is invertible. Therefore there exists a unique solution as follows:

$$\begin{aligned} W_L = &-\langle \nabla\bar{\sigma}_L(x)\Delta x, \nabla\bar{\sigma}_L(x)\Delta x\rangle^{-1} \\ &\times \left\langle Q(x) + 2\int_0^{u(x)} \Phi^{-\mathrm{T}}(\bar{U}^{-1}s)\bar{U}Rds, \nabla\bar{\sigma}_L(x)\Delta x \right\rangle, \end{aligned} \tag{2.165}$$

and the control can be derived as

$$u = \bar{U}\Phi\Big(-\frac{1}{2}(\bar{U}R)^{-1}g^{\mathrm{T}}\nabla\sigma_L^{\mathrm{T}}W_L\Big). \tag{2.166}$$

For reducing computation, the integration in (2.165) is approximated by the definition of Riemann integration [4].

A mesh of points over the integral region can be introduced on Ω, with the size δx chosen as small as possible. Moreover, p is required to be larger than L, and the activation functions are linearly independent to guarantee $(A^{\mathrm{T}}A)$ invertible. The specific expressions are given as follows:

$$A = [\nabla\bar{\sigma}_L(x)\Delta x|_{x=x_1}, \ldots, \nabla\bar{\sigma}_L(x)\Delta x|_{x=x_p}]^{\mathrm{T}}, \tag{2.167}$$

$$B = \begin{bmatrix} Q(x) + 2\int_0^{u(x)} \Phi^{-\mathrm{T}}(\bar{U}^{-1}s)\bar{U}Rds\Big|_{x=x_1} \\ \vdots \\ Q(x) + 2\int_0^{u(x)} \Phi^{-\mathrm{T}}(\bar{U}^{-1}s)\bar{U}Rds\Big|_{x=x_p} \end{bmatrix}, \tag{2.168}$$

$$\langle \nabla\bar{\sigma}_L(x)\Delta x, \nabla\bar{\sigma}_L(x)\Delta x\rangle = \lim_{\|\delta x\|\to 0} (A^{\mathrm{T}}A)\cdot\delta x, \tag{2.169}$$

$$\left\langle Q(x)+2\int_0^{u(x)} \Phi^{-\mathrm{T}}(\bar{U}^{-1}s)\bar{U}R\,ds, \nabla\bar{\sigma}_L(x)\Delta x\right\rangle = \lim_{\|\delta x\|\to 0}(A^{\mathrm{T}}B)\cdot\delta x. \quad (2.170)$$

Therefore, we get

$$W_L = -(A^{\mathrm{T}}A)^{-1}A^{\mathrm{T}}B. \quad (2.171)$$

The design procedure of the optimal constrained controller of nonlinear discrete-time systems with actuator saturation is given below:

1. Using a neural network to approximate $V(x)$, i.e., we have $V(x)=\sum_{j=1}^{L}\times w_j\sigma_j(x)$.
2. Select an initial admissible control $u^{[0]}$, then solve GHJB$(V^{[0]},u^{[0]})=0$ by applying the least-square method to obtain $\mathrm{W}^{[0]}$, and accordingly $V^{[0]}$ is computed.
3. For $i=0,1,2,\ldots$, update the control $u^{[i+1]}=\bar{U}\Phi(-\frac{1}{2}(\bar{U}R)^{-1}g^{\mathrm{T}}\nabla V^{[i]})$.
4. For $i=0,1,2,\ldots$, solve GHJB$(V^{[i+1]},u^{[i+1]})=0$ by the least-square method to obtain W^{i+1}, and then we can get $V^{[i+1]}$.
5. If $V^{[i]}(0)-V^{[i+1]}(0)\le\varepsilon$, where ε is a small positive constant, then $J^*=V^{[i]}$, stop; else $i=i+1$, go back to step 3 and go on.
6. After J^* being solved off-line, the optimal state feedback control $u^*=\bar{U}\Phi(-\frac{1}{2}(\bar{U}R)^{-1}g^{\mathrm{T}}\nabla J^*)$ will be implemented on-line.

2.4.3 Simulations

In order to demonstrate the effectiveness of the method developed in this section, an example is presented in this subsection.

Example 2.28 Consider the following affine nonlinear discrete-time system with actuator saturation:

$$x(k+1)=f(x(k))+g(x(k))u(k), \quad (2.172)$$

where

$$f(x(k))=\begin{bmatrix}-0.8x_2(k)\\ \sin(0.8x_1(k)-x_2(k))+1.8x_2(k)\end{bmatrix},$$

$$g(x(k))=\begin{bmatrix}0\\ -x_2(k)\end{bmatrix},$$

and the upper bound $\bar{U}$ of actuator saturation is 0.35.

The control objective is to design an optimal controller with bound less than 0.35. Define the cost functional as

$$J(x(0),u)=\sum_{k=0}^{\infty}\left\{x^{\mathrm{T}}(k)Qx(k)+2\int_0^{u(x(k))}\tanh^{-\mathrm{T}}(\bar{U}^{-1}s)\bar{U}R\,ds\right\}, \quad (2.173)$$

where the weight matrices are chosen as $Q=\left[\begin{smallmatrix}1&0\\0&1\end{smallmatrix}\right]$ and $R=1$.

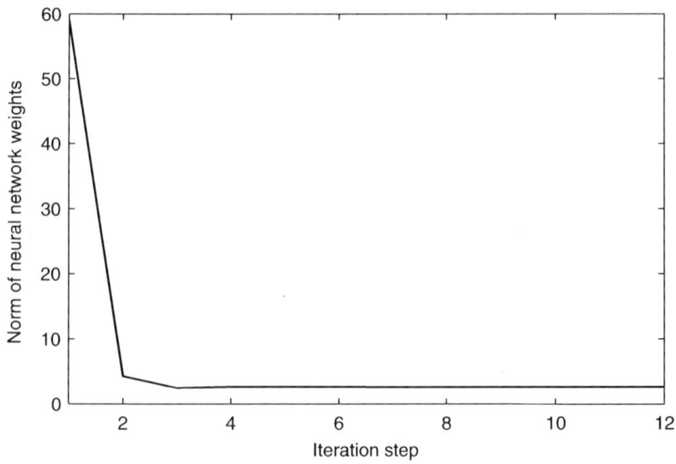

Fig. 2.20 Norm of neural-network weights at each step

To find a nearly optimal controller, a Volterra neural network is used to approximate the value function of the system as follows:

$$\begin{aligned} V(x) = {} & w_1 x_1^2 + w_2 x_2^2 + w_3 x_1 x_2 + w_4 x_1^4 + w_5 x_2^4 + w_6 x_1^3 x_2 \\ & + w_7 x_1^2 x_2^2 + w_8 x_1 x_2^3 + w_9 x_1^6 + w_{10} x_2^6 + w_{11} x_1^5 x_2 \\ & + w_{12} x_1^4 x_2^2 + w_{13} x_1^3 x_2^3 + w_{14} x_1^2 x_2^4 + w_{15} x_1 x_2^5. \end{aligned} \tag{2.174}$$

The algorithm is implemented over the region Ω defined by $|x_1| \le 0.5$, $|x_2| \le 0.5$. Select the initial control $u_0(k) = x_1(k) + 1.5x_2(k)$, which is admissible, and then update the control by $u^{[i+1]} = \bar{U} \tanh(-(\bar{U} R)^{-1} g^{\mathrm{T}} \nabla V^{[i]}/2)$, where $u^{[i]}$ and $V^{[i]}$ satisfy the following GHJB equation:

$$\begin{aligned} & \nabla V^{[i]\mathrm{T}}(x)(f(x) + g(x)u^{[i]}(x) - x) + x^{\mathrm{T}} Q x \\ & \quad + 2 \int_0^{u^{[i]}(x)} \tanh^{-\mathrm{T}}(\bar{U}^{-1} s) \bar{U} R ds = 0. \end{aligned} \tag{2.175}$$

In the simulation, the parameters are chosen as follows: the mesh size $\delta x = 0.01$, the small positive constant $\varepsilon = 0.01$, and the initial states $x_1(0) = x_2(0) = 0.5$. Figure 2.20 shows the trajectory of the norm of neural-network weights at each iteration step. Figure 2.21 shows that the value function converges to a constant very rapidly. After 12 successive iterative steps, the nearly optimal saturated control can be obtained off-line. Then, the controller is applied to the system with given initial states for 100 time steps. Figure 2.22 shows the control trajectory and Fig. 2.23 shows the state trajectories, whereas Figs. 2.24 and 2.25 illustrate the control trajectory and state trajectories without considering actuator saturation, respectively. By comparison, we can see that saturated control and corresponding state trajectories have fewer oscillations, and saturation has been overcome successfully.

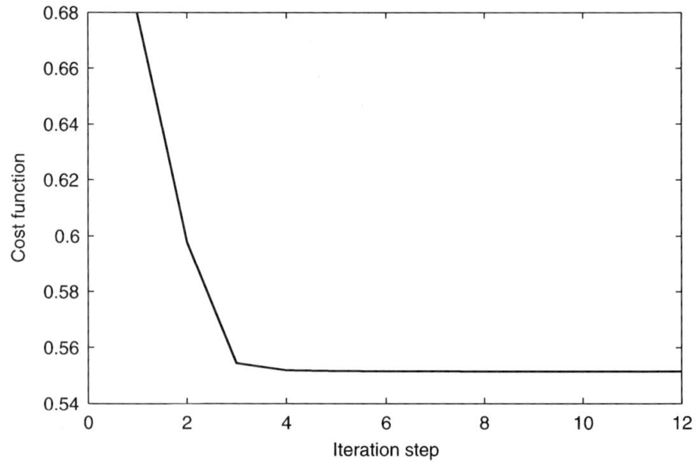

Fig. 2.21 The value function at each iteration step

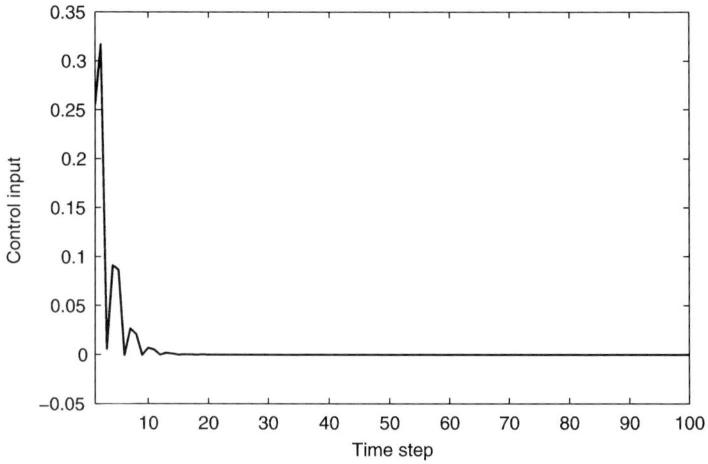

Fig. 2.22 The control trajectory

2.5 Finite-Horizon Optimal State Feedback Control Based on HDP

In this section, we will develop a new ADP scheme for the finite-horizon optimal control problem. We will study the optimal control problem with an ε-error bound using ADP algorithms. First, the HJB equation for finite-horizon optimal control of discrete-time systems is derived. In order to solve this HJB equation, a new iterative ADP algorithm is developed with convergence and optimality proofs. Second, the difficulties of obtaining the optimal solution using the iterative ADP algorithm is presented and then the ε-*optimal* control algorithm is derived based on the iterative

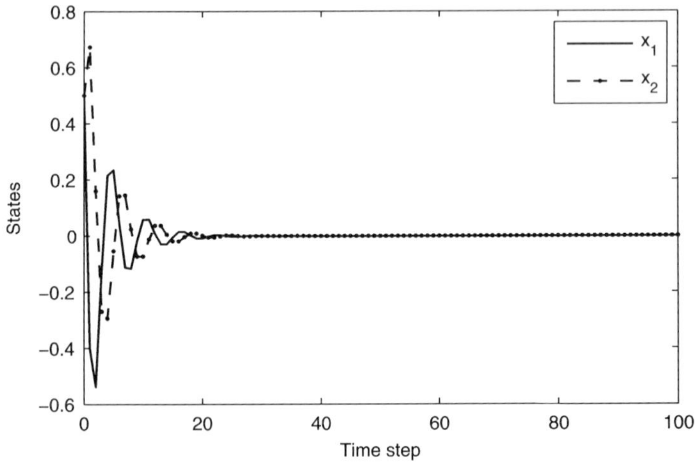

Fig. 2.23 The state trajectories

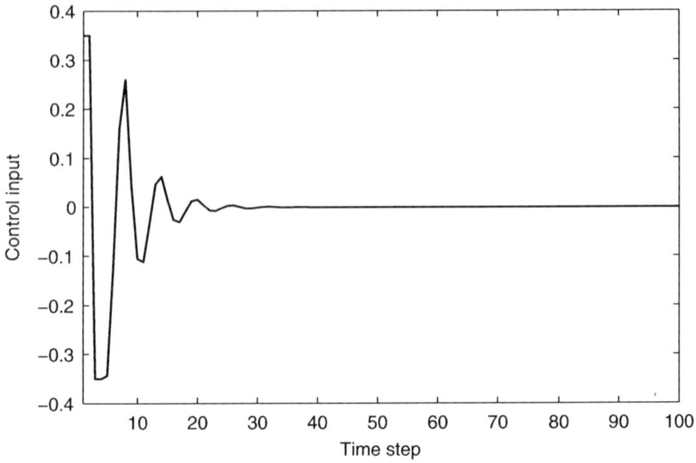

Fig. 2.24 The control trajectory without considering the actuator saturation in the controller design

ADP algorithm. Next, it will be shown that the ε-optimal control algorithm can obtain suboptimal control solutions within a fixed finite number of control steps that make the value function converge to its optimal value with an ε-error. Furthermore, in order to facilitate the implementation of the iterative ADP algorithms, we use NNs to obtain the iterative value function and the optimal control policy. Finally, an ε-optimal state feedback controller is obtained for the finite-horizon optimal control problem.

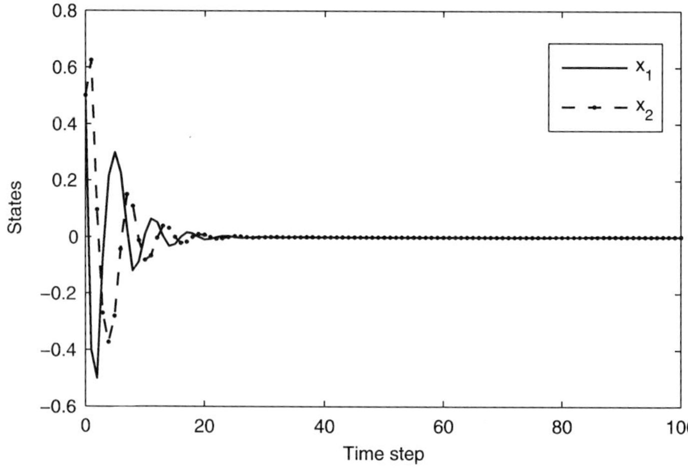

Fig. 2.25 The state trajectories without considering the actuator saturation in the controller design

2.5.1 Problem Formulation

In this section, we will study the following deterministic discrete-time systems:

$$x(k+1) = F(x(k), u(k)),\ k = 0, 1, 2, \dots, \tag{2.176}$$

where $x(k) \in \mathbb{R}^n$ is the state and $u(k) \in \mathbb{R}^m$ is the control vector. Let $x(0)$ be the initial state. The system function $F(x(k), u(k))$ is continuous for $\forall x(k), u(k)$ and $F(0, 0) = 0$. Hence, $x = 0$ is an equilibrium state of system (2.176) under the control $u = 0$. The cost function for state $x(0)$ under the control sequence $\underline{u}_0^{N-1} = (u(0), u(1), \dots, u(N-1))$ is defined as

$$J\big(x(0), \underline{u}_0^{N-1}\big) = \sum_{i=0}^{N-1} l(x(i), u(i)), \tag{2.177}$$

where l is the utility function, $l(0, 0) = 0$, and $l(x(i), u(i)) \geq 0$ for $\forall x(i), u(i)$.

The sequence $\underline{u}_0^{N-1}$ defined above is a finite sequence of controls. Using this sequence of controls, system (2.176) gives a trajectory starting from $x(0)$: $x(1) = F(x(0), u(0))$, $x(2) = F(x(1), u(1))$, $\dots$, $x(N) = F(x(N-1), u(N-1))$. We call the number of elements in the control sequence $\underline{u}_0^{N-1}$ the length of $\underline{u}_0^{N-1}$ and denote it as $|\underline{u}_0^{N-1}|$. Then, $|\underline{u}_0^{N-1}| = N$. The length of the associated trajectory $\underline{x}_0^N = (x(0), x(1), \dots, x(N))$ is $N + 1$. We denote the final state of the trajectory as $x^{(f)}(x(0), \underline{u}_0^{N-1})$, i.e., $x^{(f)}(x(0), \underline{u}_0^{N-1}) = x_N$. Then, for $\forall k \geq 0$, the finite control sequence starting at k can be written as $\underline{u}_k^{k+i-1} = (u(k), u(k+1), \dots, u(k+i-1))$, where $i \geq 1$ is the length of the control sequence. The final state can be written as $x^{(f)}(x(k), \underline{u}_k^{k+i-1}) = x(k+i)$.

We note that the cost function defined in (2.177) does not have the term associated with the final state since in the present study we specify the final state $x(N) = F(x(N-1), u(N-1))$ to be at the origin, i.e., $x(N) = x^{(f)} = 0$. For the present finite-horizon optimal control problem, the feedback controller $u(k) = u(x(k))$ must not only drive the system state to zero within a finite number of time steps but also guarantee the cost function (2.177) to be finite, i.e., $\underline{u}_k^{N-1} = (u(x(k)), u(x(k+1)), \ldots, u(x(N-1)))$ must be a finite-horizon admissible control sequence, where $N > k$ is a finite integer.

Definition 2.29 A control sequence $\underline{u}_k^{N-1}$ is said to be finite-horizon admissible for a state $x(k) \in \mathbb{R}^n$, if $x^{(f)}(x(k), \underline{u}_k^{N-1}) = 0$ and $J(x(k), \underline{u}_k^{N-1})$ is finite, where $N > k$ is a finite integer.

A state $x(k)$ is said to be finite-horizon controllable (controllable for brief) if there is a finite-horizon admissible control sequence associated with this state.

Let $\underline{u}_k$ be an arbitrary finite-horizon admissible control sequence starting at k and let

$$\mathfrak{A}_{x(k)} = \left\{\underline{u}_k : x^{(f)}\left(x(k), \underline{u}_k\right) = 0\right\}$$

be the set of all finite-horizon admissible control sequences of $x(k)$. Let

$$\mathfrak{A}_{x(k)}^{(i)} = \left\{\underline{u}_k^{k+i-1} : x^{(f)}\left(x(k), \underline{u}_k^{k+i-1}\right) = 0, \ \left|\underline{u}_k^{k+i-1}\right| = i\right\}$$

be the set of all finite-horizon admissible control sequences of $x(k)$ with length i. Then, $\mathfrak{A}_{x(k)} = \bigcup_{1 \le i < \infty} \mathfrak{A}_{x(k)}^{(i)}$. In this notation, a state $x(k)$ is controllable if and only if $\mathfrak{A}_{x(k)} \neq \emptyset$.

For any given system state $x(k)$, the objective of the present finite-horizon optimal control problem is to find a finite-horizon admissible control sequence $\underline{u}_k^{N-1} \in \mathfrak{A}_{x(k)}^{(N-k)} \subseteq \mathfrak{A}_{x(k)}$ to minimize the cost $J(x(k), \underline{u}_k^{N-1})$. The control sequence $\underline{u}_k^{N-1}$ has finite length. However, before it is determined, we do not know its length which means that the length of the control sequence $|\underline{u}_k^{N-1}| = N - k$ is unspecified. This kind of optimal control problems have been called finite-horizon problems with unspecified terminal time [3] (but in the present case, with fixed terminal state $x^{(f)} = 0$).

Define the optimal value function as

$$J^*(x(k)) = \inf_{\underline{u}_k} \left\{J(x(k), \underline{u}_k) : \underline{u}_k \in \mathfrak{A}_{x(k)}\right\}. \tag{2.178}$$

Then, according to Bellman's principle of optimality, $J^*(x(k))$ satisfies the discrete-time HJB equation

$$J^*(x(k)) = \min_{u_k} \left\{l(x(k), u(k)) + J^*(F(x(k), u(k)))\right\}. \tag{2.179}$$

Now, define the law of optimal control sequence starting at k by

$$\underline{u}^*(x(k)) = \arg\inf_{\underline{u}_k} \left\{ J(x(k), \underline{u}_k) \colon \underline{u}_k \in \mathfrak{A}_{x(k)} \right\},$$

and define the law of optimal control vector by

$$u^*(x(k)) = \arg\min_{u(k)} \left\{ l(x(k), u(k)) + J^*(F(x(k), u(k))) \right\}.$$

In other words, $\underline{u}^*(x(k)) = \underline{u}_k^*$ and $u^*(x(k)) = u_k^*$. Hence, we have

$$J^*(x(k)) = l(x(k), u_k^*) + J^*(F(x(k), u_k^*)).$$

2.5.2 Finite-Horizon Optimal State Feedback Control Based on HDP

In this subsection, a new iterative ADP algorithm is developed to obtain the finite-horizon optimal controller for nonlinear systems. The goal of the present iterative ADP algorithm is to construct an optimal control policy $u^*(x(k))$, $k = 0, 1, \ldots,$ which drives the system from an arbitrary initial state $x(0)$ to the singularity 0 within a finite time, and simultaneously minimizes the performance index function. Convergence proofs will also be given to show that the performance index function will indeed converge to the optimum.

2.5.2.1 Derivation and Properties of the Iterative ADP Algorithm

We first consider the case where for any state $x(k)$, there exists a control vector $u(k)$ such that $F(x(k), u(k)) = 0$, i.e., we can control the state of system (2.176) to zero in one step from any initial state. For the case where $F(x(k), u(k)) = 0$ does not hold, we will discuss and solve the problem later in the subsection.

In the iterative ADP algorithm, the value function and control policy are updated by recursive iterations, with the iteration index number i increasing from 0 and with the initial performance index function $V_0(x) = 0$ for $\forall x \in \mathbb{R}^n$.

The value function for $i = 1$ is computed as

$$\begin{aligned} V_1(x(k)) &= \min_{u(k)} \{ l(x(k), u(k)) + V_0(F(x(k), u(k))) \} \\ &\qquad \text{subject to } \; F(x(k), u(k)) = 0 \\ &= \min_{u(k)} l(x(k), u(k)) \;\; \text{subject to } \; F(x(k), u(k)) = 0 \\ &= l(x(k), u_k^*(x(k))), \end{aligned} \tag{2.180}$$

where $V_0(F(x(k),u(k))) = 0$ and $F(x(k),u_k^*(x(k))) = 0$. The control vector $v_1(x(k))$ for $i = 1$ is chosen as $v_1(x(k)) = u_k^*(x(k))$. Therefore, (2.180) can also be written as

$$\begin{aligned} V_1(x(k)) &= \min_{u(k)} l(x(k),u(k)) \text{ subject to } F(x(k),u(k)) = 0 \\ &= l(x(k), v_1(x(k))), \end{aligned} \tag{2.181}$$

where

$$v_1(x(k)) = \arg\min_{u(k)} l(x(k),u(k)) \text{ subject to } F(x(k),u(k)) = 0. \tag{2.182}$$

For $i = 2, 3, 4, \ldots$, the iterative ADP algorithm will be implemented as follows:

$$\begin{aligned} V_i(x(k)) &= \min_{u(k)} \{l(x(k),u(k)) + V_{i-1}(F(x(k),u(k)))\} \\ &= l(x(k), v_i(x(k))) + V_{i-1}(F(x(k), v_i(x(k)))), \end{aligned} \tag{2.183}$$

where

$$\begin{aligned} v_i(x(k)) &= \arg\min_{u(k)} \{l(x(k),u(k)) + V_{i-1}(x(k+1))\} \\ &= \arg\min_{u(k)} \{l(x(k),u(k)) + V_{i-1}(F(x(k),u(k)))\}. \end{aligned} \tag{2.184}$$

Equations (2.181)–(2.184) form the iterative ADP algorithm.

Remark 2.30 Equations (2.181)–(2.184) in the iterative ADP algorithm are similar to the HJB equation (2.179), but they are not the same. There are at least two obvious differences:

1. For any finite time k, if $x(k)$ is the state at k, then the optimal value function in HJB equation (2.179) is unique, i.e., $J^*(x(k))$, while in the iterative ADP equations (2.181)–(2.184), the value function is different for each iteration index i, i.e., $V_i(x(k)) \neq V_j(x(k))$ for $\forall i \neq j$ in general.
2. For any finite time k, if $x(k)$ is the state at k, then the optimal control law obtained by HJB equation (2.179) possesses the unique optimal control expression, i.e., $u_k^* = u^*(x(k))$, while the control law solved by the iterative ADP algorithm (2.181)–(2.184) is different from each other for each iteration index i, i.e., $v_i(x(k)) \neq v_j(x(k))$ for $\forall i \neq j$ in general.

Remark 2.31 According to (2.177) and (2.183), we have

$$V_{i+1}(x(k)) = \min_{\underline{u}_k^{k+i}} \left\{ J\left(x(k), \underline{u}_k^{k+i}\right) : \underline{u}_k^{k+i} \in \mathfrak{A}_{x(k)}^{(i+1)} \right\}. \tag{2.185}$$

Since

$$\begin{aligned}V_{i+1}(x(k)) &= \min_{u(k)} \{l(x(k), u(k)) + V_i(x(k+1))\}\\ &= \min_{u(k)} \Big\{ l(x(k), u(k)) + \min_{u(k+1)} \Big\{ l(x(k+1), u(k+1))\\ &\quad + \min_{u(k+2)} \big\{ l(x(k+2), u(k+2)) + \cdots\\ &\quad + \min_{u(k+i-1)} \{l(x(k+i-1), u(k+i-1))\\ &\quad + V_1(x(k+i))\} \cdots \big\} \Big\} \Big\},\end{aligned}$$

where

$$\begin{aligned}V_1(x(k+i)) = \min_{u(k+i)} l(x(k+i), u(k+i))\\ \text{subject to } F(x(k+i), u(k+i)) = 0,\end{aligned}$$

we obtain

$$\begin{aligned}V_{i+1}(x(k)) &= \min_{\underline{u}_k^{k+i}} \{l(x(k), u(k)) + l(x(k+1), u(k+1))\\ &\quad + \cdots + l(x(k+i), u(k+i))\}\\ &\quad \text{subject to } F(x(k+i), u(k+i)) = 0,\\ &= \min_{\underline{u}_k^{k+i}} \left\{ J\big(x(k), \underline{u}_k^{k+i}\big) : \underline{u}_k^{k+i} \in \mathfrak{A}_{x(k)}^{(i+1)} \right\}.\end{aligned}$$

Using the notation in (2.184), we can also write

$$V_{i+1}(x(k)) = \sum_{j=0}^{i} l\big(x(k+j), v_{i+1-j}(x(k+j))\big). \tag{2.186}$$

In the above, we can see that the value function $J^*(x(k))$ solved by HJB equation (2.179) is replaced by a sequence of iterative value functions $V_i(x(k))$ and the optimal control law $u^*(x(k))$ is replaced by a sequence of iterative control law $v_i(x(k))$, where $i \geq 1$ is the index of iteration. We can prove that $J^*(x(k))$ defined in (2.178) is the limit of $V_i(x(k))$ as $i \to \infty$.

Theorem 2.32 *Let $x(k)$ be an arbitrary state vector. Suppose that $\mathfrak{A}_{x(k)}^{(1)} \neq \emptyset$. Then, the value function $V_i(x(k))$ obtained by* (2.181)–(2.184) *is a monotonically nonincreasing sequence for $\forall i \geq 1$, i.e., $V_{i+1}(x(k)) \leq V_i(x(k))$ for $\forall i \geq 1$.*

Proof We prove this by mathematical induction. First, we let $i = 1$. Then, we have $V_1(x(k))$ given as in (2.181) and the finite-horizon admissible control sequence is $\hat{\underline{u}}_k^k = (v_1(x(k)))$.

Next, we show that there exists a finite-horizon admissible control sequence $\hat{\underline{u}}_k^{k+1}$ with length 2 such that $J(x(k), \hat{\underline{u}}_k^{k+1}) = V_1(x(k))$. The trajectory starting from $x(k)$ under the control of $\hat{\underline{u}}_k^k = (v_1(x(k)))$ is $x(k+1) = F(x(k), v_1(x(k))) = 0$. Then, we create a new control sequence $\hat{\underline{u}}_k^{k+1}$ by adding a 0 to the end of sequence $\hat{\underline{u}}_k^k$ to obtain the control sequence $\hat{\underline{u}}_k^{k+1} = (\hat{\underline{u}}_k^k, 0)$. Obviously, $|\hat{\underline{u}}_k^{k+1}| = 2$. The state trajectory under the control of $\hat{\underline{u}}_k^{k+1}$ is $x(k+1) = F(x(k), v_1(x(k))) = 0$ and $x(k+2) = F(x(k+1), \hat{u}(k+1))$, where $\hat{u}(k+1) = 0$. Since $x(k+1) = 0$ and $F(0,0) = 0$, we have $x(k+2) = 0$. So, $\hat{\underline{u}}_k^{k+1}$ is a finite-horizon admissible control sequence. Furthermore,

$$\begin{aligned} J(x(k), \hat{\underline{u}}_k^{k+1}) &= l(x(k), v_1(x(k))) + l(x(k+1), \hat{u}(k+1)) \\ &= l(x(k), v_1(x(k))) \\ &= V_1(x(k)) \end{aligned}$$

since $l(x(k+1), \hat{u}(k+1)) = l(0,0) = 0$. On the other hand, according to Remark 2.31, we have

$$V_2(x(k)) = \min_{\underline{u}_k^{k+1}} \left\{ J\big(x(k), \underline{u}_k^{k+1}\big) : \underline{u}_k^{k+1} \in \mathfrak{A}_{x(k)}^{(2)} \right\}.$$

Then, we obtain

$$\begin{aligned} V_2(x(k)) &= \min_{\underline{u}_k^{k+1}} \left\{ J\big(x(k), \underline{u}_k^{k+1}\big) : \underline{u}_k^{k+1} \in \mathfrak{A}_{x(k)}^{(2)} \right\} \\ &\le J\big(x(k), \hat{\underline{u}}_k^{k+1}\big) \\ &= V_1(x(k)). \end{aligned} \tag{2.187}$$

Therefore, the theorem holds for $i = 1$.

Assume that the theorem holds for any $i = q$, where $q > 1$. From (2.186), we have

$$V_q(x(k)) = \sum_{j=0}^{q-1} l\big(x(k+j), v_{q-j}(x(k+j))\big).$$

The corresponding finite-horizon admissible control sequence is $\hat{\underline{u}}_k^{k+q-1} = (v_q(x(k)), v_{q-1}(x(k+1)), \ldots, v_1(x(k+q-1)))$.

For $i = q+1$, we create a control sequence

$$\hat{\underline{u}}_k^{k+q} = \big(v_q(x(k)), v_{q-1}(x(k+1)), \ldots, v_1(x(k+q-1)), 0\big)$$

with length $q+1$. Then, the state trajectory under the control of $\underline{\hat{u}}_k^{k+q}$ is $x(k)$, $x(k+1) = F(x(k), v_q(x(k)))$, $x(k+2) = F(x(k+1), v_{q-1}(x(k+1))), \ldots, x(k+q) = F(x(k+q-1), v_1(x(k+q-1))) = 0$, $x(k+q+1) = F(x(k+q), 0) = 0$. So, $\underline{\hat{u}}_k^{k+q}$ is a finite-horizon admissible control sequence. The value function under this control sequence is

$$\begin{aligned} J(x(k), \underline{\hat{u}}_k^{k+q}) &= l(x(k), v_q(x(k))) + l(x(k+1), v_{q-1}(x(k+1))) \\ &\quad + \cdots + l(x(k+q-1), v_1(x(k+q-1))) + l(x(k+q), 0) \\ &= \sum_{j=0}^{q-1} l\big(x(k+j), v_{q-j}(x(k+j))\big) \\ &= V_q(x(k)) \end{aligned}$$

since $l(x(k+q), 0) = l(0, 0) = 0$.

On the other hand, we have

$$V_{q+1}(x(k)) = \min_{\underline{u}_k^{k+q}} \left\{ J\left(x(k), \underline{u}_k^{k+q}\right) : \underline{u}_k^{k+q} \in \mathfrak{A}_{x(k)}^{(q+1)} \right\}.$$

Thus, we obtain

$$\begin{aligned} V_{q+1}(x(k)) &= \min_{\underline{u}_k^{k+q}} \left\{ J\left(x(k), \underline{u}_k^{k+q}\right) : \underline{u}_k^{k+q} \in \mathfrak{A}_{x(k)}^{(q+1)} \right\} \\ &\le J\left(x(k), \underline{\hat{u}}_k^{k+q}\right) \\ &= V_q(x(k)), \end{aligned}$$

which completes the proof. □

From Theorem 2.32, we know that the value function $V_i(x(k)) \ge 0$ is a monotonically nonincreasing sequence and is bounded below for iteration index $i = 1, 2, \ldots$. Now, we can derive the following theorem.

Theorem 2.33 *Let $x(k)$ be an arbitrary state vector. Define the performance index function $V_\infty(x(k))$ as the limit of the iterative function $V_i(x(k))$, i.e.,*

$$V_\infty(x(k)) = \lim_{i \to \infty} V_i(x(k)). \tag{2.188}$$

Then, we have

$$V_\infty(x(k)) = \min_{u(k)} \{ l(x(k), u(k)) + V_\infty(x(k+1)) \}.$$

Proof Let $\eta_k = \eta(x(k))$ be any admissible control vector. According to Theorem 2.32, for $\forall i$, we have

$$V_\infty(x(k)) \le V_{i+1}(x(k)) \le l(x(k), \eta_k) + V_i(x(k+1)).$$

Let $i \to \infty$, we have

$$V_\infty(x(k)) \le l(x(k), \eta_k) + V_\infty(x(k+1)),$$

which is true for $\forall\, \eta_k$. Therefore,

$$V_\infty(x(k)) \le \min_{u(k)}\{l(x(k), u(k)) + V_\infty(x(k+1))\}. \tag{2.189}$$

Let $\varepsilon > 0$ be an arbitrary positive number. Since $V_i(x(k))$ is nonincreasing for $i \ge 1$ and $\lim_{i\to\infty} V_i(x(k)) = V_\infty(x(k))$, there exists a positive integer p such that

$$V_p(x(k)) - \varepsilon \le V_\infty(x(k)) \le V_p(x(k)).$$

From (2.183), we have

$$\begin{aligned} V_p(x(k)) &= \min_{u(k)}\{l(x(k), u(k)) + V_{p-1}(F(x(k), u(k)))\} \\ &= l(x(k), v_p(x(k))) + V_{p-1}(F(x(k), v_p(x(k)))). \end{aligned}$$

Hence,

$$\begin{aligned} V_\infty(x(k)) &\ge l(x(k), v_p(x(k))) + V_{p-1}(F(x(k), v_p(x(k)))) - \varepsilon \\ &\ge l(x(k), v_p(x(k))) + V_\infty(F(x(k), v_p(x(k)))) - \varepsilon \\ &\ge \min_{u(k)}\{l(x(k), u(k)) + V_\infty(x(k+1))\} - \varepsilon. \end{aligned}$$

Since ε is arbitrary, we have

$$V_\infty(x(k)) \ge \min_{u(k)}\{l(x(k), u(k)) + V_\infty(x(k+1))\}. \tag{2.190}$$

Combining (2.189) and (2.190), we prove the theorem. □

Next, we will prove that the iterative value function $V_i(x(k))$ converges to the optimal value function $J^*(x(k))$ as $i \to \infty$.

Theorem 2.34 (cf. [14]) *Let $V_\infty(x(k))$ be defined in* (2.188). *If the system state $x(k)$ is controllable, then we have the value function $V_\infty(x(k))$ equal to the optimal value function $J^*(x(k))$, i.e.,*

$$\lim_{i\to\infty} V_i(x(k)) = J^*(x(k)),$$

where $V_i(x(k))$ is defined in (2.183).

Proof According to (2.178) and (2.185), we have

$$J^*(x(k)) \le \min_{\underline{u}_k^{k+i-1}} \left\{ J(x(k), \underline{u}_k^{k+i-1}) \colon \underline{u}_k^{k+i-1} \in \mathfrak{A}_{x(k)}^{(i)} \right\} = V_i(x(k)).$$

Then, let $i \to \infty$, and we obtain

$$J^*(x(k)) \le V_\infty(x(k)). \tag{2.191}$$

Next, we show that

$$V_\infty(x(k)) \le J^*(x(k)). \tag{2.192}$$

For any $\omega > 0$, by the definition of $J^*(x(k))$ in (2.178), there exists $\underline{\eta}_k \in \mathfrak{A}_{x(k)}$ such that

$$J(x(k), \underline{\eta}_k) \le J^*(x(k)) + \omega. \tag{2.193}$$

Suppose that $|\underline{\eta}_k| = p$. Then, $\underline{\eta}_k \in \mathfrak{A}_{x(k)}^{(p)}$. So, by Theorem 2.32 and (2.185), we have

$$\begin{aligned}
V_\infty(x(k)) &\le V_p(x(k)) \\
&= \min_{\underline{u}_k^{k+p-1}} \left\{ J(x(k), \underline{u}_k^{k+p-1}) \colon \underline{u}_k^{k+p-1} \in \mathfrak{A}_{x(k)}^{(p)} \right\} \\
&\le J(x(k), \underline{\eta}_k) \\
&\le J^*(x(k)) + \omega.
\end{aligned}$$

Since ω is chosen arbitrarily, we know that (2.192) is true. Therefore, from (2.191) and (2.192), we have proven the theorem. □

We can now present the following corollary.

Corollary 2.35 *Let the value function $V_i(x(k))$ be defined by* (2.183). *If the system state $x(k)$ is controllable, then the iterative control law $v_i(x(k))$ converges to the optimal control law $u^*(x(k))$, i.e.,*

$$\lim_{i \to \infty} v_i(x(k)) = u^*(x(k))$$

Remark 2.36 Generally speaking, for the *finite-horizon* optimal control problem, the optimal value function depends not only on state $x(k)$ but also on the time left (see [7] and [11]). For the finite-horizon optimal control problem with unspecified terminal time, we have proved that the iterative value functions converge to the optimal as the iterative index i reaches infinity. Then, the time left is negligible and we say that the optimal value function $J^*(x(k))$ is only a function of the state $x(k)$ which is like the case of infinite-horizon optimal control problems.

According to Theorem 2.34 and Corollary 2.35, we know that if $x(k)$ is controllable, then, as $i \to \infty$, the iterative value function $V_i(x(k))$ converges to the optimal value function $J^*(x(k))$ and the iterative control law $v_i(x(k))$ also converges to the optimal control law $u^*(x(k))$. So, it is important to note that for controllable state

$x(k)$, the iterative value functions $V_i(x(k))$ are well defined for all i under the iterative control law $v_i(x(k))$.

Let $\mathcal{T}_0 = \{0\}$. For $i = 1, 2, \ldots$, define

$$\mathcal{T}_i = \{x(k) \in \mathbb{R}^n \mid \exists u(k) \in \mathbb{R}^m \text{ s.t. } F(x(k), u(k)) \in \mathcal{T}_{i-1}\}. \tag{2.194}$$

Next, we prove the following theorem.

Theorem 2.37 *Let $\mathcal{T}_0 = \{0\}$ and $\mathcal{T}_i$ be defined in* (2.194). *Then, for $i = 0, 1, \ldots$, we have $\mathcal{T}_i \subseteq \mathcal{T}_{i+1}$.*

Proof We prove the theorem by mathematical induction. First, let $i = 0$. Since $\mathcal{T}_0 = \{0\}$ and $F(0, 0) = 0$, we know that $0 \in \mathcal{T}_1$. Hence, $\mathcal{T}_0 \subseteq \mathcal{T}_1$.

Next, assume that $\mathcal{T}_{i-1} \subseteq \mathcal{T}_i$ holds. Now, if $x(k) \in \mathcal{T}_i$, we have $F(x(k), \eta_{i-1}(x(k))) \in \mathcal{T}_{i-1}$ for some $\eta_{i-1}(x(k))$. Hence, $F(x(k), \eta_{i-1}(x(k))) \in \mathcal{T}_i$ by the assumption of $\mathcal{T}_{i-1} \subseteq \mathcal{T}_i$. So, $x(k) \in \mathcal{T}_{i+1}$ by (2.194). Thus, $\mathcal{T}_i \subseteq \mathcal{T}_{i+1}$, which proves the theorem. □

According to Theorem 2.37, we have

$$\{0\} = \mathcal{T}_0 \subseteq \mathcal{T}_1 \subseteq \cdots \subseteq \mathcal{T}_{i-1} \subseteq \mathcal{T}_i \subseteq \cdots.$$

We can see that by introducing the sets $\mathcal{T}_i$, $i = 0, 1, \ldots$, the state $x(k)$ can be classified correspondingly. According to Theorem 2.37, the properties of the ADP algorithm can be derived in the following theorem.

Theorem 2.38

(i) *For any i, $x(k) \in \mathcal{T}_i \Leftrightarrow \mathfrak{A}_{x(k)}^{(i)} \neq \emptyset \Leftrightarrow V_i(x(k))$ is defined at $x(k)$.*

(ii) *Let $\mathcal{T}_\infty = \bigcup_{i=1}^{\infty} \mathcal{T}_i$. Then, $x(k) \in \mathcal{T}_\infty \Leftrightarrow \mathfrak{A}_{x(k)} \neq \emptyset \Leftrightarrow J^*(x(k))$ is defined at $x(k) \Leftrightarrow x(k)$ is controllable.*

(iii) *If $V_i(x(k))$ is defined at $x(k)$, then $V_j(x(k))$ is defined at $x(k)$ for every $j \geq i$.*

(iv) *$J^*(x(k))$ is defined at $x(k)$ if and only if there exists an i such that $V_i(x(k))$ is defined at $x(k)$.*

2.5.2.2 The ε-Optimal Control Algorithm

In the previous subsection, we have proved that the iterative value function $V_i(x(k))$ converges to the optimal value function $J^*(x(k))$ and $J^*(x(k)) = \min_{\underline{u}_k}\{J(x(k), \underline{u}_k),\ \underline{u} \in \mathfrak{A}_{x(k)}\}$ satisfies the Bellman's equation (2.179) for any controllable state $x(k) \in \mathcal{T}_\infty$.

To obtain the optimal value function $J^*(x(k))$, a natural strategy is to run the iterative ADP algorithm (2.181)–(2.184) until $i \to \infty$. But unfortunately, it is not practical to do so. In many cases, we cannot find the equality $J^*(x(k)) = V_i(x(k))$ for any finite i. That is, for any admissible control sequence $\underline{u}_k$ with finite length, the cost starting from $x(k)$ under the control of $\underline{u}_k$ will be larger than, not equal to, $J^*(x(k))$. On the other hand, by running the iterative ADP algorithm (2.181)–(2.184), we can obtain a control vector $v_\infty(x(k))$ and then construct a control sequence $\underline{u}_\infty(x(k)) = (v_\infty(x(k)), v_\infty(x(k+1)), \dots, v_\infty(x(k+i)), \dots)$, where $x(k+1) = F(x(k), v_\infty(x(k))), \dots, x(k+i) = F(x(k+i-1), v_\infty(x(k+i-1)))$, ... In general, $\underline{u}_\infty(x(k))$ has infinite length. That is, the controller $v_\infty(x(k))$ cannot control the state to reach the target in finite number of steps. To overcome this difficulty, a new ε-optimal control method using iterative ADP algorithm will be developed.

First, we will introduce our method of iterative ADP with the consideration of the length of control sequences. For different $x(k)$, we will consider different length i for the optimal control sequence. For a given error bound $\varepsilon > 0$, the number i will be chosen so that the error between $J^*(x(k))$ and $V_i(x(k))$ is within the bound.

Let $\varepsilon > 0$ be any small number and $x(k) \in \mathcal{T}_\infty$ be any controllable state. Let the value function $V_i(x(k))$ be defined by (2.183) and $J^*(x(k))$ be the optimal value function. According to Theorem 2.34, given $\varepsilon > 0$, there exists a finite i such that

$$|V_i(x(k)) - J^*(x(k))| \le \varepsilon. \tag{2.195}$$

Definition 2.39 (cf. [14]) Let $x(k) \in \mathcal{T}_\infty$ be a controllable state vector. Let $\varepsilon > 0$ be a small positive number. The approximate length of optimal control sequence with respect to ε is defined as

$$K_\varepsilon(x(k)) = \min\{i : |V_i(x(k)) - J^*(x(k))| \le \varepsilon\}. \tag{2.196}$$

Given a small positive number ε, for any state vector $x(k)$, the number $K_\varepsilon(x(k))$ gives a suitable length of control sequence for optimal control starting from $x(k)$. For $x(k) \in \mathcal{T}_\infty$, since $\lim_{i\to\infty} V_i(x(k)) = J^*(x(k))$, we can always find i such that (2.195) is satisfied. Therefore, $\{i : |V_i(x(k)) - J^*(x(k))| \le \varepsilon\} \neq \emptyset$ and $K_\varepsilon(x(k))$ is well defined.

We can see that an error ε between $V_i(x(k))$ and $J^*(x(k))$ is introduced into the iterative ADP algorithm which makes the value function $V_i(x(k))$ converge within a finite number of iteration steps. In this part, we will show that the corresponding control is also an effective control that drives the value function to within error bound ε from its optimal.

From Definition 2.39, we can see that all the states $x(k)$ that satisfy (2.196) can be classified into one set. Motivated by the definition in (2.194), we can further classify this set using the following definition.

Definition 2.40 (cf. [14]) Let ε be a positive number. Define $\mathcal{T}_0^{(\varepsilon)} = \{0\}$ and for $i = 1, 2, \dots$, define

$$\mathcal{T}_i^{(\varepsilon)} = \{x(k) \in \mathcal{T}_\infty \colon K_\varepsilon(x(k)) \le i\}.$$

Accordingly, when $x(k) \in \mathcal{T}_i^{(\varepsilon)}$, to find the optimal control sequence which has value less than or equal to $J^*(x(k)) + \varepsilon$, one only needs to consider the control sequences $\underline{u}_k$ with length $|\underline{u}_k| \le i$. The sets $\mathcal{T}_i^{(\varepsilon)}$ have the following properties.

Theorem 2.41 (cf. [14]) *Let* $\varepsilon > 0$ *and* $i = 0, 1, \ldots$. *Then*:

(i) $x(k) \in \mathcal{T}_i^{(\varepsilon)}$ *if and only if* $V_i(x(k)) \le J^*(x(k)) + \varepsilon$
(ii) $\mathcal{T}_i^{(\varepsilon)} \subseteq \mathcal{T}_i$
(iii) $\mathcal{T}_i^{(\varepsilon)} \subseteq \mathcal{T}_{i+1}^{(\varepsilon)}$
(iv) $\bigcup_i \mathcal{T}_i^{(\varepsilon)} = \mathcal{T}_\infty$
(v) *If* $\varepsilon > \delta > 0$, *then* $\mathcal{T}_i^{(\varepsilon)} \supseteq \mathcal{T}_i^{(\delta)}$

Proof (i) Let $x(k) \in \mathcal{T}_i^{(\varepsilon)}$. By Definition 2.40, $K_\varepsilon(x(k)) \le i$. Let $j = K_\varepsilon(x(k))$. Then, $j \le i$ and by Definition 2.39, $|V_j(x(k)) - J^*(x(k))| \le \varepsilon$. So, $V_j(x(k)) \le J^*(x(k)) + \varepsilon$. By Theorem 2.32, $V_i(x(k)) \le V_j(x(k)) \le J^*(x(k)) + \varepsilon$. On the other hand, if $V_i(x(k)) \le J^*(x(k)) + \varepsilon$, then $|V_i(x(k)) - J^*(x(k))| \le \varepsilon$. So, $K_\varepsilon(x(k)) = \min\{j : |V_j(x(k)) - J^*(x(k))| \le \varepsilon\} \le i$, which implies that $x(k) \in \mathcal{T}_i^{(\varepsilon)}$.

(ii) If $x(k) \in \mathcal{T}_i^{(\epsilon)}$, $K_\epsilon(x(k)) \le i$ and $|V_i(x(k)) - J^*(x(k))| \le \varepsilon$. So, $V_i(x(k))$ is defined at $x(k)$. According to Theorem 2.38 (i), we have $x(k) \in \mathcal{T}_i$. Hence, $\mathcal{T}_i^{(\epsilon)} \subseteq \mathcal{T}_i$.

(iii) If $x(k) \in \mathcal{T}_i^{(\varepsilon)}$, $K_\varepsilon(x(k)) \le i < i + 1$. So, $x(k) \in \mathcal{T}_{i+1}^{(\varepsilon)}$. Thus, $\mathcal{T}_i^{(\varepsilon)} \subseteq \mathcal{T}_{i+1}^{(\varepsilon)}$.

(iv) Obviously, $\bigcup_i \mathcal{T}_i^{(\varepsilon)} \subseteq \mathcal{T}_\infty$ since $\mathcal{T}_i^{(\varepsilon)}$ are subsets of $\mathcal{T}_\infty$. For any $x(k) \in \mathcal{T}_\infty$, let $p = K_\varepsilon(x(k))$. Then, $x(k) \in \mathcal{T}_p^{(\varepsilon)}$. So, $x(k) \in \bigcup_i \mathcal{T}_i^{(\varepsilon)}$. Hence, $\mathcal{T}_\infty \subseteq \bigcup_i \mathcal{T}_i^{(\varepsilon)} \subseteq \mathcal{T}_\infty$, and we obtain, $\bigcup_i \mathcal{T}_i^{(\varepsilon)} = \mathcal{T}_\infty$.

(v) If $x(k) \in \mathcal{T}_i^{(\delta)}$, $V_i(x(k)) \le J^*(x(k)) + \delta$ by part (i) of this theorem. Clearly, $V_i(x(k)) \le J^*(x(k)) + \varepsilon$ since $\delta < \varepsilon$. This implies that $x(k) \in \mathcal{T}_i^{(\varepsilon)}$. Therefore, $\mathcal{T}_i^{(\varepsilon)} \supseteq \mathcal{T}_i^{(\delta)}$. □

According to Theorem 2.41(i), $\mathcal{T}_i^{(\varepsilon)}$ is just the region where $V_i(x(k))$ is close to $J^*(x(k))$ with error less than ε. This region is a subset of $\mathcal{T}_i$ according to Theorem 2.41(ii). As stated in Theorem 2.41(iii), when i is large, the set $\mathcal{T}_i^{(\varepsilon)}$ is also large. That means that, when i is large, we have a large region where we can use $V_i(x(k))$ as the approximation of $J^*(x(k))$ under certain error. On the other hand, we claim that if $x(k)$ is far away from the origin, we have to choose long control sequence to approximate the optimal control sequence. Theorem 2.41(iv) means that for every controllable state $x(k) \in \mathcal{T}_\infty$, we can always find a suitable control sequence with length i to approximate the optimal control. The size of the set $\mathcal{T}_i^{(\varepsilon)}$ depends on the value of ε. Smaller value of ε gives smaller set $\mathcal{T}_i^{(\varepsilon)}$ which is indicated by Theorem 2.41(v).

Let $x(k) \in \mathcal{T}_\infty$ be an arbitrary controllable state. If $x(k) \in \mathcal{T}_i^{(\varepsilon)}$, the iterative value function satisfies (2.195) under the control $v_i(x(k))$, we call this control the

ε-optimal control and denote it as $\mu_\varepsilon^*(x(k))$, i.e.,

$$\mu_\varepsilon^*(x(k)) = v_i(x(k)) = \arg\min_{u(k)} \{l(x(k), u(k)) + V_{i-1}(F(x(k), u(k)))\}. \quad (2.197)$$

We have the following corollary.

Corollary 2.42 (cf. [14]) *Let $\mu_\varepsilon^*(x(k))$ be expressed in* (2.197) *that makes the value function satisfy* (2.195) *for $x(k) \in \mathcal{T}_i^{(\varepsilon)}$. Then, for any $x'_k \in \mathcal{T}_i^{(\varepsilon)}$, $\mu_\varepsilon^*(x'_k)$ guarantees*

$$|V_i(x'_k) - J^*(x'_k)| \le \varepsilon. \quad (2.198)$$

Proof The corollary can be proved by contradiction. Assume that the conclusion is not true. Then, the inequality (2.198) is false under the control $\mu_\varepsilon^*(\cdot)$ for some $x''_k \in \mathcal{T}_i^{(\varepsilon)}$.

As $\mu_\varepsilon^*(x(k))$ makes the value function satisfy (2.195) for $x(k) \in \mathcal{T}_i^{(\varepsilon)}$, we have $K_\varepsilon(x(k)) \le i$. Using the ε-optimal control law $\mu_\varepsilon^*(\cdot)$ at the state x''_k, according to the assumption, we have $|V_i(x''_k) - J^*(x''_k)| > \varepsilon$. Then, $K_\varepsilon(x''_k) > i$ and $x''_k \notin \mathcal{T}_i^{(\varepsilon)}$. It is in contradiction with the assumption $x''_k \in \mathcal{T}_i^{(\varepsilon)}$. Therefore, the assumption is false and (2.198) holds for any $x'_k \in \mathcal{T}_i^{(\varepsilon)}$. □

Remark 2.43 Corollary 2.42 is very important for neural-network implementation of the iterative ADP algorithm. It shows that we do not need to obtain the optimal control law by searching the entire subset $\mathcal{T}_i^{(\varepsilon)}$. Instead, we can just find one point of $\mathcal{T}_i^{(\varepsilon)}$, i.e., $x(k) \in \mathcal{T}_i^{(\varepsilon)}$, to obtain the ε-optimal control $\mu_\varepsilon^*(x(k))$ which will be effective for any other state $x'_k \in \mathcal{T}_i^{(\varepsilon)}$. This property not only makes the computational complexity much reduced but also makes the optimal control law easily obtained using neural networks.

Theorem 2.44 (cf. [14]) *Let $x(k) \in \mathcal{T}_i^{(\varepsilon)}$ and let $\mu_\varepsilon^*(x(k))$ be expressed in* (2.197). *Then, $F(x(k), \mu_\varepsilon^*(x(k))) \in \mathcal{T}_{i-1}^{(\varepsilon)}$. In other words, if $K_\varepsilon(x(k)) = i$, then $K_\varepsilon(F(x(k), \mu_\varepsilon^*(x(k)))) \le i - 1$.*

Proof Since $x(k) \in \mathcal{T}_i^{(\varepsilon)}$, by Theorem 2.41 (i) we know that

$$V_i(x(k)) \le J^*(x(k)) + \varepsilon. \quad (2.199)$$

According to (2.183) and (2.197), we have

$$V_i(x(k)) = l(x(k), \mu_\varepsilon^*(x(k))) + V_{i-1}(F(x(k), \mu_\varepsilon^*(x(k)))). \quad (2.200)$$

Combining (2.199) and (2.200), we have

$$\begin{aligned} V_{i-1}(F(x(k), \mu_\varepsilon^*(x(k)))) &= V_i(x(k)) - l(x(k), \mu_\varepsilon^*(x(k))) \\ &\le J^*(x(k)) + \varepsilon - l(x(k), \mu_\varepsilon^*(x(k))). \end{aligned} \quad (2.201)$$

On the other hand, we have

$$J^*(x(k)) \le l(x(k), \mu_\varepsilon^*(x(k))) + J^*(F(x, \mu_\varepsilon^*(x(k)))). \tag{2.202}$$

Putting (2.202) into (2.201), we obtain

$$V_{i-1}(F(x(k), \mu_\varepsilon^*(x(k)))) \le J^*(F(x(k), \mu_\varepsilon^*(x(k)))) + \varepsilon.$$

By Theorem 2.41 (i), we have

$$F(x(k), \mu_\varepsilon^*(x(k))) \in \mathcal{T}_{i-1}^{(\varepsilon)}. \tag{2.203}$$

So, if $K_\varepsilon(x(k)) = i$, we know that $x(k) \in \mathcal{T}_i^{(\varepsilon)}$ and $F(x, \mu_\varepsilon^*(x(k))) \in \mathcal{T}_{i-1}^{(\varepsilon)}$ according to (2.203). Therefore, we have

$$K_\varepsilon(F(x(k), \mu_\varepsilon^*(x(k)))) \le i - 1,$$

which proves the theorem. □

Remark 2.45 From Theorem 2.44 we can see that the parameter $K_\varepsilon(x(k))$ gives an important property of the finite-horizon ADP algorithm. It not only gives an optimal condition of the iterative process, but also gives an optimal number of control steps for the finite-horizon ADP algorithm. For example, if $|V_i(x(k)) - J^*(x(k))| \le \varepsilon$ for small ε, then we have $V_i(x(k)) \approx J^*(x(k))$. According to Theorem 2.44, we can get $N = k + i$, where N is the number of control steps to drive the system to zero. The whole control sequence $\underline{u}_0^{N-1}$ may not be ε-optimal but the control sequence $\underline{u}_k^{N-1}$ is ε-optimal control sequence. If $k = 0$, we have $N = K_\varepsilon(x(0)) = i$. Under this condition, we say that the iteration index $K_\varepsilon(x(0))$ denotes the number of ε-optimal control steps.

Corollary 2.46 *Let $\mu_\varepsilon^*(x(k))$ be expressed in* (2.197) *that makes the value function satisfy* (2.195) *for $x(k) \in \mathcal{T}_i^{(\varepsilon)}$. Then, for any $x_k' \in \mathcal{T}_j^{(\varepsilon)}$, where $0 \le j \le i$, $\mu_\varepsilon^*(x_k')$ guarantees*

$$|V_i(x_k') - J^*(x_k')| \le \varepsilon. \tag{2.204}$$

Proof The proof is similar to Corollary 2.42 and is omitted here. □

Remark 2.47 Corollary 2.46 shows that the ε-optimal control $\mu_\varepsilon^*(x(k))$ obtained for $\forall x(k) \in \mathcal{T}_i^{(\varepsilon)}$ is effective for any state $x_k' \in \mathcal{T}_j^{(\epsilon)}$, where $0 \le j \le i$. This means that for $\forall x_k' \in \mathcal{T}_j^{(\varepsilon)}$, $0 \le j \le i$, we can use a same ε-optimal control $\mu_\varepsilon^*(x_k')$ to control the system.

According to Theorem 2.41(iii) and Corollary 2.42, the ε-optimal control $\mu_\varepsilon^*(x(k))$ obtained for an $x(k) \in \mathcal{T}_i^{(\varepsilon)}$ is effective for any state $x_k' \in \mathcal{T}_{i-1}^{(\varepsilon)}$ (which

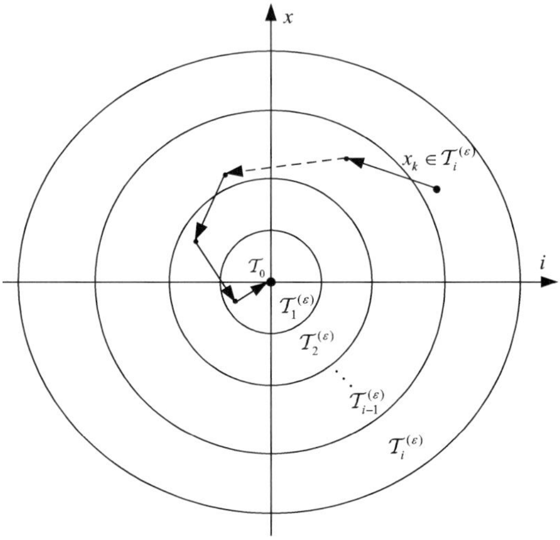

Fig. 2.26 The control process of the controllable sate $x(k) \in \mathcal{T}_i^{(\varepsilon)}$ using the iterative ADP algorithm

is also stated in Corollary 2.46). That is to say: in order to obtain effective ε-optimal control, the iterative ADP algorithm only needs to run at some state $x(k) \in \mathcal{T}_\infty$. In order to obtain an effective ε-optimal control law $\mu_\varepsilon^*(x(k))$, we should choose the state $x(k) \in \mathcal{T}_i^{(\varepsilon)} \backslash \mathcal{T}_{i-1}^{(\varepsilon)}$ for each i to run the iterative ADP algorithm. The control process using the iterative ADP algorithm is illustrated in Fig. 2.26.

From the iterative ADP algorithm (2.181)–(2.184), we can see that for any state $x(k) \in \mathbb{R}^n$, there exists a control $u(k) \in \mathbb{R}^m$ that drives the system to zero in one step. In other words, for $\forall x(k) \in \mathbb{R}^n$, there exists a control $u(k) \in \mathbb{R}^m$ such that $x(k+1) = F(x(k), u(k)) = 0$ holds. A large class of systems possesses this property; for example, all linear systems of the type $x(k+1) = Ax(k) + Bu(k)$ when B is invertible and the affine nonlinear systems with the type $x(k+1) = f(x(k)) + g(x(k))u(k)$ when the inverse of $g(x(k))$ exists. But there are also other classes of systems for which there does not exist any control $u(k) \in \mathbb{R}^m$ that drives the state to zero in one step for some $x(k) \in \mathbb{R}^n$, i.e., $\exists\, x(k) \in \mathbb{R}^n$ such that $F(x(k), u(k)) = 0$ is not possible for $\forall u(k) \in \mathbb{R}^m$. In the following part, we will discuss the situation where $F(x(k), u(k)) \neq 0$ for some $x(k) \in \mathbb{R}^m$.

Since $x(k)$ is controllable, there exists a finite-horizon admissible control sequence $\underline{u}_k^{k+i-1} = (u_k, u(k+1), \ldots, u_{k+i-1}) \in \mathfrak{A}_{x(k)}^{(i)}$ that makes $x^{(f)}(x(k), \underline{u}_k^{k+i-1}) = x(k+i) = 0$. Let $N = k+i$ be the terminal time. Assume that for $k+1, k+2, \ldots, N-1$, the optimal control sequence $\underline{u}_{k+1}^{(N-1)*} = (u^*(k+1), u_{k+2}^*, \ldots, u_{N-1}^*) \in \mathfrak{A}_{x(k+1)}^{(N-k-1)}$ has been determined. Denote the value function for $x(k+1)$ as $J(x(k+1), \underline{u}_{k+1}^{(N-1)*}) = V_0(x(k+1))$. Now, we use the iterative ADP algorithm to determine the optimal control sequence for the state $x(k)$.

The value function for $i = 1$ is computed as

$$V_1(x(k)) = l(x(k), v_1(x(k))) + V_0(F(x(k), v_1(x(k)))), \tag{2.205}$$

where

$$v_1(x(k)) = \arg\min_{u(k)}\{l(x(k), u(k)) + V_0(F(x(k), u(k)))\}. \tag{2.206}$$

Note that the initial condition used in the above expression is the value function V_0 which is obtained previously for $x(k+1)$ and now applied at $F(x(k), u(k))$. For $i = 2, 3, 4, \dots$, the iterative ADP algorithm will be implemented as follows:

$$V_i(x(k)) = l(x(k), v_i(x(k))) + V_{i-1}(F(x(k), v_i(x(k)))), \tag{2.207}$$

where

$$v_i(x(k)) = \arg\min_{u(k)} \{l(x(k), u(k)) + V_{i-1}(F(x(k), u(k)))\}. \tag{2.208}$$

Theorem 2.48 *Let $x(k)$ be an arbitrary controllable state vector. Then, the value function $V_i(x(k))$ obtained by* (2.205)–(2.208) *is a monotonically nonincreasing sequence for $\forall i \ge 0$, i.e., $V_{i+1}(x(k)) \le V_i(x(k))$ for $\forall i \ge 0$.*

Proof It can easily be proved by following the proof of Theorem 2.32, and the proof is omitted here. □

Theorem 2.49 *Let the value function $V_i(x(k))$ be defined by* (2.207). *If the system state $x(k)$ is controllable, then the value function $V_i(x(k))$ obtained by* (2.205)–(2.208) *converges to the optimal value function $J^*(x(k))$ as $i \to \infty$, i.e.,*

$$\lim_{i\to\infty} V_i(x(k)) = J^*(x(k)).$$

Proof This theorem can be proved following similar steps to the proof of Theorem 2.34 and the proof is omitted here. □

We can see that the iterative ADP algorithm (2.205)–(2.208) is an expansion from of the previous one (2.181)–(2.184). So, the properties of the iterative ADP algorithm (2.181)–(2.184) is also effective for the current one (2.205)–(2.208). But there also exist differences. From Theorem 2.32, we can see that $V_{i+1}(x(k)) \le V_i(x(k))$ for all $i \ge 1$, which means that $V_1(x(k)) = \max\{V_i(x(k)) : i = 0, 1, \dots\}$. While Theorem 2.48 shows that $V_{i+1}(x(k)) \le V_i(x(k))$ for all $i \ge 0$ which means that $V_0(x(k)) = \max\{V_i(x(k)) : i = 0, 1, \dots\}$. This difference is caused by the difference of the initial conditions of the two iterative ADP algorithms.

In the previous iterative ADP algorithm (2.181)–(2.184), it begins with the initial value function $V_0(x(k)) = 0$ since $F(x(k), u(k)) = 0$ can be solved. While in the current iterative ADP algorithm (2.205)–(2.208), it begins with the value function V_0 for the state $x(k+1)$ which is determined previously. This also causes the difference between the proofs of Theorems 2.32 and 2.34 and the corresponding results in Theorems 2.48 and 2.49. But the difference of the initial conditions of the iterative performance index function does not affect the convergence property of the two iterative ADP algorithms.

For the iterative ADP algorithm, the optimal criterion (2.195) is very difficult to verify because the optimal value function $J^*(x(k))$ is unknown in general. So, an equivalent criterion is established to replace (2.195).

If $|V_i(x(k)) - J^*(x(k))| \le \varepsilon$ holds, we have $V_i(x(k)) \le J^*(x(k)) + \varepsilon$ and $J^*(x(k)) \le V_{i+1}(x(k)) \le V_i(x(k))$. These imply that

$$0 \le V_i(x(k)) - V_{i+1}(x(k)) \le \varepsilon, \tag{2.209}$$

or

$$|V_i(x(k)) - V_{i+1}(x(k))| \le \varepsilon.$$

On the other hand, according to Theorem 2.49, $|V_i(x(k)) - V_{i+1}(x(k))| \to 0$ implies that $V_i(x(k)) \to J^*(x(k))$. Therefore, for any given small ε, if $|V_i(x(k)) - V_{i+1}(x(k))| \le \varepsilon$ holds, we have $|V_i(x(k)) - J^*(x(k))| \le \varepsilon$ if i is sufficiently large.

We will use inequality (2.209) as the optimal criterion instead of the optimal criterion (2.195).

Let $\widehat{\underline{u}}_0^{K-1} = (u(0), u(1), \ldots, u(K-1))$ be an arbitrary finite-horizon admissible control sequence and the corresponding state sequence be $\widehat{\underline{x}}_0^{K} = (x(0), x(1), \ldots, x(K))$ where $x(K) = 0$.

We can see that the initial control sequence $\widehat{\underline{u}}_0^{K-1} = (u(0), u(1), \ldots, u(K-1))$ may not be optimal, which means that the initial number of control steps K may not be optimal. So, the iterative ADP algorithm must complete two kinds of optimization. One is to optimize the number of control steps. The other is to optimize the control law. In the following, we will show how the number of control steps and the control law are optimized simultaneously in the iterative ADP algorithm.

For the state $x(K-1)$, we have $F(x(K-1), u(K-1)) = 0$. Then, we run the iterative ADP algorithm (2.181)–(2.184) at $x(K-1)$ as follows. The value function for $i = 1$ is computed as

$$\begin{aligned} V_1^1(x(K-1)) = & \min_{u(K-1)} \{l(x(K-1), u(K-1)) + V_0(F(x(K-1), u(K-1)))\} \\ & \text{subject to } \; F(x(K-1), u(K-1)) = 0 \\ = & l(x(K-1), v_1^1(x(K-1))), \end{aligned} \tag{2.210}$$

where

$$\begin{aligned} v_1^1(x(K-1)) = & \arg \min_{u(K-1)} l(x(K-1), u(K-1)) \\ & \text{subject to } \; F(x(K-1), u(K-1)) = 0, \end{aligned} \tag{2.211}$$

and $V_0(F(x(K-1), u(K-1))) = 0$. The iterative ADP algorithm will be implemented as follows for $i = 2, 3, 4, \ldots$:

$$V_i^1(x(K-1)) = l(x(K-1), v_i^1(x(K-1)))$$

$$+ V_{i-1}^{1}(F(x(K-1), v_i^1(x(K-1)))), \tag{2.212}$$

where

$$v_i^1(x(K-1)) = \arg\min_{u(K-1)} \big\{ l(x(K-1), u(K-1)) \\ + V_{i-1}^{1}(F(x(K-1), u(K-1))) \big\}, \tag{2.213}$$

until the inequality

$$\left| V_{l_1}^{1}(x(K-1)) - V_{l_1+1}^{1}(x(K-1)) \right| \le \varepsilon \tag{2.214}$$

is satisfied for $l_1 > 0$. This means that $x(K-1) \in \mathcal{T}_{l_1}^{(\varepsilon)}$ and the optimal number of control steps is $K_\varepsilon(x(K-1)) = l_1$.

Considering $x(K-2)$, we have $F(x(K-2), u(K-2)) = x(K-1)$. Put $x(K-2)$ into (2.214). If $|V_{l_1}^{1}(x(K-2)) - V_{l_1+1}^{1}(x(K-2))| \le \varepsilon$ holds, then according to Theorem 2.41(i), we know that $x(K-2) \in \mathcal{T}_{l_1}^{(\varepsilon)}$. Otherwise, if $x(K-2) \notin \mathcal{T}_{l_1}^{(\varepsilon)}$, we will run the iterative ADP algorithm as follows. Using the value function $V_{l_1}^{1}$ as the initial condition, we compute, for $i = 1$,

$$V_1^2(x(K-2)) = l(x(K-2), v_1^2(x(K-2))) \\ + V_{l_1}^{1}(F(x(K-2), v_1^2(x(K-2)))), \tag{2.215}$$

where

$$v_1^2(x(K-2)) = \arg\min_{u(K-2)} \{ l(x(K-2), u(K-2)) \\ + V_{l_1}^{1}(F(x(K-2), u(K-2))) \}. \tag{2.216}$$

For $i = 2, 3, 4, \ldots$, the iterative ADP algorithm will be implemented as follows:

$$V_i^2(x(K-2)) = l(x(K-2), v_i^2(x(K-2))) \\ + V_{i-1}^{2}(F(x(K-2), v_i^2(x(K-2)))), \tag{2.217}$$

where

$$v_i^2(x(K-2)) = \arg\min_{u(K-2)} \big\{ l(x(K-2), u_{K-2}) \\ + V_{i-1}^{2}(F(x(K-2), u(K-2))) \big\}, \tag{2.218}$$

until the inequality

$$\left| V_{l_2}^{2}(x(K-2)) - V_{l_2+1}^{2}(x(K-2)) \right| \le \varepsilon \tag{2.219}$$

is satisfied for $l_2 > 0$. We then obtain $x(K-2) \in \mathcal{T}_{l_2}^{(\varepsilon)}$, and the optimal number of control steps is $K_\varepsilon(x(K-2)) = l_2$.

Next, assume that $j \geq 2$ and $x(K-j+1) \in \mathcal{T}_{l_{j-1}}^{(\varepsilon)}$, i.e.,

$$\left| V_{l_{j-1}}^{j-1}(x(K-j+1)) - V_{l_{j-1}+1}^{j-1}(x(K-j+1)) \right| \leq \varepsilon \tag{2.220}$$

holds. Considering $x(K-j)$, we have $F(x(K-j), u(K-j)) = x(K-j+1)$. Putting $x(K-j)$ into (2.220) and if

$$\left| V_{l_{j-1}}^{j-1}(x(K-j)) - V_{l_{j-1}+1}^{j-1}(x(K-j)) \right| \leq \varepsilon \tag{2.221}$$

holds, then we know that $x(K-j) \in \mathcal{T}_{l_{j-1}}^{(\varepsilon)}$. Otherwise, if $x(K-j) \notin \mathcal{T}_{l_{j-1}}^{(\varepsilon)}$, then we run the iterative ADP algorithm as follows. Using the performance index function $V_{l_{j-1}}^{j-1}$ as the initial condition, we compute for $i = 1$,

$$\begin{aligned} V_1^j(x(K-j)) = {} & l(x(K-j), v_1^j(x(K-j))) \\ & + V_{l_{j-1}}^{j-1}(F(x(K-j), v_1^j(x(K-j)))), \end{aligned} \tag{2.222}$$

where

$$\begin{aligned} v_1^j(x(K-j)) = \arg \min_{u(K-j)} \{ & l(x(K-j), u(K-j)) \\ & + V_{l_{j-1}}^{j-1}(F(x(K-j), u(K-j)))\}. \end{aligned} \tag{2.223}$$

For $i = 2, 3, 4, \ldots$, the iterative ADP algorithm will be implemented as follows:

$$\begin{aligned} V_i^j(x(K-j)) = {} & l(x(K-j), v_i^j(x(K-j))) \\ & + V_{i-1}^j(F(x(K-j), v_i^j(x(K-j)))), \end{aligned} \tag{2.224}$$

where

$$\begin{aligned} v_i^j(x(K-j)) = \arg \min_{u(K-j)} \big\{ & l(x(K-j), u(K-j)) \\ & + V_{i-1}^j(F(x(K-j), u(K-j)))\big\}, \end{aligned} \tag{2.225}$$

until the inequality

$$\left| V_{l_j}^j(x(K-j)) - V_{l_j+1}^j(x(K-j)) \right| \leq \varepsilon \tag{2.226}$$

is satisfied for $l_j > 0$. We then obtain $x(K-j) \in \mathcal{T}_{l_j}^{(\varepsilon)}$, and the optimal number of control steps is $K_\varepsilon(x(K-j)) = l_j$.

Finally, considering $x(0)$, we have $F(x(0), u(0)) = x(1)$. If

$$\left|V_{l_{K-1}}^{K-1}(x(0)) - V_{l_{K-1}+1}^{K-1}(x(0))\right| \leq \varepsilon$$

holds, then we know that $x(0) \in \mathcal{T}_{l_{K-1}}^{(\varepsilon)}$. Otherwise, if $x(0) \notin \mathcal{T}_{l_{K-1}}^{(\varepsilon)}$, then we run the iterative ADP algorithm as follows. Using the performance index function $V_{l_{K-1}}^{K-1}$ as the initial condition, we compute, for $i = 1$,

$$V_1^K(x(0)) = l(x(0), v_1^K(x(0))) + V_{l_{K-1}}^{K-1}(F(x(0), v_1^K(x(0)))), \tag{2.227}$$

where

$$v_1^K(x(0)) = \arg\min_{u(0)}\{l(x(0), u(0)) + V_{l_{K-1}}^{K-1}(F(x(0), u(0)))\}. \tag{2.228}$$

For $i = 2, 3, 4, \ldots$, the iterative ADP algorithm will be implemented as follows:

$$V_i^K(x(0)) = l(x(0), v_i^K(x(0))) + V_{i-1}^K(F(x(0), v_i^K(x(0)))), \tag{2.229}$$

where

$$v_i^K(x(0)) = \arg\min_{u(0)}\left\{l(x(0), u(0)) + V_{i-1}^K(F(x(0), u(0)))\right\}, \tag{2.230}$$

until the inequality

$$\left|V_{l_K}^K(x(0)) - V_{l_K+1}^K(x(0))\right| \leq \varepsilon \tag{2.231}$$

is satisfied for $l_K > 0$. Therefore, we obtain $x(0) \in \mathcal{T}_{l_K}^{(\varepsilon)}$, and the optimal number of control steps is $K_\varepsilon(x(0)) = l_K$.

Starting from the initial state $x(0)$, the optimal number of control steps is l_K according to our ADP algorithm.

Remark 2.50 For the case where there are some $x(k) \in \mathbb{R}^n$, there does not exist a control $u(k) \in \mathbb{R}^m$ that drives the system to zero in one step; the computational complexity of the iterative ADP algorithm is strongly related to the original finite-horizon admissible control sequence $\widehat{\underline{u}}_0^{K-1}$. First, we repeat the iterative ADP algorithm at $x(K-1), x(K-2), \ldots, x(1), x(0)$, respectively. It is related to the control steps K of $\widehat{\underline{u}}_0^{K-1}$. If K is large, it means that $\widehat{\underline{u}}_0^{K-1}$ takes large number of control steps to drive the initial state $x(0)$ to zero and then the number of times needed to repeat the iterative ADP algorithm will be large. Second, the computational complexity is also related to the quality of control results of $\widehat{\underline{u}}_0^{K-1}$. If $\widehat{\underline{u}}_0^{K-1}$ is close to the optimal control sequence $\underline{u}_0^{(N-1)*}$, then it will take less computation to make (2.226) hold for each j.

Now, we summarize the iterative ADP algorithm as follows:

Step 1. Choose an error bound ε and choose randomly an array of initial states $x(0)$.

Step 2. Obtain an initial finite-horizon admissible control sequence $\underline{\widehat{u}}_0^{K-1} = (u(0), u(1), \ldots, u(K-1))$ and obtain the corresponding state sequence

$$\underline{\widehat{x}}_0^{K} = (x(0), x(1), \ldots, x(K)),$$

where $x(K) = 0$.

Step 3. For the state $x(K-1)$ with $F(x(K-1), u(K-1)) = 0$, run the iterative ADP algorithm (2.210)–(2.213) at $x(K-1)$ until (2.214) holds.

Step 4. Record $V_{l_1}^1(x(K-1))$, $v_{l_1}^1(x_{K-1})$ and $K_\varepsilon(x(K-1)) = l_1$.

Step 5. For $j = 2, 3, \ldots, K$, if for $x(K-j)$ the inequality (2.221) holds, go to Step 7; otherwise, go to Step 6.

Step 6. Using the value function $V_{l_{j-1}}^{j-1}$ as the initial condition, run the iterative ADP algorithm (2.222)–(2.225) until (2.226) is satisfied.

Step 7. If $j = K$, then we have obtained the optimal value function $V^*(x(0)) = V_{l_K}^K(x(0))$, the law of the optimal control sequence $u^*(x(0)) = v_{l_K}^K(x(0))$ and the number of optimal control steps $K_\varepsilon(x(0)) = l_K$; otherwise, set $j = j+1$, and go to Step 5.

Step 8. Stop.

2.5.3 Simulations

To evaluate the performance of our iterative ADP algorithm, we choose two examples with quadratic utility functions for numerical experiments.

Example 2.51 Our first example is chosen from [16]. We consider the following nonlinear system:

$$x(k+1) = f(x(k)) + g(x(k))u(k),$$

where $x(k) = [x_1(k) \;\; x_2(k)]^{\mathrm{T}}$ and $u(k) = [u_1(k) \;\; u_2(k)]^{\mathrm{T}}$ are the state and control variables, respectively. The system functions are given as

$$f(x(k)) = \begin{bmatrix} 0.2x_1(k)\exp(x_2^2(k)) \\ 0.3x_2^3(k) \end{bmatrix}, \quad g(x(k)) = \begin{bmatrix} -0.2 & 0 \\ 0 & -0.2 \end{bmatrix}.$$

The initial state is $x(0) = [1 \;\; -1]^{\mathrm{T}}$. The value function is in quadratic form with finite time horizon expressed as

$$J\big(x(0), \underline{u}_0^{N-1}\big) = \sum_{k=0}^{N-1} \big(x^{\mathrm{T}}(k)Qx(k) + u^{\mathrm{T}}(k)Ru(k)\big),$$

where the matrix $Q = R = I$, and I denotes the identity matrix with suitable dimensions.

The error bound of the iterative ADP is chosen as $\varepsilon = 10^{-5}$. Neural networks are used to implement the iterative ADP algorithm and the neural-network structure can

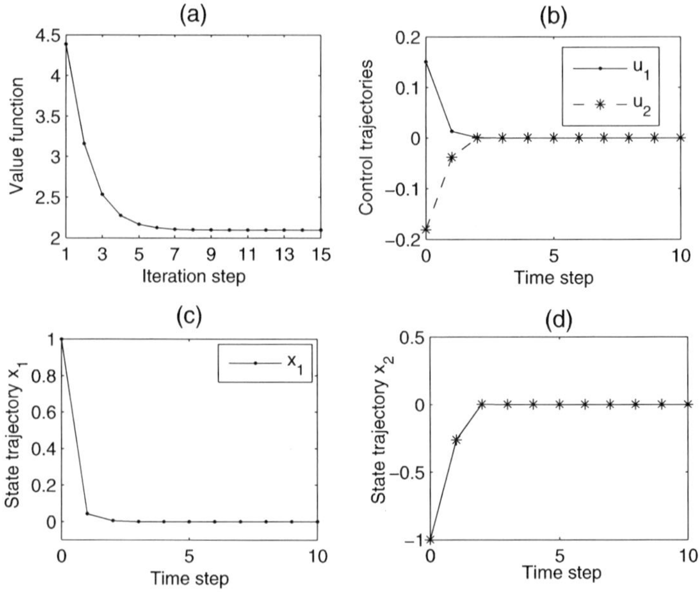

Fig. 2.27 Simulation results for Example 2.51. (**a**) The convergence of value function. (**b**) The ε-optimal control vectors. (**c**) and (**d**) The corresponding state trajectories

be seen in [13, 16]. The critic network and the action network are chosen as three-layer BP neural networks with the structures of 2–8–1 and 2–8–2, respectively. The model network is also chosen as a three-layer BP neural network with the structure of 4–8–2. The critic network is used to approximate the iterative value functions which are expressed by (2.210), (2.212), (2.215), (2.217), (2.222), (2.224), (2.227), and (2.229). The action network is used to approximate the optimal control laws which are expressed by (2.211), (2.213), (2.216), (2.218), (2.223), (2.225), (2.228), and (2.230). The training rules of the neural networks can be seen in [12]. For each iteration step, the critic network and the action network are trained for 1000 iteration steps using the learning rate of $\alpha = 0.05$, so that the neural-network training error becomes less than 10^{-8}. Enough iteration steps should be implemented to guarantee the iterative value functions and the control law to converge sufficiently. We let the algorithm run for 15 iterative steps to obtain the optimal value function and optimal control law. The convergence curve of the value function is shown in Fig. 2.27(a). Then, we apply the optimal control law to the system for $T_f = 10$ time steps and obtain the following results. The ε-optimal control trajectories are shown in Fig. 2.27(b) and the corresponding state curves are shown in Figs. 2.27(c) and (d).

After seven iteration steps, we have $|V_6(x(0)) - V_7(x(0))| \le 10^{-5} = \varepsilon$. Then, we obtain the optimal number of control steps $K_\varepsilon(x(0)) = 6$. We can see that after six time steps, the state variable becomes $x_6 = [0.912 \times 10^{-6},\ 0.903 \times 10^{-7}]^{\mathrm{T}}$. The entire computation process takes about 10 seconds before satisfactory results are obtained.

Example 2.52 The second example is chosen from [9] with some modifications. We consider the following system:

$$x(k+1) = F(x(k), u(k)) = x(k) + \sin(0.1x(k)^2 + u(k)), \tag{2.232}$$

where $x(k), u(k) \in \mathbb{R}$, and $k = 0, 1, 2, \ldots$. The cost functional is defined as in Example 2.51 with $Q = R = 1$. The initial state is $x(0) = 1.5$. Since $F(0, 0) = 0$, $x(k) = 0$ is an equilibrium state of system (2.232). But since $(\partial F(x(k), u(k))/\partial x(k))(0, 0) = 1$, system (2.232) is marginally stable at $x(k) = 0$ and the equilibrium $x(k) = 0$ is not attractive.

We can see that, for the fixed initial state $x(0)$, there does not exist a control $u(0) \in \mathbb{R}$ that makes $x(1) = F(x(0), u(0)) = 0$. The error bound of the iterative ADP algorithm is chosen as $\varepsilon = 10^{-4}$. The critic network, the action network, and the model network are chosen as three-layer BP neural networks with the structures of 1–3–1, 1–3–1, and 2–4–1, respectively. According to (2.232), the control can be expressed by

$$u(k) = -0.1x(k)^2 + \sin^{-1}(x(k+1) - x(k)) + 2\lambda\pi, \tag{2.233}$$

where $\lambda = 0, \pm 1, \pm 2, \ldots$.

To show the effectiveness of our algorithm, we choose two initial finite-horizon admissible control sequences.

Case 1. The control sequence is $\underline{\hat{u}}_0^1 = (-0.225 - \sin^{-1}(0.7),\ -0.064 - \sin^{-1}(0.8))$ and the corresponding state sequence is $\underline{\hat{x}}_0^2 = (1.5, 0.8, 0)$.

For the initial finite-horizon admissible control sequences in this case, run the iterative ADP algorithm at the states $x(k) = 0.8$ and 1.5, respectively. For each iterative step, the critic network and the action network are trained for 1000 iteration steps using the learning rate of $\alpha = 0.05$ so that the neural-network training accuracy of 10^{-8} is reached. After the algorithm has run for 15 iterative steps, we obtain the performance index function trajectories shown in Figs. 2.28(a) and (b), respectively. The ε-optimal control and state trajectories are shown in Figs. 2.28(c) and (d), respectively, for 10 time steps. We obtain $K_\varepsilon(0.8) = 5$ and $K_\varepsilon(1.5) = 8$.

Case 2. The control sequence is $\underline{\hat{u}}_0^3 = (-0.225 - \sin^{-1}(0.01),\ 2\pi - 2.2201 - \sin^{-1}(0.29),\ -0.144 - \sin^{-1}(0.5),\ -\sin^{-1}(0.7))$ and the corresponding state sequence is $\underline{\hat{x}}_0^4 = (1.5, 1.49, 1.2, 0.7, 0)$.

For the initial finite-horizon admissible control sequence in this case, run the iterative ADP algorithm at the states $x(k) = 0.7$, 1.2, and 1.49, respectively. For each iteration step, the critic network and the action network are also trained for 1000 iteration steps using the learning rate of $\alpha = 0.05$, so that the neural-network training accuracy of 10^{-8} is reached. Then we obtain the value function trajectories shown in Figs. 2.29(a)–(c), respectively. We have $K_\varepsilon(0.7) = 4$, $K_\varepsilon(1.2) = 6$, and $K_\varepsilon(1.49) = 8$.

After 25 iteration steps, the value function $V_i(x(k))$ is sufficiently convergent at $x(k) = 1.49$, with $V_8^3(1.49)$ as the value function. For the state $x(k) = 1.5$, we have $|V_8^3(1.5) - V_9^3(1.5)| = 0.52424 \times 10^{-7} < \varepsilon$. Therefore, the optimal value function

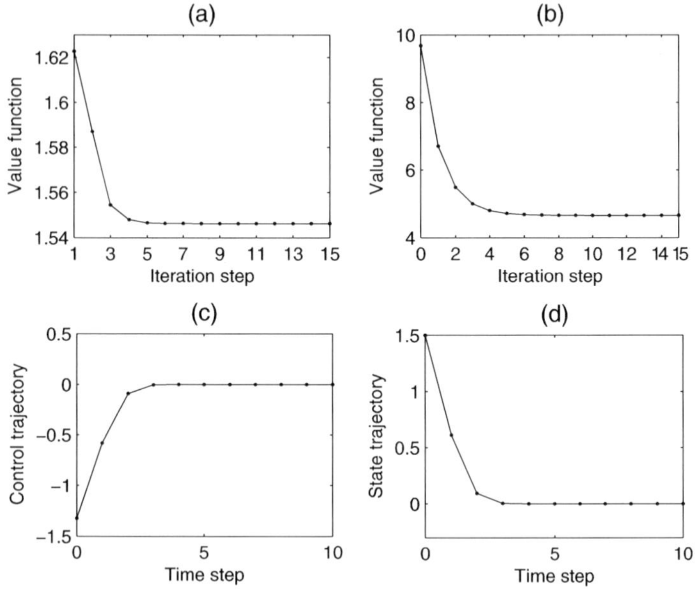

Fig. 2.28 Simulation results for Case 1 of Example 2.52. (**a**) The convergence of value function at $x(k) = 0.8$. (**b**) The convergence of value function at $x(k) = 1.5$. (**c**) The ε-optimal control trajectory. (**d**) The corresponding state trajectory

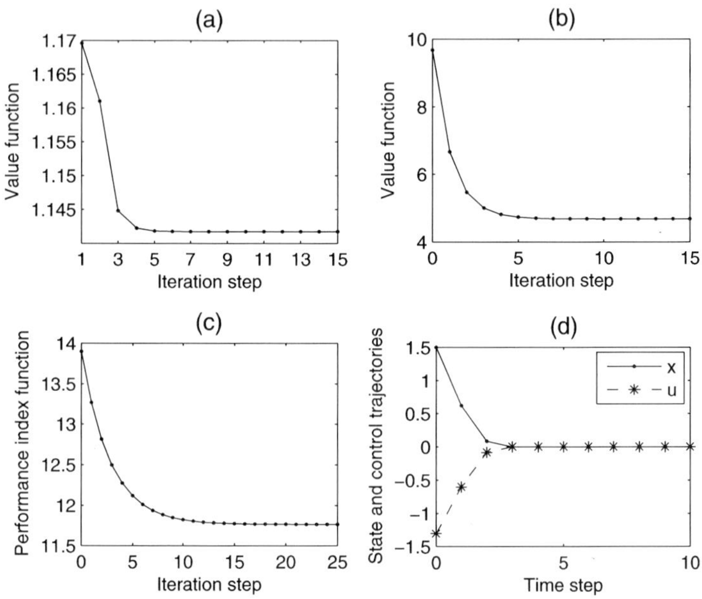

Fig. 2.29 Simulation results for Case 2 of Example 2.52. (**a**) The convergence of value function at $x(k) = 0.7$. (**b**) The convergence of value function at $x(k) = 1.2$. (**c**) The convergence of performance index function at $x(k) = 1.49$. (**d**) The ε-optimal control trajectory and the corresponding state trajectory

at $x(k) = 1.5$ is $V_8^3(1.5)$ and, thus, we have $x(k) = 1.5 \in \mathcal{T}_8^{(\varepsilon)}$ and $K_\varepsilon(1.5) = 8$. The whole computation process takes about 20 seconds and then satisfactory results are obtained.

Then we apply the optimal control law to the system for $T_f = 10$ time steps. The ε-optimal control and state trajectories are shown in Fig. 2.29(d).

We can see that the ε-optimal control trajectory in Fig. 2.29(d) is the same as the one in Fig. 2.28(c). The corresponding state trajectory in Fig. 2.29(d) is the same as the one in Fig. 2.28(d). Therefore, the optimal control law is not dependent on the initial control law. The initial control sequence $\underline{\hat{u}}_0^{K-1}$ can arbitrarily be chosen as long as it is finite-horizon admissible.

Remark 2.53 If the number of control steps of the initial admissible control sequence is larger than the number of control steps of the optimal control sequence, then we will find some of the states in the initial sequence to possess the same number of optimal control steps. For example, in Case 2 of Example 2.52, we see that the two states $x = 1.49$ and $x = 1.5$ possess the same number of optimal control steps, i.e., $K_\varepsilon(1.49) = K_\varepsilon(1.5) = 8$. Thus, we say that the control $u = -0.225 - \sin^{-1}(0.01)$ that makes $x = 1.5$ run to $x = 1.49$ is an unnecessary control step. After the unnecessary control steps are identified and removed, the number of control steps will reduce to the optimal number of control steps and, thus, the initial admissible control sequence does not affect the final optimal control results.

2.6 Summary

In this chapter, several infinite-time and finite-time optimal control schemes have been developed to solve the corresponding control problems for several kinds of nonlinear system. In Sects. 2.2, 2.3, and 2.4, the presented optimal controllers were all infinite-time optimal state feedback controllers though the developed ADP algorithms. The optimal control objective can be achieved when the number of control steps tends to infinity. In Sect. 2.5, an effective iterative ADP algorithm has been developed for the ε-optimal controller for a class of discrete-time nonlinear systems, where the optimal control objective can be achieved with a finite number of control steps.

References

1. Al-Tamimi A, Lewis FL (2007) Discrete-time nonlinear HJB solution using approximate dynamic programming: convergence proof. In: Proceedings of IEEE international symposium on approximate dynamic programming and reinforcement learning, Honolulu, HI, pp 38–43
2. Bagnell J, Kakade S, Ng A, Schneider J (2003) Policy search by dynamic programming. In: Proceedings of 17th annual conference on neural information processing systems, Vancouver, Canada, vol 16, pp 831–838

3. Bryson AE, Ho YC (1975) Applied optimal control: optimization, estimation, and control. Hemisphere–Wiley, New York
4. Burk F (1998) Lebesgue measure and integration. Wiley, New York
5. Chen Z, Jagannathan S (2008) Generalized Hamilton–Jacobi–Bellman formulation-based neural network control of affine nonlinear discrete-time systems. IEEE Trans Neural Netw 19(1):90–106
6. Cui LL, Zhang HG, Liu D, Kim YS (2009) Constrained optimal control of affine nonlinear discrete-time systems using GHJB method. In: Proceedings of IEEE international symposium on adaptive dynamic programming and reinforcement learning, Nashville, USA, pp 16–21
7. Han D, Balakrishnan SN (2002) State-constrained agile missile control with adaptive-critic-based neural networks. IEEE Trans Control Syst Technol 10(4):481–489
8. Haykin S (1999) Neural networks: a comprehensive foundation. Prentice Hall, Upper Saddle River
9. Jin N, Liu D, Huang T, Pang Z (2007) Discrete-time adaptive dynamic programming using wavelet basis function neural networks. In: Proceedings of the IEEE symposium on approximate dynamic programming and reinforcement learning, Honolulu, HI, pp 135–142
10. Liu D, Wang D, Zhao D, Wei Q, Jin N (2012) Neural-network-based optimal control for a class of unknown discrete-time nonlinear systems using globalized dual heuristic programming. IEEE Trans Autom Sci Eng 9(3):628–634
11. Plumer ES (1996) Optimal control of terminal processes using neural networks. IEEE Trans Neural Netw 7(2):408–418
12. Si J, Wang YT (2001) On-line learning control by association and reinforcement. IEEE Trans Neural Netw 12(2):264–276
13. Wang FY, Zhang HG, Liu D (2009) Adaptive dynamic programming: an introduction. IEEE Comput Intell Mag 4(2):39–47
14. Wang FY, Jin N, Liu D, Wei Q (2011) Adaptive dynamic programming for finite-horizon optimal control of discrete-time nonlinear systems with ε-error bound. IEEE Trans Neural Netw 22(1):24–36
15. Zhang HG, Xie X (2011) Relaxed stability conditions for continuous-time T-S fuzzy-control systems via augmented multi-indexed matrix approach. IEEE Trans Fuzzy Syst 19(3):478–492
16. Zhang HG, Wei QL, Luo YH (2008) A novel infinite-time optimal tracking control scheme for a class of discrete-time nonlinear systems via the greedy HDP iteration algorithm. IEEE Trans Syst Man Cybern, Part B, Cybern 38(4):937–942
17. Zhang HG, Luo YH, Liu D (2009) Neural-network-based near-optimal control for a class of discrete-time affine nonlinear systems with control constraints. IEEE Trans Neural Netw 20(9):1490–1503

Chapter 3
Optimal Tracking Control for Discrete-Time Systems

3.1 Introduction

In the last chapter, we have studied the optimal state feedback control problem for discrete-time systems. As a matter of fact, the optimal tracking control problem carries the same weight as the state feedback problem. Therefore, the subject of the optimal tracking control problem has been investigated over the past several decades [2, 3, 7, 10–12]. Traditional optimal tracking control is mostly effective in the neighborhood of the equilibrium point [4, 8], and the earlier optimal control for nonlinear system is mostly open-loop control [1]. Hence, our aim of this chapter is to present some direct methods that are adapted to the closed-loop optimal tracking control problem.

In this chapter, we first study the infinite-horizon optimal tracking control for discrete-time affine nonlinear systems in Sect. 3.2. A new type of cost functional is defined. The iterative HDP algorithm is introduced to deal with the optimal tracking control problem. It is worth noting that many practical applications have nonlinear structures not only in the states but also in the control input [9]. Due to the high complexity of controlling the nonaffine nonlinear system, the results deriving from the affine nonlinear systems may not applicable to the nonaffine case [17]. Therefore, in Sect. 3.3, we focus on a class of infinite-horizon discrete-time nonaffine nonlinear systems. It is also noticed that most real-world systems need to be effectively controlled within a finite time horizon. Hence, based on the above research, we will study how to solve the finite-horizon optimal tracking control problem using the ADP approach in Sect. 3.4. The iterative ADP algorithm is introduced to obtain the finite-horizon optimal tracking controller.

3.2 Infinite-Horizon Optimal Tracking Control Based on HDP

In this section, the infinite-horizon optimal tracking control problem for a class of discrete-time nonlinear systems is studied. We will first transform the tracking con-

H. Zhang et al., *Adaptive Dynamic Programming for Control*, Communications and Control Engineering, DOI 10.1007/978-1-4471-4757-2_3,

trol problem into an optimal regulation problem, and then the iterative HDP algorithm can be properly introduced to deal with this regulation problem. Three neural networks are used to approximate the value function, compute the optimal control policy, and model the nonlinear system respectively for facilitating the implementation of the HDP iteration algorithm.

3.2.1 Problem Formulation

Consider a class of affine nonlinear systems of the form

$$x(k+1) = f(x(k)) + g(x(k))u(k), \tag{3.1}$$

where $x(k) \in \mathbb{R}^n$ is the state vector, $u(k) \in \mathbb{R}^m$ is the input vector, $f(x(k)) \in \mathbb{R}^n$, $g(x(k)) \in \mathbb{R}^{n\times m}$. Here assume that the system is strongly controllable on $\Omega \subset \mathbb{R}^n$.

For the infinite-horizon optimal tracking control problem, the control objective is to design optimal control $u(k)$ for system (3.1) such that the state $x(k)$ tracks the specified desired trajectory $x_d(k) \in \mathbb{R}^n$, $k = 0, 1, \dots$. It is assumed that there exists a function $\delta\colon \mathbb{R}^n \to \mathbb{R}^n$ satisfying $x_d(k+1) = \delta(x_d(k))$. We further assume that the mapping between the state $x(k)$ and the desired trajectory $x_d(k)$ is one-to-one.

Define the tracking error as

$$e(k) = x(k) - x_d(k). \tag{3.2}$$

For time-invariant optimal tracking control problems of linear systems, the cost functional is generally defined as the following quadratic form:

$$\begin{aligned} J(e(0), u) = \sum_{k=0}^{\infty} \Big\{ & e^{\mathrm{T}}(k) Q_1 e(k) + (u(k+1) - u(k))^{\mathrm{T}} R \\ & \times (u(k+1) - u(k)) \Big\}, \end{aligned} \tag{3.3}$$

where $Q_1 \in \mathbb{R}^{n\times n}$ and $R \in \mathbb{R}^{m\times m}$ are two diagonal positive definite matrices.

However, for the time-variant tracking case in a nonlinear environment, the problem is much more complex and the aforementioned cost functional may be invalid, i.e., $J(e(0), u)$ calculated by (3.3) may be infinite, because the control $u(k)$ depends on the desired trajectory $x_d(k)$.

To solve this problem, we present the following cost functional similar to the framework of references [12, 15]:

$$J(e(0), w) = \sum_{k=0}^{\infty} \Big\{ e^{\mathrm{T}}(k) Q_1 e(k) + (w(k) - w(k-1))^{\mathrm{T}} R$$

$$\times (w(k) - w(k-1))\Big\}, \tag{3.4}$$

where $w(k)$ is defined as follows:

$$w(k) = u(k) - u_e(k), \tag{3.5}$$

and $w(k) = 0$ for $k < 0$. $u_e(k)$ denotes the expected control, which can be given as follows:

$$u_e(k) = g^{-1}(x_d(k))(\delta(x_d(k)) - f(x_d(k))), \tag{3.6}$$

where $g^{-1}(x_d(k))g(x_d(k)) = I$, $I \in \mathbb{R}^{m\times m}$ is the identity matrix.

In (3.4), the first term means the tracking error and the second term means the difference of the control error. However, there still exists the problem that if only the tracking error and the difference of the control error are considered in the cost functional, the system may oscillate. For example, the difference of the control error may be small but the error between the real tracking control and the expected tracking control may be large. Therefore, it is necessary to define a new type of cost functional for solving this kind of infinite horizon optimal tracking control problem. We propose a new type of quadratic cost functional as follows:

$$\begin{aligned} J(e(0), u, w) = \sum_{k=0}^{\infty} \Big\{ & e^{\mathrm{T}}(k) Q_1 e(k) + (u(k) - u_e(k))^{\mathrm{T}} S (u(k) - u_e(k)) \\ & + (w(k) - w(k-1))^{\mathrm{T}} R (w(k) - w(k-1)) \Big\}, \end{aligned} \tag{3.7}$$

where $S \in \mathbb{R}^{m\times m}$ is diagonal positive definite matrix and other parameters are the same as those of (3.4).

Compared with (3.4), a new term $(u(k) - u_e(k))^{\mathrm{T}} S(u(k) - u_e(k))$ is added to represent the error between the real and the expected tracking control. Via this cost functional, not only the tracking error and the difference of the control error are considered, but also the error between the real and expected tracking control; therefore, the oscillation can well be prevented.

After the definition of such a new type of cost functional, we will mainly discuss the optimal tracking control scheme based on the iterative HDP algorithm in the following.

3.2.2 *Infinite-Horizon Optimal Tracking Control Based on HDP*

3.2.2.1 System Transformation

For simplicity, denote $v(k) = w(k) - w(k-1)$. Noticing that $v(0) = w(0)$, we obtain

$$w(k) = v(k) + v(k-1) + \cdots + v(0). \tag{3.8}$$

Define $z(k)=[e^{\mathrm{T}}(k), x_d^{\mathrm{T}}(k)]^{\mathrm{T}}$. Thus, the cost functional (3.7) can be written as follows:

$$\begin{aligned} J(z(k), v) = \sum_{i=k}^{\infty} \Big\{ & z^{\mathrm{T}}(i)Qz(i) + v^{\mathrm{T}}(i)Rv(i) \\ & + [v(i)+v(i-1)+\cdots+v(0)]^{\mathrm{T}} \\ & \times S\,[v(i)+v(i-1)+\cdots+v(0)] \Big\}, \end{aligned} \tag{3.9}$$

where $Q=\left[\begin{smallmatrix} Q_1 & 0^{n\times n} \\ 0^{n\times n} & 0^{n\times n}\end{smallmatrix}\right]$. According to (3.5) and (3.6), we obtain

$$\begin{aligned} e(k+1) &= x(k+1) - x_d(k+1) \\ &= -\delta(x_d(k)) + f(e(k)+x_d(k)) \\ &\quad + g(e(k)+x_d(k))\,(v(k)+v(k-1)+\cdots+v(0)) \\ &\quad - g(e(k)+x_d(k))g^{-1}(x_d(k))(f(x_d(k))-\delta(x_d(k))). \end{aligned} \tag{3.10}$$

Let

$$\begin{aligned} &\Big[\Big(-\phi\,(x_d(k)) - g\,(\bar{z}(k)+x_d(k))\,g^{-1}\,(x_d(k))\,(f\,(x_d(k)) - \phi\,(x_d(k))) \\ &\qquad + f\,(\bar{z}(k)+x_d(k))\Big)^{\mathrm{T}}, \phi^{\mathrm{T}}\,(x_d(k))\Big]^{\mathrm{T}} \\ &\quad = \bar{f}(z(k)), \end{aligned}$$

and $[g^{\mathrm{T}}\,(\bar{z}(k)+x_d(k))\,, 0^{m\times n}]^{\mathrm{T}} = \bar{g}\,(z(k))$. Considering $z(k+1)=[e^{\mathrm{T}}(k+1), x_d^{\mathrm{T}}(k+1)]^{\mathrm{T}}$, we have

$$z(k+1) = \bar{f}(z(k)) + \bar{g}(z(k))(v(k)+\cdots+v(0)). \tag{3.11}$$

Therefore, the optimal tracking control problem of (3.1) is transformed into the optimal regulation problem of (3.11) with respect to (3.9). Here, the optimal controller is designed by the state feedback of the transformed system (3.11). Thus, our next work is to design a feedback controller $v(z(k))$ for (3.11) to regulate $e(k)$ and simultaneously to guarantee (3.9) is finite, i.e., admissible control.

3.2.2.2 Derivation of the Iterative HDP Algorithm

Here, we will design the optimal tracking controller using iterative HDP algorithm.

According to the Bellman optimality principle and (3.9), the HJB equation becomes

$$\begin{aligned} J^*(z(k)) = \min_{v(k)} \Big\{ & z^{\mathrm{T}}(k)Qz(k) + v^{\mathrm{T}}(k)Rv(k) \\ & + [v(k) + v(k-1) + \cdots + v(0)]^{\mathrm{T}} \\ & \times S\,[v(k) + v(k-1) + \cdots + v(0)] + J^*(z(k+1)) \Big\}, \end{aligned} \tag{3.12}$$

where $J^*(z(k))$ is the optimal value function which is defined as $J^*(z(k)) = \min_{v(k)} J(z(k), v(\cdot))$.

In the iterative HDP algorithm, the value function and control policy are updated by recurrent iteration, with the iteration number increasing from 0 to ∞.

First, for $i = 0$, we start with initial value function $V_0(z(k)) = 0$, and the control $v_0(k)$ can be computed as follows:

$$\begin{aligned} v_0(k) = \arg \min_{v(k)} \Big\{ & z^{\mathrm{T}}(k)Qz(k) + v^{\mathrm{T}}(k)Rv(k) \\ & + [v(k) + v(k-1) + \cdots + v(0)]^{\mathrm{T}} \\ & \times S\,[v(k) + v(k-1) + \cdots + v(0)] + V_0(z(k+1)) \Big\}. \end{aligned} \tag{3.13}$$

As there is no constraint for the system (3.11), we obtain

$$v_0(k) = -\,(R + S)^{-1}\, S\,(v(k-1) + \cdots + v(0)). \tag{3.14}$$

Then, we update the value function as follows:

$$\begin{aligned} V_1(z(k)) = {} & z^{\mathrm{T}}(k)Qz(k) + (v(k-1) + \cdots + v(0))^{\mathrm{T}} \\ & \times S\,(R+S)^{-1}\, R\,(R+S)^{-1}\, S\,(v(k-1) + \cdots + v(0)) \\ & + \Big(-\,(R+S)^{-1} S\,(v(k-1) + \cdots + v(0)) \\ & \quad + v(k-1) + \cdots + v(0) \Big)^{\mathrm{T}} \\ & \times S \Big(-\,(R+S)^{-1} S\,(v(k-1) + \cdots + v(0)) \\ & \quad + v(k-1) + \cdots + v(0) \Big). \end{aligned} \tag{3.15}$$

Next, for $i = 1, 2, \ldots$, the iterative HDP algorithm can be used to implement the iteration between

$$v_i(k) = -\frac{1}{2}(R+S)^{-1}\Bigg(2S\left(v(k-1)+\cdots+v(0)\right) + \bar{g}^{\mathrm{T}}(z(k))\frac{\partial V_i(z(k+1))}{\partial z(k+1)}\Bigg) \tag{3.16}$$

and

$$\begin{aligned} V_{i+1}(z(k)) = {} & z^{\mathrm{T}}(k)Qz(k) \\ & + \frac{1}{4}\Bigg(2S\left(v(k-1)+\cdots+v(0)\right) + \bar{g}^{\mathrm{T}}(z(k))\frac{\partial V_i(z(k+1))}{\partial z(k+1)}\Bigg)^{\mathrm{T}} \\ & \times (R+S)^{-1}R(R+S)^{-1} \\ & \times \Bigg(2S\left(v(k-1)+\cdots+v(0)\right) + \bar{g}^{\mathrm{T}}(z(k))\frac{\partial V_i(z(k+1))}{\partial z(k+1)}\Bigg) \\ & + \Bigg(-\frac{1}{2}(R+S)^{-1}\Bigg(2S\left(v(k-1)+\cdots+v(0)\right) \\ & \qquad + \bar{g}^{\mathrm{T}}(z(k))\frac{\partial V_i(z(k+1))}{\partial z(k+1)}\Bigg) + v(k-1)+\cdots+v(0)\Bigg)^{\mathrm{T}} \\ & \times S\Bigg(-\frac{1}{2}(R+S)^{-1}\Bigg(2S\left(v(k-1)+\cdots+v(0)\right) \\ & \qquad + \bar{g}^{\mathrm{T}}(z(k))\frac{\partial V_i(z(k+1))}{\partial z(k+1)}\Bigg) + v(k-1)+\cdots+v(0)\Bigg) \\ & + V_i(z(k+1)). \end{aligned} \tag{3.17}$$

In the following, we will present a proof of convergence of the iteration for (3.16) and (3.17), with the value function $V_i(z(k)) \to J^*(z(k))$ and $v_i(k) \to u^*(k)$ as $i \to \infty$, $\forall k$.

Lemma 3.1 *Let $\tilde{v}_i(k), k = 0, 1\ldots$ be any sequence of control, and $v_i(k)$ is expressed as* (3.16). *Define $V_{i+1}(z(k))$ as* (3.17) *and $\Lambda_{i+1}(z(k))$ as*

$$\begin{aligned} \Lambda_{i+1}(z(k)) = {} & z^{\mathrm{T}}(k)Qz(k) + \tilde{v}_i^{\mathrm{T}}(k)R\tilde{v}_i(k) \\ & + [\tilde{v}_i(k) + v(k-1)+\cdots+v(0)]^{\mathrm{T}} \end{aligned}$$

$$\times S\left[\tilde{v}_i(k) + v(k-1) + \cdots + v(0)\right] + \Lambda_i(z(k+1)). \tag{3.18}$$

If $V_0(z(k)) = \Lambda_0(z(k)) = 0$, *then* $V_i(z(k)) \le \Lambda_i(z(k))$, $\forall i$.

In order to prove the convergence of the value function, the following theorem is also necessary.

Theorem 3.2 (cf. [20]) *Let the sequence* $\{V_i(z(k))\}$ *be defined by* (3.17). *If* $e(k)$ *for the system* (3.10) *is strongly controllable, then there is an upper bound* Y *such that* $0 \le V_i(z(k)) \le Y$, $\forall i$.

Proof Let $\bar{v}(k)$, $k = 0, 1, \ldots$ be any admissible control input. Define a new sequence $\{P_i(z(k))\}$ as follows:

$$\begin{aligned} P_{i+1}(z(k)) = {} & z^{\mathrm{T}}(k)Qz(k) + \bar{v}^{\mathrm{T}}(k)R\bar{v}(k) \\ & + [\bar{v}(k) + v(k-1) + \cdots + v(0)]^{\mathrm{T}} \\ & \times S[\bar{v}(k) + v(k-1) + \cdots + v(0)] + P_i(z(k+1)), \end{aligned} \tag{3.19}$$

with $P_0(z(k)) = V_0(z(k)) = 0$, $V_i(z(k))$ is updated by (3.17). Thus, we obtain

$$\begin{aligned} P_{i+1}(z(k)) - P_i(z(k)) &= P_i(z(k+1)) - P_{i-1}(z(k+1)) \\ &\ \ \vdots \\ &= P_1(z(k+i)) - P_0(z(k+i)). \end{aligned} \tag{3.20}$$

Because $P_0(z(k+i)) = 0$, we have

$$\begin{aligned} P_{i+1}(z(k)) &= P_1(z(k+i)) + P_i(z(k)) \\ &= P_1(z(k+i)) + P_1(z(k+i-1)) + P_{i-1}(z(k)) \\ &= P_1(z(k+i)) + P_1(z(k+i-1)) \\ &\quad + P_1(z(k+i-2)) + \cdots + P_1(z(k)) \\ &= \sum_{j=0}^{i} P_1(z(k+j)). \end{aligned} \tag{3.21}$$

According to (3.19), (3.21) can be written as

$$\begin{aligned} P_{i+1}(z(k)) = \sum_{j=0}^{i} \Big\{ & z^{\mathrm{T}}(k+j)Qz(k+j) + \bar{v}^{\mathrm{T}}(k+j)R\bar{v}(k+j) \\ & + [\bar{v}(k+j) + \bar{v}(k+j-1) + \cdots + v(0)]^{\mathrm{T}} \\ & \times S[\bar{v}(k+j) + \bar{v}(k+j-1) + \cdots + v(0)] \Big\} \end{aligned}$$

$$\begin{aligned}
&\leq \sum_{j=0}^{\infty} \Big\{ z^{\mathrm{T}}(k+j) Q z(k+j) + \bar{v}^{\mathrm{T}}(k+j) R \bar{v}(k+j) \\
&\quad + [\bar{v}(k+j) + \bar{v}(k+j-1) + \cdots + v(0)]^{\mathrm{T}} \\
&\quad \times S[\bar{v}(k+j) + \bar{v}(k+j-1) + \cdots + v(0)] \Big\}.
\end{aligned} \tag{3.22}$$

Noting that the control input $\bar{v}(k), k = 0, 1, \ldots$ is an admissible control, we obtain

$$\forall i: \; P_{i+1}(z(k)) \leq \sum_{j=0}^{\infty} P_1(z(k+j)) \leq Y. \tag{3.23}$$

From Lemma 3.1, we have

$$\forall i: \; V_{i+1}(z(k)) \leq P_{i+1}(z(k)) \leq Y. \tag{3.24}$$

This completes the proof. □

With Lemma 3.1 and Theorem 3.2, the following main theorem can be derived.

Theorem 3.3 (cf. [20]) *Define the sequence* $\{V_i(z(k))\}$ *as* (3.17), *with* $V_0(z(k)) = 0$. *Then,* $\{V_i(z(k))\}$ *is a nondecreasing sequence in which* $V_{i+1}(z(k)) \geq V_i(z(k))$, $\forall i$, *and converges to the optimal value function of discrete-time HJB equation, i.e.,* $V_i(z(k)) \to J^*(z(k))$ *as* $i \to \infty$.

Proof For the convenience of analysis, define a new sequence $\{\Phi_i(z(k))\}$ as follows:

$$\begin{aligned}
\Phi_{i+1}(z(k)) &= z^{\mathrm{T}}(k) Q z(k) + v_{i+1}^{\mathrm{T}}(k) R v_{i+1}(k) \\
&\quad + [v_{i+1}(k) + v(k-1) + \cdots + v(0)]^{\mathrm{T}} \\
&\quad \times S[v_{i+1}(k) + v(k-1) + \cdots + v(0)] + \Phi_i(z(k+1)),
\end{aligned} \tag{3.25}$$

where $v_i(k)$ obtained by (3.16) and $\Phi_0(z(k)) = V_0(z(k)) = 0$.

In the following part, we prove $\Phi_i(z(k)) \leq V_{i+1}(z(k))$ by mathematical induction.

First, we prove that it holds for $i = 0$. Noting that

$$\begin{aligned}
V_1(z(k)) - \Phi_0(z(k)) &= z^{\mathrm{T}}(k) Q z(k) + v_0^{\mathrm{T}}(k) R v_0(k) \\
&\quad + [v_0(k) + v(k-1) + \cdots + v(0)]^{\mathrm{T}} \\
&\quad \times S[v_0(k) + v(k-1) + \cdots + v(0)] \\
&\geq 0.
\end{aligned} \tag{3.26}$$

Thus, for $i = 0$, we get

$$V_1(z(k)) \geq \Phi_0(z(k)). \tag{3.27}$$

Second, we assume it holds for $i-1$, i.e., $V_i(z(k)) \geq \Phi_{i-1}(z(k))$, $\forall z(k)$. Then, for i, because

$$\begin{aligned}\Phi_i(z(k)) = {} & z^{\mathrm{T}}(k)Qz(k) + v_i^{\mathrm{T}}(k)Rv_i(k) \\ & + [v_i(k) + v(k-1) + \cdots + v(0)]^{\mathrm{T}} \\ & \times S[v_i(k) + v(k-1) + \cdots + v(0)] + \Phi_{i-1}(z(k+1))\end{aligned} \tag{3.28}$$

and

$$\begin{aligned}V_{i+1}(z(k)) = {} & z^{\mathrm{T}}(k)Qz(k) + v_i^{\mathrm{T}}(k)Rv_i(k) \\ & + [v_i(k) + v(k-1) + \cdots + v(0)]^{\mathrm{T}} \\ & \times S[v_i(k) + v(k-1) + \cdots + v(0)] + V_i(z(k+1)),\end{aligned} \tag{3.29}$$

we obtain

$$V_{i+1}(z(k)) - \Phi_i(z(k)) = V_i(z(k)) - \Phi_{i-1}(z(k)) \geq 0, \tag{3.30}$$

i.e.,

$$\Phi_i(z(k)) \leq V_{i+1}(z(k)). \tag{3.31}$$

Therefore, the mathematical induction proof is completed.

Moreover, from Lemma 3.1, we know that $V_i(z(k)) \leq \Phi_i(z(k))$, and we obtain

$$V_i(z(k)) \leq \Phi_i(z(k)) \leq V_{i+1}(z(k)). \tag{3.32}$$

It is proved that $\{V_i(z(k))\}$ is a nondecreasing sequence bounded by (3.24). Hence, we conclude that $V_i(z(k)) \to J^*(z(k))$ as $i \to \infty$. □

3.2.2.3 Summary of the Algorithm

Now we summarize the iterative HDP algorithm for the nonlinear optimal tracking control problem as follows:

1. Give $x(0)$, $i_{\max}$, computation accuracy ε, desired trajectory $x_d(k)$, and control sequence $u(0), u(1), \ldots, u(k-1)$.
2. Compute $z(k)$ according to (3.11) and $v(0), \ldots, v(k-1)$ according to (3.5) and (3.8).
3. Set $i = 0$, $V_0(z(k)) = 0$.
4. Compute $v_0(k)$ by (3.14) and the value function $V_1(z(k))$ by (3.15).
5. Set $i = i + 1$.
6. Compute $v_i(k)$ by (3.16) and the corresponding value function $V_{i+1}(z(k))$ by (3.17).
7. If $|V_{i+1}(z(k)) - V_i(z(k))| < \varepsilon$, then go to Step 9; else go to Step 8.
8. If $i > i_{\max}$ then go to Step 9; otherwise, go to Step 5.
9. Stop.

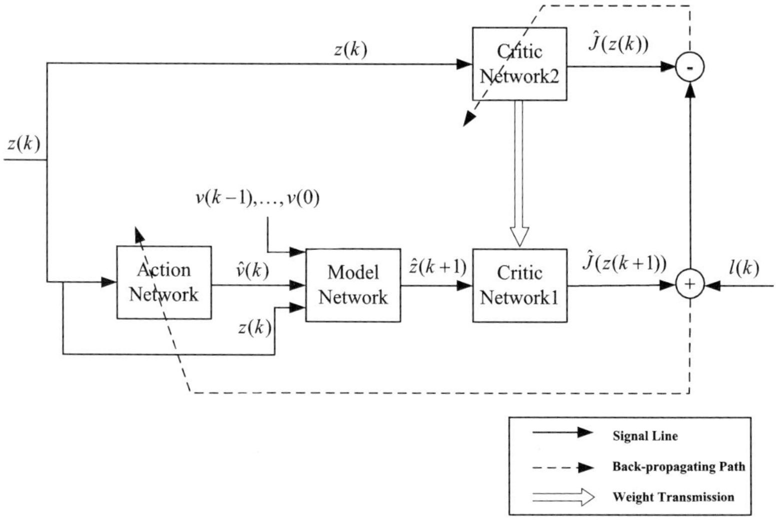

Fig. 3.1 The structure diagram of the iterative HDP algorithm

If the optimal tracking control policy $v(k)$ is obtained under the given accuracy ε. Then, we can compute tracking control input for the original system (3.1) by $u(k) = v(k) + v(k-1) + \cdots + v(0) - g^{-1}(x_d(k))(f(x_d(k)) - x_d(k+1))$.

3.2.2.4 Neural-Network Implementation for the Tracking Control Scheme

In the case of linear systems, the value function is quadratic and the control policy is linear. In the nonlinear case, this is not necessarily true, and therefore we use neural networks to approximate $v_i(k)$ and $V_i(z(k))$ [19, 21–24]. There are three neural networks needed to implement the algorithm, which are critic, model, and action networks. All the neural networks are chosen as three-layer feedforward networks. The structure diagram for running the iterative HDP algorithm is shown in Fig. 3.1. The utility term in the figure denotes $z^{\mathrm{T}}(k)Qz(k) + \hat{v}^{\mathrm{T}}(k)R\hat{v}(k) + [\hat{v}(k) + v(k-1) + \cdots + v(0)]^{\mathrm{T}}S\left[\hat{v}(k) + v(k-1) + \cdots + v(0)\right]$. The gradient descent rule is adopted for the weight update rules of each neural network, and the details can be seen in [14], and for the analysis of the neural network one may refer to [16]; it is omitted here.

3.2.3 Simulations

Example 3.4 An illustrative example is provided to demonstrate the effectiveness of the present tracking control scheme.

Consider the following affine nonlinear system:

$$x(k+1) = f(x(k)) + g(x(k))u(k), \tag{3.33}$$

where $x(k) = [x_1(k) \;\; x_2(k)]^{\mathrm{T}}$, $u(k) = [u_1(k) \;\; u_2(k)]^{\mathrm{T}}$,

$$f(x(k)) = \begin{bmatrix} 0.2x_1(k)\exp(x_2^2(k)) \\ 0.3x_2^3(k) \end{bmatrix}, \quad g(x(k)) = \begin{bmatrix} -0.2 & 0 \\ 0 & -0.2 \end{bmatrix}.$$

The given initial state is $x(0) = [1.5 \; 1]^{\mathrm{T}}$ and the desired trajectory is set to $x_d(k) = [\sin(k/20 + \pi/2) \; 0.5\cos(k/20)]^{\mathrm{T}}$. In order to demonstrate the advantage of (3.7), a comparison on the tracking performance for two different performance indices is presented. For ease of comparison, we define an evaluation function by PER $= \sum_0^{T_f} e^{\mathrm{T}}(k)e(k)$, where T_f is the running time steps.

Case 1: The cost functional is defined by (3.4) where $Q_1 = \left[\begin{smallmatrix} 1 & 0 \\ 0 & 1 \end{smallmatrix}\right]$ and $R = \left[\begin{smallmatrix} 0.2 & 0 \\ 0 & 0.2 \end{smallmatrix}\right]$. The system is first transformed as (3.11). We implement the algorithm at the time instant $k = 0$. The critic network, the action network, and the model network are chosen as three-layer neural networks with the structure 4–8–1, 4–8–2 and 6–8–4, respectively. The initial weights of action network, critic network, and model network are all set to be random in $[-1, 1]$. We take 1000 groups of sampling data to train the model network. For each group of data with the learning rate 0.01, we train 4000 steps to reach the given accuracy $\varepsilon = 10^{-6}$. After the training of the model network is completed, the weights keep unchanged. Then, the critic network and the action network are trained for 10000 iteration steps with a learning rate of 0.01, so that the given accuracy $\varepsilon = 10^{-6}$ is reached. Then, we apply the optimal tracking control policy to the system for $T_f = 250$ time steps. The convergence trajectory of the value function is shown in Fig. 3.2. The state trajectories are given in Figs. 3.3 and 3.4, the corresponding control trajectories are given in Fig. 3.5. In this case, we can obtain the evaluation function value of the present tracking control scheme as PER $= 4.2958$.

Case 2: Define the cost functional as (3.7) where $S = \left[\begin{smallmatrix} 1 & 0 \\ 0 & 1 \end{smallmatrix}\right]$. All the other parameters are set as the same as Case 1. We also implement the algorithm at the time instant $k = 0$. The convergence process of the value function is shown in Fig. 3.6. The state trajectories are given in Figs. 3.7 and 3.8, and the corresponding control trajectories are given in Fig. 3.9. In this case, with $T_f = 250$, we can obtain the evaluation function value of the present tracking control scheme as PER $= 2.1424$.

It can be seen that the tracking performance of Case 2 is much better than that of Case 1, though in both cases, the system states finally track the desired trajectories. In Case 1, both the states and control inputs oscillate seriously. While in Case 2, the oscillation is much slighter, and the evaluation function value is much smaller than that in Case 1. Hence, the control scheme developed in this section does solve the time-variant tracking control problem quite satisfyingly, and the obtained optimal tracking controller has shown excellent performance.

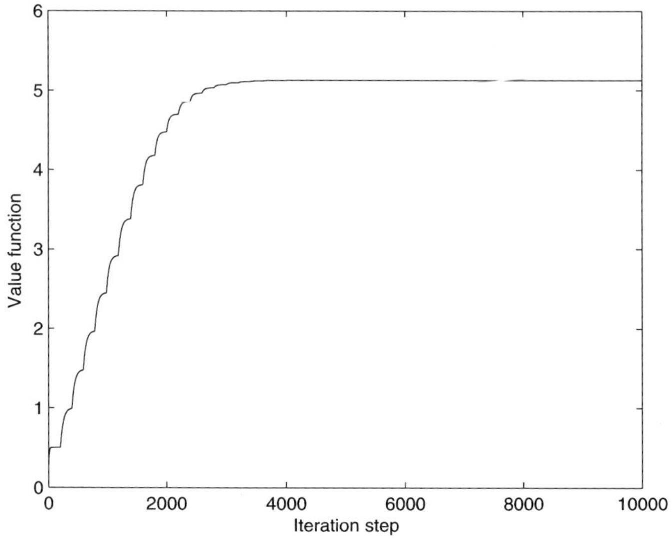

Fig. 3.2 The convergence of value function

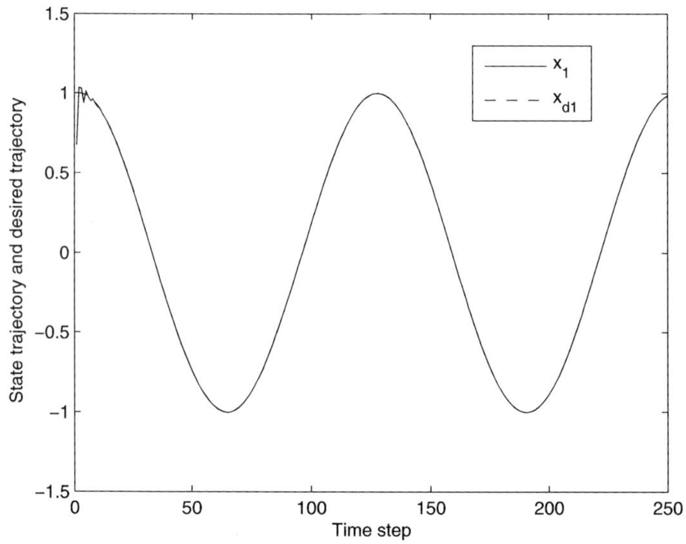

Fig. 3.3 The state trajectory x_1 and desired trajectory x_{d1}

3.3 Infinite-Horizon Optimal Tracking Control Based on GDHP

In this section an infinite-horizon optimal tracking control scheme is presented for a class of discrete-time non-affine systems, where the optimal tracking controller is

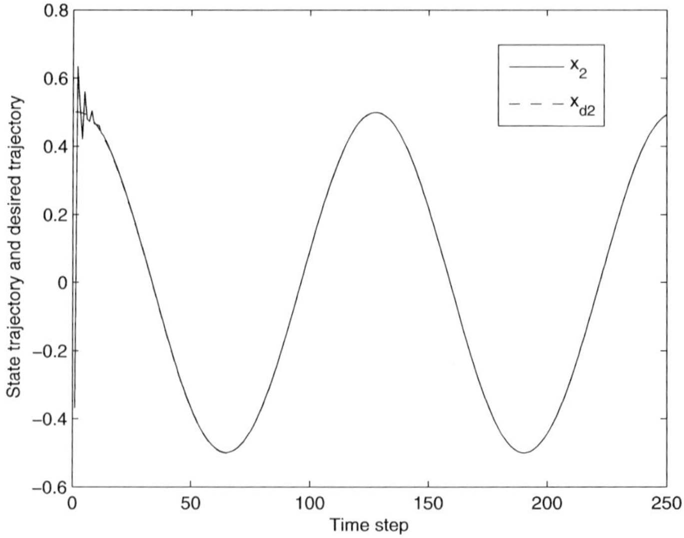

Fig. 3.4 The state trajectory x_2 and desired trajectory x_{d2}

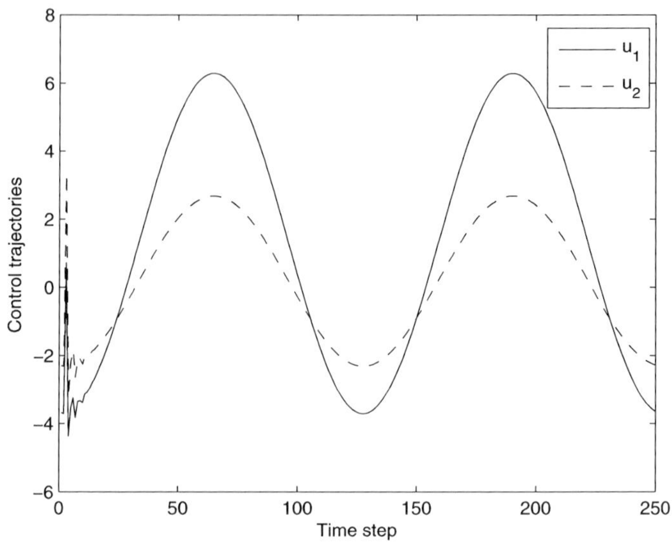

Fig. 3.5 The optimal tracking control trajectories

composed of two sub-controllers: the feedforward controller and the feedback controller. The feedforward controller is designed by implicit function theorem, while

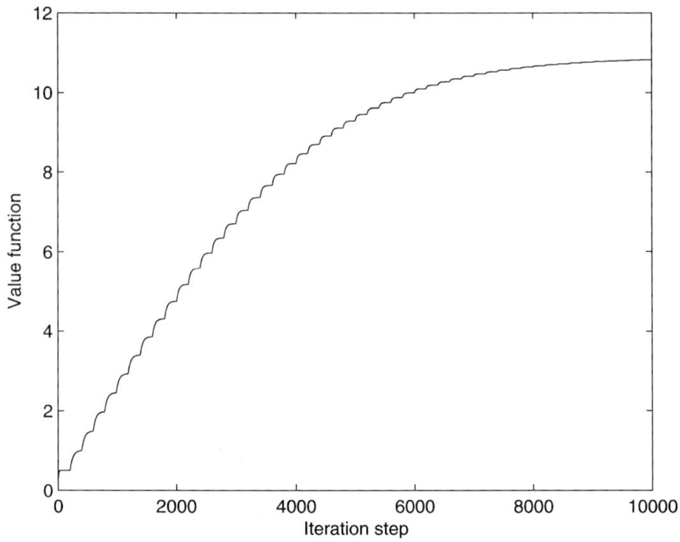

Fig. 3.6 The convergence of value function

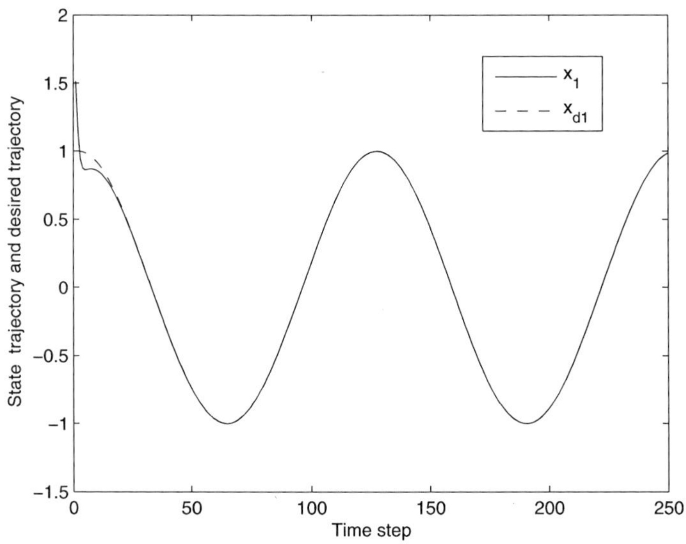

Fig. 3.7 The state trajectory x_1 and desired trajectory x_{d1}

the feedback controller is realized by the GDHP iteration algorithm. To facilitate the implementation of the algorithm, three neural networks are adopted.

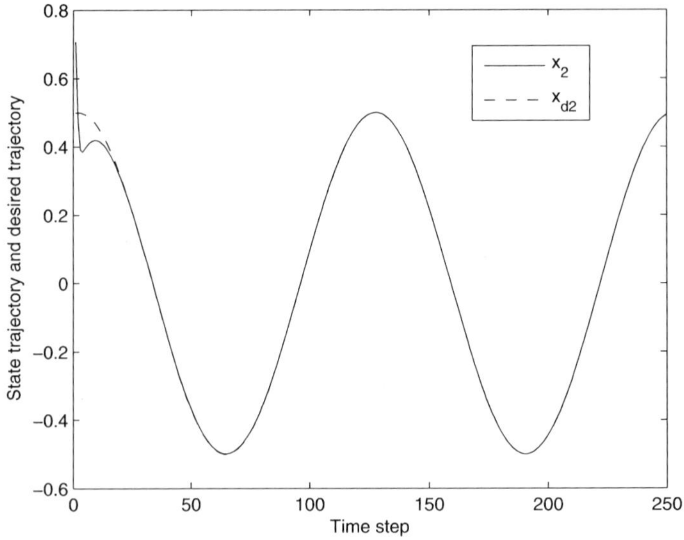

Fig. 3.8 The state trajectory x_2 and desired trajectory x_{d2}

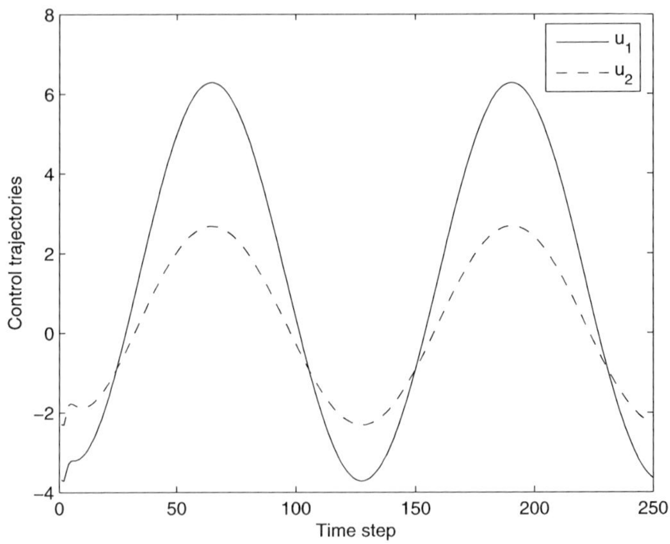

Fig. 3.9 The optimal tracking control trajectories

3.3.1 Problem Formulation

Consider a class of non-affine MIMO nonlinear systems as follows:

$$
\begin{cases}
x_1^1(k+1) = x_2^1(k) \\
x_2^1(k+1) = x_3^1(k) \\
\vdots \\
x_{n_1}^1(k+1) = f_1(x(k), u(k)) \\
\vdots \\
x_1^m(k+1) = x_2^m(k) \\
x_2^m(k+1) = x_3^m(k) \\
\vdots \\
x_{n_m}^m(k+1) = f_m(x(k), u(k)) \\
y_1(k) = x_1^1(k) \\
\vdots \\
y_m(k) = x_1^m(k),
\end{cases}
\tag{3.34}
$$

where $x(k) \in \mathbb{R}^n$, $n = \sum_{i=1}^m n_i$ denotes the state vector, $u(k) \in \mathbb{R}^m$ denotes the input vector, $u(k) = [u_1(k), u_2(k), \ldots, u_m(k)]^{\mathrm{T}}$, and $y(k) \in \mathbb{R}^m$ denotes the output vector of the system, $y(k) = [y_1(k), y_2(k), \ldots, y_m(k)]^{\mathrm{T}}$. $f_i : \mathbb{R}^n \times \mathbb{R}^m \to \mathbb{R}$ are assumed to be unknown but analytic functions. It is further assumed that $f_i(0, 0) = 0$, and meanwhile the system is strongly controllable on a compact set $\Omega_x \subset \mathbb{R}^n$ in the sense that there exists an analytic control policy that renders the system's output track specific desired trajectories.

Assumption 3.5 The system satisfies the result that $f_i : \mathbb{R}^n \times \mathbb{R}^m \to \mathbb{R}$ is C^1 for all $x(k) \in \Omega_x$ and $u(k) \in \mathbb{R}^m$.

For optimal tracking control problem, the control objective is to design an optimal control $u(k)$ for the system (3.34) such that the output vector $y(k)$ tracks the specified desired trajectories $y_d(k)$. Specifically, it is assumed that for $k = 0, 1, \ldots,$ the specified desired output trajectories are set as

$$
y_d(k) = [y_{1d}(k), y_{2d}(k), \ldots, y_{md}(k)]^{\mathrm{T}} \in \mathbb{R}^m, \tag{3.35}
$$

which lead to the desired state trajectories $x_d(k)$ as follows:

$$
x_d(k) = [x_d^1(k), x_d^2(k), \ldots, x_d^m(k)]^{\mathrm{T}}, \tag{3.36}
$$

where $x_d^i(k) = [y_{id}(k), y_{id}(k+1), \ldots, y_{id}(k+n_i-1)]$, with $i = 1, 2, \ldots, m$. It can be seen that the obtained desired state trajectories $x_d(k)$ satisfy $x_d(k) \in \Omega_d \subset \Omega_x$.

The state tracking error can be expressed as

$$
e(k) = x(k) - x_d(k). \tag{3.37}
$$

As is known, for infinite-horizon state feedback control problem, the cost functional is generally defined as the following quadratic form:

$$J(e(0), u) = \sum_{k=0}^{\infty} \left\{ x^{\mathrm{T}}(k) Q_1 x(k) + u^{\mathrm{T}}(k) R u(k) \right\}, \tag{3.38}$$

where $Q_1 \in \mathbb{R}^{n \times n}$ and $R \in \mathbb{R}^{m \times m}$ are positive definite matrices.

However, for the infinite-horizon tracking control problem, the cost functional cannot similarly be defined as

$$J(e(0), u) = \sum_{k=0}^{\infty} \left\{ e^{\mathrm{T}}(k) Q_1 e(k) + u^{\mathrm{T}}(k) R u(k) \right\}. \tag{3.39}$$

Because zero tracking error may not necessarily lead to zero control signals $u(k)$, which implies that this cost functional $J(e(0), u)$ may become infinite.

To overcome the difficulty related to the cost functional, we need to find an ideal control, required for maintaining the tracking in *steady-state stage*. Then, via subtracting this ideal control from the control signal $u(k)$, the remaining control signal is allowed to vanish in the steady-state stage. Denote the ideal control signal $u_f(k)$; it will be called the feedforward control in the sequel. Further, denote the remaining control signal as $v(k)$, calling it the feedback control. Therefore, the controller $u(k)$ becomes $u(k) = v(k) + u_f(k)$, where $u_f(k)$ is to maintain the tracking error close to zero at the steady-state stage, while $v(k)$ is to stabilize the state tracking error dynamics at *transient stage* and vanish in the steady-state stage. In this way, by using $v^{\mathrm{T}}(k) R v(k)$ to replace $u^{\mathrm{T}}(k) R u(k)$ in the cost functional (3.39), the obtained new cost functional can be finite, and accordingly the optimal tracking control problem can be appropriately defined.

Moreover, in order to further consider the control signal in the varying process, we add a new term $(v(k) - v(k-1))^{\mathrm{T}} S (v(k) - v(k-1))$ to the cost functional (3.39). Specifically, a new cost functional is defined as

$$\begin{aligned} J(e(k), v) = \sum_{i=k}^{\infty} \big\{ & e^{\mathrm{T}}(i) Q_1 e(i) + v^{\mathrm{T}}(i) R v(i) \\ & + (v(i) - v(i-1))^{\mathrm{T}} S (v(i) - v(i-1)) \big\}, \end{aligned} \tag{3.40}$$

where $Q_1 \in \mathbb{R}^{n \times n}$, $R \in \mathbb{R}^{m \times m}$ and $S \in \mathbb{R}^{m \times m}$ are all positive definite, $v(k)$ is the feedback control satisfying $v(k) = 0$ for $k < 0$.

After defining the cost functional (3.40), in the following development, we will focus on the optimal tracking controller design. It can be seen that the optimal controller is composed of two sub-controllers, i.e., the feedforward control $u_f(k)$ and the feedback control $v(k)$. In order to design these sub-controllers, we first derive the error system.

For convenience of analysis, the system (3.34) can be rewritten as

$$x(k+1) = \mathbb{F}(x(k), u(k)), \tag{3.41}$$

where $\mathbb{F}\colon \mathbb{R}^n \times \mathbb{R}^m \to \mathbb{R}^n$ is the vector function describing the nonlinear dynamics of the system (3.34).

In view of

$$e(k+1) = x(k+1) - x_d(k+1), \tag{3.42}$$

one has

$$e(k+1) = \mathbb{F}(x(k), u(k)) - x_d(k+1). \tag{3.43}$$

Since $u(k) = v(k) + u_f(k)$, the error equation can be reformulated as

$$e(k+1) = \mathbb{F}\big(e(k) + x_d(k), v(k) + u_f(k)\big) - x_d(k+1). \tag{3.44}$$

Keeping (3.42) in mind, once the error system (3.44) is stabilized, i.e., $e \to 0$, the state vector of the original system will track the desired trajectories simultaneously. Since the feedforward control $u_f(k)$ is designed beforehand to keep the state vector of the plant to the desired trajectories at the steady-state stage, it first is required to stabilize the error dynamic by choosing an appropriate feedback control $v(k)$.

Therefore, based on the above analysis, the optimal tracking control problem of (3.34) has been converted into two problems: one is to compute the feedforward control $u_f(k)$, and the other is to design the optimal feedback control $v(k)$ for stabilizing the error equation (3.44). In the following development, we will discuss the feedforward controller design and the optimal feedback controller design in detail.

3.3.2 Infinite-Horizon Optimal Tracking Control Based on GDHP

3.3.2.1 Design and Implementation of Feedforward Controller

As mentioned above, in order to make a plant follow specific desired trajectories, it is necessary to design the feedforward controller $u_f(k)$ so that the tracking error can keep zero at the steady-state stage. Therefore, by substituting the desired state $x_d(k)$ and the ideal feedforward control $u_f(k)$ in place of $x(k)$ and $u(k)$ in (3.34), the ideal steady-state equation can be written as

$$\begin{cases} x^1_{n_1 d}(k+1) = f_1(x_d(k), u_f(k)) \\ x^2_{n_2 d}(k+1) = f_2(x_d(k), u_f(k)) \\ \vdots \\ x^m_{n_m d}(k+1) = f_m(x_d(k), u_f(k)), \end{cases} \tag{3.45}$$

where the ideal feedforward control $u_f(k)$ is left to be solved in the following part.

Let

$$x_{nd}(k+1) = [x^1_{n_1 d}(k+1), x^2_{n_2 d}(k+1), \ldots, x^m_{n_m d}(k+1)]^{\mathrm{T}}. \tag{3.46}$$

Then the ideal steady-state equation (3.45) can further be written as

$$x_{nd}(k+1) = F(x_d(k), u_f(k)), \tag{3.47}$$

where $F\colon \mathbb{R}^n \times \mathbb{R}^m \to \mathbb{R}^m$ is the vector function defined as

$$F(\cdot) = \left[f_1(\cdot), f_2(\cdot), \ldots, f_m(\cdot)\right]^{\mathrm{T}}.$$

To solve the ideal feedforward control $u_f(k)$ uniquely from the above equation (3.47), we need to make sure that the implicit function theorem holds around the desired trajectories $x_d(k)$. Therefore, we first present an assumption as follows.

Assumption 3.6 The Jacobian matrix $\varUpsilon$ of the system (3.34) is nonsingular for all $x(k) \in \varOmega_d$, $u(k) \in \mathbb{R}^m$, where

$$\varUpsilon = \begin{bmatrix} \frac{\partial f_1(x(k),u(k))}{\partial u_1(k)} & \frac{\partial f_1(x(k),u(k))}{\partial u_2(k)} & \cdots & \frac{\partial f_1(x(k),u(k))}{\partial u_m(k)} \\ \frac{\partial f_2(x(k),u(k))}{\partial u_1(k)} & \frac{\partial f_2(x(k),u(k))}{\partial u_2(k)} & \cdots & \frac{\partial f_2(x(k),u(k))}{\partial u_m(k)} \\ \vdots & \vdots & & \vdots \\ \frac{\partial f_m(x(k),u(k))}{\partial u_1(k)} & \frac{\partial f_m(x(k),u(k))}{\partial u_2(k)} & \cdots & \frac{\partial f_m(x(k),u(k))}{\partial u_m(k)} \end{bmatrix}. \tag{3.48}$$

Based on Assumptions 3.5 and 3.6, the implicit function theorem is obviously ensured to hold. Therefore, the ideal feedforward control input $u_f(k)$ uniquely exists around the specific desired state trajectories $x_d(k) \in \varOmega_d$ and can further be expressed as follows:

$$u_f(k) = G(x_d(k), x_{nd}(k+1)), \tag{3.49}$$

where $G\colon \mathbb{R}^n \times \mathbb{R}^m \to \mathbb{R}$ is a unique one-to-one differentiable vector function satisfying $x_{nd}(k+1) = F(x_d(k), G(x_d(k), x_{nd}(k+1)))$.

3.3.2.2 Design and Implementation of Optimal Feedback Controller

The feedforward control has already been presented in the above subsection, which is the unique control input required at steady-state stage. The crucial problem left to solve is to design the optimal feedback control by the iterative algorithm after performing some transformation on the error system (3.44).

System Transformation

With the feedforward control $u_f(k)$ in (3.49), we substitute it into the error equation (3.44) and obtain

$$e(k+1) = \mathbb{F}(e(k) + x_d(k), v(k) + G(x_d(k), x_{nd}(k+1))) - x_d(k+1). \tag{3.50}$$

From (3.36), we can know that $x_d(k)$ is only denoting the desired trajectory. It is assumed that the desired trajectory is generated by the following relationship:

$$x_d(k+1) = \delta(x_d(k)), \tag{3.51}$$

where $\delta(\cdot) = [\delta_1(\cdot), \delta_2(\cdot), \ldots, \delta_n(\cdot)]^{\mathrm{T}}$ is an analytic vector function. Then, it is straightforward to show from (3.51) that

$$x_{nd}(k+1) = [\delta_{n_1}(x_d(k)), \delta_{n_1+n_2}(x_d(k)), \ldots, \delta_n(x_d(k))].$$

Therefore, the error system (3.50) can further be written as

$$e(k+1) = \Gamma(e(k), x_d(k), v(k)). \tag{3.52}$$

In the above equation, $e(k)$ is the states of the error equation and $x_d(k)$ can be treated as signals. However, they are time-varying. To the best of our knowledge, it is still not clear in the literature how to obtain the analytical solution for the optimal control of the above systems with time-varying signals. Inspired by the *dimension augmentation* technique, in this section we will construct an augmented system to equivalently solve the optimal feedback control $v(k)$.

Define $z_1(k) = e(k)$ and $z_2(k) = x_d(k)$, and then the error system (3.52) and the desired state trajectories equation (3.51) can be rewritten as

$$\begin{cases} z_1(k+1) = \Gamma(z_1(k), z_2(k), v(k)) \\ z_2(k+1) = \delta(z_2(k)), \end{cases} \tag{3.53}$$

which is called the augmented system of the error system (3.50). Here, the arguments $z_1(k)$ and $z_2(k)$ are viewed as two new state vectors of the augmented system. By this augmentation, the augmented system (3.53) has higher dimension than the original error system (3.52), and there is no time variable explicitly appearing in (3.53).

Accordingly, for this augmented system, the cost functional (3.40) can be redefined as

$$\begin{aligned} J(z(k), v) = \sum_{t=k}^{\infty} \big\{ & z^{\mathrm{T}}(t) Q z(t) + v^{\mathrm{T}}(t) R v(t) \\ & + (v(t) - v(t-1))^{\mathrm{T}} S (v(t) - v(t-1)) \big\}, \end{aligned} \tag{3.54}$$

where $z(t) = [e^{\mathrm{T}}(t)\ x_d^{\mathrm{T}}(t)]^{\mathrm{T}} \in \mathbb{R}^{2n}$ and $Q = \begin{bmatrix} Q_1 & 0_{n\times n} \\ 0_{n\times n} & 0_{n\times n} \end{bmatrix} \in \mathbb{R}^{2n\times 2n}$. This cost functional (3.54) is equivalent to (3.40) in the sense that the objectives of the two cost functionals are the same.

With this augmented model (3.53), it is now possible to design the optimal feedback controller to guarantee that the cost functional (3.54) is finite.

The Optimal Feedback Controller Design Based on GDHP Algorithm

In order to solve the optimal feedback control problem, the admissible control is defined similarly as Definition 2.1. However, the state vector is $z(k)$ and the cost functional is $J(z(0), v)$.

Actually the cost functional (3.54) can be expanded as

$$\begin{aligned} J(z(k), v) = {} & z^{\mathrm{T}}(k) Q z(k) + v^{\mathrm{T}}(k) R v(k) \\ & + (v(k) - v(k-1))^{\mathrm{T}} S (v(k) - v(k-1)) \\ & + J(z(k+1), v). \end{aligned} \tag{3.55}$$

From Bellman's optimality principle, the HJB equation is

$$\begin{aligned} J^*(z(k)) = {} & \min_{v(k)} \Big\{ z^{\mathrm{T}}(k) Q z(k) + v^{\mathrm{T}}(k) R v(k) \\ & + (v(k) - v(k-1))^{\mathrm{T}} S \big(v(k) - v(k-1)\big) \\ & + J^*(z(k+1)) \Big\}, \end{aligned} \tag{3.56}$$

where $J^*(z(k))$ is the optimal value function.

The corresponding optimal control can be defined as

$$\begin{aligned} u^*(k) = {} & \arg \min_{v(k)} \Big\{ z^{\mathrm{T}}(k) Q z(k) + v^{\mathrm{T}}(k) R v(k) \\ & + (v(k) - v(k-1))^{\mathrm{T}} S \big(v(k) - v(k-1)\big) \\ & + J^*(z(k+1)) \Big\}. \end{aligned} \tag{3.57}$$

It should be noted that if the optimal value function $J^*(z(k))$ can be solved from HJB equation (3.56), the optimal feedback control $u^*(k)$ can be computed. However, it is not straightforward to solve this HJB equation in most cases. Thus, in this section, we propose a new iterative algorithm named GDHP algorithm to solve the optimal control by introducing a new vector $\lambda_i(z(k))$.

First, we start with the initial value function $V_0(\cdot) = 0$, initial costate function $\lambda_0(\cdot) = 0$, and then for $i = 0, 1, \ldots$ we implement the iteration between

$$\begin{aligned} v_i(k) = {} & \arg \min_{v(k)} \Big\{ z^{\mathrm{T}}(k) Q z(k) + v^{\mathrm{T}}(k) R v(k) \\ & + (v(k) - v(k-1))^{\mathrm{T}} S (v(k) - v(k-1)) \\ & + V_i(z(k+1)) \Big\} \end{aligned} \tag{3.58}$$

i.e.,

$$v_i(k) = -\frac{1}{2}(R+S)^{-1}\left(-2Sv(k-1) + \left(\frac{dz(k+1)}{dv_i(k)}\right)^{\mathrm{T}} \lambda_i(z(k+1))\right) \tag{3.59}$$

and

$$\begin{aligned} V_{i+1}(z(k)) = {} & z^{\mathrm{T}}(k)Qz(k) + v_i^{\mathrm{T}}(k)Rv_i(k) \\ & + (v_i(k) - v(k-1))^{\mathrm{T}} S\,(v_i(k) - v(k-1)) \\ & + V_i\big(z(k+1)\big), \end{aligned} \tag{3.60}$$

$$\lambda_{i+1}(z(k)) = 2Qz(k) + \left(\frac{\partial z(k+1)}{\partial z(k)}\right)^{\mathrm{T}} \lambda_i(z(k+1)). \tag{3.61}$$

From the above analysis, we can find that via the costate function sequence (3.61), the optimal control sequence can be directly obtained by (3.59), which can avoid computing $\partial V_i(z(k+1))/\partial z(k+1)$ at each iterative step. This explains the working principle for the present GDHP algorithm.

Convergence Analysis of GDHP Algorithm

In the following we will present a concise convergence analysis of the iteration among (3.58), (3.60) and (3.61) with $V_i(z(k)) \to J^*(z(k))$, $\lambda_i(z(k)) \to \lambda^*(z(k))$ and the control policy $v_i(k) \to u^*(k)$ as $i \to \infty$.

Lemma 3.7 *Let μ_i be any sequence of control policies, and v_i is the optimal policies by* (3.58). *Define $V_{i+1}(z(k))$ as* (3.60) *and $\Lambda_{i+1}(z(k))$ as*

$$\begin{aligned} \Lambda_{i+1}(z(k)) = {} & z^{\mathrm{T}}(k)Qz(k) + \mu_i^{\mathrm{T}}(k)R\mu_i(k) \\ & + (\mu_i(k) - v(k-1))^{\mathrm{T}} S\,(\mu_i(k) - v(k-1)) \\ & + \Lambda_i(z(k+1)). \end{aligned} \tag{3.62}$$

If $V_0(\cdot) = \Lambda_0(\cdot) = 0$, then $V_i(z(k)) \le \Lambda_i(z(k))$, $\forall i$.

Lemma 3.8 *For the sequence $\{V_i(z(k))\}$ defined by* (3.60), *if the system* (3.52) *is controllable, then there is an upper bound Y such that $0 \le V_i \le Y$, $\forall i$.*

With Lemma 3.7 and Lemma 3.8, we are now ready to present the main theorem.

Theorem 3.9 *Define the sequence $\{V_i(z(k))\}$ as* (3.60), *with $V_0(\cdot) = 0$, the sequence $\{\lambda_i(z(k))\}$ as* (3.61), *with $\lambda_0(\cdot) = 0$. Then, $\{V_i(z(k))\}$ is a nondecreasing* *sequence satisfying* $V_{i+1}(z(k)) \geq V_i(z(k))$, $\forall i$, *and converges to the optimal value function of the discrete-time HJB equation* (3.56), *i.e.*, $V_i(z(k)) \to J^*(z(k))$ *as* $i \to \infty$. *Meanwhile, the costate function and the control input sequences are convergent, with* $\lambda_i(z(k)) \to \lambda^*(z(k))$ *and* $v_i(k) \to u^*(k)$ *as* $i \to \infty$.

Proof For the convenience of analysis, define a new sequence $\{\Phi_i(z(k))\}$ as follows:

$$\begin{aligned}\Phi_{i+1}(z(k)) = {} & z^{\mathrm{T}}(k)Qz(k) + v_{i+1}^{\mathrm{T}}(k)Rv_{i+1}(k) \\ & + (v_{i+1}(k) - v(k-1))^{\mathrm{T}}S(v_{i+1}(k) - v(k-1)) \\ & + \Phi_i(z(k+1)),\end{aligned} \tag{3.63}$$

where $v_i(k)$ defined by (3.58) and $\Phi_0(\cdot) = V_0(\cdot) = 0$.

In the following, we prove $\Phi_i(z(k)) \leq V_{i+1}(z(k))$ by mathematical induction. First, we prove that it holds for $i = 0$. Noting that

$$\begin{aligned}V_1(z(k)) - \Phi_0(z(k)) = {} & z^{\mathrm{T}}(k)Qz(k) + v_0^{\mathrm{T}}(k)Rv_0(k) \\ & + (v_0(k) - v(k-1))^{\mathrm{T}}S(v_0(k) - v(k-1)) \\ \geq {} & 0,\end{aligned} \tag{3.64}$$

for $i = 0$, we get

$$V_1(z(k)) \geq \Phi_0(z(k)). \tag{3.65}$$

Second, we assume that it holds for $i-1$, i.e., $V_i(z(k)) \geq \Phi_{i-1}(z(k))$, $\forall z(k)$. Then, for i, since

$$\begin{aligned}\Phi_i(z(k)) = {} & z^{\mathrm{T}}(k)Qz(k) + v_i^{\mathrm{T}}(k)Rv_i(k) \\ & + (v_i(k) - v(k-1))^{\mathrm{T}}S(v_i(k) - v(k-1)) \\ & + \Phi_{i-1}(z(k+1)),\end{aligned} \tag{3.66}$$

and

$$\begin{aligned}V_{i+1}(z(k)) = {} & z^{\mathrm{T}}(k)Qz(k) + v_i^{\mathrm{T}}(k)Rv_i(k) \\ & + (v_i(k) - v(k-1))^{\mathrm{T}}S(v_i(k) - v(k-1)) \\ & + V_i(z(k+1)),\end{aligned} \tag{3.67}$$

we obtain

$$V_{i+1}(z(k)) - \Phi_i(z(k)) = V_i(z(k)) - \Phi_{i-1}(z(k)) \geq 0, \tag{3.68}$$

i.e.,

$$\Phi_i(z(k)) \leq V_{i+1}(z(k)). \tag{3.69}$$

Therefore, the mathematical induction proof is completed.

Moreover, from Lemma (3.7), we know that $V_i(z(k)) \le \Phi_i(z(k))$; therefore we obtain

$$V_i(z(k)) \le \Phi_i(z(k)) \le V_{i+1}(z(k)), \tag{3.70}$$

which proves that $\{V_i(z(k))\}$ is a bounded nondecreasing sequence. Hence we conclude that $V_i(z(k)) \to J^*(z(k))$ as $i \to \infty$. With $\lambda_i(z(k)) = \partial V_i(z(k))/\partial z(k)$ and $\lambda^*(z(k)) = \partial J^*(z(k))/\partial z(k)$, the corresponding costate sequence $\{\lambda_i(z(k))\}$ is also convergent with $\lambda_i(z(k)) \to \lambda^*(z(k))$ as $i \to \infty$. Since the costate functions are convergent, we conclude that the corresponding control sequence $v_i(k)$ converges to the optimal control $u^*(k)$ as $i \to \infty$.

This completes the proof. □

Neural-Network Implementation of the Optimal Feedback Controller

For facilitating the implementation of the GDHP algorithm, we use three neural networks, named model network, critic network, and action network, respectively.

Model Network

Before carrying out the iterative process, we first construct the model network to approximate the nonlinear dynamic of the augmented system. The model network predicts $z(k+1)$ given $z(k)$ and $v(k)$, i.e., it learns the mapping given in (3.53). The output of the model network is denoted

$$\hat{z}(k+1) = W_m^{\mathrm{T}} \phi(V_m^{\mathrm{T}} I_m(k)), \tag{3.71}$$

where $I_m(k) = \mathrm{col}\{v(k),\ z(k)\}$ is the input vector of the model network.

Define the output error of the network as

$$e_m(k) = \hat{z}(k+1) - z(k+1). \tag{3.72}$$

The weights in the model network are updated to minimize the following performance error measure:

$$E_m(k) = \frac{1}{2} e_m^{\mathrm{T}}(k) e_m(k). \tag{3.73}$$

The weight update rule for model network is chosen as a gradient-based adaptation given by

$$W_m(j+1) = W_m(j) - \alpha_m(j) \left[\frac{\partial E_m(k)}{\partial W_m} \right], \tag{3.74}$$

$$V_m(j+1) = V_m(j) - \alpha_m(j) \left[\frac{\partial E_m(k)}{\partial V_m} \right], \tag{3.75}$$

where $\alpha_m(j)$ is the learning rate of the model network.

After the model network is trained, its weights are kept unchanged.

Critic Network

The output of the critic network is composed of two parts; one part is used to approximate the index function $V_{i+1}(k)$, and the other part is used to approximate the costate function $\lambda_{i+1}(k)$. Define the output of the critic network as $O_{c(i+1)}(k)$, which can be described by

$$O_{c(i+1)}(k) = W_{c(i+1)}^{\mathrm{T}} \phi(V_{c(i+1)}^{\mathrm{T}} z(k)). \tag{3.76}$$

Define the output target of the critic network as $D_{c(i+1)}(k)$, which is also composed of two parts, i.e.,

$$D_{c(i+1)}(k) = [V_{i+1}(k) \quad \lambda_{i+1}(k)]^{\mathrm{T}}, \tag{3.77}$$

where $V_{i+1}(k)$ is the target of value function computed by (3.60); $\lambda_{i+1}(k)$ is the target of the costate function computed by (3.61), respectively.

Define the error function for the critic network as

$$e_{c(i+1)}(k) = O_{c(i+1)}(k) - D_{c(i+1)}(k). \tag{3.78}$$

Also, define the objective function to be minimized in the critic network as

$$E_{c(i+1)}(k) = \frac{1}{2} e_{c(i+1)}^{\mathrm{T}}(k) e_{c(i+1)}(k). \tag{3.79}$$

Choose the weight update rule for the critic network as a gradient-based rule given by

$$W_{c(i+1)}(j+1) = W_{c(i+1)}(j) - \alpha_c(j) \left[\frac{\partial E_{c(i+1)}(k)}{\partial W_{c(i+1)}} \right], \tag{3.80}$$

$$V_{c(i+1)}(j+1) = V_{c(i+1)}(j) - \alpha_c(j) \left[\frac{\partial E_{c(i+1)}(k)}{\partial V_{c(i+1)}} \right], \tag{3.81}$$

where $\alpha_c(j) > 0$ is the learning rate of critic network.

Action Network

In the action network the augmented vector $z(k)$ is used as input to create the optimal feedback control as the output of the network. The output can be formulated as

$$\hat{v}_i(k) = W_{ai}^{\mathrm{T}} \phi(V_{ai}^{\mathrm{T}} z(k)). \tag{3.82}$$

The output target $v_i(k)$ is given by (3.59). Thus, we can define the output error of the action network as

$$e_{ai}(k) = \hat{v}_i(k) - v_i(k). \tag{3.83}$$

The weights in the action network are updated to minimize the following performance error measure:

$$E_{ai}(k) = \frac{1}{2} e_{ai}^{\mathrm{T}}(k) e_{ai}(k). \tag{3.84}$$

The update algorithm is then similar to the one in the critic network. By the gradient descent rule, we have

$$W_{ai}(j+1) = W_{ai}(j) - \alpha_a(j) \left[\frac{\partial E_{ai}(k)}{\partial W_{ai}} \right], \tag{3.85}$$

$$V_{ai}(j+1) = V_{ai}(j) - \alpha_a(j) \left[\frac{\partial E_{ai}(k)}{\partial V_{ai}} \right], \tag{3.86}$$

where $\alpha_a(j) > 0$ is the learning rate of action network.

Design Procedure of the Optimal Tracking Controller

The procedures of the optimal tracking control scheme based on the GI-GDHP algorithm are summarized as follows:

1. Given $x(0), u(0), u(1), \dots, u(k-1), i_{\max}, \varepsilon_0$, and the desired trajectory $y_d(k)$.
2. Compute $x_d(k), x_d(k+1), z(k)$ according to (3.36) and (3.37).
3. Compute the feedforward control $\hat{u}_f(k)$ by solving the implicit equation $x_{nd}(k+1) = F(x_d(k), u_f(k))$.
4. Construct the model network $\hat{z}(k+1) = W_m^{\mathrm{T}} \phi(V_m^{\mathrm{T}} I_m(k))$ with the initial weight parameters chosen randomly from $[-0.1, 0.1]$ and train the model network with a random input vector uniformly distributed in the interval $[-1, 1]$ and an arbitrary initial state vector till the given accuracy ε_0 is reached.
5. Set the iterative step $i = 0$, the initial value function $V_i(\cdot) = 0$ and costate function $\lambda_i(\cdot) = 0$. Construct both the critic network and the action network with the initial weight parameters chosen randomly from $[-1, 1]$.
6. Compute the output target by (3.59) and update the weight parameters of the action network by (3.85) and (3.86).
7. Compute the output target by (3.77) and update the weight parameters of the critic network by (3.80) and (3.81).
8. If

$$\|\lambda_{i+1}(z(k)) - \lambda_i(z(k))\|^2 + \|V_{i+1}(z(k)) - V_i(z(k))\|^2 < \varepsilon_0,$$

go to Step 9; otherwise $i = i + 1$, go to Step 6.

9. If $i > i_{\max}$, $v(k) = v_i(k)$ and go to Step 11; otherwise $i = i + 1$, go to Step 6.
10. Compute the approximate optimal control $u(k) = v(k) + u_f(k)$.
11. Stop.

As stated in the last subsection, the iterative algorithm will be convergent with $V_i(z(k)) \to J^*(z(k))$, $\lambda_i(z(k)) \to \lambda^*(z(k))$ and the control sequence $v_i(k) \to v^*(k)$ as $i \to \infty$. However, in the practical application, we cannot implement the iteration till $i \to \infty$. Actually, we iterate the algorithm for a max number $i_{\max}$ or with a tolerance value ε_0 to test the convergence of the algorithm. In the above procedures, the parameters ε_0 and $i_{\max}$ are chosen by the designer. The smaller ε_0 is set, the more accurate the costate function; the value function can be found in this way. If the condition set in Step 8 is satisfied, it implies that the costate function and the value function are convergent to the pre-specified accuracy. The larger the $i_{\max}$ is set, the more accurate the controller obtained will be at the price of increasing the computing burden.

3.3.2.3 Convergence Characteristics of the Neural-Network Approximation Process

Based on the insight in [6, 14], in this part we are ready to provide some convergence analysis for the neural-network approximation process in an averaged sense by using the stochastic approximation algorithm. In [13], a recursive stochastic procedure is presented to find the root of a real-valued function $r(\omega)$ of a real variable ω.

A function $r(\omega)$ with the form $r(\omega) = E\{h(\omega)\}$ ($E\{\cdot\}$ is the expectation operator) is called a regression function of $h(\omega)$, and, conversely, $h(\omega)$ is called a sample function of $r(\omega)$. According to the Robbins–Monro algorithm, the following iterative algorithm can be used to solve the root ω^* of the function $r(\omega)$:

$$\omega(j+1) = \omega(j) - q(j)h[\omega(j)], \tag{3.87}$$

where $q(j)$ is a sequence of positive numbers which satisfy the following conditions:

$$\begin{aligned} &1)\ \sum_{j=0}^{\infty} q(j) = \infty \\ &2)\ \sum_{j=0}^{\infty} q^2(j) < \infty \\ &3)\ \lim_{j\to\infty} q(j) = 0. \end{aligned} \tag{3.88}$$

Conditions (N1) through (N3) should be satisfied.

(N1) $r(\omega)$ has a single root ω^*, $r(\omega^*) = 0$ and

$$\begin{cases} r(\omega) < 0 & \text{if } \omega < \omega^*, \\ r(\omega) > 0 & \text{if } \omega > \omega^*. \end{cases} \tag{3.89}$$

(N2) The variance of $h(\omega)$ from $r(\omega)$ is finite

$$\sigma^2(\omega) = E\{r(\omega) - h(\omega)\}^2 < \infty. \tag{3.90}$$

(N3) The real-valued function $r(\omega)$ satisfies

$$| r(\omega) | < M_0 | \omega - \omega^* | + M_1 < \infty. \tag{3.91}$$

If the iterative algorithm (3.87) is implemented under the conditions (N1) through (N3), the obtained $\omega(j)$ will converge toward the true root ω^* in the mean square error sense with probability one, i.e.,

$$\lim_{j\to\infty} E\left\{\| \omega(j) - \omega^* \|^2\right\} = 0 \tag{3.92}$$

$$\text{Prob}\left\{\lim_{j\to\infty} \omega(j) = \omega^*\right\} = 1. \tag{3.93}$$

In this section, similar to the framework in [14], the Robbins–Monro algorithm is applied by setting $r(\omega) = \partial E/\partial\omega$, where E is an objective function to be optimized. If E has a local optimum at ω^*, $r(\omega)$ will satisfy the condition (N1) locally at ω^*. If E has a quadratic form, $r(\omega)$ will satisfy the condition (N1) globally.

First, we are ready to investigate the learning process of action network. Recall that the instantaneous error measure for the action network is of the following form:

$$e_{ai}(k) = \hat{v}_i(k) - v_i(k). \tag{3.94}$$

The performance error measure is defined as

$$E_{ai}(k) = \frac{1}{2} e_{ai}^{\mathrm{T}}(k) e_{ai}(k). \tag{3.95}$$

Let

$$\tilde{E}_{ai} = E\{E_{ai}\}, \tag{3.96}$$

and assume that the expectation is well defined over the discrete state measurements. The derivative of $\tilde{E}_{ai}$ with respect to the weights of the action network is then of the form

$$\frac{\partial \tilde{E}_{ai}}{\partial \omega_{ai}} = E\left\{\left(\frac{\partial e_{ai}}{\partial \omega_{ai}}\right)^{\mathrm{T}} e_{ai}\right\}. \tag{3.97}$$

According to the Robbins–Monro algorithm, the root (it may be a local root) of $\partial \tilde{E}_{ai}/\partial \omega_{ai}$ as a function of ω_{ai} can be obtained by the following recursive procedure, if the root exists and if the step size $\alpha_a(j)$ meets all the requirements described in (3.88):

$$\omega_{ai}(j+1) = \omega_{ai}(j) - \alpha_a(j)\left[\left(\frac{\partial e_{ai}}{\partial \omega_{ai}}\right)^{\mathrm{T}} e_{ai}\right]. \tag{3.98}$$

Recalling the update rule for the action network, we find that (3.98) is equivalent to (3.83)–(3.86). In this sense, the action network update rule of (3.83)–(3.86) is actually converging to a (local) minimum of the performance measure (3.96) in a statistical average sense.

In a similar framework, the convergence analysis of critic network and model network in the optimal feedback scheme can be derived, except that the definitions of the error measures are different. The details are therefore omitted here to save space.

3.3.3 Simulations

Example 3.10 An example is provided to demonstrate the present effectiveness of the optimal tracking control scheme.

Consider the following nonaffine nonlinear system:

$$\begin{cases} x_1^1(k+1) = x_2^1(k) \\ x_2^1(k+1) = 0.4x_2^2(k) - 0.3\cos(x_2^1(k)) + 2u_1 + 2\tanh(u_2) \\ x_1^2(k+1) = x_2^2(k) \\ x_2^2(k+1) = x_1^1(k) - 2u_2 + 3\sin(x_2^2(k))u_1 \\ y_1(k) = x_1^1(k) \\ y_2(k) = x_1^2(k), \end{cases} \tag{3.99}$$

where $x(k) = [x_1^1(k)\ x_2^1(k)\ x_1^2(k)\ x_2^2(k)] \in \mathbb{R}^4$ denotes the state vector, $u(k) = [u_1(k)\ u_2(k)] \in \mathbb{R}^2$ is the control vector, $y(k) = [y_1(k)\ y_2(k)] \in \mathbb{R}^2$ is the output vector.

The cost functional is defined as (3.54), the weight matrices are set as $Q_1 = 5I_4$, $R = 2I_2$ and $S = 2I_2$, where I_n denotes the $n \times n$ identity matrix. The desired output trajectories are set as $y_{1d}(k) = 0.5\sin[2\pi k/100]$, $y_{2d}(k) = 0.5\cos[2\pi k/100]$.

We would like to mention that the above test system (3.99) satisfies Assumptions 3.5 and 3.6 for the desired trajectories $y_{1d}(k)$, $y_{2d}(k)$. Therefore, the optimal tracking control scheme developed above can be used to design the optimal controller for this plant.

Here, the model network, the critic network, and the action network are set as the feedforward neural networks with the structures of 10–8–8, 8–8–9, and 8–8–2, respectively. All the neural networks are trained with a certain number of training cycles, and in each training cycle there are a certain number of iterative steps. For the optimal feedback controller design, we should first train the model network. The learning rate is chosen as $\alpha_m(j) = 0.1(1 - j/N_m)$, where N_m is the iterative steps of each training cycle which is set to $N_m = 5000$. Denote the training cycle number of all the neural networks as L. We train the model network for $L = 1000$ cycles with a random input vector uniformly distributed in the interval $[-1, 1]$ and an arbitrary initial state vector to get the predefined accuracy $\varepsilon_m = 10^{-4}$. Then, keeping the weights of the model network fixed, both the critic network and the action network are trained for $L = 200$ cycles to reach the predefined accuracy $\varepsilon_0 = 10^{-3}$. We should mention that in the training of the critic network and the action

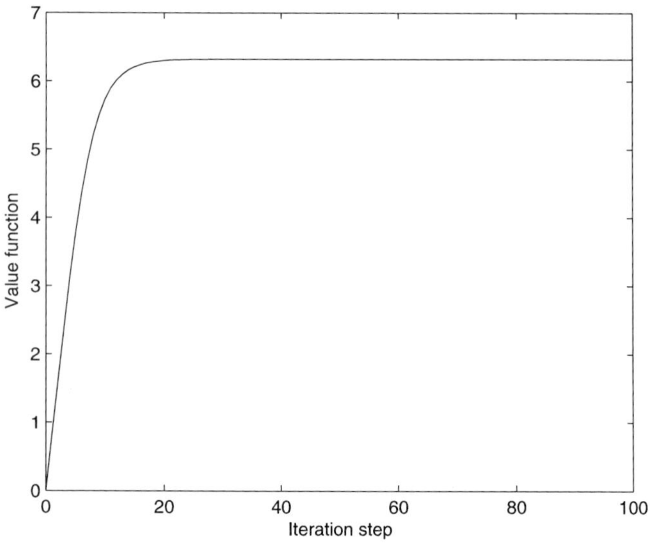

Fig. 3.10 The convergence trajectory of value function

network, the training cycle number L denotes the outer iterative number i in the GDHP algorithm. The learning rate of the critic network is set as $\alpha_c(j) = 0.1(1 - j/N_c)$, where N_c is the iterative steps of each training cycle in critic network. The learning rate of action network is set as $\alpha_a(j) = 0.1(1 - j/N_a)$, where N_a is the iterative steps of each training cycle in the action network. In the simulation, we set the iterative steps as $N_c = N_a = 200$. After simulation, the convergence trajectories of the value function and costate function are shown in Figs. 3.10, 3.11. In order to compare the different actions of the control policies obtained by different training cycles, for the same initial state vector $x(0) = [-0.2 \;\; 0.2314 \;\; 0.3 \;\; 0.499]$, we use the different control policies obtained for the plant and obtain the simulation results as follows. The output tracking error trajectories are shown in Figs. 3.12, 3.13, 3.14 and 3.15. It can be seen that the controlled plant's dynamics is out of order when the number of training cycles L is less than 12, and when $L > 15$ the tracking performance is improved with the training cycles increasing. When $L > 120$, the iterative process is basically convergent with slight improvement in the tracking performance.

From the simulation results, we can see that the output trajectories do track the desired trajectories quite well, so the control scheme developed in this section can solve the tracking control problem satisfactorily.

3.4 Finite-Horizon Optimal Tracking Control Based on ADP

A finite-horizon neuro-optimal tracking control scheme for a class of discrete-time nonlinear systems is developed in this section. Through a system transformation,

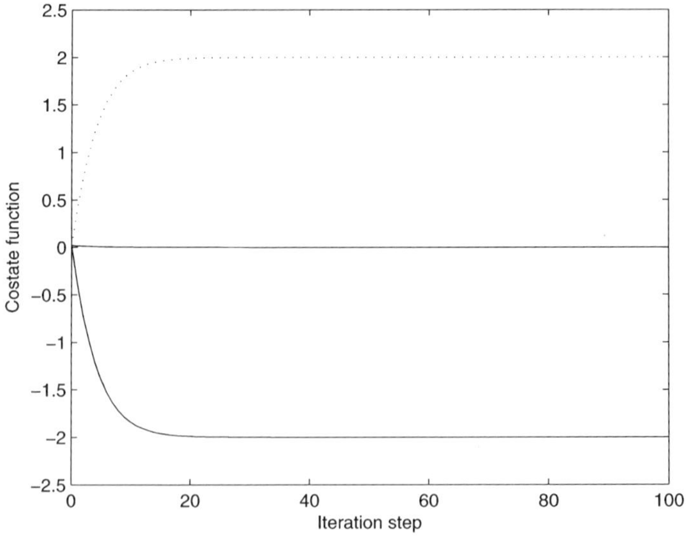

Fig. 3.11 The convergence trajectory of costate function

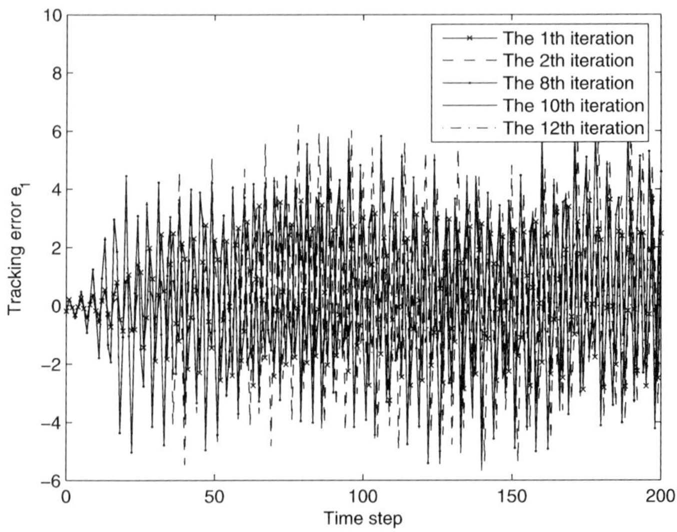

Fig. 3.12 The output tracking error e_1 for $L = 1, 2, 8, 10, 12$

the optimal tracking control problem is converted into a finite-horizon optimal regulation problem for the tracking error dynamics. Then, with convergence analysis in terms of value function and control law, the iterative ADP algorithm via the HDP technique is introduced to obtain the finite-horizon optimal tracking controller which makes the value function close to its optimal value within an ε-error bound.

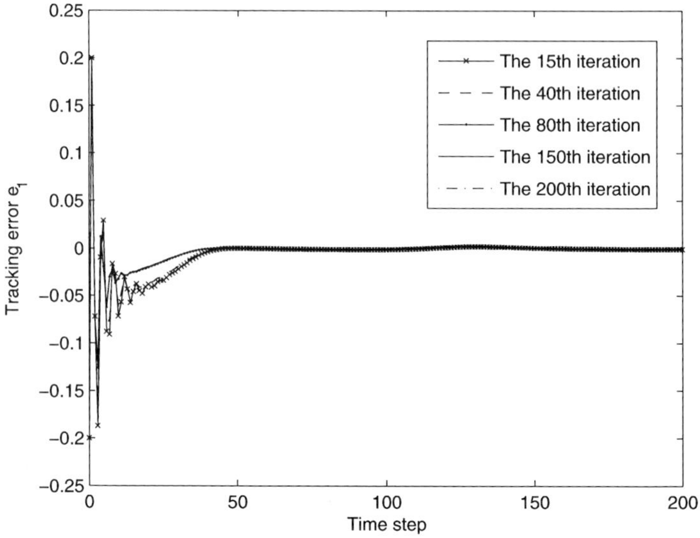

Fig. 3.13 The output tracking error e_1 for $L = 15, 40, 80, 150, 200$

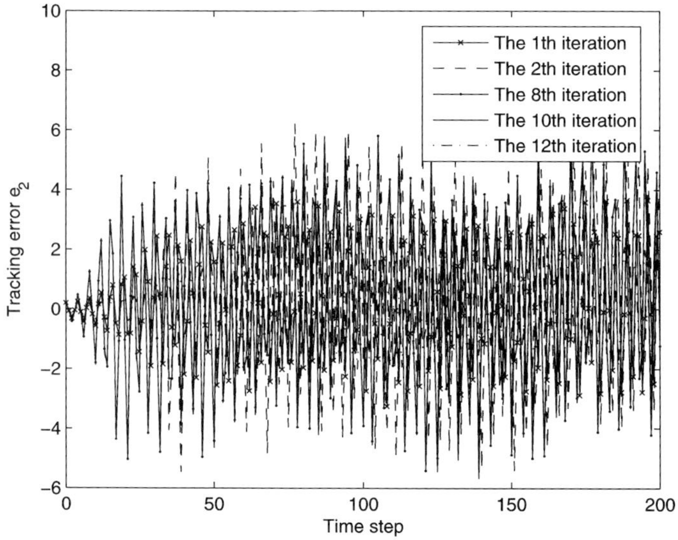

Fig. 3.14 The output tracking error e_2 for $L = 1, 2, 8, 10, 12$

Three neural networks are used as parametric structures to implement the algorithm, which aims at approximating the value function, the control law, and the error dynamics, respectively.

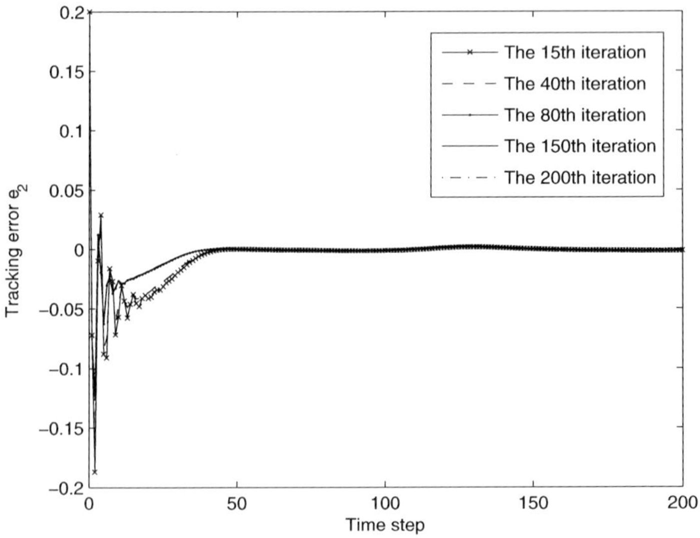

Fig. 3.15 The output tracking error e_2 for $L = 15, 40, 80, 150, 200$

3.4.1 Problem Formulation

Consider the discrete-time nonlinear systems given by

$$x(k+1) = f(x(k)) + g(x(k))u_p(k), \tag{3.100}$$

where $x(k) \in \mathbb{R}^n$ is the state, $u_p(k) \in \mathbb{R}^m$ is the control vector, $f(\cdot)$ and $g(\cdot)$ are differentiable in their argument with $f(0) = 0$. Assume that $f + gu_p$ is Lipschitz continuous on a set Ω in $\mathbb{R}^n$ containing the origin, and that the system (3.100) is controllable in the sense that there exists a continuous control on Ω that asymptotically stabilizes the system.

The objective for the optimal tracking control problem is to determine the optimal control law u_p^*, so as to make the state trajectory of the nonlinear system (3.100) to track a reference (or desired) trajectory $x_d(k)$ in an optimal manner. Here, we assume that the reference trajectory $x_d(k)$ satisfies

$$x_d(k+1) = \delta(x_d(k)), \tag{3.101}$$

where $x_d(k) \in \mathbb{R}^n$ and $\delta(x_d(k)) \in \mathbb{R}^n$. Then, we define the tracking error as

$$e(k) = x(k) - x_d(k). \tag{3.102}$$

Inspired by the work of [12, 20], we define the steady control corresponding to the reference trajectory $x_d(k)$ as

$$u_d(k) = g^{-1}(x_d(k))(\delta(x_d(k)) - f(x_d(k))), \tag{3.103}$$

where $g^{-1}(x_d(k))g(x_d(k)) = I_m$ and I_m is an $m \times m$ identity matrix.

By denoting

$$u(k) = u_p(k) - u_d(k) \tag{3.104}$$

and using (3.100)–(3.103), we obtain the new system as

$$\begin{cases} e(k+1) = f(e(k)+x_d(k)) + g(e(k)+x_d(k))g^{-1}(x_d(k))(\delta(x_d(k)) \\ \qquad\qquad - f(x_d(k))) - \delta(x_d(k)) + g(e(k)+x_d(k))u(k) \\ x_d(k+1) = \delta(x_d(k)). \end{cases} \tag{3.105}$$

Note that in system (3.105), $e(k)$ and $x_d(k)$ are regarded as the system variables, while $u(k)$ is seen as system input. The second equation of system (3.105) only gives the evolution of the reference trajectory which is not affected by the system input. Therefore, for simplicity, (3.105) can be rewritten as

$$e(k+1) = F(e(k), u(k)). \tag{3.106}$$

Now, let $e(0)$ be an initial state of system (3.106) and define $\underline{u}_0^{N-1} = (u(0), u(1), \ldots, u(N-1))$ be a control sequence with which the system (3.106) gives a trajectory starting from $e(0)$: $e(1)$, $e(2)$, ..., $e(N)$. The number of elements in the control sequence $\underline{u}_0^{N-1}$ is called the length of $\underline{u}_0^{N-1}$ and we denote it $|\underline{u}_0^{N-1}|$. Then $|\underline{u}_0^{N-1}| = N$. The final state under the control sequence $\underline{u}_0^{N-1}$ is denoted $e^{(f)}(e(0), \underline{u}_0^{N-1}) = e(N)$.

Definition 3.11 A nonlinear dynamical system is said to be stabilizable on a compact set $\Omega \in \mathbb{R}^n$, if for all initial conditions $e(0) \in \Omega$, there exists a control sequence $\underline{u}_0^{N-1} = (u(0), u(1), \ldots, u(N-1))$, $u(i) \in \mathbb{R}^m$, $i = 0, 1, \ldots, N-1$, such that the state $e^{(f)}(e(0), \underline{u}_0^{N-1}) = 0$.

Let $\underline{u}_k^{N-1} = (u(0), u(1), \ldots, u(N-1))$ be the control sequence starting at k with length $N-k$. For the finite-horizon optimal tracking control problem, it is desired to find the control sequence which minimizes the following cost function:

$$J\big(e(k), \underline{u}_k^{N-1}\big) = \sum_{i=k}^{N-1} l(e(i), u(i)), \tag{3.107}$$

where l is the utility function, $l(0,0) = 0$, $l(e(i), u(i)) \geq 0$ for $\forall e(i), u(i)$. In this section, the utility function is chosen as the quadratic form as follows:

$$l(e(i), u(i)) = e^{\mathrm{T}}(i)Qe(i) + u^{\mathrm{T}}(i)Ru(i).$$

This quadratic cost functional cannot only force the system state to follow the reference trajectory, but also force the system input to be close to the steady value in maintaining the state to its reference value. In fact, it can also be expressed as

$$l(e(i), u(i)) = \begin{bmatrix} e^{\mathrm{T}}(i) & x_d^{\mathrm{T}}(i) \end{bmatrix} \begin{bmatrix} Q & 0 \\ 0 & 0 \end{bmatrix} \begin{bmatrix} e(i) \\ x_d(i) \end{bmatrix} + u^{\mathrm{T}}(i)Ru(i)$$

from the view point of system (3.105).

In this sense, the nonlinear tracking control problem is converted into a regulation problem and the finite-horizon cost functional for tracking is written in terms of $e(i)$ and $u(i)$. Then, the problem of solving the finite-horizon optimal tracking control law u_p^* for system (3.100) is transformed into seeking the finite-horizon optimal control law u^* for system (3.106) with respect to (3.107). As a result, we will focus on how to design u^* in the following.

For the finite-horizon optimal control problem, the designed feedback control must be finite-horizon admissible, which means that it must not only stabilize the controlled system on Ω within a finite number of time steps, but also guarantee the cost functional to be finite.

Definition 3.12 A control sequence $\underline{u}_k^{N-1}$ is said to be finite horizon admissible for a state $e(k) \in \mathbb{R}^n$ with respect to (3.107) on Ω if $\underline{u}_k^{N-1}$ is continuous on a compact set $\Omega_u \in \mathbb{R}^m$, $u(0)=0$, $e^{(f)}(e(k), \underline{u}_k^{N-1})=0$ and $J(e(k), \underline{u}_k^{N-1})$ is finite.

Let $\mathfrak{A}_{e(k)} = \{\underline{u}(k) \colon e^{(f)}(e(k), \underline{u}(k)) = 0\}$ be the set of all finite-horizon admissible control sequences of e_k and

$$\mathfrak{A}_{e(k)}^{(i)} = \left\{\underline{u}_k^{k+i-1} \colon e^{(f)}\left(e(k), \underline{u}_k^{k+i-1}\right) = 0, \left|\underline{u}_k^{k+i-1}\right| = i\right\}$$

be the set of all finite-horizon admissible control sequences of $e(k)$ with length i. Define the optimal value function as

$$J^*(e(k)) = \inf_{\underline{u}(k)} \left\{J(e(k), \underline{u}(k)) \colon \underline{u}(k) \in \mathfrak{A}_{e(k)}\right\}. \tag{3.108}$$

Note that (3.107) can be written as

$$\begin{aligned} J\left(e(k), \underline{u}_k^{N-1}\right) &= e^{\mathrm{T}}(k)Qe(k) + u^{\mathrm{T}}(k)R(k)u(k) + \sum_{i=k+1}^{N-1} l(e(k), u(k)) \\ &= e^{\mathrm{T}}(k)Qe(k) + u^{\mathrm{T}}(k)R(k)u(k) + J\left(e(k+1), \underline{u}_{k+1}^{N-1}\right). \end{aligned} \tag{3.109}$$

Then, according to Bellman's optimality principle, it is known that the optimal value function $J^*(e_k)$ satisfies the discrete-time HJB equation

$$J^*(e(k)) = \min_{u(k)} \left\{e^{\mathrm{T}}(k)Qe(k) + u^{\mathrm{T}}(k)Ru(k) + J^*(e(k+1))\right\}. \tag{3.110}$$

The optimal control u^* satisfies the first-order necessary condition, which is given by the gradient of the right hand side of (3.110) with respect to $u(k)$. Then,

$$u^*(e(k)) = -\frac{1}{2}R^{-1}g^{\mathrm{T}}(e(k) + x_d(k))\frac{\partial J^*(e(k+1))}{\partial e(k+1)}. \tag{3.111}$$

By substituting (3.111) into (3.110), the discrete-time HJB equation becomes

$$\begin{aligned} J^*(e(k)) &= e^{\mathrm{T}}(k)Qe(k) + \frac{1}{4}\left(\frac{\partial J^*(e(k+1))}{\partial e(k+1)}\right)^{\mathrm{T}} g(e(k)+x_d(k))R^{-1} \\ &\quad \times g^{\mathrm{T}}(e(k)+x_d(k))\frac{\partial J^*(e(k+1))}{\partial e(k+1)} + J^*(e(k+1)), \end{aligned} \tag{3.112}$$

where $J^*(e(k))$ is the optimal value function corresponding to the optimal control law $u^*(e(k))$. Since the above discrete-time HJB equation cannot be solved exactly, we will present a novel algorithm to approximate the value function iteratively in next section. Before that, we make the following assumption.

Assumption 3.13 For system (3.105), the inverse of the control coefficient matrix $g(e(k)+x_d(k))$ exists.

Based on Assumption 3.13, for given $e(k)$ and $x_d(k)$, there exists an initial control $u(k)$ which can drive $e(k)$ to zero in one time step.

3.4.2 Finite-Horizon Optimal Tracking Control Based on ADP

3.4.2.1 Derivation of the Iterative ADP Algorithm

In this part, we present the iterative ADP algorithm, where the cost function and the control law are updated by recursive iterations.

First, we start with the initial value function $V_0(\cdot)=0$, and then solve for the law of single control vector $v_0(e_k)$ as follows:

$$\begin{aligned} &v_0(e(k)) = \arg\min_{u(k)} \left\{ l(e(k),u(k)) + V_0(e(k+1)) \right\} \\ &\text{subject to } F(e(k),u(k)) = 0. \end{aligned} \tag{3.113}$$

Once the control law $v_0(e(k))$ is determined, we update the cost function as

$$V_1(e(k)) = \min_{u(k)} \left\{ l(e(k),u(k)) + V_0(e(k+1)) \right\} = l(e(k), v_0(e(k))),$$

which can also be written as

$$\begin{aligned} V_1(e(k)) &= \min_{u(k)} l(e(k),u(k)) \text{ subject to } F(e(k),u(k)) = 0 \\ &= l(e(k), v_0(e(k))). \end{aligned} \tag{3.114}$$

Then, for $i = 1, 2, \ldots$, the iterative algorithm can be implemented for the control law

$$v_i(e_k) = \arg\min_{u(k)} \left\{ l(e(k),u(k)) + V_i(e(k+1)) \right\}$$

$$= -\frac{1}{2}R^{-1}g^{\mathrm{T}}(e(k)+x_d(k))\frac{\partial V_i(e(k+1))}{\partial e(k+1)} \tag{3.115}$$

and the value function

$$\begin{aligned} V_{i+1}(e(k)) &= \min_{u(k)}\left\{l(e(k),u(k)) + V_i(e(k+1))\right\} \\ &= l(e(k), v_i(e(k))) + V_i(F(e(k), v_i(e(k)))). \end{aligned} \tag{3.116}$$

Remark 3.14 In the iterative ADP algorithm (3.113)–(3.116), i is the iteration index of the control law and the value function, while k is the time index of the system's control and state trajectories. The value function and control law are updated until they converge to the optimal ones.

Next, we will present a convergence proof of the iteration between (3.115) and (3.116) with the value function $V_i \to J^*$ and the control law $v_i \to u^*$ as $i \to \infty$. Before that, we will see what $V_{i+1}(e_k)$ will be when it is expanded. According to (3.114) and (3.116), we obtain

$$\begin{aligned} V_{i+1}(e(k)) &= \min_{u(k)}\left\{l(e(k),u(k)) + V_i(e(k+1))\right\} \\ &= \min_{\underline{u}_k^{k+1}}\left\{l(e(k),u(k)) + l(e(k+1),u(k+1)) + V_{i-1}(e(k+2))\right\} \\ &\ \ \vdots \\ &= \min_{\underline{u}_k^{k+i-1}}\left\{l(e(k),u(k)) + l(e(k+1),u(k+1))\right. \\ &\quad \left. + \cdots + l(e(k+i-1),u(k+i-1)) + V_1(e(k+i))\right\}, \end{aligned} \tag{3.117}$$

where

$$\begin{aligned} V_1(e(k+i)) = &\min_{u(k+i)} l(e(k+i),u(k+i)) \\ &\text{subject to } F(e(k+i),u(k+i)) = 0. \end{aligned} \tag{3.118}$$

Then, we have

$$\begin{aligned} V_{i+1}(e(k)) &= \min_{\underline{u}_k^{k+i}} \sum_{j=0}^{i} l(e(k+j),u(k+j)) \\ &\qquad \text{subject to } F(e(k+i),u(k+i)) = 0 \\ &= \min_{\underline{u}_k^{k+i}}\left\{V\left(e(k),\underline{u}_k^{k+i}\right) : \underline{u}_k^{k+i} \in \mathfrak{A}_{e(k)}^{(i+1)}\right\}, \end{aligned} \tag{3.119}$$

which can also be written as

$$V_{i+1}(e(k))=\sum_{j=0}^{i} l(e(k+j), v_{i-j}(e(k+j))) \tag{3.120}$$

when using the notation in (3.115). These equations will be useful in the convergence proof of the iterative ADP algorithm.

3.4.2.2 Convergence Analysis of the Iterative ADP Algorithm

Theorem 3.15 (cf. [18]) *Suppose* $\mathfrak{A}_{e(k)}^{(1)} \neq \emptyset$. *Then, the value function sequence* $\{V_i\}$ *obtained by* (3.113)–(3.116) *is a monotonically nonincreasing sequence satisfying* $V_{i+1}(e(k)) \leq V_i(e(k))$ *for* $\forall i \geq 1$, *i.e.*, $V_1(e(k))=\max\{V_i(e(k))\colon i=1,2,\dots\}$.

Proof We prove this theorem by mathematical induction. First, we let $i=1$. The value function $V_1(e(k))$ is given in (3.114) and the finite-horizon admissible control sequence is $\underline{\hat{u}}_k^k=(v_0(e(k)))$. Now, we show that there exists a finite-horizon admissible control sequence $\underline{\hat{u}}_k^{k+1}$ with length 2 such that $V(e(k), \underline{\hat{u}}_k^{k+1})=V_1(e(k))$. Let $\underline{\hat{u}}_k^{k+1}=(\underline{\hat{u}}_k^k, 0)$; then $|\underline{\hat{u}}_k^{k+1}|=2$. Since $e(k+1)=F(e(k), v_0(e(k)))=0$ and $\hat{u}_{k+1}=0$, we have $e(k+2)=F(e(k+1), \hat{u}(k+1))=F(0,0)=0$. Thus, $\underline{\hat{u}}_k^{k+1}$ is a finite-horizon admissible control sequence. Since $l(e(k+1), \hat{u}(k+1))=U(0,0)=0$, we can obtain

$$\begin{aligned}
V\big(e(k), \underline{\hat{u}}_k^{k+1}\big) &= l(e(k), v_0(e(k)))+l(e(k+1), \hat{u}(k+1))\\
&= l(e(k), v_0(e(k)))\\
&= V_1(e(k)).
\end{aligned}$$

On the other hand, according to (3.119), we have

$$V_2(e(k))=\min_{\underline{u}_k^{k+1}}\Big\{V\big(e(k), \underline{u}_k^{k+1}\big)\colon \underline{u}_k^{k+1}\in \mathfrak{A}_{e(k)}^{(2)}\Big\},$$

which reveals that

$$V_2(e(k)) \leq V\big(e(k), \underline{\hat{u}}_k^{k+1}\big)=V_1(e(k)). \tag{3.121}$$

Therefore, the theorem holds for $i=1$.

Next, assume that the theorem holds for any $i=q$, where $q>1$. The current value function can be expressed as

$$V_q(e(k))=\sum_{j=0}^{q-1} l(e(k+j), v_{q-1-j}(e(k+j))),$$

where $\hat{\underline{u}}_k^{k+q-1} = (v_{q-1}(e(k)), v_{q-2}(e(k+1)), \ldots, v_0(e(k+q-1)))$ is the corresponding finite-horizon admissible control sequence.

Then, for $i = q+1$, we can construct a control sequence $\hat{\underline{u}}_k^{k+q} = (v_{q-1}(e(k)), v_{q-2}(e(k+1)), \ldots, v_0(e(k+q-1)), 0)$ with length $q+1$. The error trajectory is given as $e(k)$, $e(k+1) = F(e(k), v_{q-1}(e(k)))$, $e(k+2) = F(e(k+1), v_{q-2}(e(k+1))), \ldots, e(k+q) = F(e(k+q-1), v_0(e(k+q-1))) = 0$, $e(k+q+1) = F(e(k+q), \hat{u}(k+q)) = F(0,0) = 0$. This shows that $\hat{\underline{u}}_k^{k+q}$ is a finite-horizon admissible control sequence. As $l(e(k+q), \hat{u}(k+q)) = l(0,0) = 0$, we obtain

$$\begin{aligned}
V\big(e(k), \hat{\underline{u}}_k^{k+q}\big) &= l(e(k), v_{q-1}(e(k))) + l(e(k+1), v_{q-2}(e(k+1))) \\
&\quad + \cdots + l(e(k+q-1), v_0(e(k+q-1))) \\
&\quad + l(e(k+q), \hat{u}(k+q)) \\
&= \sum_{j=0}^{q-1} l(e(k+j), v_{q-1-j}(e(k+j))) \\
&= V_q(e(k)).
\end{aligned}$$

On the other hand, according to (3.119), we have

$$V_{q+1}(e(k)) = \min_{\underline{u}_k^{k+q}} \Big\{ V\big(e(k), \underline{u}_k^{k+q}\big) : \underline{u}_k^{k+q} \in \mathfrak{A}_{e(k)}^{(q+1)} \Big\},$$

which implies that

$$V_{q+1}(e(k)) \le V\big(e(k), \hat{\underline{u}}_k^{k+q}\big) = V_q(e(k)). \tag{3.122}$$

Accordingly, we complete the proof by mathematical induction. □

We have concluded that the value function sequence $\{V_i(e_k)\}$ is a monotonically nonincreasing sequence which is bounded below, and therefore, its limit exists. Here, we denote it as $V_\infty(e(k))$, i.e., $\lim_{i\to\infty} V_i(e(k)) = V_\infty(e(k))$. Next, let us consider what will happen when we make $i \to \infty$ in (3.116).

Theorem 3.16 (cf. [18]) *For any discrete time step k and tracking error $e(k)$, the following equation holds*:

$$V_\infty(e(k)) = \min_{u(k)} \big\{ l(e(k), u(k)) + V_\infty(e(k+1)) \big\}. \tag{3.123}$$

Proof For any admissible control $\tau(k) = \tau(e(k))$ and i, according to Theorem 3.15 and (3.116), we have

$$\begin{aligned}V_\infty(e(k)) &\le V_{i+1}(e(k))\\ &= \min_{u(k)}\big\{l(e(k),u(k)) + V_i(e(k+1))\big\}\\ &\le l(e(k),\tau(k)) + V_i(e(k+1)).\end{aligned}$$

Let $i \to \infty$; we get

$$V_\infty(e(k)) \le l(e(k),\tau(k)) + V_\infty(e(k+1)).$$

Note that in the above equation, $\tau(k)$ is chosen arbitrarily. Thus, we can obtain

$$V_\infty(e(k)) \le \min_{u(k)}\big\{l(e(k),u(k)) + V_\infty(e(k+1))\big\}. \tag{3.124}$$

On the other hand, let $\epsilon > 0$ be an arbitrary positive number. Then, there exists a positive integer l such that

$$V_l(e(k)) - \epsilon \le V_\infty(e(k)) \le V_l(e(k)), \tag{3.125}$$

because $V_i(e(k))$ is nonincreasing for $i \ge 1$ with $V_\infty(e(k))$ as its limit. Besides, from (3.116), we obtain

$$\begin{aligned}V_l(e(k)) &= \min_{u(k)}\big\{l(e(k),u(k)) + V_{l-1}(e(k+1))\big\}\\ &= l(e(k),v_{l-1}(e(k))) + V_{l-1}(F(e(k),v_{l-1}(e(k)))).\end{aligned}$$

Combining with (3.125), we obtain

$$\begin{aligned}V_\infty(e(k)) &\ge l(e(k),v_{l-1}(e(k))) + V_{l-1}(F(e(k),v_{l-1}(e(k)))) - \delta\\ &\ge l(e(k),v_{l-1}(e(k))) + V_\infty(F(e(k),v_{l-1}(e(k)))) - \delta\\ &\ge \min_{u(k)}\{L(e(k),u(k)) + V_\infty(e(k+1))\} - \delta,\end{aligned}$$

which reveals that

$$V_\infty(e(k)) \ge \min_{u(k)}\big\{l(e(k),u(k)) + V_\infty(e(k+1))\big\}, \tag{3.126}$$

because of the arbitrariness of ϵ. Based on (3.124) and (3.126), we conclude that (3.123) is true. □

Next, we will prove that the value function sequence $\{V_i\}$ converges to the optimal value function J^* as $i \to \infty$.

Theorem 3.17 (cf. [18]) *Define the value function sequence $\{V_i\}$ as in* (3.116) *with $V_0(\cdot) = 0$. If the system state $e(k)$ is controllable, then J^* is the limit of the value function sequence $\{V_i\}$, i.e.,*

$$V_\infty(e(k)) = J^*(e(k)).$$

Proof On one hand, in accordance with (3.108) and (3.119), we find

$$\begin{aligned} J^*(e(k)) &= \inf_{\underline{u}(k)} \left\{ V(e(k), \underline{u}(k)) \colon \underline{u}(k) \in \mathfrak{A}_{e(k)} \right\} \\ &\le \min_{\underline{u}_k^{k+i-1}} \left\{ V\left(e(k), \underline{u}_k^{k+i-1}\right) \colon \underline{u}_k^{k+i-1} \in \mathfrak{A}_{e(k)}^{(i)} \right\} \\ &= V_i(e(k)). \end{aligned}$$

Letting $i \to \infty$, we get

$$J^*(e(k)) \le V_\infty(e(k)). \tag{3.127}$$

On the other hand, according to the definition of $J^*(e(k))$, for any $\eta > 0$, there exists an admissible control sequence $\underline{\sigma}(k) \in \mathfrak{A}_{e(k)}$ such that

$$V\big(e(k), \underline{\sigma}(k)\big) \le J^*(e(k)) + \eta. \tag{3.128}$$

Now, we suppose that $|\underline{\sigma}(k)| = q$, which shows that $\underline{\sigma}(k) \in \mathfrak{A}_{e(k)}^{(q)}$. Then, we have

$$\begin{aligned} V_\infty(e(k)) &\le V_q(e(k)) \\ &= \min_{\underline{u}_k^{k+q-1}} \left\{ V\left(e(k), \underline{u}_k^{k+q-1}\right) \colon \underline{u}_k^{k+q-1} \in \mathfrak{A}_{e(k)}^{(q)} \right\} \\ &\le V\big(e(k), \underline{\sigma}(k)\big), \end{aligned}$$

using Theorem 3.15 and (3.119). Combining with (3.128), we get

$$V_\infty(e(k)) \le J^*(e(k)) + \eta.$$

Noticing that η is chosen arbitrarily in the above expression, we have

$$V_\infty(e(k)) \le J^*(e(k)). \tag{3.129}$$

Based on (3.127) and (3.129), we can conclude that $J^*(e(k))$ is the limit of the value function sequence $\{V_i\}$ as $i \to \infty$, i.e., $V_\infty(e(k)) = J^*(e(k))$. □

From Theorems 3.15–3.17, we find that the value function sequence $\{V_i(e(k))\}$ converges to the optimal value function $J^*(e(k))$ of the discrete-time HJB equation, i.e., $V_i \to J^*$ as $i \to \infty$. Then, according to (3.111) and (3.115), we can conclude the convergence of the corresponding control law sequence. Now, we present the following corollary.

Corollary 3.18 *Define the value function sequence* $\{V_i\}$ *as in* (3.116) *with* $V_0(\cdot) = 0$, *and the control law sequence* $\{v_i\}$ *as in* (3.115). *If the system state* $e(k)$ *is controllable, then the sequence* $\{v_i\}$ *converges to the optimal control law* u^* *as* $i \to \infty$, *i.e.,*

$$\lim_{i \to \infty} v_i(e(k)) = u^*(e(k)).$$

3.4.2.3 The ε-Optimal Control Algorithm

According to Theorems 3.15–3.17 and Corollary 3.18, we should run the iterative ADP algorithm (3.113)–(3.116) until $i \to \infty$ to obtain the optimal value function $J^*(e(k))$, and then to get a control vector $v_\infty(e(k))$. Based on this control vector, we can construct a control sequence $\underline{u}_\infty(e(k)) = (v_\infty(e(k)), v_\infty(e(k+1)), \ldots, v_\infty(e(k+i)), \ldots)$ to control the state to reach the target. Obviously, $\underline{u}_\infty(e(k))$ has infinite length. Though it is feasible in terms of theory, it is always not practical to do so because most real-world systems need to be effectively controlled within finite horizon. Therefore, in this section, we will propose a novel ε-optimal control strategy using the iterative ADP algorithm to deal with the problem. The idea is that, for a given error bound $\varepsilon > 0$, the iterative number i will be chosen so that the error between $V_i(e(k))$ and $J^*(e(k))$ is within the bound.

Let $\varepsilon > 0$ be any small number, $e(k)$ be any controllable state, and $J^*(e(k))$ be the optimal value of the value function sequence defined as in (3.116). From Theorem 3.17, it is clear that there exists a finite i such that

$$|V_i(e(k)) - J^*(e(k))| \le \varepsilon. \tag{3.130}$$

The length of the optimal control sequence starting from $e(k)$ with respect to ε is defined as

$$K_\varepsilon(e(k)) = \min\{i : |V_i(e(k)) - J^*(e(k))| \le \varepsilon\}. \tag{3.131}$$

The corresponding control law

$$\begin{aligned} v_{i-1}(e(k)) &= \arg\min_{u(k)} \left\{ l(e(k), u(k)) + V_{i-1}(e(k+1)) \right\} \\ &= -\frac{1}{2} R^{-1} g^{\mathrm{T}}(e(k) + x_d(k)) \frac{\partial V_{i-1}(e(k+1))}{\partial e(k+1)} \end{aligned} \tag{3.132}$$

is called the ε-optimal control and is denoted $\mu_\varepsilon^*(e(k))$.

In this sense, we can see that an error ε between $V_i(e(k))$ and $J^*(e(k))$ is introduced into the iterative ADP algorithm, which makes the value function sequence $\{V_i(e(k))\}$ converge in a finite number of iteration steps.

However, the optimal criterion (3.130) is difficult to verify because the optimal value function $J^*(e(k))$ is unknown in general. Consequently, we will use an equivalent criterion, i.e.,

$$|V_i(e(k)) - V_{i+1}(e(k))| \le \varepsilon, \tag{3.133}$$

replacing (3.130).

In fact, if $|V_i(e(k)) - J^*(e(k))| \le \varepsilon$ holds, we have $V_i(e(k)) \le J^*(e(k)) + \varepsilon$. Combining with $J^*(e(k)) \le V_{i+1}(e(k)) \le V_i(e(k))$, we find that

$$0 \le V_i(e(k)) - V_{i+1}(e(k)) \le \varepsilon,$$

which means

$$|V_i(e(k)) - V_{i+1}(e(k))| \le \varepsilon.$$

On the other hand, according to Theorem 3.17, $|V_i(e(k)) - V_{i+1}(e(k))| \to 0$ connotes that $V_i(e(k)) \to J^*(e(k))$. As a result, if $|V_i(e(k)) - V_{i+1}(e(k))| \le \varepsilon$ holds for any given small ε, we can derive the conclusion that $|V_i(e(k)) - J^*(e(k))| \le \varepsilon$ holds if i is sufficiently large.

3.4.2.4 Summary of the Algorithm

The detailed design procedure for the finite-horizon nonlinear optimal tracking control scheme using the iterative ADP algorithm will be given as follows:

1. Specify an error bound ε for the given initial state x_0. Choose $i_{\max}$, the reference trajectory $x_d(k)$, and the matrices Q and R.
2. Compute $e(k)$ according to (3.101) and (3.102).
3. Set $i = 0$, $V_0(e(k)) = 0$. Obtain the initial finite-horizon admissible vector $v_0(e(k))$ by (3.113) and update the value function $V_1(e(k))$ by (3.114).
4. Set $i = i + 1$.
5. Compute $v_i(e(k))$ by (3.115) and compute the corresponding value function $V_{i+1}(e(k))$ by (3.116).
6. If $|V_i(e(k)) - V_{i+1}(e(k))| \le \varepsilon$, then go to Step 8; otherwise, go to Step 7.
7. If $i > i_{\max}$, then go to Step 8; otherwise, go to Step 4.
8. Stop.

After the optimal control law $u^*(e(k))$ for system (3.105) is derived under the given error bound ε, we can compute the optimal tracking control input for the original system (3.100) by

$$\begin{aligned} u_p^*(k) &= u^*(e(k)) + u_d(k) \\ &= u^*(e(k)) + g^{-1}(x_d(k))(\delta(x_d(k)) - f(x_d(k))). \end{aligned} \tag{3.134}$$

3.4.2.5 Neural-Network Implementation of the Iterative ADP Algorithm via HDP Technique

Now, we implement the iterative HDP algorithm in (3.113)–(3.116) using NNs. In the iterative HDP algorithm, there are three networks, which are model network, critic network, and action network. All the networks are chosen as three-layer feedforward NNs. The input of the critic network and the action network are $e(k)$, while the input of the model network is $e(k)$ and $\hat{v}_i(e(k))$. The structure diagram of the iterative HDP algorithm is shown in Fig. 3.16.

Model Network

The purpose of designing the model network is to approximate the error dynamics. We should train the model network before carrying out the iterative HDP algorithm.

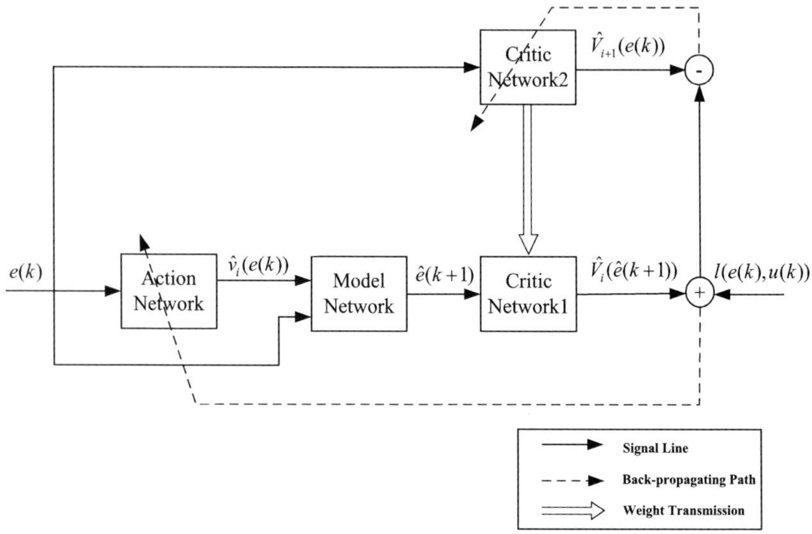

Fig. 3.16 The structure diagram of the iterative HDP algorithm

For given $e(k)$ and $\hat{v}_i(e(k))$, we can obtain the output of the model network as

$$\hat{e}(k+1) = W_m^{\mathrm{T}} \phi\big(V_m^{\mathrm{T}} z(k)\big), \tag{3.135}$$

where

$$z(k) = \big[e^{\mathrm{T}}(k) \;\; \hat{v}_i^{\mathrm{T}}(e(k))\big]^{\mathrm{T}}.$$

We define the error function of the model network as

$$e_m(k) = \hat{e}(k+1) - e(k+1). \tag{3.136}$$

The weights in the model network are updated to minimize the following performance measure:

$$E_m(k) = \frac{1}{2} e_m^{\mathrm{T}}(k) e_m(k). \tag{3.137}$$

Using the gradient-based adaptation rule, the weights can be updated as

$$W_m(j+1) = W_m(j) - \alpha_m \left[\frac{\partial E_m(k)}{\partial W_m(j)}\right], \tag{3.138}$$

$$V_m(j+1) = V_m(j) - \alpha_m \left[\frac{\partial E_m(k)}{\partial V_m(j)}\right], \tag{3.139}$$

where $\alpha_m > 0$ is the learning rate of the model network, and j is the iterative step for updating the weight parameters.

After the model network is trained, its weights are kept unchanged.

Critic Network

The critic network is used to approximate the value function $V_i(e(k))$. The output of the critic network is denoted

$$\hat{V}_i(e(k)) = W_{ci}^{\mathrm{T}}\phi\big(V_{ci}^{\mathrm{T}}e(k)\big). \tag{3.140}$$

The target function can be written as

$$V_i(e(k)) = e^{\mathrm{T}}(k)Qe(k) + v_{i-1}^{\mathrm{T}}(e(k))Rv_{i-1}(e(k)) + \hat{V}_{i-1}(\hat{e}(k+1)). \tag{3.141}$$

Then, we define the error function for the critic network as

$$e_{ci}(k) = \hat{V}_i(e(k)) - V_i(e(k)). \tag{3.142}$$

The objective function to be minimized for the critic network is

$$E_{ci}(k) = \frac{1}{2}e_{ci}^{\mathrm{T}}(k)e_{ci}(k). \tag{3.143}$$

The weight updating rule for training the critic network is also a gradient-based adaptation, given by

$$W_{ci}(j+1) = W_{ci}(j) - \alpha_c\left[\frac{\partial E_{ci}(k)}{\partial W_{ci}(j)}\right], \tag{3.144}$$

$$V_{ci}(j+1) = V_{ci}(j) - \alpha_c\left[\frac{\partial E_{ci}(k)}{\partial V_{ci}(j)}\right], \tag{3.145}$$

where $\alpha_c > 0$ is the learning rate of the critic network, and j is the inner-loop iterative step for updating the weight parameters.

Action Network

In the action network, the state $e(k)$ is used as input to obtain the optimal control. The output can be formulated as

$$\hat{v}_i(e(k)) = W_{ai}^{\mathrm{T}}\phi\big(V_{ai}^{\mathrm{T}}e(k)\big). \tag{3.146}$$

The target control input is given by

$$v_i(e(k)) = -\frac{1}{2}R^{-1}g^{\mathrm{T}}(e(k)+x_d(k))\frac{\partial\hat{V}_i(\hat{e}(k+1))}{\partial\hat{e}(k+1)}. \tag{3.147}$$

The error function of the action network can be defined as

$$e_{ai}(k) = \hat{v}_i(e(k)) - v_i(e(k)). \tag{3.148}$$

The weights of the action network are updated to minimize the following performance error measure:

$$E_{ai}(k) = \frac{1}{2} e_{ai}^{\mathrm{T}}(k) e_{ai}(k). \tag{3.149}$$

Similarly, the weight updating algorithm is

$$W_{ai}(j+1) = W_{ai}(j) - \alpha_a \left[\frac{\partial E_{ai}(k)}{\partial W_{ai}(j)} \right], \tag{3.150}$$

$$V_{ai}(j+1) = V_{ai}(j) - \alpha_a \left[\frac{\partial E_{ai}(k)}{\partial V_{ai}(j)} \right], \tag{3.151}$$

where $\alpha_a > 0$ is the learning rate of the action network, and j is the inner-loop iterative step for updating the weight parameters.

3.4.3 Simulations

In this part, two simulation examples are provided to confirm the theoretical results.

Example 3.19 The first example is derived from [5] with some modifications. Consider the following nonlinear system:

$$x(k+1) = f(x(k)) + g(x(k)) u_p(k), \tag{3.152}$$

where $x(k) = [x_1(k)\ x_2(k)]^{\mathrm{T}} \in \mathbb{R}^2$ and $u_p(k) = [u_{p1}(k)\ u_{p2}(k)]^{\mathrm{T}} \in \mathbb{R}^2$ are the state and control variables, respectively. The parameters of the value function are chosen as $Q = 0.5I$ and $R = 2I$, where I denotes the identity matrix with suitable dimensions. The state of the controlled system is initialized by $x(0) = [0.8\ \ -0.5]^{\mathrm{T}}$. The system functions are given as

$$f(x(k)) = \begin{bmatrix} \sin(0.5x_2(k)) x_1^2(k) \\ \cos(1.4x_2(k)) \sin(0.9x_1(k)) \end{bmatrix}, \quad g(x(k)) = \begin{bmatrix} 1 & 0 \\ 0 & 1 \end{bmatrix}.$$

The reference trajectory for the above system is selected as

$$x_d(k) = \begin{bmatrix} \sin(0.25(k)) \\ \cos(0.25(k)) \end{bmatrix}.$$

We set the error bound of the iterative HDP algorithm as $\varepsilon = 10^{-5}$ and implement the algorithm at time instant $k = 0$. The initial control vector of system (3.105) can be computed as $v_0(e(0)) = [0.64 \sin(0.25)\ \ -\sin(0.72)\cos(0.7)]^{\mathrm{T}}$, where $e(0) = [0.8\ -1.5]^{\mathrm{T}}$. Then, we choose three-layer feedforward NNs as model network, critic network, and action network with the structures 4–8–2, 2–8–1, 2–8–2, respectively. The initial weights of the three networks are all set to be random

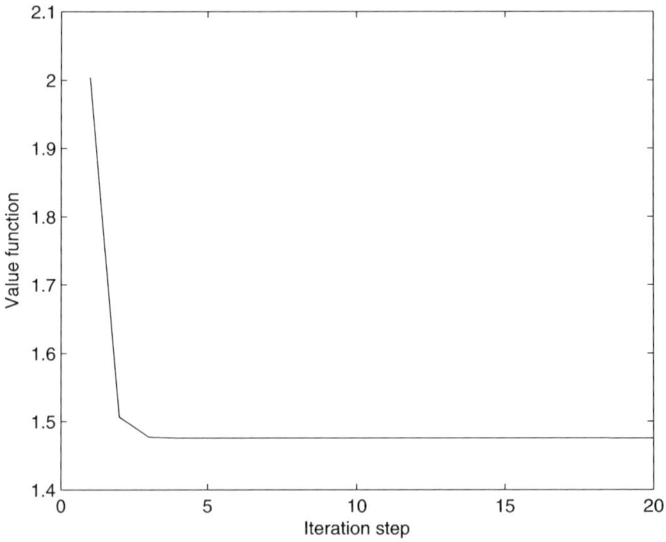

Fig. 3.17 The convergence process of the cost function

in $[-1, 1]$. It should be mentioned that the model network should be trained first. We train the model network for 1000 steps using 500 samples under the learning rate $\alpha_m = 0.1$. After the training of the model network is completed, the weights are kept unchanged. Then, we train the critic network and the action network for 20 iterations (i.e., for $i = 1, 2, \ldots, 20$) with each iteration of 2000 training steps to make sure the given error bound $\varepsilon = 10^{-5}$ is reached. In the training process, the learning rate is $\alpha_c = \alpha_a = 0.05$. The convergence process of the value function of the iterative HDP algorithm is shown in Fig. 3.17, for $k = 0$. We can see that the iterative value function sequence does converge to the optimal value function quite rapidly, which indicates the effectiveness of the iterative HDP algorithm. Therefore, we have $|V_{19}(e(0)) - V_{20}(e(0))| \le \varepsilon$, which means that the number of steps of the ε-optimal control is $K_\varepsilon(e(0)) = 19$. Besides, the ε-optimal control law $\mu_\varepsilon^*(e(0))$ for system (3.105) can also be obtained during the iterative process.

Next, we compute the near-optimal tracking control law for the original system (3.100) using (3.134) and apply it to the controlled system for 40 time steps. The obtained state trajectories are shown in Figs. 3.18 and 3.19, where the corresponding reference trajectories are also plotted to evaluate the tracking performance. The tracking control trajectories and the tracking errors are shown in Figs. 3.20 and 3.21, respectively. Besides, we can derive that the tracking error becomes $e(19) = [0.2778 \times 10^{-5} \quad -0.8793 \times 10^{-5}]^{\mathrm{T}}$ after 19 time steps. These simulation results verify the excellent performance of the tracking controller developed by the iterative ADP algorithm.

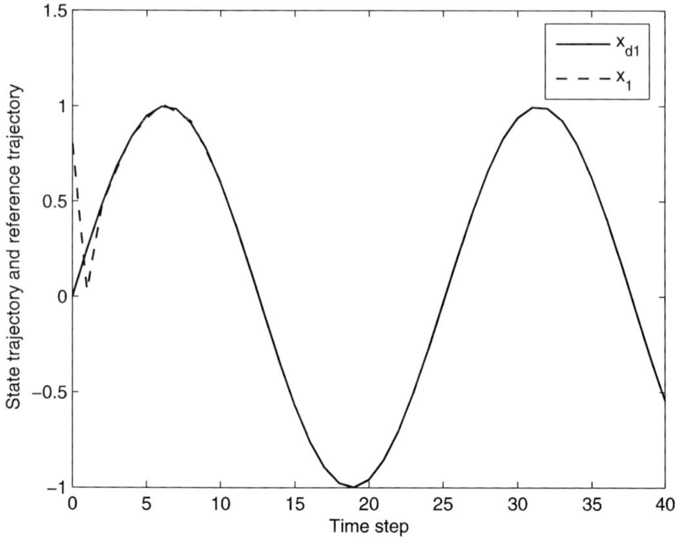

Fig. 3.18 The state trajectory x_1 and the reference trajectory r_1

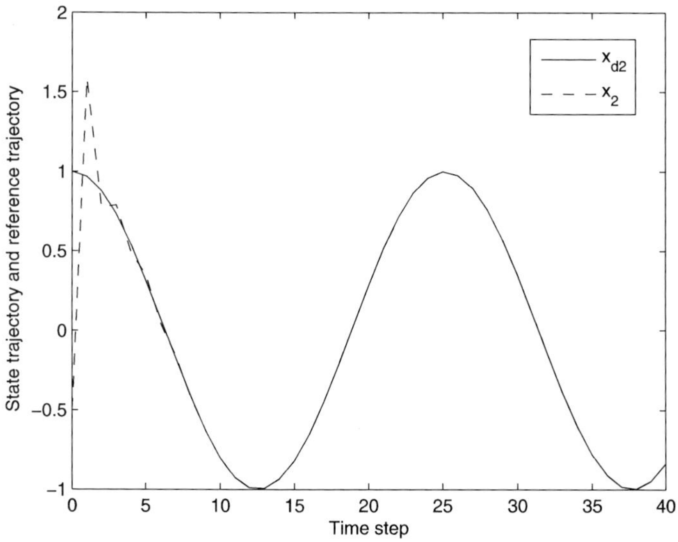

Fig. 3.19 The state trajectory x_2 and the reference trajectory r_2

Example 3.20 The second example is obtained from [20]. Consider the nonlinear discrete-time system described by (3.152) where

$$f(x(k)) = \begin{bmatrix} 0.2x_1(k)e^{x_2^2(k)} \\ 0.3x_2^3(k) \end{bmatrix}, \quad g(x(k)) = \begin{bmatrix} -0.2 & 0 \\ 0 & -0.2 \end{bmatrix}.$$

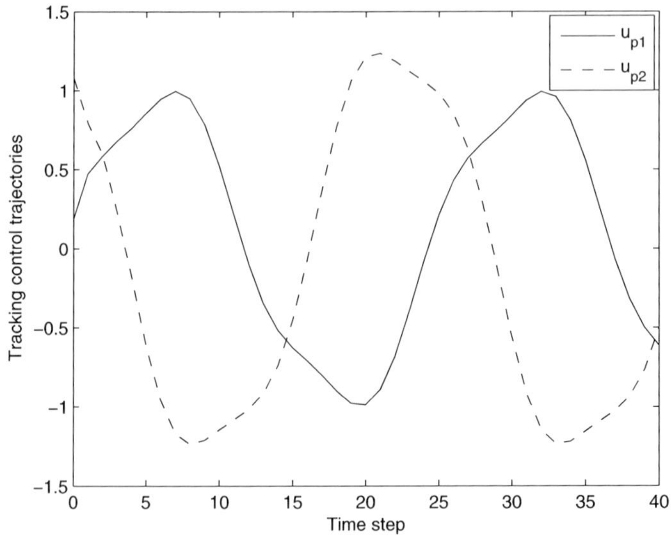

Fig. 3.20 The tracking control trajectories u_p

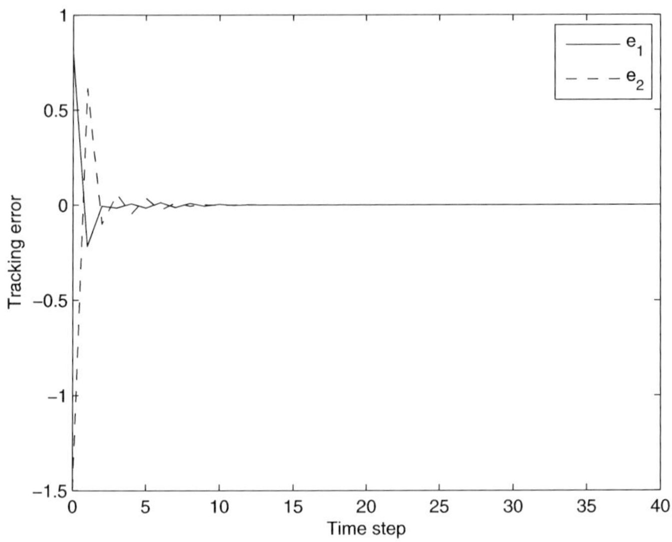

Fig. 3.21 The tracking error e

The desired trajectory is set to

$$x_d(k) = \begin{bmatrix} \sin(k + 0.5\pi) \\ 0.5 \cos k \end{bmatrix}.$$

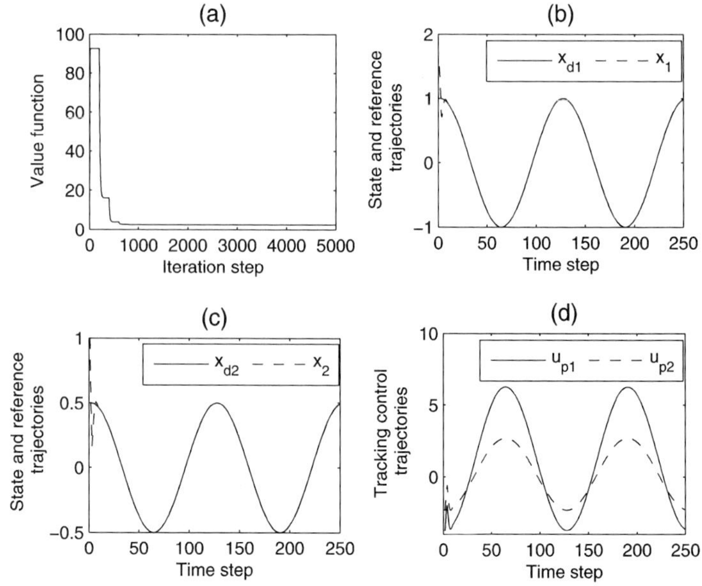

Fig. 3.22 Simulation results of Example 3.20

In the implementation of the iterative HDP algorithm, the initial weights and structures of three networks are set in the same way as Example 3.19.

Then, for the given initial state $x(0) = [1.5 \quad 1]^{\mathrm{T}}$, we train the model network for 10000 steps using 1000 data samples under the learning rate $\alpha_m = 0.05$. Besides, the critic network and the action network are trained for 5000 iterations so that the given error bound $\varepsilon = 10^{-6}$ is reached. The learning rate in the training process is also $\alpha_c = \alpha_a = 0.05$.

The convergence process of the value function of the iterative HDP algorithm is shown in Fig. 3.22(a), for $k = 0$. Then, we apply the tracking control law to the system for 250 time steps and obtain the state and reference trajectories shown in Figs. 3.22(b) and (c). Besides, the tracking control trajectories are given in Fig. 3.22(d). It is clear from the simulation results that the iterative HDP algorithm developed in this section is very effective in solving the finite-horizon tracking control problem.

3.5 Summary

In this chapter, we studied the optimal tracking control problem for discrete-time affine nonlinear systems via the ADP approach. In Sect. 3.2, the infinite-horizon optimal tracking control problem for discrete-time affine nonlinear systems was investigated. Then, we focused on a class of infinite-horizon discrete-time nonaffine nonlinear systems in Sect. 3.3. In Sect. 3.4, the finite-horizon optimal tracking control problem was studied. Several numerical simulations showed that the present methods are effective and can be used for a wide class of nonlinear systems.

References

1. Aganovic Z, Gajic Z (1994) The successive approximation procedure for finite-time optimal control of bilinear systems. IEEE Trans Autom Control 39:1932–1935
2. Cimen T, Banks SP (2004) Nonlinear optimal tracking control with application to super-tankers for autopilot design. Automatica 40:1845–1863
3. Cui LL, Zhang HG, Chen B, Zhang QL (2010) Asymptotic tracking control scheme for mechanical systems with external disturbances and friction. Neurocomputing 73:1293–1302
4. Devasiad S, Chen D, Paden B (1996) Nonlinear inversion-based output tracking. IEEE Trans Autom Control 41:930–942
5. Dierks T, Jagannathan S (2009) Optimal tracking control of affine nonlinear discrete-time systems with unknown internal dynamics. In: Proceedings of joint 48th IEEE conference on decision and control and 28th Chinese control conference, Shanghai, China, pp 6750–6755
6. Enns R, Si J (2003) Helicopter trimming and tracking control using direct neural dynamic programming. IEEE Trans Neural Netw 14:929–939
7. Gao D, Tang G, Zhang B (2006) Approximate optimal tracking control for a class of nonlinear systems with disturbances. In: Proceedings of 6th world congress on intelligent control and automation, Dalian, China, pp 521–525
8. Ha IJ, Gilbert EG (1987) Robust tracking in nonlinear systems. IEEE Trans Autom Control 32:763–771
9. Hsu CT, Chen SL (2003) Nonlinear control of a 3-pole active magnetic bearing system. Automatica 39:291–298
10. Lewis FL, Syrmos VL (1995) Optimal control. Wiley, New York
11. Mclain TW, Bailry CA, Beard RW (1999) Synthesis and experimental testing of a nonlinear optimal tracking controller. In: Proceedings of the American control conference, San Diego, CA, USA, pp 2847–2851
12. Park YM, Choi MS, Lee KY (1996) An optimal tracking neuro-controller for nonlinear dynamic systems. IEEE Trans Neural Netw 7:1009–1110
13. Robbins H, Monro S (1951) A stochastic approximation method. Ann Math Stat 22:400–407
14. Si J, Wang YT (2001) On-line learning control by association and reinforcement. IEEE Trans Neural Netw 12:264–276
15. Stefano ML (1981) Optimal design of PID regulators. Int J Control 33:601–616
16. Yang Q, Jagannathan S (2007) Online reinforcement learning neural network controller design for nanomanipulation. In: Proceedings of the IEEE symposium on approximate dynamic programming and reinforcement learning, Honolulu, HI, pp 225–232
17. Wang FY, Jin N, Liu D, Wei Q (2011) Adaptive dynamic programming for finite-horizon optimal control of discrete-time nonlinear systems with ε-error bound. IEEE Trans Neural Netw 22:24–36
18. Wang D, Liu D, Wei Q (2012) Finite-horizon neuro-optimal tracking control for a class of discrete-time nonlinear systems using adaptive dynamic programming approach. Neurocomputing 78:14–22
19. Zhang HG, Yang J, Su CY (2007) T-S fuzzy-model-based robust H-infinity design for networked control systems with uncertainties. IEEE Trans Ind Inform 3:289–301
20. Zhang HG, Wei QL, Luo YH (2008) A novel infinite-time optimal tracking control scheme for a class of discrete-time nonlinear systems via the greedy HDP iteration algorithm. IEEE Trans Syst Man Cybern, Part B, Cybern 38:937–942

21. Zhang HG, Wang ZS, Liu DR (2009) Global asymptotic stability and robust stability of a class of Cohen–Grossberg neural networks with mixed delays. IEEE Trans Circuits Syst I, Regul Pap 56:616–629
22. Zheng CD, Zhang HG (2007) Generalized multivariate rectangular matrix Pade-type approximants. IEEE Trans Circuits Syst I, Regul Pap 54:2099–2105
23. Zheng CD, Zhang HG, Tian H (2007) Generalized homogeneous multivariate matrix Pade-type approximants and Pade approximants. IEEE Trans Autom Control 52:2160–2165
24. Zheng CD, Zhang HG, Wang ZS (2009) Delay-dependent globally exponential stability criteria for static neural networks: an LMI approach. IEEE Trans Neural Netw 56:605–609

Chapter 4
Optimal State Feedback Control of Nonlinear Systems with Time Delays

4.1 Introduction

For time-delay system, much research is about decentralized control, synchronization control, and stability analysis [4–7, 10–13, 15, 16]. However, the optimal control problem is often encountered in industrial production. So in this chapter, the optimal state feedback control problems of nonlinear systems with time delays will be discussed. In general, the optimal control for the time-delay systems is an infinite-dimensional control problem [1], which is very difficult to solve. The analysis of systems with time delays is much more difficult than that of systems without delays, and there is no method strictly facing this problem for nonlinear time-delay systems.

In Sect. 4.2, the optimal state feedback control problem of nonlinear systems with time delays both in states and controls is solved. By introducing a delay matrix function, we can obtain the explicit expression of the optimal control function. Also, it is proved that the value function converges to the optimum using the present iterative ADP algorithm.

In Sect. 4.3, we study the optimal control problem of nonlinear time-delay systems with saturating actuators. First, the HJB equation for time-delay systems with saturating actuators is derived using a nonquadratic function. In order to solve this HJB equation, two optimization searching processes are developed. We also give the convergence proof for the new iterative HDP algorithm. Finally, two neural networks are used to implement the iterative HDP algorithm.

The above two sections are for the infinite-horizon optimal control problem. To the best of our knowledge, only [8] presented finite-horizon optimal control in detail. However, time-delay nonlinear systems have not been considered. So, we develop an effective algorithm to solve the finite-horizon optimal control problem for a class of time-delay systems in Sect. 4.4.

H. Zhang et al., *Adaptive Dynamic Programming for Control*,
Communications and Control Engineering, DOI 10.1007/978-1-4471-4757-2_4,

4.2 Infinite-Horizon Optimal State Feedback Control via Delay Matrix

4.2.1 Problem Formulation

We consider the following discrete-time affine nonlinear system with *time delays* in state and control variables:

$$\begin{aligned} x(k+1) = \; & f(x(k), x(k-\sigma)) + g_0(x(k), x(k-\sigma))u(k) \\ & + g_1(x(k), x(k-\sigma))u(k-\tau), \end{aligned} \tag{4.1}$$

with the initial condition given by $x(s) = \phi(s)$, $s = -\sigma, -\sigma+1, \ldots, 0$, where $x(k) \in \mathbb{R}^{\mathrm{n}}$ is the state vector, $f: \mathbb{R}^{\mathrm{n}} \times \mathbb{R}^{\mathrm{n}} \to \mathbb{R}^{\mathrm{n}}$, and $g_0, g_1: \mathbb{R}^{\mathrm{n}} \times \mathbb{R}^{\mathrm{n}} \to \mathbb{R}^{\mathrm{n}\times\mathrm{m}}$ are differentiable functions and we have the control $u(k) \in \mathbb{R}^{\mathrm{m}}$. The state and control delays σ and τ are both nonnegative integers. Assume that $f(x(k), x(k-\sigma)) + g_0(x(k), x(k-\sigma))u(k) + g_1(x(k), x(k-\sigma))u(k-\tau)$ is Lipschitz continuous on a set Ω in $\mathbb{R}^{\mathrm{n}}$ containing the origin, and that the system (4.1) is controllable in the sense that there exists a bounded control on Ω that asymptotically stabilizes the system. In this section, we mainly discuss how to design an optimal state feedback controller for this class of delayed discrete-time systems. Therefore, we desire to find the optimal control $u(x)$ satisfying $u(x(k)) = u(k)$ to minimize the generalized cost functional as follows:

$$\begin{aligned} J(x(0), u) = & \sum_{k=0}^{\infty} \big\{ x^{\mathrm{T}}(k) Q_0 x(k) + 2x^{\mathrm{T}}(k) Q_1 x(k-\sigma) \\ & + x^{\mathrm{T}}(k-\sigma) Q_2 x(k-\sigma) + u^{\mathrm{T}}(k) R_0 u(k) \\ & + 2u^{\mathrm{T}}(k) R_1 u(k-\tau) + u^{\mathrm{T}}(k-\tau) R_2 u(k-\tau) \big\}, \end{aligned} \tag{4.2}$$

where $\left[\begin{smallmatrix} Q_0 & Q_1 \\ Q_1^{\mathrm{T}} & Q_2 \end{smallmatrix}\right] \geq 0$ and $\left[\begin{smallmatrix} R_0 & R_1 \\ R_1^{\mathrm{T}} & R_2 \end{smallmatrix}\right] > 0$ and $l(x(k), x(k-\sigma), u(k), u(k-\tau)) = x^{\mathrm{T}}(k) Q_0 x(k) + 2x^{\mathrm{T}}(k) Q_1 x(k-\sigma) + x^{\mathrm{T}}(k-\sigma) Q_2 x(k-\sigma) + u^{\mathrm{T}}(k) R_0 u(k) + 2u^{\mathrm{T}}(k) R_1 u(k-\tau) + u^{\mathrm{T}}(k-\tau) R_2 u(k-\tau)$ is the utility function. Let $J^*(x)$ denote the optimal value function which satisfies

$$J^*(x) = \min_u J(x, u). \tag{4.3}$$

According to Bellman's optimality principle, we get the following HJB equation:

$$\begin{aligned} J^*(x(k)) = & \min_{u(k)} \big\{ x^{\mathrm{T}}(k) Q_0 x(k) + 2x^{\mathrm{T}}(k) Q_1 x(k-\sigma) \\ & + x^{\mathrm{T}}(k-\sigma) Q_2 x(k-\sigma) + u^{\mathrm{T}}(k) R_0 u(k) \\ & + 2u^{\mathrm{T}}(k) R_1 u(k-\tau) + u^{\mathrm{T}}(k-\tau) R_2 u(k-\tau) \\ & + J^*(x(k+1)) \big\}. \end{aligned} \tag{4.4}$$

For an optimal control problem, the state feedback control $u(x)$ must not only stabilize the system on Ω but also guarantee that (4.2) is finite, i.e., $u(x)$ must be admissible.

4.2.2 Optimal State Feedback Control Using Delay Matrix

Noting that the nonlinear delayed system (4.1) is infinite-dimensional, the control variable $u(k)$ couples with $u(k-\tau)$. It is nearly impossible to obtain the expression of the optimal control by solving the HJB equation (4.4). To overcome the difficulty, a new iterative algorithm is developed in this section.

Lemma 4.1 *For the delayed nonlinear system* (4.1) *with respect to the cost functional* (4.2), *if there exists a control* $u(k) \neq 0$ *at time step* k, *then there exists a bounded matrix function* $M(k)$ *so that*

$$u(k-\tau) = M(k)u(k). \tag{4.5}$$

Proof As $u(k)$ and $u(k-\tau)$ are bounded real vectors, so we can construct a function that satisfies

$$u(k-\tau) = h(u(k)). \tag{4.6}$$

Then, using the method of undetermined coefficients, let $M(u(k))$ satisfy

$$h(u(k)) = M(u(k))u(k). \tag{4.7}$$

Then we obtain $M(u(k))$ expressed as

$$M(u(k)) = h(u(k))u^{\mathrm{T}}(k)\left(u(k)u^{\mathrm{T}}(k)\right)^{-1}, \tag{4.8}$$

where $(u(k)u^{\mathrm{T}}(k))^{-1}$ means the generalized inverse matrix of $(u(k)u^{\mathrm{T}}(k))$. On the other hand, since $u(k)$ and $u(k-\tau)$ are both bounded real vectors, we have $h(u(k))$ and $(u(k)u^{\mathrm{T}}(k))^{-1}$ are bounded. So $M(k) = M(u(k))$ is the solution. □

According to Lemma 4.1, the HJB equation becomes

$$\begin{aligned} J^*(x(k)) &= x^{\mathrm{T}}(k)Q_0x(k) + 2x^{\mathrm{T}}(k)Q_1x(k-\sigma) + x^{\mathrm{T}}(k-\sigma)Q_2x(k-\sigma) \\ &\quad + u^{*\mathrm{T}}(k)R_0u^*(k) + 2u^{*\mathrm{T}}(k)R_1M^*(k)u^*(k) \\ &\quad + u^{*\mathrm{T}}(k)M^{*\mathrm{T}}(k)R_2M^*(k)u^*(k) + J^*(x(k+1)), \end{aligned} \tag{4.9}$$

where $u^*(k)$ is the optimal control and $u^*(k-\tau) = M^*(k)u^*(k)$.

According to Bellman's principle of optimality, we can obtain the optimal control by differentiating the HJB equation (4.9) with respect to the control u. Then we obtain the optimal control $u^*(k)$ formulated as

$$\begin{aligned}u^*(k) = & -\frac{1}{2}\left(R_0 + 2R_1 M^*(k) + M^{*\mathrm{T}}(k) R_2 M^*(k)\right)^{-1} \\ & \times \left(g_0\left(x(k), x(k-\sigma)\right) + g_1\left(x(k), x(k-\sigma)\right) M^*(k)\right)^{\mathrm{T}} \\ & \times \frac{\partial J^*(x(k+1))}{\partial x(k+1)}.\end{aligned} \tag{4.10}$$

In (4.10), the inverse of $\left(R_0 + 2R_1 M^*(k) + M^{*\mathrm{T}}(k) R_2 M^*(k)\right)$ should exist.

From (4.10), the explicit optimal control expression u^* is obtained by solving the HJB equation (4.9). We see that the optimal control u^* depends on M^* and $J^*(x)$, where $J^*(x)$ is a solution of the HJB equation (4.9). It is still an open question how to solve the HJB equation and there is currently no method for rigorously seeking for the value function of this delayed optimal control problem. Furthermore, the optimal delay matrix function M^* is also unknown, which makes the optimal control u^* more difficult to obtain. So an iterative index i is introduced into the ADP approach to obtain the optimal control iteratively.

First, for $i = 0, 1, \dots,$ let

$$u^{[i+1]}(k-\tau) = M^{[i]}(k) u^{[i+1]}(k), \tag{4.11}$$

where $M^{[0]}(k) = I$ and $u^{[0]}(k-\tau) = M^{[0]}(k) u^{[0]}(k)$. We start with an initial value function $V^{[0]}(x(k)) = 0$, and the control $u^{[0]}(k)$ can be computed as follows:

$$u^{[0]}(x(k)) = \arg\min_u \left\{\Gamma^{[0]} + V^{[0]}(x(k+1))\right\}, \tag{4.12}$$

where

$$\begin{aligned}\Gamma^{[0]} = & \, x^{\mathrm{T}}(k) Q_0 x(k) + 2x^{\mathrm{T}}(k) Q_1 x(k-\sigma) \\ & + x^{\mathrm{T}}(k-\sigma) Q_2 x(k-\sigma) + u^{[0]\mathrm{T}}(k) R_0 u^{[0]}(k) \\ & + 2u^{[0]\mathrm{T}}(k) R_1 M^{[0]}(k) u^{[0]}(k) \\ & + u^{[0]\mathrm{T}}(k) M^{[0]\mathrm{T}}(k) R_2 M^{[0]}(k) u^{[0]}(k).\end{aligned}$$

Then, the value function is updated as

$$V^{[1]}(x(k)) = \Gamma^{[0]} + V^{[0]}(x(k+1)). \tag{4.13}$$

Thus, for $i = 1, 2, \dots,$ the iterative ADP can be used to implement the iteration between

$$u^{[i]}(x(k)) = \arg\min_u \left\{\Gamma^{[i]} + V^{[i]}(x(k+1))\right\}$$

$$\begin{aligned}&= -\frac{1}{2}\left(R_0 + 2R_1 M^{[i-1]}(k) + M^{[i-1]\mathrm{T}}(k) R_2 M^{[i-1]}(k)\right)^{-1} \\ &\quad\times \left(g_0(x(k), x(k-\sigma)) + g_1(x(k), x(k-\sigma)) M^{[i-1]}(k)\right)^{\mathrm{T}} \\ &\quad\times \frac{\partial V^{[i]}(x(k+1))}{\partial x(k+1)},\end{aligned} \tag{4.14}$$

where

$$\begin{aligned}\Gamma^{[i]} &= x^{\mathrm{T}}(k) Q_0 x(k) + 2x^{\mathrm{T}}(k) Q_1 x(k-\sigma) \\ &\quad + x^{\mathrm{T}}(k-\sigma) Q_2 x(k-\sigma) + u^{[i]\mathrm{T}}(k) R_0 u^{[i]}(k) \\ &\quad + 2u^{[i]\mathrm{T}}(k) R_1 M^{[i-1]}(k) u^{[i]}(k) \\ &\quad + u^{[i]\mathrm{T}}(k) M^{[i-1]\mathrm{T}}(k) R_2 M^{[i-1]}(k) u^{[i]}(k),\end{aligned}$$

and

$$V^{[i+1]}(x(k)) = \Gamma^{[i]} + V^{[i]}(x(k+1)). \tag{4.15}$$

From (4.14) and (4.15), it can be seen that, during the iterative process, the control actions for different control steps obey different control laws. After iteration i, the obtained control law sequence is $(u^{[0]}, u^{[1]}, \dots, u^{[i]})$. For the infinite-horizon problem, both the optimal value function and the optimal control law are unique. Therefore, it is necessary to show that the iterative value function $V^{[i]}(x(k))$ will converge when iteration $i \to \infty$ under the iterative control $u^{[i]}(k)$ and this will be proved in the following.

Lemma 4.2 *Let $\tilde{u}^{[i]}(k), k = 0, 1 \dots$ be any sequence of control, and let $u^{[i]}(k)$ be expressed as* (4.14). *Define $V^{[i+1]}(x(k))$ as* (4.15) *and $\Lambda^{[i+1]}(x(k))$ as*

$$\begin{aligned}\Lambda^{[i+1]}(x(k)) &= x^{\mathrm{T}}(k) Q_0 x(k) + 2x^{\mathrm{T}}(k) Q_1 x(k-\sigma) \\ &\quad + x^{\mathrm{T}}(k-\sigma) Q_2 x(k-\sigma) + \tilde{u}^{[i]\mathrm{T}}(k) R_0 \tilde{u}^{[i]}(k) \\ &\quad + 2\tilde{u}^{[i]\mathrm{T}}(k) R_1 M^{[i-1]}(k) \tilde{u}^{[i]}(k) \\ &\quad + \tilde{u}^{[i]\mathrm{T}}(k) M^{[i-1]\mathrm{T}}(k) R_2 M^{[i-1]}(k) \tilde{u}^{[i]}(k) \\ &\quad + \Lambda^{[i]}(x(k+1)).\end{aligned} \tag{4.16}$$

If $V^{[0]}(x(k)) = \Lambda^{[0]}(x(k)) = 0$, then $V^{[i]}(x(k)) \le \Lambda^{[i]}(x(k)), \forall i$.

In order to prove the convergence of the value function, the following theorem is also necessary.

Theorem 4.3 (cf. [9]) *Let the value function $V^{[i]}(x(k))$ be defined by* (4.15). *If $x(k)$ of the system* (4.1) *is controllable, then there exists an upper bound Y such that $0 \le V^{[i]}(x(k)) \le Y, \forall i$.*

Proof As the system (4.1) is Lipschitz, $M^{[i]}(k)$ is a bounded matrix for $i=0,1,\ldots$. Define a *delay matrix function* $\bar{M}(k)$ which enforces

$$\chi^{\mathrm{T}}\left(R_0+2R_1\bar{M}(k)+\bar{M}^{\mathrm{T}}(k)R_2\bar{M}(k)\right)\chi \\ -\chi^{\mathrm{T}}\left(R_0+2R_1M^{[i]}(k)+M^{[i]\mathrm{T}}(k)R_2M^{[i]}(k)\right)\chi\geq 0 \tag{4.17}$$

for $\forall i$, where χ is any nonzero m-dimensional vector. Let $\bar{u}(k), k=0,1\ldots$ be any admissible control input. Define a new sequence $P^{[i]}(x(k))$ as follows:

$$\begin{aligned}P^{[i+1]}(x(k))=&\,x^{\mathrm{T}}(k)Q_0x(k)+2x^{\mathrm{T}}(k)Q_1x(k-\sigma)\\&+x^{\mathrm{T}}(k-\sigma)Q_2x(k-\sigma)+\bar{u}^{\mathrm{T}}(k)R_0\bar{u}(k)\\&+2\bar{u}^{\mathrm{T}}(k)R_1\bar{M}(k)\bar{u}(k)\\&+\bar{u}^{\mathrm{T}}(k)\bar{M}^{\mathrm{T}}(k)R_2\bar{M}(k)\bar{u}(k)+P^{[i]}(x(k+1)),\end{aligned} \tag{4.18}$$

where we let $P^{[0]}(x(k))=V^{[0]}(x(k))=0$ and $\bar{u}(k-\tau)=\bar{M}(k)\bar{u}(k)$. $V^{[i]}(x(k))$ is updated by (4.15). Thus, we obtain

$$\begin{aligned}P^{[i+1]}(x(k))-P^{[i]}(x(k))=&\;P^{[i]}(x(k+1))-P^{[i-1]}(x(k+1))\\&\;\vdots\\=&\;P^{[1]}(x(k+i))-P^{[0]}(x(k+i)).\end{aligned} \tag{4.19}$$

Because $P^{[0]}(x(k+i))=0$, we have

$$\begin{aligned}P^{[i+1]}(x(k))&=P^{[1]}(x(k+i))+P^{[i]}(x(k))\\&=\sum_{j=0}^{i}P^{[1]}(x(k+j)).\end{aligned} \tag{4.20}$$

According to (4.18), (4.20) can be rewritten as

$$P^{[i+1]}(x(k))=\sum_{j=0}^{i}\Xi(k+j)\leq\sum_{j=0}^{\infty}\Xi(k+j), \tag{4.21}$$

where

$$\begin{aligned}\Xi(k+j)=&\,x^{\mathrm{T}}(k+j)Q_0x(k+j)+2x^{\mathrm{T}}(k+j)Q_1x(k+j-\sigma)\\&+x^{\mathrm{T}}(k+j-\sigma)Q_2x(k+j-\sigma)+\bar{u}^{\mathrm{T}}(k+j)R_0\bar{u}(k+j)\\&+2\bar{u}^{\mathrm{T}}(k+j)R_1\bar{M}(k+j)\bar{u}(k+j)\\&+\bar{u}^{\mathrm{T}}(k+j)\bar{M}^{\mathrm{T}}(k+j)R_2\bar{M}(k+j)u(k+j).\end{aligned}$$

Noting that the control input $\bar{u}(k)$, $k = 0, 1, \ldots$, is an admissible control, we obtain

$$\forall i:\ P^{[i+1]}(x(k)) \leq \sum_{j=0}^{\infty} P^{[i+j]}(x(k+j)) \leq Y. \tag{4.22}$$

From Lemma 4.1, we have

$$\forall i:\ V^{[i+1]}(x(k)) \leq P^{[i+1]}(x(k)) \leq Y. \tag{4.23}$$

This completes the proof. □

With Lemma 4.1 and Theorem 4.3, the following main theorem can be derived.

Theorem 4.4 *Define the value function $V^{[i]}(x(k))$ as (4.15), with $V^{[0]}(x(k)) = 0$. If $x(k)$ for the system (4.1) is controllable, then $V^{[i]}(x(k))$ is a nondecreasing sequence satisfying $V^{[i]}(x(k)) \leq V^{[i+1]}(x(k))$, and $V^{[i]}(x(k))$ is convergent as $i \to \infty$.*

Proof For convenience in our analysis, define a new sequence $\Phi^{[i]}(x(k))$ as follows:

$$\begin{aligned} \Phi^{[i+1]}(x(k)) = {} & x^{\mathrm{T}}(k) Q_0 x(k) + 2x^{\mathrm{T}}(k) Q_1 x(k-\sigma) \\ & + x^{\mathrm{T}}(k-\sigma) Q_2 x(k-\sigma) + u^{[i+1]\mathrm{T}}(k) R_0 u^{[i+1]}(k) \\ & + 2u^{[i+1]\mathrm{T}}(k) R_1 M^{(i)}(k) u^{[i+1]}(k) \\ & + u^{[i+1]\mathrm{T}}(k) M^{[i]\mathrm{T}}(k) R_2 M^{[i]}(k) u^{[i+1]}(k) \\ & + \Phi^{[i]}(x(k+1)), \end{aligned} \tag{4.24}$$

where $u^{[i]}(k)$ is obtained by (4.14) and $\Phi^{[0]}(x(k)) = V^{[0]}(x(k)) = 0$.

In the following part, we prove $\Phi^{[i]}(x(k)) \leq V^{[i+1]}(x(k))$ by mathematical induction.

First, we prove that it holds for $i = 0$. Noting that

$$\begin{aligned} V^{[1]}(x(k)) - \Phi^{[0]}(x(k)) &= x^{\mathrm{T}}(k) Q_0 x(k) + 2x^{\mathrm{T}}(k) Q_1 x(k-\sigma) \\ & \quad + x^{\mathrm{T}}(k-\sigma) Q_2 x(k-\sigma) \\ & \geq 0, \end{aligned} \tag{4.25}$$

for $i = 0$, we get

$$V^{[1]}(x(k)) \geq \Phi^{[0]}(x(k)). \tag{4.26}$$

Second, we assume it holds for $i-1$, i.e., $V^{[i]}(x(k)) - \Phi^{[i-1]}(x(k)) \geq 0$, $\forall x(k)$. Then, for i, from (4.15) and (4.24), we obtain

$$\begin{aligned} V^{[i+1]}(x(k)) - \Phi^{[i]}(x(k)) &= V^{[i]}(x(k+1)) - \Phi^{[i-1]}(x(k+1)) \\ & \geq 0, \end{aligned} \tag{4.27}$$

i.e.,

$$\Phi^{[i]}(x(k)) \leq V^{[i+1]}(x(k)). \tag{4.28}$$

Thus, the mathematical induction proof is completed.

Moreover, from Lemma 4.1, we know that $V^{[i]}(x(k)) \leq \Phi^{[i]}(x(k))$ and therefore we obtain

$$V^{[i]}(x(k)) \leq \Phi^{[i]}(x(k)) \leq V^{[i+1]}(x(k)), \tag{4.29}$$

which proves that $V^{[i]}(x(k))$ is a nondecreasing sequence bounded by (4.23). Hence, we conclude that $V^{[i]}(x(k))$ is a nondecreasing convergent sequence as $i \to \infty$. □

Corollary 4.5 *If Theorem* 4.4 *holds, then the delay matrix function $M^{[i]}(k)$ is a convergent sequence as $i \to \infty$.*

According to Corollary 4.5, we define

$$M^{[\infty]}(k) = \lim_{i \to \infty} M^{[i]}(k). \tag{4.30}$$

Next, we will prove that the value function sequence $V^{[i]}(x(k))$ converges to $J^*(x(k))$ as $i \to \infty$. As $V^{[i]}(x(k))$ is a convergent sequence as $i \to \infty$, we define

$$V^{[\infty]}(x(k)) = \lim_{i \to \infty} V^{[i]}(x(k)). \tag{4.31}$$

Let $\bar{u}_l$ be the lth admissible control; then, similar to the proof of Theorem 4.3, we can construct the value function sequence $P_l^{[i]}(x)$ as follows:

$$\begin{aligned} P_l^{[i+1]}(x(k)) = {} & x^{\mathrm{T}}(k)Q_0x(k) + 2x^{\mathrm{T}}(k)Q_1x(k-\sigma) \\ & + x^{\mathrm{T}}(k-\sigma)Q_2x(k-\sigma) \\ & + \bar{u}_l^{\mathrm{T}}(k)R_0\bar{u}_l(k) + 2\bar{u}_l(k)R_1M^{[\infty]}(k)\bar{u}_l(k) \\ & + \bar{u}_l(k)M^{[\infty]\mathrm{T}}(k)R_2M^{[\infty]}(k)\bar{u}_l(k) \\ & + P_l^{[i]}(x(k+1)), \end{aligned} \tag{4.32}$$

with $P_l^{[0]}(\cdot) = 0$ and $\bar{u}_l(k) = M^{[\infty]}(k)\bar{u}_l(k-\tau)$. According to Theorem 4.3, we have

$$\begin{aligned} P_l^{[i+1]}(x(k)) = \sum_{j=0}^{i} \big\{ & x^{\mathrm{T}}(k+j)Q_0x(k+j) + 2x^{\mathrm{T}}(k+j)Q_1x(k+j-\sigma) \\ & + x^{\mathrm{T}}(k+j-\sigma)Q_2x(k+j-\sigma) + \bar{u}_l^{\mathrm{T}}(k+j)R_0\bar{u}_l(k+j) \\ & + 2\bar{u}_l^{\mathrm{T}}(k+j)R_1M^{[\infty]}(k+j)\bar{u}_l(k+j) \\ & + \bar{u}_l^{\mathrm{T}}(k+j)M^{[\infty]\mathrm{T}}(k+j)R_2M^{[\infty]}(k+j)\bar{u}_l(k+j \big\}. \end{aligned} \tag{4.33}$$

Let

$$P_l^{[\infty]}(x(k)) = \lim_{i\to\infty} P_l^{[i+1]}(x(k)); \tag{4.34}$$

then we have

$$P_l^{[i]}(x(k)) \le P_l^{[\infty]}(x(k)). \tag{4.35}$$

Theorem 4.6 *Define* $P_l^{[\infty]}(x(k))$ *as in* (4.34). *Define the value function* $V^{[i]}(x(k))$ *as in* (4.15) *with* $V^{[0]}(\cdot) = 0$. *For any state vector* $x(k)$, *define* $J^*(x(k)) = \min_l\{P_l^{[\infty]}(x(k))\}$. *Then we conclude that* $J^*(x(k))$ *is the limit of the value function* $V^{[i]}(x(k))$ *as* $i \to \infty$.

Proof For any l, there exists an upper bound Y_l such that

$$P_l^{[i+1]}(x(k)) \le P_l^{[\infty]}(x(k)) \le Y_l. \tag{4.36}$$

According to (4.23), for $\forall l$, we have

$$V^{[\infty]}(x(k)) \le P_l^{[\infty]}(x(k)) \le Y_l. \tag{4.37}$$

Since $J^*(x(k)) = \min_l\{P_l^{[\infty]}(x(k))\}$, for any $\epsilon > 0$, there exists an admissible control $\bar{u}_K$, where K is a nonnegative number such that the associated value function satisfies $P_K^{[\infty]}(x(k)) \le J^*(x(k)) + \epsilon$. According to (4.23), we have $V^{[\infty]}(x(k)) \le P_l^{[\infty]}(x(k))$ for any l. Thus, we obtain $V^{[\infty]}(x(k)) \le P_K^{[\infty]}(x(k)) \le J^*(x(k)) + \epsilon$. Noting that ϵ is chosen arbitrarily, we have

$$V^{[\infty]}(x(k)) \le J^*(x(k)). \tag{4.38}$$

On the other hand, since $V^{[i]}(x(k))$ is bounded for $\forall i$, according to the definition of admissible control, the control sequence associated with the value function $V^{[\infty]}(x(k))$ must be an admissible control, i.e., there exists an admissible control $\bar{u}_N^{[i]}$ such that $V^{[\infty]}(x(k)) = P_N^{[\infty]}(x(k))$. Combining with the definition $J^*(x(k)) = \min_l\{P_l^{[\infty]}(x(k))\}$, we obtain

$$V^{[\infty]}(x(k)) \ge J^*(x(k)). \tag{4.39}$$

Therefore, combining (4.38) and (4.39), we conclude that

$$V^{[\infty]}(x(k)) = \lim_{i\to\infty} V^{[i]}(x(k)) = J^*(x(k)), \tag{4.40}$$

i.e., $J^*(x(k))$ is the limit of the value function $V^{[i]}(x(k))$, as $i \to \infty$. □

Based on Theorem 4.6, we will prove that the value function $J^*(x(k))$ satisfies the principle of optimality, which shows that $V^{[i]}(x(k))$ can reach the optimum as $i \to \infty$.

Theorem 4.7 *Consider any state vector $x(k)$. The "optimal" value function $J^*(x(k))$ satisfies*

$$\begin{aligned} J^*(x(k)) = \min_{u(k)} \{ & x^{\mathrm{T}}(k)Q_0x(k) + 2x^{\mathrm{T}}(k)Q_1x(k-\sigma) + x^{\mathrm{T}}(k-\sigma)Q_2x(k-\sigma) \\ & + u^{\mathrm{T}}(k)R_0u(k) + 2u^{\mathrm{T}}(k)R_1M(k)u(k) \\ & + u^{\mathrm{T}}(k)M(k)R_2M(k)u(k) + J^*(x(k+1))\}, \end{aligned} \tag{4.41}$$

where $u(k-\tau) = M(k)u(k)$.

Proof For any $u(k)$ and i, based on Bellman's optimality principle, we have

$$V^{[i]}(x(k)) \leq \Upsilon^{[i-1]} + V^{[i-1]}(x(k+1)), \tag{4.42}$$

where

$$\begin{aligned} \Upsilon^{[i-1]} = {} & x^{\mathrm{T}}(k)Q_0x(k) + 2x^{\mathrm{T}}(k)Q_1x(k-\sigma) \\ & + x^{\mathrm{T}}(k-\sigma)Q_2x(k-\sigma) \\ & + u^{\mathrm{T}}(k)R_0u(k) + 2u^{\mathrm{T}}(k)R_1M^{[i-1]}(k)u(k) \\ & + u^{\mathrm{T}}(k)M^{[i-1]\mathrm{T}}(k)R_2M^{[i-1]}(k)u(k). \end{aligned}$$

As $V^{[i]}(x(k)) \leq V^{[i+1]}(x(k)) \leq V^{[\infty]}(x(k))$ and $V^{[\infty]}(x(k)) = J^*(x(k))$, we obtain

$$V^{[i]}(x(k)) \leq \Upsilon^{[i-1]} + J^*(x(k+1)). \tag{4.43}$$

Let $i \to \infty$; we have

$$J^*(x(k)) \leq \Upsilon^{[\infty]} + J^*(x(k+1)). \tag{4.44}$$

Since $u(k)$ in the above equation is chosen arbitrarily, the following equation holds:

$$J^*(x(k)) \leq \min_{u(k)} \left\{ \Upsilon^{[\infty]} + J^*(x(k+1)) \right\}. \tag{4.45}$$

On the other hand, for any i, the value function satisfies

$$V^{[i]}(x(k)) = \Omega^{[i-1]} + V^{[i-1]}(x(k+1)), \tag{4.46}$$

where

$$\begin{aligned} \Omega^{[i-1]} = {} & x^{\mathrm{T}}(k)Q_0x(k) + 2x^{\mathrm{T}}(k)Q_1x(k-\sigma) \\ & + x^{\mathrm{T}}(k-\sigma)Q_2x(k-\sigma) + u^{[i-1]\mathrm{T}}(k)R_0u^{[i-1]}(k) \\ & + 2u^{[i]\mathrm{T}}(k)R_1M^{[i-2]}(k)u^{[i-1]\mathrm{T}}(k) \\ & + u^{[i-1]\mathrm{T}}(k)M^{[i-2]\mathrm{T}}(k)R_2M^{[i-2]}(k)u^{[i-1]\mathrm{T}}(k). \end{aligned}$$

Combining with $V^{[i]}(x(k)) \le J^*(x(k)), \forall i$, we have

$$J^*(x(k)) \ge \Omega^{[i-1]} + V^{[i-1]}(x(k+1)). \tag{4.47}$$

Let $i \to \infty$; then

$$\begin{aligned} J^*(x(k)) &\ge \lim_{i\to\infty} \left\{\Omega^{[i-1]} + V^{[i-1]}(x(k+1))\right\} \\ &\ge \min_{u(k)} \left\{\Omega^{[\infty]} + J^*(x(k+1))\right\}. \end{aligned} \tag{4.48}$$

Combining (4.45) with (4.48), we have

$$\begin{aligned} J^*(x(k)) = &\min_{u(k)}\{\Omega^{[\infty]} + J^*(x(k+1))\} \\ = &\, x^{\mathrm{T}}(k)Q_0x(k) + 2x^{\mathrm{T}}(k)Q_1x(k-\sigma) \\ &+ x^{\mathrm{T}}(k-\sigma)Q_2x(k-\sigma) + u^{*\mathrm{T}}(k)R_0u^*(k) \\ &+ 2u^{*\mathrm{T}}(k)R_1M^{[\infty]}(k)u^*(k) \\ &+ u^{*\mathrm{T}}(k)M^{[\infty]\mathrm{T}}(k)R_2M^{[\infty]}(k)u^*(k) \\ &+ J^*(x(k+1)). \end{aligned} \tag{4.49}$$

Thus, we have $u^{[i]}(k) \to u^*(k)$ as $i \to \infty$. So does $u^{[i]}(k-\tau)$. On the other hand, we also have $M^{[i]}(k) \to M^{[\infty]}(k)$ and $u^{[i]}(k-\tau) = M^{[i-1]}(k)u^{[i]}(k)$. Let $i \to \infty$, and we get

$$u^*(k-\tau) = M^{[\infty]}(k)u^*(k). \tag{4.50}$$

Therefore, we have $M^{[\infty]}(k) = M^*(k)$, and (4.49) can be written as

$$\begin{aligned} J^*(x(k)) = &\, x^{\mathrm{T}}(k)Q_0x(k) + 2x^{\mathrm{T}}(k)Q_1x(k-\sigma) \\ &+ x^{\mathrm{T}}(k-\sigma)Q_2x(k-\sigma) + u^{*\mathrm{T}}(k)R_0u^*(k) \\ &+ 2u^{*\mathrm{T}}(k)R_1M^*(k)u^*(k) \\ &+ u^{*\mathrm{T}}(k)M^{*\mathrm{T}}(k)R_2M^*(k)u^*(k) \\ &+ J^*(x(k+1)), \end{aligned} \tag{4.51}$$

where $u^*(k-\tau) = M^*(k)u^*(k)$. □

Therefore, we conclude that the value function $V^{[i]}(x(k))$ converges to the optimum $J^*(x(k))$ as $i \to \infty$.

In the case of linear systems, the value function is quadratic and the control law is linear. In the nonlinear case, this is not necessarily true and therefore we use neural networks to approximate $u^{[i]}(k)$ and $V^{[i]}(x(k))$.

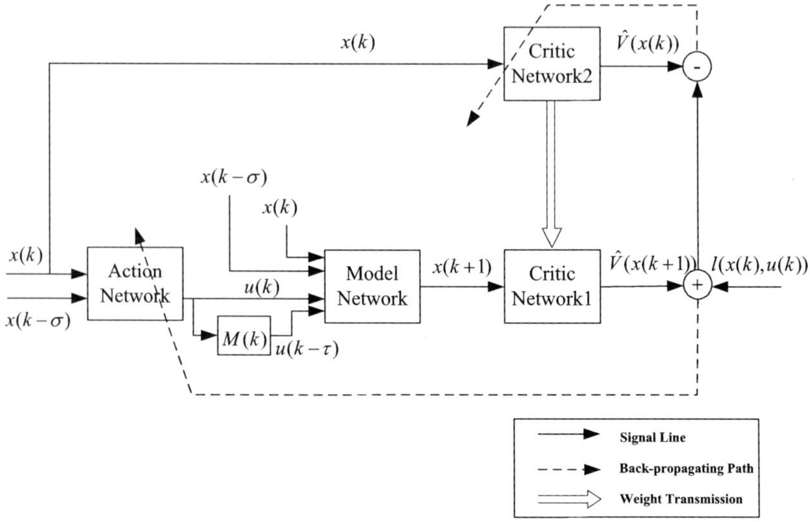

Fig. 4.1 The structure diagram of the algorithm

The number of hidden layer neurons is denoted by N_1, the weight matrix between the input layer and hidden layer is denoted by V, and the weight matrix between the hidden layer and output layer is denoted by W. Then, the output of the three-layer NN is represented by

$$\hat{F}(X, V, W) = W^{\mathrm{T}}\phi(V^{\mathrm{T}}X), \tag{4.52}$$

where $\phi(V^{\mathrm{T}}X) \in \mathbb{R}^{\mathrm{N}_1}$ is the activation function.

The NN estimation error can be expressed by

$$F(X) = F(X, V, W^*) + \varepsilon(X), \tag{4.53}$$

where V^*, W^* are the ideal weight parameters; $\varepsilon(X)$ is the reconstruction error.

In our case, there are four neural networks, which are the critic network, model network, action network, and delay matrix function network (M network), respectively. All the neural networks are chosen as three-layer feedforward network. The whole structure diagram is shown in Fig. 4.1. The utility term in the figure denotes $x^{\mathrm{T}}(k)Q_0x(k)+2x^{\mathrm{T}}(k)Q_1x(k-\sigma)+x^{\mathrm{T}}(k-\sigma)Q_2x(k-\sigma)+u^{\mathrm{T}}(k)R_0u(k)+2u^{\mathrm{T}}(k)R_1u(k-\tau)+u^{\mathrm{T}}(k-\tau)R_2u(k-\tau)$.

4.2.2.1 Model Network

The model network is to approximate the system dynamics and it should be trained before the implementation of the iterative ADP algorithm. The update rule of the model network adopts the gradient descent method. The training process is simple and general. The details can be found in [14] and they are omitted here.

After the model network is trained, its weights are kept unchanged.

4.2.2.2 The M Network

The M network is to approximate the delay matrix function $M(k)$. The output of the M network is denoted

$$\hat{u}(k-\tau) = W_M^{\mathrm{T}} \phi(V_M^{\mathrm{T}} u(k)). \tag{4.54}$$

We define the error function of the model network as

$$e_M(k) = \hat{u}(k-\tau) - u(k-\tau). \tag{4.55}$$

Define the error measure as

$$E_M(k) = \frac{1}{2} e_M^{\mathrm{T}}(k) e_M(k). \tag{4.56}$$

Then, the gradient-based weight updating rule for the M network can be described by

$$w_M(k+1) = w_M(k) + \Delta w_M(k), \tag{4.57}$$

$$\Delta w_M(k) = \alpha_M \left[-\frac{\partial E_M(k)}{\partial w_M(k)} \right], \tag{4.58}$$

where α_M is the learning rate of the M network.

4.2.2.3 Critic Network

The critic network is used to approximate the value function $V^{[i]}(x(k))$. The output of the critic network is denoted

$$\hat{V}^{[i]}(x(k)) = W_{ci}^{\mathrm{T}} \phi(V_{ci}^{\mathrm{T}} z(k)). \tag{4.59}$$

The target function can be written as

$$V^{[i+1]}(x(k)) = \Gamma^{[i]} + \hat{V}^{[i]}(x(k+1)). \tag{4.60}$$

Then, we define the error function for the critic network as

$$e_{ci}(k) = \hat{V}^{[i+1]}(x(k)) - V^{[i+1]}(x(k)). \tag{4.61}$$

The objective function to be minimized in the critic network is

$$E_{ci}(k) = \frac{1}{2} e_{ci}^{\mathrm{T}}(k) e_{ci}(k). \tag{4.62}$$

So the gradient-based weight updating rule for the critic network is given by

$$w_{c(i+1)}(k) = w_{ci}(k) + \Delta w_{ci}(k), \tag{4.63}$$

$$\Delta w_{ci}(k) = \alpha_c \left[-\frac{\partial E_{ci}(k)}{\partial w_{ci}(k)} \right], \tag{4.64}$$

$$\frac{\partial E_{ci}(k)}{\partial w_{ci}(k)} = \frac{\partial E_{ci}(k)}{\partial \hat{V}^{[i]}(x(k))} \frac{\partial \hat{V}^{[i]}(x(k))}{\partial w_{ci}(k)}, \tag{4.65}$$

where $\alpha_c > 0$ is the learning rate of critic network and $w_c(k)$ is the weight vector in the critic network.

4.2.2.4 Action Network

In the action network, the state $x(k)$ is used as input to create the optimal control. The output can be formulated as

$$\hat{u}^{[i]}(k) = W_{ai}^{\mathrm{T}} \phi(V_{ai}^{\mathrm{T}} x(k)). \tag{4.66}$$

And the output target of the action network is given by (4.14). So we define the output error of the action network as

$$e_{ai}(k) = \hat{u}^{[i]}(k) - u^{[i]}(k), \tag{4.67}$$

where $u^{[i]}(k)$ is the target function which can be described by

$$\begin{aligned} u^{[i]}(k) = & -\frac{1}{2}\Big(R_0 + 2R_1 M^{[i-1]}(k) + M^{[i-1]\mathrm{T}}(k) R_2 M^{[i-1]}(k)\Big)^{-1} \\ & \times \Big(g_0(x(k), x(k-\sigma)) + g_1(x(k), x(k-\sigma)) M^{[i-1]}(k)\Big)^{\mathrm{T}} \\ & \times \frac{\partial \hat{V}^{[i]}(x(k+1))}{\partial x(k+1)}. \end{aligned}$$

As $u^{[i]}(k-\tau) = M^{[i-1]}(k) u^{[i]}(k)$, we have $\dfrac{\partial u^{[i]}(k-\tau)}{\partial u^{[i]}(k)} = M^{[i-1]}(k)$. Then, according to (4.54), $M^{[i-1]}(k)$ can be expressed as

$$M_{ij}^{[i-1]}(k) = V_{Mi}^{\mathrm{T}} \left[1 - \left(\phi(V_M^{\mathrm{T}} u(k)) \right)_i^2 \right] W_{Mj} \tag{4.68}$$

for $i, j = 1, 2, \ldots, m$. $M_{ij}^{[i-1]}(k)$ denotes the element of row i, column j of matrix $M^{[i-1]}(k)$. V_{Mi} and W_{Mj} mean the column i and column j of the weight matrices V_M and W_M, respectively. $(\phi(V_M^{\mathrm{T}} u(k)))_i$ is the ith element of the vector $\phi(V_M^{\mathrm{T}} u(k))$.

The weights in the action network are updated to minimize the following error measure:

$$E_{ai}(k) = \frac{1}{2} e_{ai}^{\mathrm{T}}(k) e_{ai}(k). \tag{4.69}$$

The weight updating algorithm is similar to the one for the critic network. By the gradient descent rule, we obtain

$$w_{a(i+1)}(k) = w_{ai}(k) + \Delta w_{ai}(k), \tag{4.70}$$

$$\Delta w_{ai}(k) = \alpha_a \left[-\frac{\partial E_{ai}(k)}{\partial w_{ai}(k)} \right], \tag{4.71}$$

$$\frac{\partial E_{ai}(k)}{\partial w_{ai}(k)} = \frac{\partial E_{ai}(k)}{\partial e_{ai}(k)} \frac{\partial e_{ai}(k)}{\partial u^{[i]}(k)} \frac{\partial u^{[i]}(k)}{\partial w_{ai}(k)}, \tag{4.72}$$

where $\alpha_a > 0$ is the learning rate of action network.

4.2.3 Simulations

Example 4.8 Consider the following affine nonlinear system:

$$x(k+1) = f(x(k), x(k-\sigma)) + g(x(k), x(k-\sigma))u(k), \tag{4.73}$$

where $x(k) = [x_1(k)\ x_2(k)]^{\mathrm{T}}$, $u(k) = [u_1(k)\ u_2(k)]^{\mathrm{T}}$, and $f(x(k), x(k-\sigma)) = \left[\begin{smallmatrix} x_1(k)\exp(x_2^3(k))x_2(k-2) \\ x_2^3(k)x_1(k-2) \end{smallmatrix}\right]$, $g(x(k), x(k-\sigma)) = \left[\begin{smallmatrix} -0.2 & 0 \\ 0 & -0.2 \end{smallmatrix}\right]$. The time delay in the state is $\sigma = 2$ and the initial condition is $x(k) = [1 \quad -1]^{\mathrm{T}}$ for $-2 \le k \le 0$. The cost functional is defined as (4.2), where $Q_0 = Q_2 = R_0 = I$ and $Q_1 = R_1 = R_2 = 0$.

We implement the algorithm at the time instant $k = 5$. We choose three-layer neural networks as the critic network, the action network, and the model network with the structure 4–10–2, 2–10–1 and 6–10–2, respectively. The initial weights of action network, critic network, and model network are all set to be random in $[-0.5, 0.5]$. It should be mentioned that the model network should be trained first. For the given initial state, we train the model network for 3000 steps under the learning rate $\alpha_m = 0.05$. After the training of the model network is completed, the weights are kept unchanged. Then, the critic network and the action network are trained for 3000 steps, so that the given accuracy $\varepsilon = 10^{-6}$ is reached. In the training process, the learning rate $\alpha_a = \alpha_c = 0.05$. The convergence curve of the value function is shown in Fig. 4.2. Then, we apply the optimal control to the system for $T_f = 30$ time steps and obtain the following results. The state trajectories are given as Fig. 4.3 and the corresponding control curves are given as Fig. 4.4.

Example 4.9 For the second example, the control time delay is added to the system of Example 4.8 and the system becomes

$$\begin{aligned} x(k+1) = {} & f(x(k), x(k-\sigma)) + g_0(x(k), x(k-\sigma))u(k) \\ & + g_1(x(k), x(k-\sigma))u(k-\tau), \end{aligned} \tag{4.74}$$

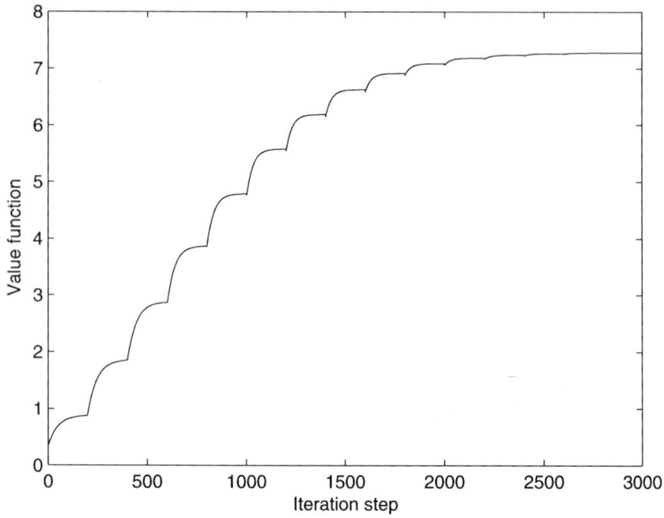

Fig. 4.2 The convergence of the value function

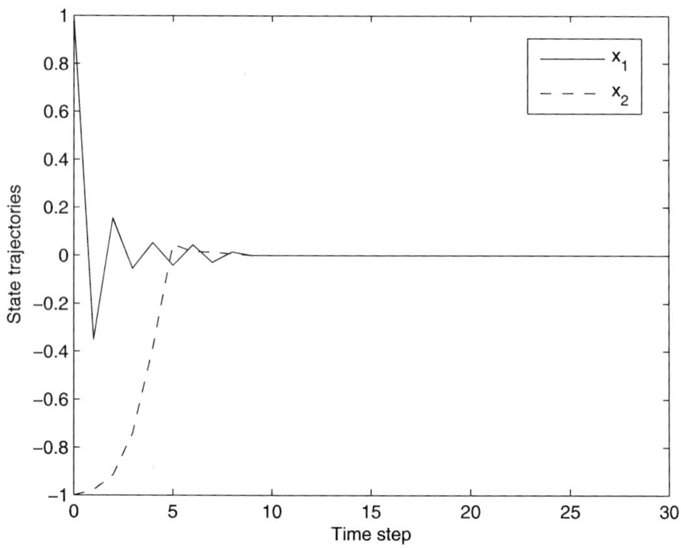

Fig. 4.3 The state variables trajectories

where $\sigma = 2$, $\tau = 1$, $x(k) = [x_1(k)\ x_2(k)]^{\mathrm{T}}$, $u(k) = [u_1(k)\ u_2(k)]^{\mathrm{T}}$, and $f(x(k), x(k-\sigma))$ is the same as Example 4.8, $g_0(x(k), x(k-\sigma)) = g_1(x(k), x(k-\sigma)) = \begin{bmatrix} -0.2 & 0 \\ 0 & -0.2 \end{bmatrix}$. The initial condition is $x(k) = [-1\ -1]^{\mathrm{T}}$ and $u(k) = 0$ for $-2 \le k \le 0$. The cost functional is defined as (4.2), where $Q_0 = Q_2 = R_0 = R_2 = I$ and $Q_1 = R_1 = 0$.

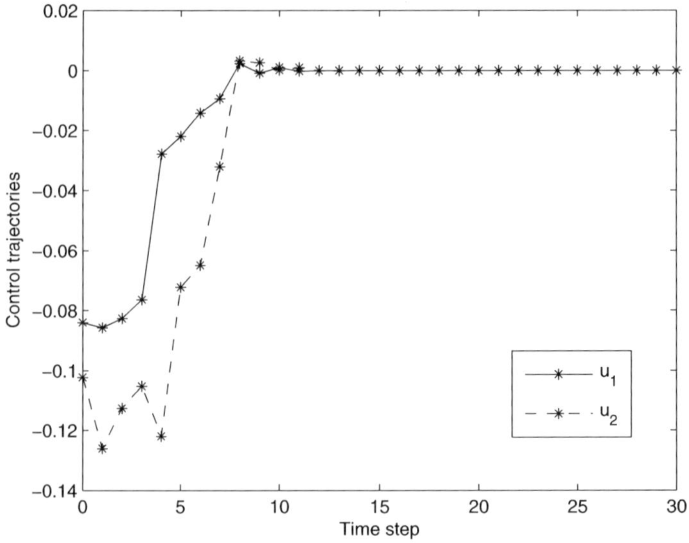

Fig. 4.4 The optimal control trajectories

We also implement the algorithm at the time instant $k = 5$. We choose three-layer neural networks as the critic network, the action network, the model network, and the M network with the structure 4–10–2, 2–10–1, 8–10–2, and 2–8–2, respectively. All the other parameters are set the same as in Example 4.8. The initial weights of action network, critic network, model network, and the M network are all set to be random in $[-0.5, 0.5]$. For the given initial state, we train the model network for 4000 steps. After the training of the model network has completed, the weights keep unchanged. Then, the critic network, the action network and the M network are trained for 3000 steps to reach the given accuracy of $\varepsilon = 10^{-6}$. The convergence curve of the value function is shown in Fig. 4.5. Then, we apply the optimal control to the system for $T_f = 30$ time steps and obtain the following results. The state trajectories are given as Fig. 4.6 and the corresponding control curves are given as Fig. 4.7.

4.3 Infinite-Horizon Optimal State Feedback Control via HDP

4.3.1 Problem Formulation

Consider a class of affine nonlinear discrete time-delay systems with saturating actuators as follows:

$$\begin{cases} x(k+1) = f(x(k-\sigma_0), \dots, x(k-\sigma_m)) + g(x(k-\sigma_0), \dots, x(k-\sigma_m))u(k), \\ x(k) = \varrho(k),\ k = -\sigma_0, -\sigma_1, \dots, -\sigma_m, \end{cases} \tag{4.75}$$

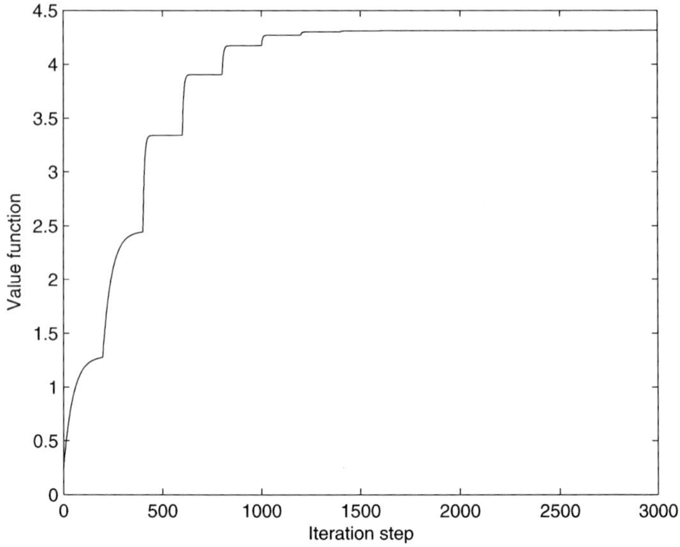

Fig. 4.5 The convergence of the value function

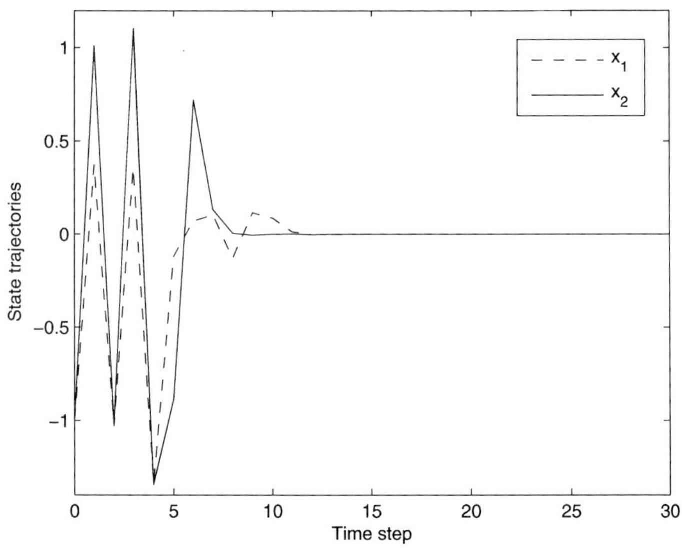

Fig. 4.6 The state variables trajectories

where $x(k-\sigma_0)$, $x(k-\sigma_1), \ldots, x(k-\sigma_m) \in \mathbb{R}^n$, $u_k \in \mathbb{R}^m$, $f(x(k-\sigma_0), x(k-\sigma_1), \ldots, x(k-\sigma_m)) \in \mathbb{R}^n$, $g(x(k-\sigma_0), x(k-\sigma_1), \ldots, x(k-\sigma_m)) \in \mathbb{R}^{n\times m}$, $\varrho(k)$ describes the initial condition. σ_i, $i = 1, \ldots, m$ is a positive integer. Set $0 =$

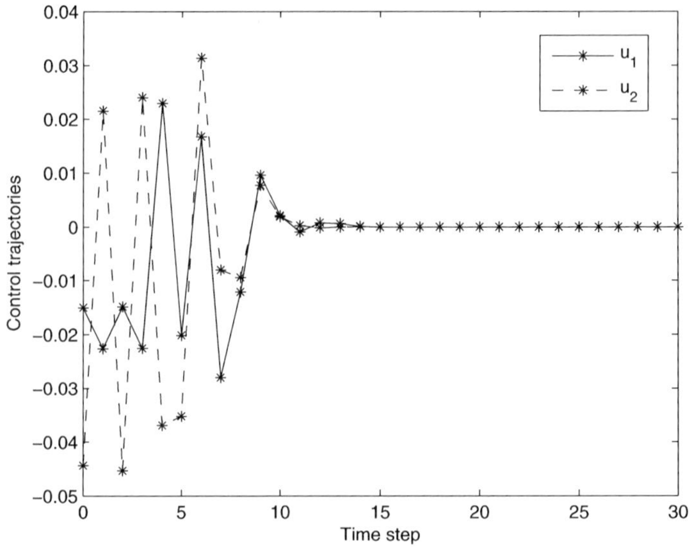

Fig. 4.7 The optimal control trajectories

$\sigma_0 < \sigma_1 < \cdots < \sigma_m$. Here assume that the system is controllable on $\Omega \subset \mathbb{R}^n$. Assume that f, g are all Lipschitz continuous functions. We denote $\Omega_u = \{u(k)|u(k) = [u_1(k)\ u_2(k)\ \ldots\ u_m(k)]^{\mathrm{T}} \in \mathbb{R}^m, |u_i(k)| \le \bar{u}_i, i = 1, \ldots, m\}$, where $\bar{u}_i$ is the saturating bound for the ith actuator. Let $\bar{U} = \mathrm{diag}\{\bar{u}_1\ \bar{u}_2\ \ldots\ \bar{u}_m\}$.

In this section, we desire to find an optimal control law for the system (4.75), which minimizes a generalized nonquadratic cost functional as follows:

$$J(x(k), u(k)) = \sum_{i=k}^{\infty} \{Q(x(i)) + W(u(i))\}, \tag{4.76}$$

where $Q(x(i)), W(u(i))$ are positive definite, and $l(x(i), u(i)) = Q(x(i)) + W(u(i))$ is the utility function.

It is noted that, for optimal control problems, the state feedback control law $u(k)$ must not only stabilize the system (4.75) on Ω, but also guarantee that (4.76) is finite. Such controls are defined to be admissible.

In this section, we mainly discuss the optimal control for a discrete time-delay system with saturating actuators. We let

$$Q(x(k)) = X^{\mathrm{T}}(k - \sigma) Q X(k - \sigma), \tag{4.77}$$

and

$$W(u(k)) = 2 \int_0^{u(k)} \varphi^{-\mathrm{T}}(\bar{U}^{-1} s) \bar{U} R \mathrm{d}s, \tag{4.78}$$

where $X(k - \sigma) = [x^{\mathrm{T}}(k - \sigma_0), x^{\mathrm{T}}(k - \sigma_1), \ldots, x^{\mathrm{T}}(k - \sigma_m)]^{\mathrm{T}}$, Q and R are positive definite. We assume that R is diagonal for simplicity in our analysis,

$s \in \mathbb{R}^m$, $\varphi \in \mathbb{R}^m$, $\varphi^{-1}(u(k)) = [\psi^{-1}(u_1(k))\ \psi^{-1}(u_2(k))\ \cdots\ \psi^{-1}(u_m(k))]^{\mathrm{T}}$, $\psi(\cdot)$ is a bounded single mapping function satisfying $|\psi(\cdot)| \le 1$ and belonging to C^p $(p \ge 1)$ and L_2. Moreover, it is a monotonically increasing odd function with its first derivative bounded by a constant M. We know that the hyperbolic tangent function $\psi(\cdot) = \tanh(\cdot)$ is one example of such functions. Noting that, $W(u(k))$ is ensured to be positive definite by the definition above, because $\psi^{-1}(\cdot)$ is a monotonic odd function, and R is positive definite.

Let $J^*(x(k)) = \min_{u(k)} J(x(k), u(k))$ denote the optimal value function, and let $u^*(k)$ denote the corresponding optimal control law. According to Bellman's principle of optimality, the optimal value function $J^*(x(k))$ should satisfy the following HJB equation:

$$\begin{aligned} J^*(x(k)) &= \min_{u(k)} \sum_{i=k}^{\infty} \{Q(x(i)) + W(u(i))\} \\ &= \min_{u(k)} \big\{Q(x(k)) + W(u(k)) + J^*(x(k+1))\big\}, \end{aligned} \tag{4.79}$$

and the optimal controller $u^*(k)$ should satisfy

$$\begin{aligned} u^*(k) &= \arg\min_{u(k)} \sum_{i=k}^{\infty} \{Q(x(i)) + W(u(i))\} \\ &= \arg\min_{u_k} \big\{Q(x(k)) + W(u(k)) + J^*(x(k+1))\big\}. \end{aligned} \tag{4.80}$$

The optimal control problem for the nonlinear discrete time-delay system with saturating actuators can be solved if the optimal value function $J^*(x(k))$ can be obtained from (4.79). However, there is currently no quite effective method for solving this value function for the nonlinear discrete time-delay system with saturating actuators. Therefore, in the following we will discuss how to utilize the iterative HDP algorithm to seek the approximate optimal control solution.

4.3.2 Optimal Control Based on Iterative HDP

First, for any given initial state $\varrho(k)$ and initial control law $\beta(k)$, we start with the initial iterative value function $V^{[0]}(\cdot) = 0$. Then, we find the control vector $u^{[0]}(k)$ as follows:

$$\begin{aligned} u^{[0]}(k) = \arg\min_{u(k)} \Bigg\{ & X^{\mathrm{T}}(k-\sigma) Q X(k-\sigma) + 2\int_0^{u(k)} \varphi^{-\mathrm{T}}(\bar{U}^{-1}s)\bar{U} R \mathrm{d}s \\ & + V^{[0]}(x(k+1)) \Bigg\}, \end{aligned} \tag{4.81}$$

and the value function is updated as

$$V^{[1]}(x(k)) = \min_{u(k)} \left\{ X^{\mathrm{T}}(k-\sigma)QX(k-\sigma) + 2\int_0^{u(k)} \varphi^{-\mathrm{T}}(\bar{U}^{-1}s)\bar{U}R\mathrm{d}s \right.$$
$$\left. + V^{[0]}(x(k+1)) \right\}, \tag{4.82}$$

where $x(1), \ldots, x(k)$ are obtained under the action $\beta(k)$, and

$$x(k+1) = f(x(k-\sigma_0), \ldots, x(k-\sigma_m))$$
$$+ g(x(k-\sigma_0), \ldots, x(k-\sigma_m))u^{[0]}(k). \tag{4.83}$$

Moreover, for $i = 1, 2, \ldots$, the iterative HDP algorithm iterates between

$$u^{[i]}(k) = \arg\min_{u(k)} \left\{ X^{\mathrm{T}}(k-\sigma)QX(k-\sigma) + 2\int_0^{u(k)} \varphi^{-\mathrm{T}}(\bar{U}^{-1}s)\bar{U}R\mathrm{d}s \right.$$
$$\left. + V^{[i]}(x(k+1)) \right\} \tag{4.84}$$

and

$$V^{[i+1]}(x(k)) = \min_{u(k)} \left\{ X^{\mathrm{T}}(k-\sigma)QX(k-\sigma) + 2\int_0^{u(k)} \varphi^{-\mathrm{T}}(\bar{U}^{-1}s)\bar{U}R\mathrm{d}s \right.$$
$$\left. + V^{[i]}(x(k+1)) \right\}, \tag{4.85}$$

where $x(1), \ldots, x(k)$ are obtained under the action $u^{[i-1]}(k)$, and

$$x(k+1) = f(x(k-\sigma_0), \ldots, x(k-\sigma_m))$$
$$+ g(x(k-\sigma_0), \ldots, x(k-\sigma_m))u^{[i]}(k). \tag{4.86}$$

We further compute the control law $u^{[i]}(k)$ from (4.84) as follows:

$$u^{[i]}(k)$$
$$= \bar{U}\varphi\left(-\frac{1}{2}(\bar{U}R)^{-1}g^{\mathrm{T}}(x(k-\sigma_0), \ldots, x(k-\sigma_m))\frac{\partial V^{[i]}(x(k+1))}{\partial x(k+1)}\right), \tag{4.87}$$

and update the value function as

$$V^{[i+1]}(x(k)) = X^{\mathrm{T}}(k-\sigma)QX(k-\sigma) + 2\int_0^{u^{[i]}(k)} \varphi^{-\mathrm{T}}(\bar{U}^{-1}s)\bar{U}R\mathrm{d}s$$
$$+ V^{[i]}(x(k+1)). \tag{4.88}$$

In the iterative HDP algorithm, the value function and control law are updated by recurrent iteration, with the iteration number i increasing from 0 to ∞.

In the following, we give the convergence analysis of the present iterative HDP algorithm for nonlinear discrete time-delay systems with saturating actuators theoretically.

Lemma 4.10 *Let $\{\mu^{[i]}(k)\}$ be an arbitrary sequence of control laws, and $\{u^{[i]}(k)\}$ be the control law sequence expressed as in* (4.84). *Let $V^{[i]}$ be as in* (4.85) *and $\Lambda^{[i]}$ as*

$$\begin{aligned}\Lambda^{[i+1]}(\tilde{x}(k)) = {} & \tilde{X}^{\mathrm{T}}(k-\sigma)Q\tilde{X}(k-\sigma) + 2\int_0^{\mu^{[i]}(k)} \varphi^{-\mathrm{T}}(\bar{U}^{-1}s)\bar{U}R\mathrm{d}s \\ & + \Lambda^{[i]}(\tilde{x}(k+1)),\end{aligned} \tag{4.89}$$

where $\tilde{x}(k+1)$ is obtained under the action of $\mu^{[i]}(k)$. For the same initial state $\varrho(k)$ and initial control law $\beta(k)$, if $V^{[0]}(\cdot) = \Lambda^{[0]}(\cdot) = 0$, then $V^{[i+1]}(x(k)) \le \Lambda^{[i+1]}(\tilde{x}(k)), \forall i$.

Proof It is straightforward from the fact that $V^{[i+1]}$ is a result of minimizing the right hand side of equation (4.85) with respect to the control input $u^{[i]}(k)$, while $\Lambda^{[i+1]}$ is the result of an arbitrary control input. □

Theorem 4.11 (cf. [3]) *Let the sequence $V^{[i+1]}(x(k))$ be defined by* (4.85). *If the system is controllable, then there is an upper bound Y such that $0 \le V^{[i+1]}(x(k)) \le Y, \forall i$.*

Proof Let $\{\gamma^{[i]}(k)\}$ be any stabilizing and admissible control input, and let $V^{[0]}(\cdot) = P^{[0]}(\cdot) = 0$, where $V^{[i+1]}(x(k))$ is updated as (4.85) and $P^{[i+1]}(\bar{x}(k))$ is updated as

$$\begin{aligned}P^{[i+1]}(\bar{x}(k)) = {} & \bar{X}^{\mathrm{T}}(k-\sigma)Q\bar{X}(k-\sigma) + 2\int_0^{\gamma^{[i]}(k)} \varphi^{-\mathrm{T}}(\bar{U}^{-1}s)\bar{U}R\mathrm{d}s \\ & + P^{[i]}(\bar{x}(k+1)),\end{aligned} \tag{4.90}$$

where $\bar{x}(1), \dots, \bar{x}(k)$ are obtained under the action $\gamma^{[i-1]}(k)$, and $\bar{x}(k+1)$ is obtained under the action of $\gamma^{[i]}(k)$. From (4.90), we further obtain

$$\begin{aligned}P^{[i]}(\bar{x}(k+1)) = {} & \bar{X}^{\mathrm{T}}(k-\sigma+1)Q\bar{X}(k-\sigma+1) \\ & + 2\int_0^{\gamma^{[i-1]}(k+1)} \varphi^{-\mathrm{T}}(\bar{U}^{-1}s)\bar{U}R\mathrm{d}s \\ & + P^{[i-1]}(\bar{x}(k+2)).\end{aligned} \tag{4.91}$$

Thus, we obtain

$$\begin{aligned}
P^{[i+1]}(\bar{x}(k)) = {} & \bar{X}^{\mathrm{T}}(k-\sigma)Q\bar{X}(k-\sigma) + 2\int_0^{\gamma^{[i]}(k)} \varphi^{-\mathrm{T}}(\bar{U}^{-1}s)\bar{U}R\mathrm{d}s \\
& + \bar{X}^{\mathrm{T}}(k-\sigma+1)Q\bar{X}(k-\sigma+1) \\
& + 2\int_0^{\gamma^{[i-1]}(k+1)} \varphi^{-\mathrm{T}}(\bar{U}^{-1}s)\bar{U}R\mathrm{d}s \\
& + \cdots \\
& + \bar{X}^{\mathrm{T}}(k-\sigma+i)Q\bar{X}(k-\sigma+i) \\
& + 2\int_0^{\gamma^{[0]}(k+i)} \varphi^{-\mathrm{T}}(\bar{U}^{-1}s)\bar{U}R\mathrm{d}s.
\end{aligned} \tag{4.92}$$

Let

$$\begin{aligned}
L^{[i-j]}(\bar{x}(k+j)) = {} & \bar{X}^{\mathrm{T}}(k-\sigma+j)Q\bar{X}(k-\sigma+j) \\
& + 2\int_0^{\gamma^{[i-j]}(k+j)} \varphi^{-\mathrm{T}}(\bar{U}^{-1}s)\bar{U}R\mathrm{d}s,
\end{aligned} \tag{4.93}$$

and then (4.92) can further be written as

$$\begin{aligned}
P^{[i+1]}(\bar{x}(k)) &= \sum_{j=0}^{i} L^{[i-j]}(\bar{x}(k+j)) \\
&= \sum_{j=0}^{i} \Bigg\{ \bar{X}^{\mathrm{T}}(k-\sigma+j)Q\bar{X}(k-\sigma+j) \\
&\quad + 2\int_0^{\gamma^{[i-j]}(k+j)} \varphi^{-\mathrm{T}}(\bar{U}^{-1}s)\bar{U}R\mathrm{d}s \Bigg\}.
\end{aligned} \tag{4.94}$$

Noticing that $\{\gamma^{[i]}(k)\}$ is an admissible control law sequence, there exists an upper bound Y such that

$$P^{[i+1]}(\bar{x}(k)) \le \lim_{i\to\infty} \sum_{j=0}^{i} L^{[i-j]}(\bar{x}(k+j)) \le Y, \ \forall i. \tag{4.95}$$

Thus, based on the definition of the cost functional and Lemma 4.10, we obtain

$$0 \le V^{[i+1]}(x(k)) \le P^{[i+1]}(\bar{x}(k)) \le Y, \ \forall i. \tag{4.96}$$

This completes the proof. □

Theorem 4.12 *Define the value function sequence* $\{V^{[i]}\}$ *as in* (4.85) *with* $V^{[0]}(\cdot)=0$, *the control law sequence* $\{u^{[i]}(k)\}$ *as in* (4.84). *Then, we conclude that* $\{V^{[i]}\}$ *is a nondecreasing sequence satisfying* $V^{[i+1]}(x(k)) \geq V^{[i]}(x(k))$, $\forall i$.

Proof We define a new sequence $\{\Phi^{[i]}(x(k))\}$ as follows:

$$\begin{aligned}\Phi^{[i]}(x(k)) &= X^{\mathrm{T}}(k-\sigma)QX(k-\sigma)+2\int_0^{u^{[i]}(k)} \varphi^{-\mathrm{T}}(\bar{U}^{-1}s)\bar{U}R\mathrm{d}s \\ &\quad + \Phi^{[i-1]}(x(k+1)),\end{aligned} \tag{4.97}$$

where $\Phi^{[0]}(\cdot)=V^{[0]}(\cdot)=0$.

In the following part, we prove $\Phi^{[i]}(x(k)) \leq V^{i+1}(x(k))$ by mathematical induction.

First,we prove that it holds for $i=0$. Noticing that

$$\begin{aligned}&V^{[1]}(x(k)) - \Phi^{[0]}(x(k)) \\ &\quad = X^{\mathrm{T}}(k-\sigma)QX(k-\sigma)+2\int_0^{u^{[0]}(k)} \varphi^{-\mathrm{T}}(\bar{U}^{-1}s)\bar{U}R\mathrm{d}s \\ &\quad \geq 0,\end{aligned} \tag{4.98}$$

for $i=0$, we get

$$V^{[1]}(x(k)) \geq \Phi^{[0]}(x(k)). \tag{4.99}$$

Second, we suppose that $V^{[i]}(x(k)) \geq \Phi^{[i-1]}(x(k))$, for $i-1$, $\forall x(k)$. Then, for i, since

$$\begin{aligned}V^{[i+1]}(x(k)) &= X^{\mathrm{T}}(k-\sigma)QX(k-\sigma) \\ &\quad +2\int_0^{u^{[i]}(k)} \varphi^{-\mathrm{T}}(\bar{U}^{-1}s)\bar{U}R\mathrm{d}s + V^{[i]}(x(k+1)),\end{aligned} \tag{4.100}$$

and

$$\begin{aligned}\Phi^{[i]}(x(k)) &= X^{\mathrm{T}}(k-\sigma)QX(k-\sigma) \\ &\quad +2\int_0^{u^{[i]}(k)} \varphi^{-\mathrm{T}}(\bar{U}^{-1}s)\bar{U}R\mathrm{d}s + \Phi^{[i-1]}(x(k+1)),\end{aligned} \tag{4.101}$$

we get

$$V^{[i+1]}(x(k)) - \Phi^{[i]}(x(k)) = V^{[i]}(x(k+1)) - \Phi^{[i-1]}(x(k+1)) \geq 0. \tag{4.102}$$

Therefore, by mathematical induction, we have

$$V^{[i+1]}(x(k)) \geq \Phi^{[i]}(x(k)), \forall i. \tag{4.103}$$

In addition, from Lemma 4.10 we know that $V^{[i]}(x(k)) \leq \Phi^{[i]}(x(k))$. Therefore, we have

$$V^{[i]}(x(k)) \leq \Phi^{[i]}(x(k)) \leq V^{[i+1]}(x(k)), \tag{4.104}$$

which proves that $\{V^{[i]}(x(k))\}$ is a nondecreasing sequence bounded by (4.96). □

Theorem 4.13 *According to Theorem* 4.11, *there is a limit for the value function sequence* $\{V^{[i]}(x(k))\}$ *when* $i \to \infty$. *Without loss of generality, we define* $\lim_{i\to\infty} V^{[i]}(x(k)) = J^*(x(k))$, *and, accordingly,* $\lim_{i\to\infty} u^{[i]}(k) = u^*(k)$. *Then, for any discrete time step k, the limit* $J^*(x(k))$ *is the "optimal" value function, i.e.,*

$$J^*(x(k)) = \min_{u(k)} \{Q(x(k)) + W(u(k)) + J^*(x(k+1))\}. \tag{4.105}$$

Proof For any i, the value function sequence satisfies

$$V^{[i+1]}(x(k)) = \min_{u(k)} \left\{Q(x(k)) + W(u(k)) + V^{[i]}(x(k+1))\right\}. \tag{4.106}$$

Combining with $V^{[i+1]}(x(k)) \leq \lim_{i\to\infty} V^{[i+1]}(x(k))$, $\forall i$, and $\lim_{i\to\infty} V^{[i+1]}(x(k)) = J^*(x(k))$, we obtain

$$\begin{aligned} J^*(x(k)) &\geq \min_{u(k)} \left\{Q(x(k)) + W(u(k)) + V^{[i]}(x(k+1))\right\} \\ &= Q(x(k)) + W(u^{[i]}(k)) + V^{[i]}(x(k+1)). \end{aligned} \tag{4.107}$$

Let $i \to \infty$, we have

$$J^*(x(k)) \geq Q(x(k)) + W(u^*(k)) + J^*(x(k+1)). \tag{4.108}$$

On the other hand, for any i and $\hat{u}^{[i]}(k)$, we have

$$V^{[i+1]}(x(k)) \leq Q(x(k)) + W(\hat{u}^{[i]}(k)) + V^{[i]}(x(k+1)). \tag{4.109}$$

As $V^{[i]}(x(k)) \leq V^{[i+1]}(x(k))$, we obtain

$$V^{[i]}(x(k)) \leq Q(x(k)) + W(\hat{u}^{[i]}(k)) + V^{[i]}(x(k+1)). \tag{4.110}$$

Let $i \to \infty$, we have

$$J^*(x(k)) \leq Q(x(k)) + W(\hat{u}^{[i]}(k)) + J^*(x(k+1)). \tag{4.111}$$

Since $\hat{u}^{[i]}(k)$ in the above equation is chosen arbitrarily, we have

$$J^*(x(k)) \leq Q(x(k)) + W(u^*(k)) + J^*(x(k+1)). \tag{4.112}$$

Thus, we further obtain

$$\begin{aligned} J^*(x(k)) &= Q(x(k)) + W(u^*(k)) + J^*(x(k+1)) \\ &= \min_{u(k)} \left\{ Q(x(k)) + W(u(k)) + J^*(x(k+1)) \right\}. \end{aligned} \tag{4.113}$$

This completes the proof. □

Therefore, we conclude that the value function sequence $\{V^{[i]}\}$ converges to the optimal value function of the discrete-time HJB equation, i.e., $V^{[i]} \to J^*$ as $i \to \infty$. Simultaneously, we conclude that the corresponding control law sequence $\{u^{[i]}(k)\}$ converges to the optimal control law $u^*(k)$ as $i \to \infty$.

4.3.3 Simulations

Example 4.14 Consider the following two-order nonlinear discrete time-delay system with saturating actuators:

$$\begin{cases} x(k+1) = f(x(k), x(k-2)) + g(x(k), x(k-2))u(k), \ k \geq 0 \\ x(k) = \varrho(k), \ k = 0, -1, -2, \end{cases} \tag{4.114}$$

where $f(x(k), x(k-2)) = \left[\begin{smallmatrix} 0.2x_1(k-2)\exp(x_2^2(k)) \\ 0.3(x_2^3(k)) \end{smallmatrix}\right]$, $g(x(k), x(k-2)) = \left[\begin{smallmatrix} 0 \\ -0.2 \end{smallmatrix}\right]$, and assume that the control constraint is set to $|u(k)| \leq 0.3$.

Define the cost functional as

$$J(x(k)) = \sum_{i=k}^{\infty} \left\{ X^{\mathrm{T}}(i-\sigma) Q X(i-\sigma) + 2\int_0^{u(i)} \tanh^{-\mathrm{T}}(\bar{U}^{-1}s)\bar{U}R\,\mathrm{d}s \right\}, \tag{4.115}$$

where $X(i-\sigma) = [x^{\mathrm{T}}(i), x^{\mathrm{T}}(i-2)]^{\mathrm{T}}$, $x(i) = [x_1(i), x_2(i)]^{\mathrm{T}}$, $Q = I_4$, $R = 0.5$. The initial condition $\varrho(k) = [1.5 \ \ 0.5]^{\mathrm{T}}$, $k = 0, -1, -2$. Moreover, we implement the algorithm at the time step $k = 2$.

In the presented iterative HDP algorithm, we choose three-layer neural networks as the critic network and the action network with the structure 2-8-1 and 4-8-1. The initial weight matrices are chosen randomly from $[-0.1, 0.1]$. The critic network and the action network are trained with the learning rates $\alpha_a = \alpha_c = 0.01$. After 5000 iteration steps we get the curves of states, control input, and value function. The state trajectories are given as Fig. 4.8, and the corresponding control curve is

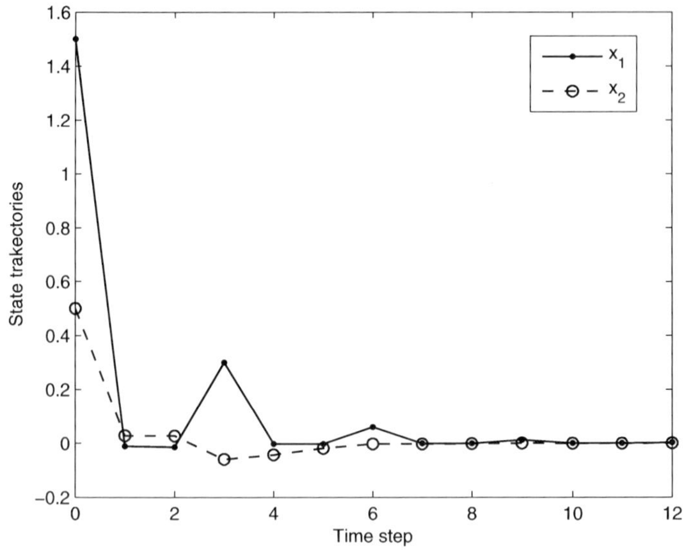

Fig. 4.8 The state variable trajectories x_1 and x_2

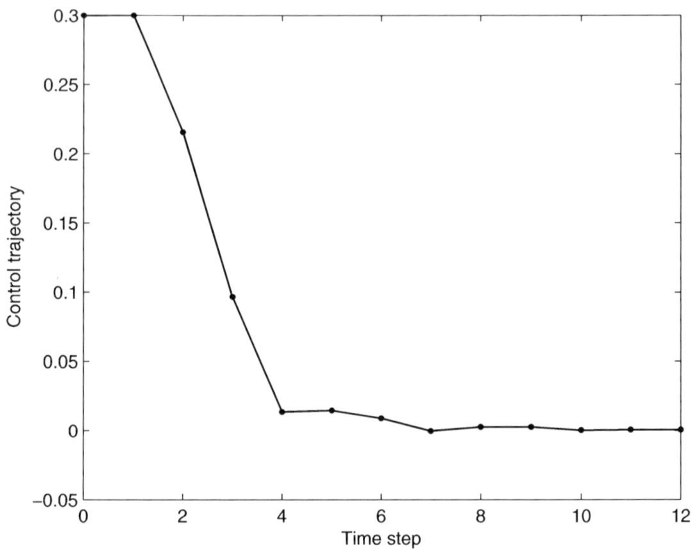

Fig. 4.9 The control input trajectory u

given as Fig. 4.9. The convergence curve of the value function is shown in Fig. 4.10.

It is clear from the simulation that the new iterative algorithm in this section is very effective.

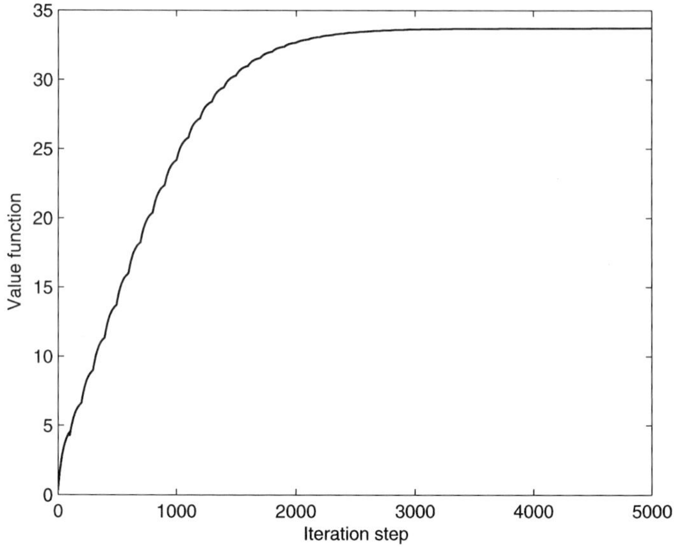

Fig. 4.10 The value function V

4.4 Finite-Horizon Optimal State Feedback Control for a Class of Nonlinear Systems with Time Delays

4.4.1 Problem Formulation

Consider a class of time-delay affine nonlinear systems

$$\begin{cases} x(t+1) = f(x(t), x(t-h)) + g(x(t), x(t-h))u(t), \\ x(t) = \varrho(t), \ -h \le t \le 0, \end{cases} \tag{4.116}$$

where $x(t)$ and $x(t-h)$ are states, and h is a nonnegative integer. We have $f(x(t), x(t-h)) \in \mathbb{R}^n$, $g(x(t), x(t-h)) \in \mathbb{R}^{n\times m}$ and the input $u(t) \in \mathbb{R}^m$. $\varrho(t)$ is the initial state. $f(x(t), x(t-h))$ and $g(x(t), x(t-h))$ are known functions. Suppose that the system is drift free, and the system (4.116) is stabilizable on a prescribed compact set $\Omega \in \mathbb{R}^n$.

For $\forall k$, $k > h$, the finite-time cost functional for state $x(k)$ under the control sequence $\underline{u}_k^{N+k-1} = (u(k), u(k+1), \ldots, u(N+k-1))$ is defined as

$$J(x(k), \underline{u}_k^{N+k-1}) = \sum_{i=k}^{N+k-1} \{x^{\mathrm{T}}(i)Qx(i) + u^{\mathrm{T}}(i)Ru(i)\}, \tag{4.117}$$

where Q and R are symmetric and positive-definite matrices. For the present finite-horizon optimal control problem, the feedback control sequence must not only drive the system state to zero within a finite number of time steps but also guarantee the cost functional (4.117) to be finite; $\underline{u}_k^{N+k-1} = (u(k), u(k+1), \ldots, u(N+$

$k-1))$ must be a finite horizon admissible control sequence [8], where $N > k$ is a finite integer, the length of the control sequence. The final state can be written as $x^{(f)}(x(k), \underline{u}_k^{N+k-1}) = x(N+k)$.

For time-delay nonlinear system, define the optimal value function as

$$J^*(x(k)) = \inf_{\underline{u}_k^{N+k-1}} \left\{ J(x(k), \underline{u}_k^{N+k-1}) \right\}. \tag{4.118}$$

From Bellman's optimality principle, it is known that, for the finite-horizon optimization case, the value function $J^*(x(k))$ satisfies the HJB equation as follows:

$$J^*(x(k)) = \inf_{u(k)} \{ x^{\mathrm{T}}(k) Q x(k) + u^{\mathrm{T}}(k) R u(k) + J^*(x(k+1)) \}. \tag{4.119}$$

The optimal control $u^*(k)$ satisfies the first-order necessary condition, which is given by the gradient of the right hand side of (4.119) with respect to $u(k)$ as

$$\begin{aligned} \frac{\partial J^*(x(k))}{\partial u(k)} &= \frac{\partial \left(x^{\mathrm{T}}(k) Q x(k) + u^{\mathrm{T}}(k) R u(k) \right)}{\partial u(k)} \\ &\quad + \left(\frac{\partial x(k+1)}{\partial u(k)} \right)^{\mathrm{T}} \frac{\partial J^*(x(k+1))}{\partial x(k+1)} \\ &= 0, \end{aligned} \tag{4.120}$$

and therefore we get

$$u^*(k) = -\frac{1}{2} R^{-1} g^{\mathrm{T}}(x(k), x(k-h)) \frac{\partial J^*(x(k+1))}{\partial x(k+1)}. \tag{4.121}$$

By substituting (4.121) into (4.119), the HJB becomes

$$\begin{aligned} J^*(x^*(k)) = x^{*\mathrm{T}}(k) Q x^*(k) &+ \frac{1}{4} \left(\frac{\partial J^*(x^*(k+1))}{\partial x^*(k+1)} \right)^{\mathrm{T}} \\ &\times g\left(x^*(k), x^*(k-h) \right) R^{-1} g^{\mathrm{T}} \left(x^*(k), x^*(k-h) \right) \\ &\times \frac{\partial J^*(x^*(k+1))}{\partial x^*(k+1)} + J^*(x^*(k+1)), \end{aligned} \tag{4.122}$$

where

$$\begin{aligned} x^*(t+1) = {} & f(x^*(t), x^*(t-h)) \\ & + g(x^*(t), x^*(t-h)) u^*(t), \quad t = 0, 1, \ldots, k, \ldots. \end{aligned} \tag{4.123}$$

4.4.2 Optimal Control Based on Improved Iterative ADP

In this section, we apply the improved ADP algorithm to solve the finite-horizon value function and obtain an optimal control sequence $(\underline{u}_k^{N+k-1})^* = (u^*(k), u^*(k+1), \ldots, u^*(N+k-1))$.

From system (4.116), we can see that for $\forall k$, and any initial state $\varrho(k)$, there exists a control vector $u(k)$, such that $x(k+1)=0$, i.e., we can control the state of system (4.116) to zero in one step from any initial state. In the iterative ADP algorithm, the value function and control law are updated by recursive iterations, with the iteration index number i increasing from 0 to ∞. The initial value function $V^{[0]}=0$, and $u^{[0]}(k)$ is an arbitrary control law.

The value function for $i=1$ is computed as

$$\begin{aligned} V^{[1]}(x^{[0]}(k)) &= \inf_{u(k)}\{x^{[0]\mathrm{T}}(k)Qx^{[0]}(k)+u^{\mathrm{T}}(k)Ru(k)\} \\ &= x^{[0]\mathrm{T}}(k)Qx^{[0]}(k)+u^{[1]\mathrm{T}}(k)Ru^{[1]}(k) \\ \text{s.t. } & f(x^{[0]}(k),x^{[0]}(k-h))+g(x^{[0]}(k),x^{[0]}(k-h))u(k)=0, \end{aligned} \tag{4.124}$$

where

$$u^{[1]}(k) = \arg\inf_{u(k)}\{x^{[0]\mathrm{T}}(k)Qx^{[0]}(k)+u^{\mathrm{T}}(k)Ru(k)\}, \tag{4.125}$$

and

$$\begin{aligned} x^{[0]}(t+1) &= \begin{cases} f(x^{[1]}(t),x^{[1]}(t-h)) \\ \quad +g(x^{[1]}(t),x^{[1]}(t-h))u^{[2]}(t), & k\le t \\ f(x^{[0]}(t),x^{[0]}(t-h)) \\ \quad +g(x^{[0]}(t),x^{[0]}(t-h))u^{[0]}(t), & 0\le t<k \end{cases} \\ x^{[0]}(t) &= \varrho(t), \ -h\le t<0. \end{aligned} \tag{4.126}$$

For $i=1,2,\ldots$, the value function $V^{[i+1]}(x^{[i]}(k))$ is updated as follows:

$$\begin{aligned} V^{[i+1]}(x^{[i]}(k)) = \inf_{u(k)}\{ & x^{[i]\mathrm{T}}(k)Qx^{[i]}(k)+u^{\mathrm{T}}(k)Ru(k) \\ & + V^{[i]}(x^{[i-1]}(k+1))\}. \end{aligned} \tag{4.127}$$

The control law is updated as follows:

$$\begin{aligned} u^{[i+1]}(k) = \arg\inf_{u(k)}\{ & x^{[i]\mathrm{T}}(k)Qx^{[i]}(k)+u^{\mathrm{T}}(k)Ru(k) \\ & + V^{[i]}(x^{[i-1]}(k+1))\}, \end{aligned} \tag{4.128}$$

where

$$x^{[i]}(t+1)=\begin{cases} f(x^{[i+1]}(t),x^{[i+1]}(t-h)) \\ \quad +g(x^{[i+1]}(t),x^{[i+1]}(t-h))u^{[i+2]}(t),\ k\le t, \\ f(x^{[i]}(t),x^{[i]}(t-h)) \\ \quad +g(x^{[i]}(t),x^{[i]}(t-h))u^{[i]}(t),\ 0\le t<k, \end{cases}$$

$$x^{[i]}(t)=\varrho(t),\ -h\le t<0. \tag{4.129}$$

In this part, we present several theorems to demonstrate that the improved iterative ADP algorithm is convergent. Furthermore, the iterative value function converges to the optimal one.

Lemma 4.15 *Let $V^{[1]}(x^{[0]}(k))$ be defined as* (4.124), *and $V^{[i+1]}(x^{[i]}(k))$ be defined as* (4.127). *Let $\underline{\hat{u}}_k^{k+i}$ be any finite-horizon admissible control sequence. Define*

$$\begin{aligned} &L^{[i+1]}(\hat{x}^{[i]}(k),\underline{\hat{u}}_k^{k+i}) \\ &\quad =\hat{x}^{[i]\mathrm{T}}(k)Q\hat{x}^{[i]}(k)+\hat{u}^{[i+1]\mathrm{T}}(k)R\hat{u}^{[i+1]}(k) \\ &\qquad +\hat{x}^{[i-1]\mathrm{T}}(k+1)Q\hat{x}^{[i-1]}(k+1)+\hat{u}^{[i]\mathrm{T}}(k+1)R\hat{u}^{[i]}(k+1)+\cdots \\ &\qquad +\hat{x}^{[0]\mathrm{T}}(k+i)Q\hat{x}^{[0]}(k+i)+\hat{u}^{[1]\mathrm{T}}(k+i)R\hat{u}^{[1]}(k+i), \\ &\quad \text{s.t. } f(\hat{x}^{[0]}(k+i),\hat{x}^{[0]}(k+i-h)) \\ &\qquad +g(\hat{x}^{[0]}(k+i),\hat{x}^{[0]}(k+i-h))\hat{u}^{[1]}(k+i)=0, \end{aligned} \tag{4.130}$$

and the final state under $\underline{u}_k^{k+i}$ satisfies

$$x^f(x^{[i]}(k),\underline{u}_k^{k+i})=0. \tag{4.131}$$

If $\hat{u}^{[0]}(k)=u^{[0]}(k)$, then we have for $\forall\hat{x}^{[i]}(k)$

$$V^{[i+1]}(x^{[i]}(k))=\inf_{\underline{u}_k^{k+i}}\left\{L^{[i+1]}\left(\hat{x}^{[i]}(k),\underline{\hat{u}}_k^{k+i}\right)\right\}. \tag{4.132}$$

Proof From (4.127) and (4.130), we easily get (4.132). □

Theorem 4.16 (cf. [2]) *For system* (4.116), *let $x(k), k>h$, be an arbitrary state vector. Then, the value function sequence $\{V^{[i+1]}(x^{[i]}(k))\}$ is a monotonically nonincreasing sequence for $i\ge 1$, i.e., $V^{[i+1]}(x^{[i]}(k))\le V^{[i]}(x^{[i-1]}(k))$ for $i\ge 1$.*

Proof In (4.130), we let $\hat{u}^{[i+1]}(k)=u^{[i]}(k)$, and then we have $\hat{x}^{[i]}(k)=x^{[i-1]}(k)$. So we further obtain

$$\begin{aligned} &L^{[i+1]}(\hat{x}^{[i]}(k),\underline{\hat{u}}_k^{k+i}) \\ &\quad =x^{[i-1]\mathrm{T}}(k)Qx^{[i-1]}(k)+u^{[i]\mathrm{T}}(k)Ru^{[i]}(k) \end{aligned}$$

$$+ x^{[i-2]\mathrm{T}}(k+1)Qx^{[i-2]}(k+1) + u^{[i-1]\mathrm{T}}(k+1)Ru^{[i-1]}(k+1) + \cdots$$
$$+ x^{[0]\mathrm{T}}(k+i-1)Qx^{[0]}(k+i-1) + u^{[1]\mathrm{T}}(k+i-1)Ru^{[1]\mathrm{T}}(k+i-1),$$
$$+ \hat{x}^{[0]\mathrm{T}}(k+i)Q\hat{x}^{[0]}(k+i) + \hat{u}^{[1]\mathrm{T}}(k+i)R\hat{u}^{[1]}(k+i). \tag{4.133}$$

From (4.124) and (4.125), we have known that $\hat{x}^{[0]}(k+i)=0$ and $\hat{u}^{[1]}(k+i)=0$. So we have

$$L^{[i+1]}(\hat{x}^{[i]}(k), \hat{\underline{u}}_k^{k+i}) = V^{[i+1]}(x^{[i]}(k)). \tag{4.134}$$

From Lemma 4.15, we have $V^{i+1}(x^{[i]}(k)) \le L^{[i+1]}(x^{[i]}(k), \hat{\underline{u}}_k^{k+i})$. So we obtain

$$V^{[i+1]}(x^{[i]}(k)) \le V^{[i]}(x^{[i-1]}(k)). \tag{4.135}$$

This completes the proof. □

From Theorem 4.16, we can see that the value function $V^{[i+1]}(x^{[i]}(k))$ is a monotonically nonincreasing sequence and the lower bound is 0. In the following part, we will demonstrate that the limit of $V^{[i+1]}(x^{[i]}(k))$ satisfies HJB. The following lemma is necessary for the further proof.

Lemma 4.17 *Let $V^{[i+1]}(x^{[i]}(k))$ and $u^{[i]}(k)$ be defined as* (4.127) *and* (4.128). *Let $\{\mu^{[i]}(k)\}$ be an arbitrary sequence of the control law, $\Lambda^{[i+1]}(\tilde{x}^{[i]}(k))$ be defined by*

$$\Lambda^{[i+1]}(\tilde{x}^{[i]}(k)) = \tilde{x}^{[i]\mathrm{T}}(k)Q\tilde{x}^{[i]} + \mu^{[i+1]\mathrm{T}}(k)R\mu^{[i+1]}(k) + \Lambda^{[i]}(\tilde{x}^{[i-1]}(k+1)), i>0, \tag{4.136}$$

with the first iteration expressed as follows:

$$\Lambda^{[1]}(\tilde{x}^{[0]}(k)) = \tilde{x}^{[0]\mathrm{T}}(k)Q\tilde{x}^{[0]}(k) + \mu^{[1]\mathrm{T}}(k)R\mu^{[1]}(k)$$
$$\text{s.t. } f(\tilde{x}^{[0]}(k), \tilde{x}^{[0]}(k-h)) + g(\tilde{x}^{[0]}(k), \tilde{x}^{[0]}(k-h))\mu^{[1]}(k) = 0, \tag{4.137}$$

where

$$\tilde{x}^{[0]}(t+1) = \begin{cases} f(\tilde{x}^{[1]}(t), \tilde{x}^{[1]}(t-h)) \\ \quad +g(\tilde{x}^{[1]}(t), \tilde{x}^{[1]}(t-h))\mu^{[2]}(t), \ t \ge k \\ f(\tilde{x}^{[0]}(t), \tilde{x}^{[0]}(t-h)) \\ \quad +g(\tilde{x}^{[0]}(t), \tilde{x}^{[0]}(t-h))\mu^{[0]}(t), \ 0 \le t < k \end{cases}$$
$$\tilde{x}^{[0]}(t) = \varrho(t), \ -h \le t \le 0, \tag{4.138}$$

and for $i \geq 0$, the state is updated as follows:

$$\tilde{x}^{[i]}(t+1) = \begin{cases} f(\tilde{x}^{[i+1]}(t), \tilde{x}^{[i+1]}(t-h)) \\ \quad +g(\tilde{x}^{[i+1]}(t), \tilde{x}^{[i+1]}(t-h))\mu^{[i+2]}(t), \ t \geq k \\ f(\tilde{x}^{[i]}(t), \tilde{x}^{[i]}(t-h)) \\ \quad +g(\tilde{x}^{[i]}(t), \tilde{x}^{[i]}(t-h))\mu^{[i]}(t), \ 0 \leq t < k \end{cases}$$

$$\tilde{x}^{[i]}(t) = \varrho(t), \ -h \leq t \leq 0. \tag{4.139}$$

If $u^{[0]}(k) = \mu^{[0]}(k)$, *and* $V^{[0]} = \Lambda^{[0]} = 0$, *then* $V^{[i+1]}(x^{[i-1]}(k)) \leq \Lambda^{[i+1]}(\tilde{x}^{[i-1]}(k)), \forall i$.

Proof It can be straightforwardly seen from the fact that $V^{[i+1]}(x^{[i]}(k))$ is the result of minimizing the right hand side of (4.127) with respect to the control input $u^{[i+1]}(k)$, while $\Lambda^{[i+1]}(\tilde{x}^{[i]}(k))$ is the result of arbitrary control input. □

Theorem 4.18 *For system* (4.116), *let* $x(k), k > h$, *be an arbitrary state vector. Let* $u^{\infty}(k) = \lim_{i\to\infty} u^{[i]}(k)$, *and* $V^{\infty}(x^{\infty}(k)) = \lim_{i\to\infty} V^{[i+1]}(x^{[i]}(k))$, *where* $x^{\infty}(k)$ *is the state under the action of* $u^{\infty}(k)$. *Then, we have*

$$\begin{aligned} V^{\infty}(x^{\infty}(k)) &= \inf_{u(k)} \{x^{\infty\mathrm{T}}(k)Qx^{\infty}(k) + u^{\mathrm{T}}(k)Qu(k) + V^{\infty}(x^{\infty}(k+1))\} \\ &= x^{\infty\mathrm{T}}(k)Qx^{\infty}(k) + u^{\infty\mathrm{T}}(k)Qu^{\infty}(k) \\ &\quad + V^{\infty}(x^{\infty}(k+1)), \end{aligned} \tag{4.140}$$

where

$$u^{\infty}(k) = \arg\inf_{u(k)} \{x^{\infty\mathrm{T}}(k)Qx^{\infty}(k) + u^{\mathrm{T}}(k)Qu(k) + V^{\infty}(x^{\infty}(k+1))\}. \tag{4.141}$$

Proof Let $\{\mu^{[i]}(k)\}$ be arbitrary sequence of control law, and $\Lambda^{[i+1]}(\tilde{x}^{[i]}(k))$ be defined as (4.136). Then, from Lemma 4.17, we get

$$\begin{aligned} V^{[i+1]}(x^{[i]}(k)) \leq{} & \tilde{x}^{[i]\mathrm{T}}(k)Q\tilde{x}^{[i]}(k) + \mu^{[i+1]\mathrm{T}}(k)R\mu^{[i+1]}(k) \\ & + \Lambda^{[i]}(\tilde{x}^{[i-1]}(k+1)). \end{aligned} \tag{4.142}$$

Let $\{\mu^{[i]}(k)\} = \{u^{[i]}(k)\}$; then (4.142) can be changed to

$$\begin{aligned} V^{[i+1]}(x^{[i]}(k)) &\leq x^{[i]\mathrm{T}}(k)Qx^{[i]}(k) + u^{[i+1]\mathrm{T}}(k)Ru^{[i+1]}(k) + V^{[i]}(x^{[i-1]}(k+1)) \\ &= \inf_{u(k)} \{x^{[i]\mathrm{T}}(k)Qx^{[i]}(k) + u^{\mathrm{T}}(k)Ru(k) \\ &\quad + V^{[i]}(x^{[i-1]}(k+1))\}. \end{aligned} \tag{4.143}$$

Let $i \to \infty$ in (4.143), we obtain

$$V^{\infty}(x^{\infty}(k)) \leq \inf_{u(k)}\{x^{\infty\mathrm{T}}(k)Qx^{\infty}(k)+u^{\mathrm{T}}(k)Qu(k) \\ +V^{\infty}(x^{\infty}(k+1))\}. \tag{4.144}$$

According to Theorem 4.16, we have $V^{[i+1]}(x^{[i]}(k)) \leq V^{[i]}(x^{[i-1]}(k))$, i.e.,

$$V^{[i]}(x^{[i-1]}(k)) \geq \inf_{u(k)}\{x^{[i]\mathrm{T}}(k)Qx^{[i]}(k)+u^{\mathrm{T}}(k)Ru(k) \\ +V^{[i]}(x^{[i-1]}(k+1))\}. \tag{4.145}$$

Let $i \to \infty$ in (4.145), and we obtain

$$V^{\infty}(x^{\infty}(k)) \geq \inf_{u(k)}\{x^{\infty\mathrm{T}}(k)Qx^{\infty}(k)+u^{\mathrm{T}}(k)Qu(k) \\ +V^{\infty}(x^{\infty}(k+1))\}. \tag{4.146}$$

Thus, combining (4.144) and (4.146), we have

$$V^{\infty}(x^{\infty}(k)) = \inf_{u(k)}\{x^{\infty\mathrm{T}}(k)Qx^{\infty}(k)+u^{\mathrm{T}}(k)Qu(k) \\ +V^{\infty}(x^{\infty}(k+1))\}. \tag{4.147}$$

This completes the proof. □

Theorem 4.19 *Let* $V^{\infty}(x^{\infty}(k)) = \lim_{i\to\infty} V^{[i+1]}(x^{[i]}(k))$, *and let* $V^{[i+1]}(x^{[i]}(k))$ *be defined as* (4.127). *Then we have* $V^{\infty}(x^{\infty}(k)) = J^{*}(x^{*}(k))$.

Proof From the definition of $J^{*}(x^{*}(k))$, we get

$$J^{*}(x^{*}(k)) \leq V^{[i+1]}(x^{[i]}(k)). \tag{4.148}$$

Let $i \to \infty$ in (4.148); we have

$$J^{*}(x^{*}(k)) \leq V^{\infty}(x^{\infty}(k)). \tag{4.149}$$

On the other hand, according to the definition $J^{*}(x^{*}(k))$ in (4.119), for any $\theta > 0$, there exists a sequence of control law $\mu^{[i+1]}(k)$, such that the associated value function $\Lambda^{[i+1]}(\tilde{x}^{[i]}(k))$ defined as (4.136) satisfies $\Lambda^{[i+1]}(\tilde{x}^{[i]}(k)) \leq J^{*}(x^{*}(k))+\theta$. From Lemma 4.17, we get

$$V^{[i+1]}(x^{[i]}(k)) \leq \Lambda^{[i+1]}(\tilde{x}^{[i]}(k)) \leq J^{*}(x^{*}(k))+\theta. \tag{4.150}$$

Let $i \to \infty$; we obtain

$$V^{\infty}(x^{\infty}(k)) \leq J^{*}(x^{*}(k))+\theta. \tag{4.151}$$

Note that θ is chosen arbitrarily, and we have

$$V^{\infty}(x^{\infty}(k)) \leq J^*(x^*(k)). \tag{4.152}$$

Thus, combining (4.149) and (4.152), we have

$$V^{\infty}(x^{\infty}(k)) = J^*(x^*(k)). \tag{4.153}$$

This completes the proof. □

From the previous part, we proved that the iterative value function $V^{[i+1]}(x^{[i]})(k)$ converges to the optimal value function $J^*(x^*(k))$ until $i \to \infty$. But, unfortunately, the infinite iteration is not practical to do so. According to Theorem 4.19, we have $\lim_{i\to\infty} V^{[i]}(x^{[i-1]}(k)) = V^*(x^*(k))$, i.e., $\forall \gamma > 0$, $\exists I \in \mathbb{N}$, such that $\forall i > I$,

$$\left| V^{[i]}(x^{[i-1]}(k)) - V^*(x^*(k)) \right| \leq \gamma. \tag{4.154}$$

Thus,

$$\begin{aligned} u_\delta^*(k) = u^{[i]}(k) = {} & \arg\min_{u(k)} \{x^{[i-1]\mathrm{T}}(k) Q x^{[i-1]}(k) \\ & + u^{\mathrm{T}}(k) R u(k) + V^{[i-1]}(x^{[i-2]}(k+1))\}. \end{aligned} \tag{4.155}$$

While the optimal value function $J^*(x^*(k))$ is unknown in general, the optimality criterion (4.154) is very difficult to verify. So the following theorem is necessary.

Theorem 4.20 *Let $V^{[i]}(x^{[i-1]}(k))$ be defined as* (4.127), *and let $J^*(x^*(k))$ be defined as* (4.122). *Then, $\forall \delta > 0$, $\exists I \in \mathbb{N}$, such that $\forall i > I$,*

$$\left| V^{[i]}(x^{[i-1]}(k)) - J^*(x^*(k)) \right| \leq \delta, \tag{4.156}$$

which is equivalent to

$$\left| V^{[i+1]}(x^{[i]}(k)) - V^{[i]}(x^{[i-1]}(k)) \right| \leq \delta. \tag{4.157}$$

Proof From (4.156), we see that

$$V^{[i]}(x^{[i-1]}(k)) \leq J^*(x^*(k)) + \delta. \tag{4.158}$$

Furthermore, according to Theorem 4.16, we get

$$J^*(x^*(k)) \leq V^{[i+1]}(x^{[i]}(k)) \leq V^{[i]}(x^{[i-1]}(k)). \tag{4.159}$$

So we obtain

$$\begin{aligned} V^{[i]}(x^{[i-1]}(k)) - V^{[i+1]}(x^{[i]}(k)) &\leq V^{[i]}(x^{[i-1]}(k)) - J^*(x^*(k)) \\ &\leq J^*(x^*(k)) + \delta - J^*(x^*(k)) \\ &= \delta. \end{aligned} \tag{4.160}$$

As $V^{[i]}(x^{[i-1]}(k)) - V^{[i+1]}(x^{[i]}(k)) \geq 0$, we have $|V^{[i+1]}(x^{[i]}(k)) - V^{[i]}(x^{[i-1]}(k))| \leq \delta$.

On the other hand, from (4.154), we have known that $|V^{[i]}(x^{[i-1]}(k)) - J^*(x^*(k))| \leq \gamma$. Here we let $\gamma = \delta$, so for $\forall i > I$, we have

$$\left| V^{[i]}(x^{[i-1]}(k)) - J^*(x^*(k)) \right| \leq \delta. \tag{4.161}$$

This completes the proof. □

In this part, two neural networks are used to approximate the iterative value function and optimal control. The process of implementation is as follows:

Step 1. Give the accuracy δ, the initial state ϱ and the initial control law $u^{[0]}$. Give the initial time step k.
Step 2. Set $i = 0$, according to (4.126), we can obtain the state sequence $x^{[0]}(1), \ldots, x^{[0]}(k)$. So we can get $V^{[1]}(x^{[0]}(k))$ and $u^{[1]}(k)$ from (4.124) and (4.125).
Step 3. Set $i = 1$, we have the state sequence $x^{[i]}(1), \ldots, x^{[i]}(k)$ from (4.129). According to (4.127) and (4.128), we obtain $V^{[i+1]}(x^{[i]}(k))$ and $u^{[i+1]}(k)$. If $|V^{[i+1]}(x^{[i]}(k)) - V^{[i]}(x^{[i-1]}(k))| \leq \delta$, then go to Step 5. Otherwise, go to Step 4.
Step 4. $i = i + 1$; go to Step 3.
Step 5. Stop.

4.4.3 Simulations

Example 4.21 Consider the following nonlinear time-delay system:

$$\begin{aligned} x(t+1) &= f(x(t), x(t-2)) + g(x(t), x(t-2))u(t) \\ x(t) &= \varrho(t), -2 \leq t \leq 0, \end{aligned} \tag{4.162}$$

where $f(x(t), x(t-2)) = \left[\begin{smallmatrix} 0.2x_1(t)\exp(x_2(t))^2 \\ 0.3x_2^3(t-2) \end{smallmatrix}\right]$ and $g(x(t), x(t-2)) = \left[\begin{smallmatrix} -0.2 & 0 \\ 0 & -0.2 \end{smallmatrix}\right]$.

In system (4.162), the initial state $\varrho(t) = [0.5 \ \ -0.5]^{\mathrm{T}}$, $-2 \leq t \leq 0$. The initial control law $u^{[0]}(t) = -2x(t)$. The initial time step is $k = 3$. In this section, we adopt BP neural networks to approximate $V^{[i+1]}(x^{[i]}(k))$ and $u^{[i+1]}(k)$. The initial weights are chosen randomly from $[-0.1, 0.1]$, and the learning rate is 0.05. We select $Q = R = I_2$.

The state trajectories are given as Fig. 4.11. The control trajectories are shown in Fig. 4.12. From the two figures we can see that the system (4.162) can be stabilized using the iteration algorithm developed in this section. The value function iteration curve is given in Fig. 4.13. It is clear that the value function is convergent as $i \to \infty$. As $|V^{[15]}(x^{[14]}) - V^{[14]}(x^{[13]})| < 10^{-5}$, we see that system (4.162) is stabilized in $N = 15$ steps with accuracy error $\delta = 10^{-5}$.

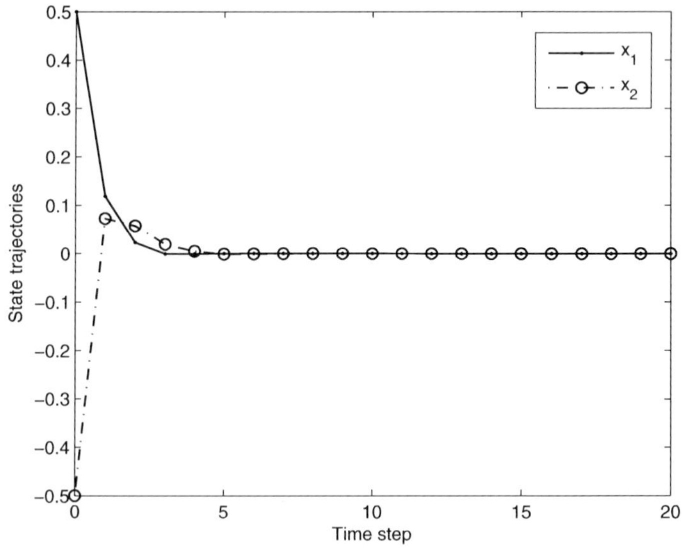

Fig. 4.11 The state variable trajectories x_1 and x_2

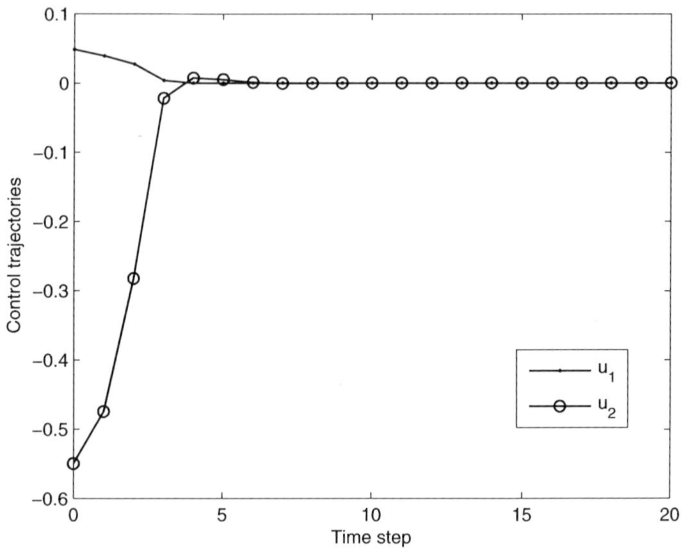

Fig. 4.12 The control variable trajectories u_1 and u_2

4.5 Summary

In this chapter, we developed some effective algorithms based on iterative ADP to solve optimal control problems for nonlinear systems with time delays. A delay matrix function was introduced and the explicit expression of optimal control was

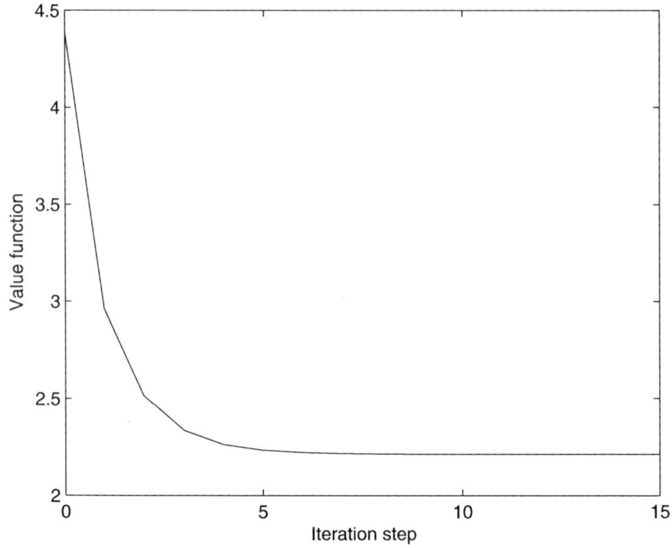

Fig. 4.13 The value function

obtained. We then developed an effective algorithm to solve the optimal control problems for nonlinear time-delay systems with saturating actuators. Finally, an effective algorithm to solve the finite-horizon optimal control problems was developed for a class of time-delay systems. Simulation studies successfully demonstrated the outstanding performance of the present optimal control schemes for time-delay nonlinear systems.

References

1. Malek-Zavarei M, Jashmidi M (1987) Time-delay systems: analysis, optimization and applications. North-Holland, Amsterdam, pp 80–96
2. Song RZ, Zhang HG (2011) The finite horizon optimal control for a class of time-delay affine nonlinear system. Neural Comput Appl. doi:10.1007/s00521-011-0706-3
3. Song RZ, Zhang HG, Wei QL, Luo YH (2010) Optimal control laws for time-delay systems with saturating actuators based on heuristic dynamic programming. Neurocomputing 73:3020–3027
4. Tong SC, Liu CL, Li YM, Zhang HG (2011) Adaptive fuzzy decentralized control for large-scale nonlinear systems with time-varying delays and unknown high-frequency gain sign. IEEE Trans Syst Man Cybern, Part B, Cybern 41:474–485
5. Wang ZS, Zhang HG (2010) Global asymptotic stability of reaction–diffusion Cohen–Grossberg neural networks with continuously distributed delays. IEEE Trans Neural Netw 21:39–49
6. Wang ZS, Zhang HG, Li P (2010) An LMI approach to stability analysis of reaction–diffusion Cohen–Grossberg neural networks concerning Dirichlet boundary conditions and distributed delays. IEEE Trans Syst Man Cybern, Part B, Cybern 40:1596–1606

7. Wang YC, Zhang HG, Wang XY, Yang DS (2010) Networked synchronization control of coupled dynamic networks with time-varying delay. IEEE Trans Syst Man Cybern, Part B, Cybern 40:1468–1479
8. Wang FY, Jin N, Liu DR, Wei Q (2011) Adaptive dynamic programming for finite-horizon optimal control of discrete-time nonlinear systems with ϵ-error bound. IEEE Trans Neural Netw 22:24–36
9. Wei QL, Zhang HG, Liu DR, Zhao Y (2010) An optimal control scheme for a class of discrete-time nonlinear systems with time delays using adaptive dynamic programming. Acta Autom Sin 36:121–129
10. Zhang HG, Wang YC (2008) Stability analysis of Markovian jumping stochastic Cohen–Grossberg neural networks with mixed time delays. IEEE Trans Neural Netw 19:366–370
11. Zhang HG, Xie YH, Wang ZL, Zheng CD (2007) Adaptive synchronization between two different chaotic neural networks with time delay. IEEE Trans Neural Netw 18:1841–1845
12. Zhang HG, Wang ZS, Liu DR (2008) Robust stability analysis for interval Cohen–Grossberg neural networks with unknown time-varying delays. IEEE Trans Neural Netw 19:1942–1955
13. Zhang HG, Wang YC, Liu DR (2008) Delay-dependent guaranteed cost control for uncertain stochastic fuzzy systems with multiple tire delays. IEEE Trans Syst Man Cybern, Part B, Cybern 38:126–140
14. Zhang HG, Wei QL, Luo YH (2008) A novel infinite-time optimal tracking control scheme for a class of discrete-time nonlinear systems via the greedy HDP iteration algorithm. IEEE Trans Syst Man Cybern, Part B, Cybern 38:937–942
15. Zheng CD, Zhang HG, Wang ZS (2009) New delay-dependent global exponential stability criterion for cellular-type neural networks with time-varying delays. IEEE Trans Circuits Syst II, Express Briefs 56:250–254
16. Zheng CD, Zhang HG, Wang ZS (2010) Improved robust stability criteria for delayed cellular neural networks via the LMI approach. IEEE Trans Circuits Syst II, Express Briefs 57:41–45

Chapter 5
Optimal Tracking Control of Nonlinear Systems with Time Delays

5.1 Introduction

Chapter 4 discussed the optimal state feedback control problems of nonlinear systems with time delay. In this chapter, we will solve the optimal tracking control problem of nonlinear systems with time delay based on the HDP algorithm. First, the HJB equation for a discrete time-delay system is derived which is based on state error and control error. In order to solve this HJB equation, a novel iterative HDP algorithm containing the iterations of state, control law, and cost functional is developed. We also give the convergence proof for the novel iterative HDP algorithm. Finally, the critic network and action network are used to approximate the value function and the corresponding control law, respectively. The main contributions of this chapter can be summarized as follows:

1. It is the first time that one solves the optimal tracking control problem of nonlinear systems with time delays using the HDP algorithm.
2. For the novel HDP algorithm, besides the value function iteration, the state is also updated according to the iterative control law in every iteration step.
3. In the state update, we adopt "*backward iteration*". For the $(i+1)$st cost functional iteration, the state at time step $k+1$ is updated according to the states before time step $k+1$ in the ith iteration and the control law in the ith iteration.

5.2 Problem Formulation

Consider a class of discrete-time affine nonlinear systems with time delays:

$$\begin{aligned} x(k+1) &= f(x(k-\sigma_0), x(k-\sigma_1), \dots, x(k-\sigma_m)) \\ &\quad + g(x(k-\sigma_0), x(k-\sigma_1), \dots, x(k-\sigma_m))u(k), \\ x(k) &= \varrho_1(k), \ -\sigma_m \le k \le 0, \end{aligned} \tag{5.1}$$

H. Zhang et al., *Adaptive Dynamic Programming for Control*,
Communications and Control Engineering, DOI 10.1007/978-1-4471-4757-2_5,

where $x(k-\sigma_0), x(k-\sigma_1), \ldots, x(k-\sigma_m) \in \mathbb{R}^n$ and $x(k-\sigma_1), \ldots, x(k-\sigma_m)$ are states of time delays. $f(x(k-\sigma_0), x(k-\sigma_1), \ldots, x(k-\sigma_m)) \in \mathbb{R}^n$, $g(x(k-\sigma_0), x(k-\sigma_1), \ldots, x(k-\sigma_m)) \in \mathbb{R}^{n\times m}$ and the input $u(k) \in \mathbb{R}^m$. $\varrho_1(k)$ is the initial state, σ_i is the time delay. We set $0 = \sigma_0 < \sigma_1 < \cdots < \sigma_m$, and these are nonnegative integer numbers. $f(x(k-\sigma_0), x(k-\sigma_1), \ldots, x(k-\sigma_m))$ and $g(x(k-\sigma_0), x(k-\sigma_1), \ldots, x(k-\sigma_m))$ are known functions, and $g(x(k-\sigma_0), x(k-\sigma_1), \ldots, x(k-\sigma_m))$ is analytic. The system (5.1) is controllable and reachable in $\Omega \in \mathbb{R}^n$ [8].

In this chapter, we define the state error as follows:

$$e(k) = x(k) - x_d(k), \tag{5.2}$$

where the reference orbit $x_d(k)$ is generated by the n-dimensional autonomous system as follows:

$$\begin{aligned} x_d(k+1) &= S(x_d(k)), \\ x_d(0) &= \varrho_2(k), -\sigma_m \le k \le 0, \end{aligned} \tag{5.3}$$

where $x_d(k) \in \mathbb{R}^n$, $S(x_d(k)) \in \mathbb{R}^n$ and ϱ_2 is the initial state, $\varrho_2(-\sigma_m) = \cdots = \varrho_2(-\sigma_1) = 0$.

The objective in this chapter is to design an optimal state feedback control law $u(k)$ based on any given $\varrho_1(k)(-\sigma_m \le k \le 0)$ and initial control law $\beta(k)$, which not only renders the state $x(k)$ asymptotically tracking the reference orbit, i.e., $e(k)$ asymptotically approaches zero; it also minimizes the cost functional as follows:

$$J(e(k), v(k)) = \sum_{i=k}^{\infty} \left\{ e^{\mathrm{T}}(i) Q e(i) + v^{\mathrm{T}}(i) R v(i) \right\}, \tag{5.4}$$

where Q and R are symmetric and positive-definite matrices. We divide $u(k)$ into two parts, i.e., $v(k)$ and $u_s(k)$. So we have

$$v(k) = u(k) - u_s(k), \tag{5.5}$$

where $u_s(k)$ denotes the steady control input corresponding to the desired trajectory $x_d(k)$. In fact, $v(k)$ is the error between actual control $u(k)$ of system (5.1) and the steady control $u_s(k)$.

We see that the problem of solving the optimal tracking control law $u(k)$ of system (5.1) is converted into solving the optimal control law $v(k)$. In the following section, we will discuss how to design $v(k)$.

5.3 Optimal Tracking Control Based on Improved Iterative ADP Algorithm

In this section, we focus on designing the optimal control law to handle the optimal tracking control problem. We first give a representation of the steady control input

$u_s(k)$. Inspired by the paper of [13], we define the steady control as follows:

$$\begin{aligned} u_s(k) &= g^{-1}(x_d(k-\sigma_0), \ldots, x_d(k-\sigma_m)) \\ &\quad \times (x_d(k+1) - f(x_d(k-\sigma_0), \ldots, x_d(k-\sigma_m))), \end{aligned} \tag{5.6}$$

where $g^{-1}(\cdot)$ denotes the inversion of $g(\cdot)$.

Remark 5.1 If $g(\cdot)$ is invertible, then u_s can be obtained directly by (5.6). So u_s exists, when $g(\cdot)$ is invertible. If $g(\cdot)$ is noninvertible, u_s also exists. Because of the numerical solution of $g^{-1}(\cdot)$ can be solved at least by one of the three methods as follows:

1. Moore–Penrose pseudoinverse technique [4]:

   The Moore–Penrose pseudoinverse is a matrix G of the same dimensions as g, satisfying four conditions: $gGg = g$, $GgG = G$, Gg is Hermitian, and gG is Hermitian.

   In [4], it is proved that the matrix G exists and is unique, if $g \neq 0$. In this chapter, $g \neq 0$ obviously, because we supposed that the system (5.1) is controllable and reachable. So we can say that $g^{-1} = G$. Furthermore, in Matlab 7.5, the Moore–Penrose pseudoinverse of g can be obtained by the Matlab function $G = \mathrm{pinv}(g)$.
2. Least square method [2]:

   As $g(\cdot)g^{-1}(\cdot) = I$, we have

$$g^{-1}(\cdot) = (g^{\mathrm{T}}(\cdot)g(\cdot))^{-1}g^{\mathrm{T}}(\cdot). \tag{5.7}$$

   Introducing $E(0, r^2) \in \mathbb{R}^{n\times m}$ into $g(\cdot)$, $\bar{g}(\cdot)$ can be expressed as follows:

$$\bar{g}(\cdot) = g(\cdot) + E(0, r^2), \tag{5.8}$$

   where every element in $E(0, r^2)$ is zero-mean Gaussian noise. So we get

$$g^{-1}(\cdot) = (\bar{g}^{\mathrm{T}}(\cdot)\bar{g}(\cdot))^{-1}g^{\mathrm{T}}(\cdot). \tag{5.9}$$

   Then, we sample K times, every time we take nm Gaussian points, so we have $E_i(0, r^2)$, $i \in \{1, 2, \ldots, K\}$. Every element in E_i is a Gaussian sample point. Thus, we have $G(\cdot) = [\bar{g}_1; \ldots; \bar{g}_K]$, where $\bar{g}_i = g + E_i(0, r^2)$, $i \in \{1, 2, \ldots, K\}$.

   So we get

$$g^{-1}(\cdot) = (G^{\mathrm{T}}(\cdot)G(\cdot))^{-1}g^{\mathrm{T}}(\cdot). \tag{5.10}$$

3. Neural network method [11]:

   First, we let the BP neural network be expressed as $\hat{F}(X, V, W) = W^{\mathrm{T}}\sigma(V^{\mathrm{T}}X)$, where W and V are the weights of the neural network and σ is the activation function. Second, we use the output $\hat{x}(k+1)$ of BP neural network to approximate $x(k+1)$. Then, we have $\hat{x}(k+1) = W^{\mathrm{T}}\sigma(V^{\mathrm{T}}X)$, where

$X = [x(k), u(k)]$. We already know that the system (5.1) is an affine nonlinear system. So we get $g = \partial\hat{x}(k+1)/\partial u$. The equation $g^{-1} = \partial u/\partial\hat{x}(k+1)$ can also be established.

So we can see that u_s is existent and can be obtained by (5.6). In this chapter, we adopt the Moore–Penrose pseudoinverse technique to get $g^{-1}(\cdot)$ in our simulation section.

According to (5.1), and in light of (5.2) with respect to $x(k)$ and (5.5) with respect to $u_s(k)$, we easily obtain

$$\begin{aligned} e(k+1) &= f(e(k-\sigma_0) + x_d(k-\sigma_0), \ldots, e(k-\sigma_m) \\ &\quad + x_d(k-\sigma_m)) + g(e(k-\sigma_0) + x_d(k-\sigma_0), \\ &\quad \ldots, e(k-\sigma_m) + x_d(k-\sigma_m)) \\ &\quad \times (g^{-1}(x_d(k-\sigma_0), \ldots, x_d(k-\sigma_m)) \\ &\quad \times (S(x_d(k)) - f(x_d(k-\sigma_0), \ldots, x_d(k-\sigma_m))) \\ &\quad + v(k)) - S(x_d(k)), \\ e(k) &= \varrho(k), -\sigma_m \le k \le 0, \end{aligned} \tag{5.11}$$

where $\varrho(k) = \varrho_1(k) - \varrho_2(k)$.

So the aim for this chapter is changed to getting an optimal control law not only making the system (5.11) asymptotically stable but also making the cost functional (5.4) minimal. To solve the optimal tracking control problem in this chapter, the following definition and assumption are required.

Definition 5.2 (cf. [10]) (Asymptotic Stability) An equilibrium state $e = 0$ for system (5.11) is asymptotically stable if:

1. It is stable, i.e., given any positive numbers k_0 and ϵ, there exists $\delta > 0$, such that every solution of system (5.11) satisfies $\max_{k_0 \le k \le k_0+\sigma_m} |e(k)| \le \delta$ and $\max_{k_0 \le k \le \infty} |e(k)| \le \epsilon$.
2. For each $k_0 > 0$ there is a $\delta > 0$ such that every solution of system (5.11) satisfies $\max_{k_0 \le k \le k_0+\sigma_m} |e(k)| \le \delta$ and $\lim_{k\to\infty} e(k) = 0$.

Assumption 5.3 Given the system (5.11), for the infinite-time horizon problem, there exists a control law $v(k)$, which satisfies:

1. $v(k)$ is continuous on Ω; if $e(k-\sigma_0) = e(k-\sigma_1) = \cdots = e(k-\sigma_m) = 0$, then $v(k) = 0$;
2. $v(k)$ stabilizes system (5.11);
3. $\forall e(-\sigma_0), e(-\sigma_1), e(-\sigma_m) \in \mathbb{R}^n$, $J(e(0), v(0))$ is finite.

Actually, for nonlinear systems without time delays, if $v(k)$ satisfies Assumption 5.3, then we can say that $v(k)$ is an admissible control. The definition of an admissible control can be found in [3].

In system (5.11), for time step k, $J^*(e(k))$ is used to denote the optimal value function, i.e., $J^*(e(k)) = \inf_{v(k)} J(e(k), v(k))$, and the corresponding control input can be formulated as $v^*(k) = \arg\inf_{v(k)} J(e(k), v(k))$. Thus, we see that $u^*(k) = v^*(k) + u_s(k)$ is the optimal control for system (5.1). Let $e^*(k) = x^*(k) - x_d(k)$, where $x^*(k)$ is used to denote the state under the action of the optimal tracking control law $u^*(k)$.

According to Bellman's principle of optimality [10], $J^*(e(k))$ should satisfy the following HJB equation:

$$J^*(e(k)) = \inf_{v(k)} \left\{ e^{\mathrm{T}}(k) Q e(k) + v^{\mathrm{T}}(k) R v(k) + J^*(e(k+1)) \right\}, \tag{5.12}$$

i.e., the optimal controller $v^*(k)$ should satisfy

$$v^*(k) = \arg \inf_{v(k)} \left\{ e^{\mathrm{T}}(k) Q e(k) + v^{\mathrm{T}}(k) R v(k) + J^*(e(k+1)) \right\}. \tag{5.13}$$

Here we define $e^*(k) = x^*(k) - x_d(k)$, and

$$\begin{aligned} x^*(k+1) &= f(x^*(k-\sigma_0), \ldots, x^*(k-\sigma_m)) \\ &\quad + g(x^*(k-\sigma_0), \ldots, x^*(k-\sigma_m)) u^*(k), \quad k = 0, 1, 2, \ldots, \\ x^*(k) &= \varrho_1(k), \ k = -\sigma_m, -\sigma_{m-1}, \ldots, -\sigma_0. \end{aligned} \tag{5.14}$$

Then the HJB equation is written as follows:

$$J^*(e^*(k)) = e^{*\mathrm{T}}(k) Q e^*(k) + v^{*\mathrm{T}}(k) R v^*(k) + J^*(e^*(k+1)). \tag{5.15}$$

Remark 5.4 Of course, one can reduce system (5.1) to a system without time delay by defining a new $(\sigma_m + 1)n$-dimensional state vector $y(k) = (x(k - \sigma_0), x(k - \sigma_1), \ldots, x(k - \sigma_m))$. However, Chyung has pointed out that there are two major disadvantages of this method in [6]. First, the resulting new system is a $(\sigma_m + 1)n$-dimensional system increasing the dimension of the system by $(\sigma_m + 1)$ fold. Second, the new system may not be controllable, even if the original system is controllable. This causes the set of attainability to have an empty interior which, in turn, introduces additional difficulties [5–7].

Therefore, in the following part, a direct method for time-delay systems is developed for the optimal tracking control law. In the novel algorithm, we emphasize that the states are regulated by the designed controller in each iteration. The detailed iteration process is as follows.

First, we start with the initial value function $V^{[0]}(\cdot) = 0$ which is not necessarily the optimal value function. Then, for any given state $\varrho_1(t)$ $(-\sigma_m \le t \le 0)$, and initial control $\beta(t)$ in system (5.1), at any current time k, we start the iterative algorithm from $i = 0$ to find the control law $v^{[0]}(k)$ as follows:

$$v^{[0]}(k) = \arg \inf_{v(k)} \left\{ e^{\mathrm{T}}(k) Q e(k) + v^{\mathrm{T}}(k) R v(k) \right\}, \tag{5.16}$$

and the value function is updated as follows:

$$V^{[1]}(e(k)) = \inf_{v(k)} \left\{e^{\mathrm{T}}(k)Qe(k) + v^{\mathrm{T}}(k)Rv(k)\right\}, \tag{5.17}$$

where $e(k) = x(k) - x_d(k)$, and

$$\begin{aligned} x(t+1) &= f(x(t-\sigma_0), \ldots, x(t-\sigma_m)) \\ &\quad + g(x(t-\sigma_0), \ldots, x(t-\sigma_m))\beta(t) \\ x(t) &= \varrho_1(t). \ -\sigma_m \le t \le 0. \end{aligned} \tag{5.18}$$

We notice that the states are generated by the designed controller, so we further get the value function iteration as follows:

$$V^{[1]}(e^{[0]}(k)) = e^{[0]\mathrm{T}}(k)Qe^{[0]}(k) + v^{[0]\mathrm{T}}(k)Rv^{[0]}(k). \tag{5.19}$$

For $i = 1, 2, \ldots$, the iterative HDP algorithm performs the iterations between

$$\begin{aligned} v^{[i]}(k) = \arg \inf_{v(k)} \{&e^{\mathrm{T}}(k)Qe(k) + v^{\mathrm{T}}(k)Rv(k) \\ &+ V^{[i]}(e(k+1))\} \end{aligned} \tag{5.20}$$

and

$$\begin{aligned} V^{[i+1]}(e(k)) = \inf_{v(k)} \{&e^{\mathrm{T}}(k)Qe(k) + v^{\mathrm{T}}(k)Rv(k) \\ &+ V^{[i]}(e(k+1))\}. \end{aligned} \tag{5.21}$$

It should be emphasized that the states in each iteration are regulated by the designed control law, so we further obtain

$$\begin{aligned} V^{[i+1]}(e^{[i]}(k)) = e^{[i]\mathrm{T}}(k)Qe^{[i]}(k) &+ v^{[i]\mathrm{T}}(k)Rv^{[i]}(k) \\ &+ V^{[i]}(e^{[i-1]}(k+1)), \end{aligned} \tag{5.22}$$

where $e^{[i]}(k) = x^{[i]}(k) - x_d(k)$, $i = 0, 1, 2, \ldots$, and the states are updated as follows:

$$\begin{aligned} x^{[i]}(t+1) &= \begin{cases} f(x^{[i+1]}(t-\sigma_0), \ldots, x^{[i+1]}(t-\sigma_m)) \\ \quad +g(x^{[i+1]}(t-\sigma_0), \ldots, x^{[i+1]}(t-\sigma_m)) \\ \quad \times u^{[i+1]}(t), \ t \ge k, \\ f(x^{[i]}(t-\sigma_0), \ldots, x^{[i]}(t-\sigma_m)) \\ \quad +g(x^{[i]}(t-\sigma_0), \ldots, x^{[i]}(t-\sigma_m)) \\ \quad \times u^{[i]}(t), \ 0 \le t < k, \end{cases} \\ x^{[i]}(t) &= \varrho_1(t), \ -\sigma_m \le t \le 0, \end{aligned} \tag{5.23}$$

where $u^{[i]}(t) = v^{[i]}(t) + u_s(t)$.

Remark 5.5 From (5.23) we find that the state $x^{[i]}(t+1), t \geq k$, is related to $x^{[i+1]}(t-\sigma_0), \ldots, x^{[i+1]}(t-\sigma_m)$ and $u^{[i+1]}(t)$. It reflects the "backward iteration" of state at $t, t \geq k$.

Remark 5.6 One important property we must point out is that the HDP algorithm developed in this chapter is different from the algorithm presented in [13]:

1. The HDP algorithm presented in [13] deals with nonlinear systems without time delays, while the HDP algorithm developed in this chapter deals with time-delay nonlinear systems.
2. For the HDP algorithm presented in [13], only the value function is updated according to the control law iteration in every iteration step. For the HDP algorithm developed in this chapter, besides the value function iteration, the state is also updated according to the control law iteration.

Based on the analysis above, our algorithm is a novel HDP algorithm. It is a development of the HDP algorithm presented in [13].

In the following part, we present the convergence analysis of the iteration of (5.16)–(5.23).

Lemma 5.7 *Let $v^{[i]}(k)$ and $V^{[i+1]}(e^{[i]}(k))$ be expressed as* (5.20) *and* (5.22), *and let $\{\mu^{[i]}(k)\}$ be arbitrary sequence of control law, and let $\Lambda^{[i+1]}$ be defined by*

$$\Lambda^{[i+1]}(e^{[i]}(k)) = e^{[i]\mathrm{T}}(k)Qe^{[i]}(k) + \mu^{[i]\mathrm{T}}(k)R\mu^{[i]}(k) + \Lambda^{[i]}(e^{[i-1]}(k+1)), i \geq 0. \tag{5.24}$$

Thus, if $V^{[0]} = \Lambda^{[0]} = 0$, then $V^{[i+1]}(e^{[i]}(k)) \leq \Lambda^{[i+1]}(e^{[i]}(k)), \forall e^{[i]}(k)$.

Proof For given $\varrho_1(t)$ $(-\sigma_m \leq t \leq 0)$ of system (5.1), it can be straightforwardly seen from the fact that $V^{[i+1]}(e^{[i]}(k))$ is obtained by the control input $v^{[i]}(k)$, while $\Lambda^{[i+1]}(e^{[i]}(k))$ is the result of arbitrary control input. □

Lemma 5.8 *Let the sequence $V^{[i+1]}(e^{[i]}(k))$ be defined by* (5.22); *$v^{[i]}(k)$ is the control law expressed as* (5.20). *There is an upper bound Y, such that $0 \leq V^{[i+1]}(e^{[i]}(k)) \leq Y, \forall e^{[i]}(k)$.*

Proof Let $\gamma^{[i]}(k)$ be any control law that satisfies Assumption 5.3. Let $V^{[0]} = P^{[0]} = 0$, and let $P^{[i+1]}(e^{[i]}(k))$ be defined as follows:

$$P^{[i+1]}(e^{[i]}(k)) = e^{[i]\mathrm{T}}(k)Qe^{[i]}(k) + (\gamma^{[i]}(k))^{\mathrm{T}}R\gamma^{[i]}(k) + P^{[i]}(e^{[i-1]}(k+1)). \tag{5.25}$$

From (5.25), we obtain

$$P^{[i]}(e^{[i-1]}(k+1)) = e^{[i-1]\mathrm{T}}(k+1)Qe^{[i-1]}(k+1)$$

$$+ \gamma^{[i-1]\mathrm{T}}(k+1)R\gamma^{[i-1]}(k+1)$$
$$+ P^{[i-1]}(e^{[i-2]}(k+2)). \tag{5.26}$$

Thus, we get

$$\begin{aligned} P^{[i+1]}(e^{[i]}(k)) = {} & e^{[i]\mathrm{T}}(k)Qe^{[i]}(k) \\ & + \gamma^{[i]\mathrm{T}}(k)R\gamma^{[i]}(k) \\ & + e^{[i-1]\mathrm{T}}(k+1)Qe^{[i-1]}(k+1) \\ & + \gamma^{[i-1]\mathrm{T}}(k+1)R\gamma^{[i-1]}(k+1) \\ & + \cdots \\ & + e^{[0]\mathrm{T}}(k+i)Qe^{[0]}(k+i) \\ & + \gamma^{[0]\mathrm{T}}(k+i)R\gamma^{[0]}(k+i). \end{aligned} \tag{5.27}$$

For $j = 0, \ldots, i$, we let

$$\begin{aligned} L(k+j) = {} & e^{[i-j]\mathrm{T}}(k+j)Qe^{[i-j]}(k+j) \\ & + \gamma^{[i-j]\mathrm{T}}(k+j)R\gamma^{[i-j]}(k+j). \end{aligned} \tag{5.28}$$

Then (5.27) can further be written as follows:

$$P^{[i+1]}(e^{[i]}(k)) = \sum_{j=0}^{i} L(k+j). \tag{5.29}$$

Furthermore, we see that

$$\forall i: \ P^{[i+1]}(e^{[i]}(k)) \le \lim_{i\to\infty} \sum_{j=0}^{i} L(k+j). \tag{5.30}$$

Noting that $\{\gamma^{[i]}(k)\}$ satisfies Assumption 5.3, according to the third condition of Assumption 5.3, there exists an upper bound Y such that

$$\lim_{i\to\infty} \sum_{j=0}^{i} L(k+j) \le Y. \tag{5.31}$$

From Lemma 5.7 we obtain

$$\forall i: \ V^{[i+1]}(e^{[i]}(k)) \le P^{[i+1]}(e^{[i]}(k)) \le Y. \tag{5.32}$$

This completes the proof. □

Theorem 5.9 (cf. [12]) *For system* (5.1), *the iterative algorithm is as in* (5.16)–(5.23); *then we have* $V^{[i+1]}(e^{[i]}(k)) \geq V^{[i]}(e^{[i]}(k))$, $\forall e^{[i]}(k)$.

Proof For convenience in our analysis, define a new sequence $\{\Phi^{[i]}(e(k))\}$ as follows:

$$\begin{aligned}\Phi^{[i]}(e(k)) &= e^{\mathrm{T}}(k)Qe(k) + v^{[i]\mathrm{T}}(k)Rv^{[i]}(k)\\ &\quad + \Phi^{[i-1]}(e(k+1)),\end{aligned} \tag{5.33}$$

with $v^{[i]}(k)$ defined by (5.20), $\Phi^{[0]}(e(k)) = 0$.

In the following part, we will prove $\Phi^{[i]}(e(k)) \leq V^{[i+1]}(e(k))$ by mathematical induction.

First, we prove that it holds for $i = 0$. Notice that

$$\begin{aligned}V^{[1]}(e(k)) - \Phi^{[0]}(e(k)) &= e^{\mathrm{T}}(k)Qe(k)v^{[0]\mathrm{T}}(k)Rv^{[0]}(k)\\ &\geq 0,\end{aligned} \tag{5.34}$$

thus, for $i = 0$, we have

$$V^{[1]}(e(k)) \geq \Phi^{[0]}(e(k)). \tag{5.35}$$

Second, we assume that it holds for i, i.e., $V^{[i]}(e(k)) \geq \Phi^{[i-1]}(e(k))$, for $\forall e(k)$. Then, from (5.21) and (5.33), we get

$$\begin{aligned}V^{[i+1]}(e(k)) - \Phi^{[i]}(e(k)) &= V^{[i]}(e(k+1)) - \Phi^{[i-1]}(e(k+1))\\ &\geq 0,\end{aligned} \tag{5.36}$$

i.e., the following inequality holds:

$$\Phi^{[i]}(e(k)) \leq V^{[i+1]}(e(k)). \tag{5.37}$$

Therefore, (5.37) is proved by mathematical induction, for $\forall e(k)$.

On the other hand, from Lemma 5.7, we have $\forall e^{[i]}(k)$, $V^{[i+1]}(e^{[i]}(k)) \leq \Phi^{[i+1]}(e^{[i]}(k))$. So we have $V^{[i]}(e(k)) \leq \Phi^{[i]}(e(k))$.

Therefore for any $e(k)$, we have

$$V^{[i]}(e(k)) \leq \Phi^{[i]}(e(k)) \leq V^{[i+1]}(e(k)). \tag{5.38}$$

So, for any $e^{[i]}(k)$, we have

$$V^{[i+1]}(e^{[i]}(k)) \geq V^{[i]}(e^{[i]}(k)). \tag{5.39}$$

□

We let $V^L(e^L(k)) = \lim_{i\to\infty} V^{[i+1]}(e^{[i]}(k))$. Accordingly, $v^L(k) = \lim_{i\to\infty} \times v^{[i]}(k)$ and $e^L(k) = \lim_{i\to\infty} e^{[i]}(k)$ are the corresponding states. Let $x^L(k) = e^L(k) + \eta(k)$ and $u^L(k) = v^L(k) + u_s(k)$.

In the following part, we will show that the value function $J^L(e^L(k))$ satisfies the corresponding HJB function, and that it is the optimal value function.

Theorem 5.10 *If the value function* $V^{i+1}(e^i(k))$ *is defined by* (5.22). *Let* $V^L(e^L(k)) = \lim_{i\to\infty} V^{[i+1]}(e^{[i]}(k))$. *Let* $v^L(k) = \lim_{i\to\infty} v^{[i]}(k)$ *and* $e^L(k) = \lim_{i\to\infty} e^{[i]}(k)$ *be the corresponding states. Then, the following equation can be established*:

$$V^L(e^L(k)) = e^{L\mathrm{T}}(k)Qe^L(k) + v^{L\mathrm{T}}(k)Rv^L(k) + V^L(e^L(k+1)). \tag{5.40}$$

Proof First, according to Theorem 5.9, we have $V^{[i+1]}(e^{[i]}(k)) \geq V^{[i]}(e^{[i]}(k))$, $\forall e^{[i]}(k)$. So, we obtain

$$\begin{aligned} V^{[i+1]}(e^{[i]}(k)) \geq{}& e^{[i]\mathrm{T}}(k)Qe^{[i]}(k) + v^{[i-1]\mathrm{T}}(k)Rv^{[i-1]}(k) \\ &+ V^{[i-1]}(e^{[i-1]}(k+1)). \end{aligned} \tag{5.41}$$

Let $i \to \infty$; we have $v^L(k) = \lim_{i\to\infty} v^{[i]}(k)$ and

$$\begin{aligned} V^L(e^L(k)) \geq{}& e^{L\mathrm{T}}(k)Qe^L(k) + v^{L\mathrm{T}}(k)Rv^L(k) \\ &+ V^L(e^L(k+1)). \end{aligned} \tag{5.42}$$

On the other hand, for any i and arbitrary control policy $\{\mu^{[i]}(k)\}$, let $\Lambda^{[i+1]}(e^{[i]}(k))$ be as in (5.24). By Lemma 5.7, we have

$$\begin{aligned} V^{[i+1]}(e^{[i]}(k)) \leq{}& e^{[i]\mathrm{T}}(k)Qe^{[i]}(k) + \mu^{[i]\mathrm{T}}(k)R\mu^{[i]}(k) \\ &+ \Lambda^{[i]}(e^{[i-1]}(k+1)). \end{aligned} \tag{5.43}$$

Since $\mu^{[i]}(k)$ in (5.43) is chosen arbitrarily, we let the control policy be $\{\mu^{[i]}(k)\} = \{v^{[i]}(k)\}, \forall i$. So we get

$$\begin{aligned} V^{[i+1]}(e^{[i]}(k)) \leq{}& e^{[i]\mathrm{T}}(k)Qe^{[i]}(k) + v^{[i]\mathrm{T}}(k)Rv^{[i]}(k) \\ &+ V^{[i]}(e^{[i-1]}(k+1)). \end{aligned} \tag{5.44}$$

Let $i \to \infty$, and then we have

$$\begin{aligned} V^L(e^L(k)) \leq{}& e^{L\mathrm{T}}(k)Qe^L(k) + v^{L\mathrm{T}}(k)Rv^L(k) \\ &+ V^L(e^L(k+1)). \end{aligned} \tag{5.45}$$

Thus, combining (5.42) with (5.45), we have

$$\begin{aligned} V^L(e^L(k)) ={}& e^{L\mathrm{T}}(k)Qe^L(k) + v^{L\mathrm{T}}(k)Rv^L(k) \\ &+ V^L(e^L(k+1)). \end{aligned} \tag{5.46}$$

□

Next, we give a theorem to demonstrate $V^L(e^L(k)) = J^*(e^*(k))$.

Theorem 5.11 *Let the value function $V^{[i+1]}(e^{[i]}(k))$ be defined as (5.22) and $V^L(e^L(k)) = \lim_{i\to\infty} V^{[i+1]}(e^{[i]}(k))$. $J^*(e^*(k))$ is defined as in (5.15). Then, we have $V^L(e^L(k)) = J^*(e^*(k))$.*

Proof According to the definition $J^*(e(k)) = \inf_{v(k)}\{J(e(k), v(k))\}$, we know that

$$V^{[i+1]}(e(k)) \geq J^*(e(k)). \tag{5.47}$$

So for $\forall e^{[i]}(k)$, we have

$$V^{[i+1]}(e^{[i]}(k)) \geq J^*(e^{[i]}(k)). \tag{5.48}$$

Let $i \to \infty$, and then we have

$$V^L(e^L(k)) \geq J^*(e^L(k)). \tag{5.49}$$

On the other hand, according to $J^*(e(k)) = \inf_{v(k)}\{J(e(k), v(k))\}$, for any $\theta > 0$ there exists a sequence of control policy $\mu^{[i]}(k)$, such that the associated value function $\Lambda^{[i+1]}(e(k))$ similar to (5.24) satisfies $\Lambda^{[i+1]}(e(k)) \leq J^*(e(k)) + \theta$. From Lemma 5.7, we get

$$V^{[i+1]}(e(k)) \leq \Lambda^{[i+1]}(e(k)) \leq J^*(e(k)) + \theta. \tag{5.50}$$

So we have

$$V^{[i+1]}(e^{[i]}(k)) \leq J^*(e^{[i]}(k)) + \theta. \tag{5.51}$$

Let $i \to \infty$, and then we obtain

$$V^L(e^L(k)) \leq J^*(e^L(k)) + \theta. \tag{5.52}$$

Noting that θ is chosen arbitrarily, we have

$$V^L(e^L(k)) \leq J^*(e^L(k)). \tag{5.53}$$

From (5.49) and (5.53), we get

$$V^L(e^L(k)) = J^*(e^L(k)). \tag{5.54}$$

From Theorem 5.10, we see that $V^L(e^L(k))$ satisfies the HJB equation, so we have $v^L(k) = v^*(k)$. From (5.14) and (5.23), we have $e^L(k) = e^*(k)$. Then, we draw the conclusion that $V^L(e^L(k)) = J^*(e^*(k))$. □

After that, we give a theorem to demonstrate that the state error system (5.11) is asymptotically stable, i.e., the system state $x(k)$ follows $x_d(k)$ asymptotically.

The following lemma is necessary for the proof of the stability property.

Lemma 5.12 *Define the value function sequence* $\{V^{[i+1]}(e^{[i]}(k))\}$ *as* (5.22) *with* $V^{[0]}=0$, *and the control law sequence* $\{v^{[i]}(k)\}$ *as* (5.20). *Then, we find that for* $\forall i=0,1,\ldots$, *the value function* $V^{[i+1]}(e^{[i]}(k))$ *is a positive definite function.*

Proof The lemma can be proved by the following three steps.

1. *Show that zero is an equilibrium point for the system* (5.11).

For the autonomous system of the system (5.11), let $e(k-\sigma_0)=e(k-\sigma_1)=\cdots=e(k-\sigma_m)=0$; we have $e(k+1)=0$. According to the first condition of Assumption 5.3, when $e(k-\sigma_0)=e(k-\sigma_1)=\cdots=e(k-\sigma_m)=0$, we have $v(k)=0$. So according to the definition of the equilibrium state in [10], we can say that zero is an equilibrium point for the system (5.11).

2. *Show that for* $\forall i$, *the value function* $V^{[i+1]}(e^{[i]}(k))=0$ *at the equilibrium point.*

This conclusion can be proved by mathematical induction.

For $i=0$, we have $V^{[0]}=0$, and

$$V^{[1]}(e^{[0]}(k))=e^{[0]\mathrm{T}}(k)Qe^{[0]}(k)+v^{[0]\mathrm{T}}(k)Rv^{[0]}(k). \tag{5.55}$$

Let $e^{[0]}(k-\sigma_0)=e^{[0]}(k-\sigma_1)=\cdots=e^{[0]}(k-\sigma_m)=0$ at the equilibrium point; hence we have $v^{[0]}(k)=0$ according to Assumption 5.3. Then we get $V^{[1]}(e^{[0]}(k))=0$ at the equilibrium point.

Assume that for any i, $V^{[i]}(e^{[i-1]}(k+1))=0$ holds at the equilibrium point. For $i+1$, we have

$$\begin{aligned}
V^{[i+1]}(e^{[i]}(k)) &= e^{[i]\mathrm{T}}(k)Qe^{[i]}(k)+v^{[i]\mathrm{T}}(k)Rv^{[i]}(k)+V^{[i]}(e^{[i-1]}(k+1))\\
&= e^{[i]\mathrm{T}}(k)Qe^{[i]}(k)+v^{[i]\mathrm{T}}(k)Rv^{[i]}(k)\\
&\quad + V^{[i]}(f(e^{[i]}(k-\sigma_0)+x_d(k-\sigma_0),\ldots,e^{[i]}(k-\sigma_m)\\
&\quad + x_d(k-\sigma_0))\\
&\quad + g(e^{[i]}(k-\sigma_0)+x_d(k-\sigma_0),\ldots,e^{[i]}(k-\sigma_m)\\
&\quad + x_d(k-\sigma_m))(v^{[i]}(k)+u_s(k))-S(x_d(k))).
\end{aligned} \tag{5.56}$$

Let $e^{[i]}(k-\sigma_0)=e^{[i]}(k-\sigma_1)=\cdots=e^{[i]}(k-\sigma_m)=0$ at the equilibrium point. Then, according to the first condition of Assumption 5.3, we have $v^{[i]}(k)=0$. So we obtain $V^{[i+1]}(e^{[i]}(k))=0$ at the equilibrium point.

3. *Show that the iterative value function* $V^{[i+1]}(e^{[i]}(k))$, $i=0,1,\ldots$, *is a positive definite function.*

From (5.22), we get

$$\begin{aligned}
V^{[i+1]}(e^{[i]}(k)) &= e^{[i]\mathrm{T}}(k)Qe^{[i]}(k)+v^{[i]\mathrm{T}}(k)Rv^{[i]}(k)\\
&\quad + e^{[i-1]\mathrm{T}}(k+1)Qe^{[i-1]}(k+1)
\end{aligned}$$

$$+ v^{[i-1]\mathrm{T}}(k+1)Rv^{[i-1]}(k+1) + \dots$$
$$+ e^{[0]\mathrm{T}}(k+i)Qe^{[0]}(k+i) + v^{[0]\mathrm{T}}(k+i)Rv^{[0]}(k+i). \quad (5.57)$$

Then, we have $V^{[i+1]}(e^{[i]}(k)) > 0$ for $e^{[i]}(k) \neq 0, \forall i$. On the other hand, as $e^{[i]}(k) \to \infty$, we have $V^{[i+1]}(e^{[i]}(k)) \to \infty$. So we can say that the value function $V^{[i+1]}(e^{[i]}(k))$ is a positive definite function. □

Now, we give the theorem of the stability property.

Theorem 5.13 *Let the optimal control* $v^*(k)$ *be expressed as* (5.13) *and let the value function* $J^*(e^*(k))$ *be expressed as* (5.15). *Then, we find that the optimal control* $v^*(k)$ *stabilizes the system* (5.11) *asymptotically.*

Proof We already know that $V^L(e^L(k))$ is a positive definite function. Furthermore, we have proved $V^L(e^L(k)) = J^*(e^*(k))$ in Theorem 5.11. So we see that $J^*(e^*(k))$ is a positive definite function.

Furthermore, according to the HJB equation (5.15), we have

$$\begin{aligned} J^*(e^*(k+1)) - J^*(e^*(k)) &= -\left\{e^{*\mathrm{T}}(k)Qe^*(k) + v^{*\mathrm{T}}(k)Rv^*(k)\right\} \\ &\leq 0. \end{aligned} \quad (5.58)$$

According to the definition of a Lyapunov function [9], we find that $J^*(e^*(k))$ is a Lyapunov function, which proves the conclusion. □

Therefore, we conclude that the value function $\{V^{[i+1]}(e^{[i]}(k))\}$ is convergent, based on the HDP algorithm developed in this chapter. Furthermore, the limit of $\{V^{[i+1]}(e^{[i]}(k))\}$ satisfies the HJB equation and is the optimal one.

5.4 Simulations

Example 5.14 Consider the following nonlinear time-delay system which is the example in [1, 13] with modification:

$$\begin{aligned} x(k+1) &= f(x(k), x(k-1), x(k-2)) \\ &\quad + g(x(k), x(k-1), x(k-2))u(k), \\ x(k) &= \varrho_1(k), -2 \leq k \leq 0, \end{aligned} \quad (5.59)$$

where

$$f(x(k), x(k-1), x(k-2)) = \begin{bmatrix} 0.2x_1(k)\exp\left(x_2^2(k)\right) & x_2(k-2) \\ 0.3x_2^2(k) & x_1(k-1) \end{bmatrix},$$

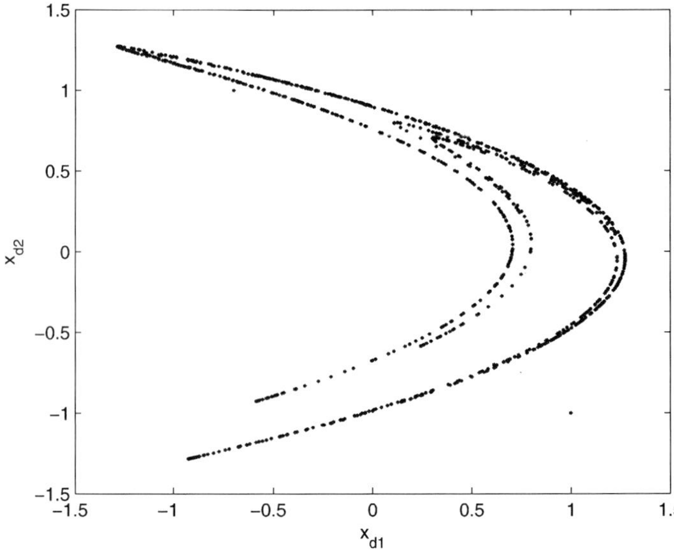

Fig. 5.1 The Hénon chaos orbits

and

$$g(x(k), x(k-1), x(k-2)) = \begin{bmatrix} -0.2 & 0 \\ 0 & -0.2 \end{bmatrix}.$$

The desired signal is the well-known Hénon mapping, as follows:

$$\begin{bmatrix} x_{d1}(k+1) \\ x_{d2}(k+1) \end{bmatrix} = \begin{bmatrix} 1 + bx_{d2}(k) - ax_{d1}^2(k) \\ x_{d2}(k) \end{bmatrix}, \tag{5.60}$$

where $a = 1.4$, $b = 0.3$, $\left[\begin{smallmatrix} x_{d1}(0) \\ x_{d2}(0) \end{smallmatrix}\right] = \left[\begin{smallmatrix} 0.1 \\ -0.1 \end{smallmatrix}\right]$. The chaotic signal orbits are given as Fig. 5.1.

Based on the implementation of the present HDP algorithm, we first give the initial states as $\varrho_1(-2) = \varrho_1(-1) = \varrho_1(0) = [0.5\ -0.5]^{\mathrm{T}}$, and the initial control law as $\beta(k) = -2x(k)$. The implementation of the algorithm is at the time instant $k = 3$. The maximal iteration step $i_{\max}$ is 50. We choose three-layer BP neural networks as the critic network and the action network with the structure 2–8–1 and 6–8–2, respectively. The iteration time of the weights updating for two neural networks is 100. The initial weights are chosen randomly from $[-0.1, 0.1]$, and the learning rate is $\alpha_a = \alpha_c = 0.05$. We select $Q = R = I_2$.

The state trajectories are given as Figs. 5.2 and 5.3. The solid lines in the two figures are the system states, and the dashed lines are the trajectories of the desired chaotic signal. From the two figures we see that the state trajectories of system (5.59) follow the chaotic signal satisfactorily. The corresponding control error curves are given as Fig. 5.4. From Lemma 5.8, Theorem 5.9 and Theorem 5.11, the value func-

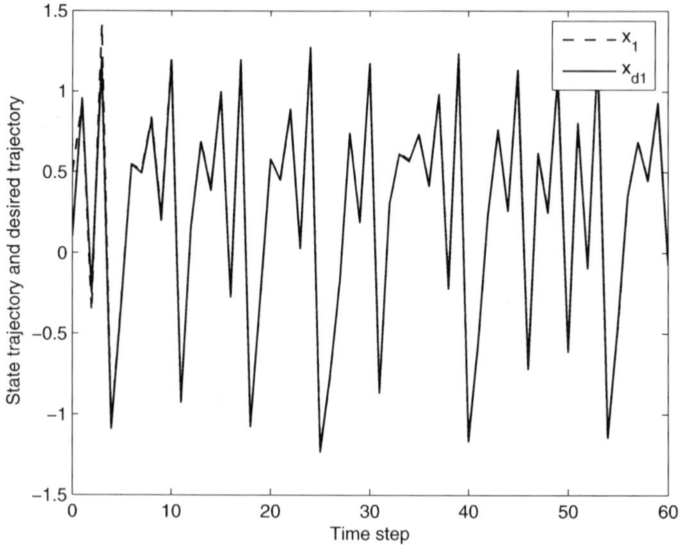

Fig. 5.2 The state variable trajectory x_1 and desired trajectory x_{d1}

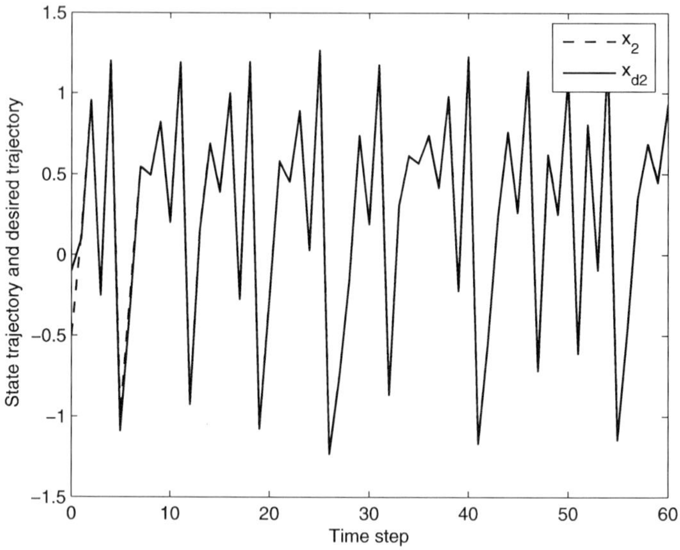

Fig. 5.3 The state variable trajectory x_2 and desired trajectory x_{d2}

tion sequence $\{V^{[i+1]}\}$ is bounded and nondecreasing. Furthermore, it converges to the optimal value function J^* as $i \to \infty$. The curve in Fig. 5.5 shows the properties of the value function sequence. According to Theorem 5.13, we know that

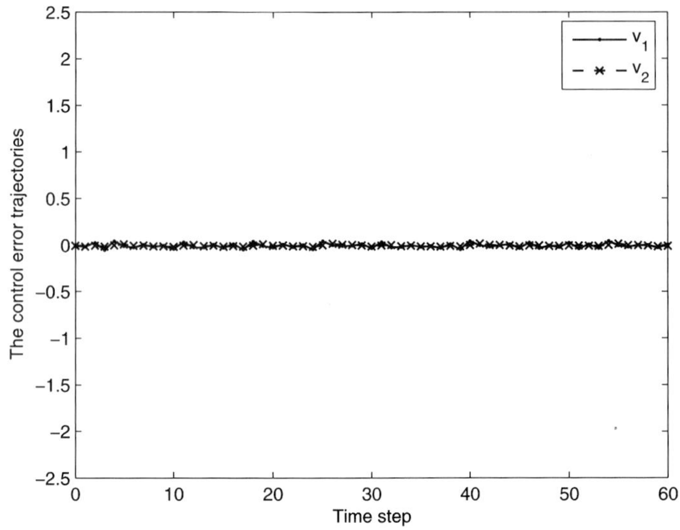

Fig. 5.4 The control error variable trajectory v_1 and v_2

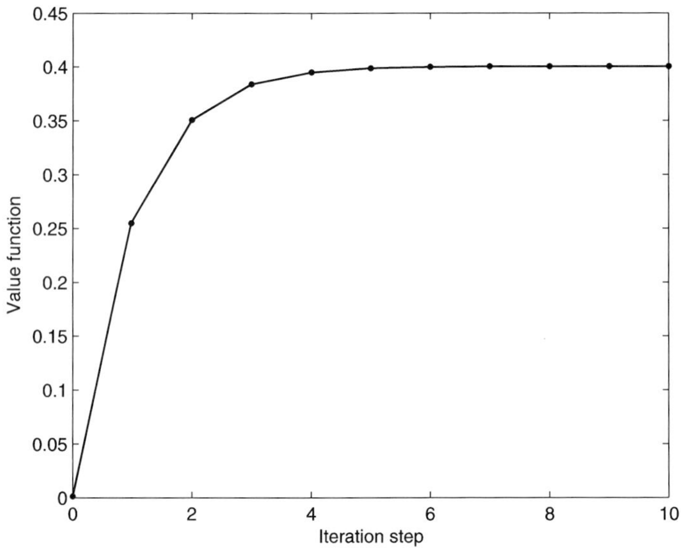

Fig. 5.5 The convergence of value function

system (5.11) is asymptotically stable. The error trajectories for the system (5.59) and Hénon chaotic signal are presented in Fig. 5.6, and they converge to zero asymptotically. It is clear that the new HDP iteration algorithm is very feasible.

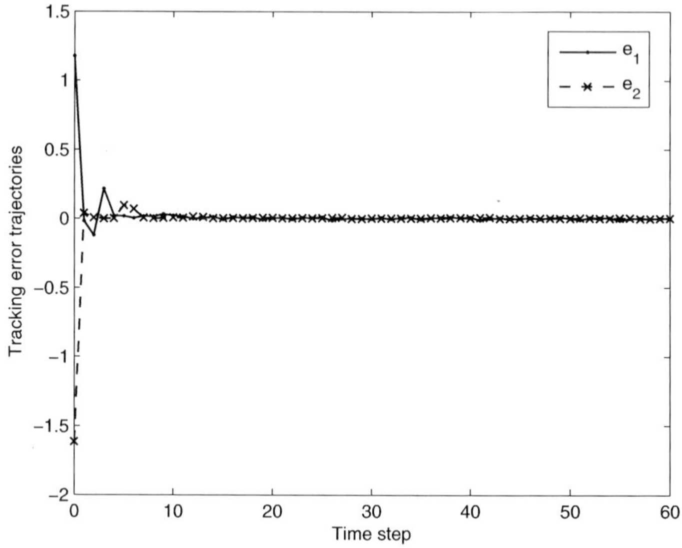

Fig. 5.6 The tracking error trajectories e_1 and e_2

Example 5.15 Consider the following nonlinear time-delay system:

$$\begin{aligned} x(k+1) &= f(x(k), x(k-1), x(k-2)) \\ &\quad + g(x(k), x(k-1), x(k-2))u(k), \\ x(k) &= \varrho_1(k), -2 \leq k \leq 0, \end{aligned} \tag{5.61}$$

where

$$f(x(k), x(k-1), x(k-2)) = \begin{bmatrix} 0.2x_1(k)\exp\left(x_2^2(k)\right) & x_2(k-2) \\ 0.3x_2^2(k) & x_1(k-1) \end{bmatrix},$$

and

$$g(x(k), x(k-1), x(k-2)) = \begin{bmatrix} x_2(k-2) & 0 \\ 0 & 1 \end{bmatrix}.$$

From system (5.61), we know that $g(\cdot)$ is not always invertible. Here the Moore–Penrose pseudoinverse technique is used to obtain $g^{-1}(\cdot)$.

The desired orbit $x_d(k)$ is generated by the following exosystem:

$$x_d(k+1) = Ax_d(k), \tag{5.62}$$

with

$$A = \begin{bmatrix} \cos wT & \sin wT \\ -\sin wT & \cos wT \end{bmatrix}, \qquad x_d(0) = \begin{bmatrix} 0 \\ 1 \end{bmatrix},$$

where $T = 0.1s$, $w = 0.8\pi$.

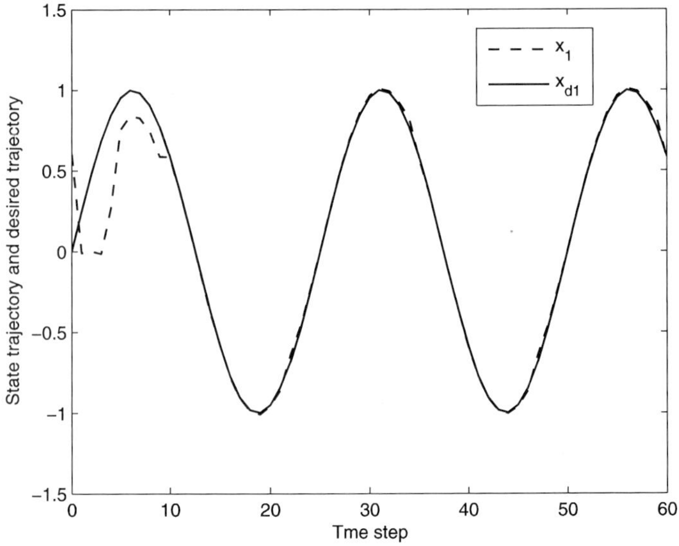

Fig. 5.7 The state variable trajectory x_1 and desired trajectory x_{d1}

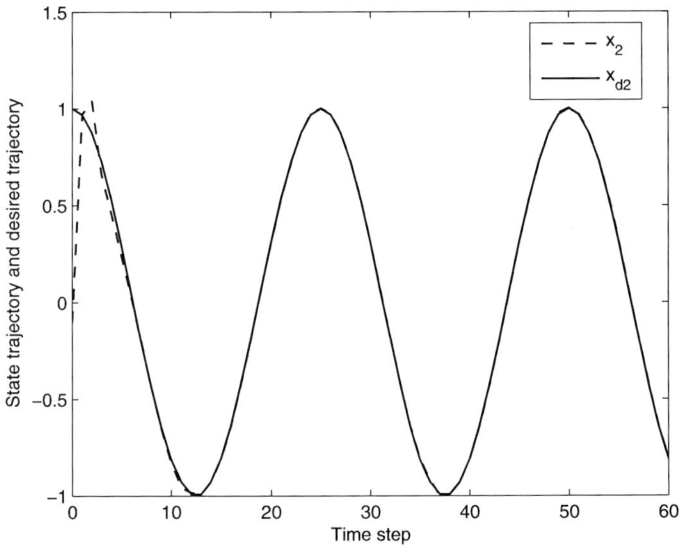

Fig. 5.8 The state variable trajectory x_2 and desired trajectory x_{d2}

First, we give the initial states as $\varrho_1(-2) = \varrho_1(-1) = \varrho_1(0) = [0.5 \quad -0.2]^{\mathrm{T}}$, and the initial control law as $\beta(k) = -2x(k)$. We also implement the present HDP algorithm at the time instant $k = 3$. The maximal iteration step $i_{\max}$ is 60. The learning rate is $\alpha_a = \alpha_c = 0.01$. We select $Q = R = I_2$.

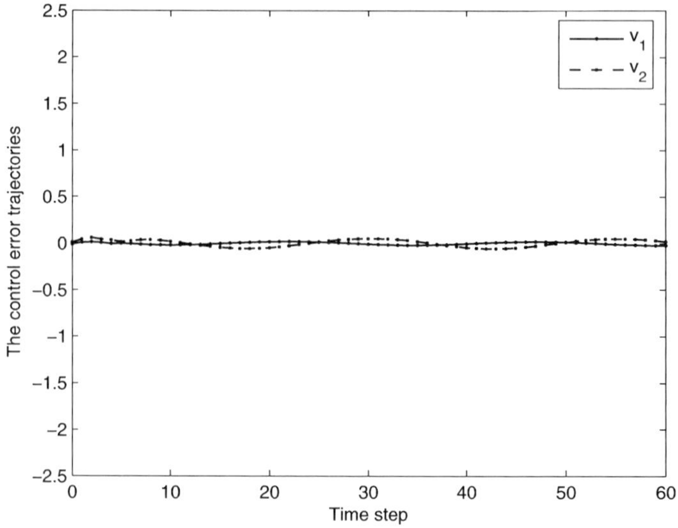

Fig. 5.9 The control error variable trajectory v_1 and v_2

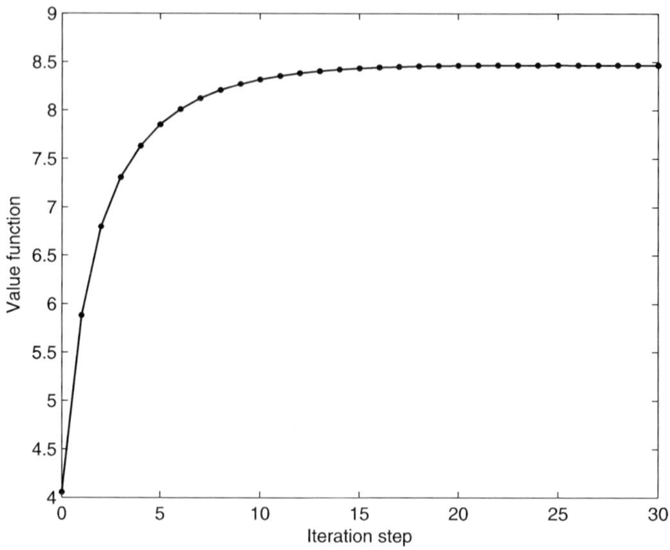

Fig. 5.10 The convergence of the value function

The trajectories of the system states are presented in Figs. 5.7 and 5.8. In the two figures, the solid lines are the system states, and the dashed lines are the desired trajectories. The corresponding control error curves are displayed in Fig. 5.9. In addition, we give the curve of the value function sequence in Fig. 5.10. It is bounded and convergent. It verifies Theorem 5.9 and Theorem 5.11 as well. The tracking er-

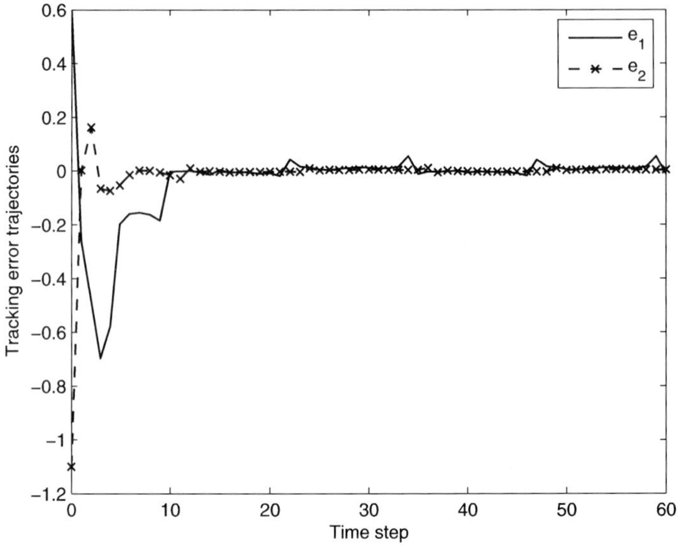

Fig. 5.11 The tracking error trajectories e_1 and e_2

rors are shown in Fig. 5.11, and they converge to zero asymptotically. It is clear that the tracking performance is satisfactory, and the new iterative algorithm developed in this chapter is very effective.

5.5 Summary

In this chapter, we developed an effective HDP algorithm to solve optimal tracking control problems for a class of nonlinear discrete-time systems with time delays. First, we defined a cost functional for time-delay systems. Then, a novel iterative HDP algorithm was developed to solve the optimal tracking control problem. Two neural networks are used to facilitate the implementation of the iterative algorithm. Simulation examples demonstrated the effectiveness of the present optimal tracking control algorithm.

References

1. Al-Tamimi A, Lewis FL (2007) Discrete-time nonlinear HJB solution using approximate dynamic programming: convergence proof. In: Proceedings of IEEE international symposium on approximate dynamic programming and reinforcement learning, Honolulu, HI, pp 38–43
2. Al-Tamimi A, Abu-Khalaf M, Lewis FL (2007) Adaptive critic designs for discrete-time zero-sum games with application to H∞ control. IEEE Trans Syst Man Cybern, Part B, Cybern 37:240–247

3. Al-Tamimi A, Lewis FL, Abu-Khalaf M (2008) Discrete-time nonlinear HJB solution using approximate dynamic programming: convergence proof. IEEE Trans Syst Man Cybern, Part B, Cybern 38:943–949
4. Ben-Israel A, Greville TNE (2002) Generalized inverse: theory and applications. Springer, New York
5. Chyung DH (1967) Discrete systems with time delay. Presented at the 5th annals Allerton conference on circuit and system theory, Urbana
6. Chyung DH (1968) Discrete optimal systems with time delay. IEEE Trans Autom Control 13:117
7. Chyung DH, Lee EB (1966) Linear optimal systems with time delays. SIAM J Control 4(3):548–575
8. Isidori A (2005) Nonlinear control systems II. Springer, Berlin
9. Liao X, Wang L, Yu P (2007) Stability of dynamical systems. Elsevier, Amsterdam
10. Manu MZ, Mohammad J (1987) Time-delay systems analysis, optimization and applications. North-Holland, New York
11. Wei QL, Zhang HG, Liu DR, Zhao Y (2010) An optimal control scheme for a class of discrete-time nonlinear systems with time delays using adaptive dynamic programming. Acta Autom Sin 36:121–129
12. Zhang HG, Song RZ (2011) Optimal tracking control for a class of nonlinear discrete-time systems with time delays based on heuristic dynamic programming. IEEE Trans Neural Netw 22(12):1851–1862
13. Zhang HG, Wei QL, Luo YH (2008) A novel infinite-time optimal tracking control scheme for a class of discrete-time nonlinear systems via the greedy HDP iteration algorithm. IEEE Trans Syst Man Cybern, Part B, Cybern 38:937–942

Chapter 6
Optimal Feedback Control for Continuous-Time Systems via ADP

6.1 Introduction

In this chapter, we will study how to design a controller for continuous-time systems via the ADP method. Although many ADP methods have been proposed for continuous-time systems [1, 6, 9, 10, 12, 13, 15, 17–19], a suitable framework in which the optimal controller can be designed for a class of unknown general continuous-time systems is still not available. Therefore, in Sect. 6.2, we will develop a novel optimal robust feedback control scheme for a class of unknown general continuous-time systems using ADP method. The merit of present method is that we require only the availability of input/output data instead of exact system model. Moreover, the obtained control input can be guaranteed to be close to the optimal control input within a small bound.

As is known, in the real world, many practical control systems are described by nonaffine structure, such as chemical reactions, dynamic model in pendulum control, etc. The difficulty associated with ADP for nonaffine nonlinear system is that the nonlinear function is an implicit function with respect to the control variable. To overcome this difficulty, in Sect. 6.3, we will extend the ADP method to a class of nonaffine nonlinear systems. Through the present two methods, optimal control problems of a quite wide class of continuous-time nonlinear systems can be solved.

6.2 Optimal Robust Feedback Control for Unknown General Nonlinear Systems

In this section, a robust approximate optimal tracking control scheme is developed for a class of unknown general nonlinear systems by using the ADP method. In the design of the controller, only available input/output data are required instead of known system dynamics. First, a data-based model is established by a *recurrent neural network* (RNN) to reconstruct the unknown system dynamics using available input/output data. Then, based on the obtained data-based model, the ADP method is utilized to design the approximate optimal tracking controller.

H. Zhang et al., *Adaptive Dynamic Programming for Control*,
Communications and Control Engineering, DOI 10.1007/978-1-4471-4757-2_6,

6.2.1 Problem Formulation

Consider the following general continuous-time nonlinear systems:

$$\dot{x}(t) = f(x(t), u(t)), \tag{6.1}$$

where $x(t) = [x_1(t), x_2(t), \dots, x_n(t)]^{\mathrm{T}} \in \mathbb{R}^n$ is the state vector, $u(t) = [u_1(t), u_2(t), \dots, u_m(t)]^{\mathrm{T}} \in \mathbb{R}^m$ is the input vector, and $f(\cdot,\cdot)$ is an unknown smooth nonlinear function with respect to $x(t)$ and $u(t)$.

In this section, our control objective is to design an optimal controller for (6.1), which ensures that the state vector $x(t)$ tracks the specified trajectory $x_d(t)$ while minimizing the infinite-horizon cost functional as follows:

$$J(e(t), u) = \int_t^{\infty} l(e(\tau), u(\tau)) d\tau, \tag{6.2}$$

where $e(t) = x(t) - x_d(t)$ denotes the state tracking error, $l(e(t), u(t)) = e^{\mathrm{T}}(t)Qe(t) + u^{\mathrm{T}}(t)Ru(t)$ is the utility function, and Q and R are symmetric positive definite matrices with appropriate dimensions.

Since the system dynamics is completely unknown, we cannot apply existing ADP methods to (6.1) directly. Therefore, it is now desirable to propose a novel control scheme that does not need the exact system dynamics but only the input/output data which can be obtained during the operation of the system. Therefore, we propose a *data-based* optimal robust tracking control scheme using ADP method for unknown general nonlinear continuous-time systems. Specifically, the design of the present controller is divided into two steps:

1. Establishing a data-based model based on an RNN by using available input/output data to reconstruct the unknown system dynamics, and
2. Designing the robust approximate optimal tracking controller based on the obtained data-based model

In the following, the establishment of the data-based model and the controller design will be discussed in detail.

6.2.2 Data-Based Robust Approximate Optimal Tracking Control

Although we cannot obtain the exact system model in general, fortunately, we can access input–output data of the unknown general nonlinear systems in many practical control processes. So it is desirable to use available input–output data in the design of the controller. The historical input–output data could be incorporated indirectly in the form of a data-based model. The data-based model could extract useful information contained in the input–output data and capture input–output mapping. Markov models, neural network models, well structured filters, wavelet models, and other function approximation models can be regarded as data-based models [14, 20–24, 26]. In this section, we develop a data-based model based on a recurrent neural network (RNN) to reconstruct the unknown system dynamics by using available input–output data.

To begin with the development, the system dynamics (6.1) is rewritten in the form of an RNN as follows [16]:

$$\dot{x}(t) = C^{*\mathrm{T}} x(t) + A^{*\mathrm{T}} h(x(t)) + C_u^{*\mathrm{T}} u(t) + A_u^{*\mathrm{T}} + \varepsilon_m(t), \tag{6.3}$$

where $\varepsilon_m(t)$ is assumed to be bounded, $C^{*\mathrm{T}}$, $A^{*\mathrm{T}}$, $C_u^{*\mathrm{T}}$, and $A_u^{*\mathrm{T}}$ are unknown ideal weight matrices. The activation function $h(\cdot)$ is selected as a monotonically increasing function and satisfies

$$0 \le h(x) - h(y) \le k(x - y), \tag{6.4}$$

for any $x, y \in \mathbb{R}$ and $x \ge y$, $k > 0$, such as $h(x) = \tanh(x)$.

Based on (6.3), the data-based model is then constructed as

$$\dot{\hat{x}}(t) = \hat{C}^{\mathrm{T}}(t)\hat{x}(t) + \hat{A}^{\mathrm{T}}(t) h(\hat{x}(t)) + \hat{C}_u^{\mathrm{T}}(t) u(t) + \hat{A}_u^{\mathrm{T}}(t) - \upsilon(t), \tag{6.5}$$

where $\hat{x}(t)$ is the estimated system state vector, $\hat{C}(t)$, $\hat{A}(t)$, $\hat{C}_u(t)$, and $\hat{A}_u(t)$ are the estimates of the ideal weight matrices C^*, A^*, C_u^*, and A_u^*, respectively, and $\upsilon(t)$ is defined as

$$\upsilon(t) = S e_m(t) + \frac{\hat{\theta}(t) e_m(t)}{e_m^{\mathrm{T}}(t) e_m(t) + \eta}, \tag{6.6}$$

where $e_m(t) = x(t) - \hat{x}(t)$ is the system modeling error, $S \in \mathbb{R}^{n\times n}$ is a design matrix, $\hat{\theta}(t) \in \mathbb{R}$ is an additional tunable parameter, and $\eta > 1$ is a constant.

Assumption 6.1 The term $\varepsilon_m(t)$ is assumed to be upper bounded by a function of modeling error such that

$$\varepsilon_m^{\mathrm{T}}(t)\varepsilon_m(t) \le \varepsilon_M(t) = \theta^* e_m^{\mathrm{T}}(t) e_m(t), \tag{6.7}$$

where θ^* is the bounded constant target value.

The modeling error dynamics is written as

$$\begin{aligned}
\dot{e}_m(t) = {} & C^{*\mathrm{T}} e_m(t) + \tilde{C}^{\mathrm{T}}(t)\hat{x}(t) + A^{*\mathrm{T}} \tilde{h}(e_m(t)) \\
& + \tilde{A}^{\mathrm{T}}(t) h(\hat{x}(t)) + \tilde{C}_u^{\mathrm{T}}(t) u(t) + \tilde{A}_u^{\mathrm{T}}(t) + \varepsilon_a(t) \\
& + S e_m(t) - \frac{\tilde{\theta}(t) e_m(t)}{e_m^{\mathrm{T}}(t) e_m(t) + \eta} + \frac{\theta^* e_m(t)}{e_m^{\mathrm{T}}(t) e_m(t) + \eta},
\end{aligned} \tag{6.8}$$

where $\tilde{C}(t) = C^* - \hat{C}(t)$, $\tilde{A}(t) = A^* - \hat{A}(t)$, $\tilde{C}_u(t) = C_u^* - \hat{C}_u(t)$, $\tilde{A}_u(t) = A_u^* - \hat{A}_u(t)$, $\tilde{h}(e_m(t)) = h(x(t)) - h(\hat{x}(t))$, and $\tilde{\theta}(t) = \theta^* - \hat{\theta}(t)$.

Theorem 6.2 (cf. [25]) *The modeling error $e_m(t)$ will be asymptotically convergent to zero as $t \to \infty$ if the weight matrices and the tunable parameter of the data-based model* (6.5) *are updated through the following equations*:

$$\begin{aligned}\dot{\hat{C}}(t) &= \Gamma_1 \hat{x}(t) e_m^{\mathrm{T}}(t),\\ \dot{\hat{A}}(t) &= \Gamma_2 f(\hat{x}(t)) e_m^{\mathrm{T}}(t),\\ \dot{\hat{C}}_u(t) &= \Gamma_3 u(t) e_m^{\mathrm{T}}(t),\\ \dot{\hat{A}}_u(t) &= \Gamma_4 e_m^{\mathrm{T}}(t),\\ \dot{\hat{\theta}}(t) &= -\Gamma_5 \frac{e_m^{\mathrm{T}}(t) e_m(t)}{e_m^{\mathrm{T}}(t) e_m(t) + \eta},\end{aligned} \tag{6.9}$$

where Γ_i is a positive definite matrix such that $\Gamma_i = \Gamma_i^{\mathrm{T}} > 0$, $i = 1, 2, \ldots, 5$.

Proof Choose the following Lyapunov function candidate:

$$J_3(t) = J_1(t) + J_2(t), \tag{6.10}$$

where

$$\begin{aligned}J_1(t) &= \frac{1}{2} e_m^{\mathrm{T}}(t) e_m(t),\\ J_2(t) &= \frac{1}{2} \mathrm{tr}\{\tilde{C}^{\mathrm{T}}(t) \Gamma_1^{-1} \tilde{C}(t) + \tilde{A}^{\mathrm{T}}(t) \Gamma_2^{-1} \tilde{A}(t)\\ &\quad + \tilde{C}_u^{\mathrm{T}}(t) \Gamma_3^{-1} \tilde{C}_u(t) + \tilde{A}_u^{\mathrm{T}}(t) \Gamma_4^{-1} \tilde{A}_u(t)\} + \frac{1}{2} \tilde{\theta}^{\mathrm{T}}(t) \Gamma_5^{-1} \tilde{\theta}(t).\end{aligned}$$

Then, the time derivative of the *Lyapunov function* candidate (6.10) along the trajectories of the error system (6.8) is computed as

$$\begin{aligned}\dot{J}_1(t) &= e_m^{\mathrm{T}}(t) C^{*\mathrm{T}} e_m(t) + e_m^{\mathrm{T}}(t) \tilde{C}^{\mathrm{T}}(t) \hat{x}(t)\\ &\quad + e_m^{\mathrm{T}}(t) A^{*\mathrm{T}} \tilde{h}(e_m(t)) + e_m^{\mathrm{T}}(t) \tilde{A}^{T}(t) h(\hat{x}(t))\\ &\quad + e_m^{\mathrm{T}}(t) \tilde{C}_u^{\mathrm{T}}(t) u(t) + e_m^{\mathrm{T}}(t) \tilde{A}_u^{\mathrm{T}}(t) + e_m^{\mathrm{T}}(t) \varepsilon_m(t)\\ &\quad + e_m^{\mathrm{T}}(t) S e_m(t) - \frac{e_m^{\mathrm{T}}(t) \tilde{\theta}(t) e_m(t)}{e_m^{\mathrm{T}}(t) e_m(t) + \eta} + \frac{e_m^{\mathrm{T}}(t) \theta^{*} e_m(t)}{e_m^{\mathrm{T}}(t) e_m(t) + \eta}.\end{aligned} \tag{6.11}$$

From (6.4), we can obtain

$$e_m^{\mathrm{T}}(t) A^{*\mathrm{T}} \tilde{h}(e_m(t)) \le \frac{1}{2} e_m^{\mathrm{T}}(t) A^{*\mathrm{T}} A^{*} e_m(t) + \frac{1}{2} k^2 e_m^{\mathrm{T}}(t) e_m(t). \tag{6.12}$$

According to Assumption 6.1, we have

$$\begin{aligned} e_m^{\mathrm{T}}(t)\varepsilon_a(t) &\le \frac{1}{2}e_m^{\mathrm{T}}(t)e_m(t) + \frac{1}{2}\varepsilon_m^{\mathrm{T}}(t)\varepsilon_m(t) \\ &\le \frac{1}{2}e_m^{\mathrm{T}}(t)e_m(t) + \frac{1}{2}\theta^* e_m^{\mathrm{T}}(t)e_m(t). \end{aligned} \tag{6.13}$$

Therefore, (6.11) can be rewritten as

$$\begin{aligned} \dot{J}_1(t) \le{}& e_m^{\mathrm{T}}(t)C^{*\mathrm{T}}e_m(t) + e_m^{\mathrm{T}}(t)\tilde{C}^{\mathrm{T}}(t)\hat{x}(t) \\ &+ \frac{1}{2}e_m^{\mathrm{T}}(t)A^{*\mathrm{T}}A^*e_m(t) + \left(\frac{1}{2} + \frac{1}{2}\theta^* + \frac{1}{2}k^2\right)e_m^{\mathrm{T}}(t)e_m(t) \\ &+ e_m^{\mathrm{T}}(t)\tilde{A}^{\mathrm{T}}(t)h(\hat{x}(t)) + e_m^{\mathrm{T}}(t)\tilde{C}_u^{\mathrm{T}}(t)u(t) + e_m^{\mathrm{T}}(t)\tilde{A}_u^{\mathrm{T}}(t) \\ &+ e_m^{\mathrm{T}}(t)Se_m(t) - \frac{e_m^{\mathrm{T}}(t)\tilde{\theta}(t)e_m(t)}{e_m^{\mathrm{T}}(t)e_m(t)+\eta} + \frac{e_m^{\mathrm{T}}(t)\theta^*e_m(t)}{e_m^{\mathrm{T}}(t)e_m(t)+\eta}. \end{aligned} \tag{6.14}$$

By computing the time derivative of $J_2(t)$, we have

$$\begin{aligned} \dot{J}_2(t) ={}& \mathrm{tr}\{\tilde{C}^{\mathrm{T}}(t)\Gamma_1^{-1}\dot{\hat{C}}(t) + \tilde{A}^{\mathrm{T}}(t)\Gamma_2^{-1}\dot{\hat{A}}(t) \\ &+ \tilde{C}_u^{\mathrm{T}}(t)\Gamma_3^{-1}\dot{\hat{C}}_u(t) + \tilde{A}_u^{\mathrm{T}}(t)\Gamma_4^{-1}\dot{\hat{A}}_u(t)\} + \tilde{\theta}^{\mathrm{T}}(t)\Gamma_5^{-1}\dot{\hat{\theta}}(t). \end{aligned} \tag{6.15}$$

Combining (6.14) with (6.15), we have

$$\begin{aligned} \dot{J}_3(t) \le{}& e_m^{\mathrm{T}}(t)C^{*\mathrm{T}}e_m(t) + \frac{1}{2}e_m^{\mathrm{T}}(t)A^{*\mathrm{T}}A^*e_m(t) \\ &+ e_m^{\mathrm{T}}(t)\left(\left(\frac{1}{2} + \frac{1}{2}\theta^* + \frac{1}{2}k^2\right)I_n + S\right)e_m(t) + \frac{e_m^{\mathrm{T}}(t)\theta^*e_m(t)}{e_m^{\mathrm{T}}(t)e_m(t)+\eta} \\ \le{}& e_m^{\mathrm{T}}(t)\Xi e_m(t), \end{aligned} \tag{6.16}$$

where I_n denotes a $n \times n$ identity matrix and

$$\Xi = C^{*\mathrm{T}} + \frac{1}{2}A^{*\mathrm{T}}A^* + \left(\frac{1}{2} + \frac{3}{2}\theta^* + \frac{1}{2}k^2\right)I_n + S,$$

and S is selected to make $\Xi < 0$. Therefore, it can be concluded that $\dot{J}_3(t) < 0$. Since $J_3(t) > 0$, it follows from [3] that $e_m(t) \to 0$ as $t \to \infty$.

This completes the proof. □

Remark 6.3 According to the results of Theorem 6.2, since $e_m(t) \to 0$ as $t \to \infty$, the term $\upsilon(t) \to 0$ as $t \to \infty$. In addition, $\dot{\hat{C}}(t) \to 0$, $\dot{\hat{A}}(t) \to 0$, $\dot{\hat{C}}_u(t) \to 0$, and $\dot{\hat{A}}_u(t) \to 0$ as $e_m(t) \to 0$. It means that $\hat{C}(t)$, $\hat{A}(t)$, $\hat{C}_u(t)$, and $\hat{A}_u(t)$ all tend to be constant matrices which are denoted C, A, C_u, and A_u, respectively.

Consequently, the nonlinear system (6.1) can be rewritten as

$$\dot{x}(t) = C^{\mathrm{T}}x(t) + A^{\mathrm{T}}h(x(t)) + C_u^{\mathrm{T}}u(t) + A_u^{\mathrm{T}}. \tag{6.17}$$

In this way, the original optimal tracking control problem of (6.1) is transformed into the optimal tracking control problem of (6.17). Next, the controller design based on (6.17) will be given in detail.

It is assumed that the desired trajectory $x_d(t)$ has the following form:

$$\dot{x}_d(t) = C^{\mathrm{T}}x_d(t) + A^{\mathrm{T}}h(x_d(t)) + C_u^{\mathrm{T}}u_d(t) + A_u^{\mathrm{T}}, \tag{6.18}$$

where $u_d(t)$ is the control input of the desired system.

By using (6.17) and (6.18), the error system can be formulated as

$$\dot{e}(t) = C^{\mathrm{T}}e(t) + A^{\mathrm{T}}h_e(t) + C_u^{\mathrm{T}}u_e(t), \tag{6.19}$$

where $h_e(t) = h(x(t)) - h(x_d(t))$ and $u_e(t) = u(t) - u_d(t)$. It is noted that the controller $u(t)$ is composed of two parts, the steady-state controller $u_d(t)$ and the feedback controller $u_e(t)$.

The steady-state controller $u_d(t)$ can be obtained from (6.18) as follows:

$$u_d(t) = C_u^{-\mathrm{T}}(\dot{x}_d(t) - C^{\mathrm{T}}x_d(t) - A^{\mathrm{T}}h(x_d(t)) - A_u^{\mathrm{T}}), \tag{6.20}$$

where C_u^{-1} stands for the pseudo-inverse of C_u. The steady-state controller is used to maintain the tracking error close to zero at the steady-state stage.

Next, the feedback controller $u_e(t)$ will be designed to stabilize the state tracking error dynamics at transient stage in an optimal manner. In the following, for brevity, the denotations $e(t)$, $u_d(t)$, $u_e(t)$, $u(t)$, and $V(e(t))$ are rewritten as e, u_d, u_e, u, and $V(e)$.

The infinite-horizon cost functional (6.2) is transformed into

$$J(e,u) = \int_t^{\infty} l(e(\tau), u_e(\tau))\mathrm{d}\tau, \tag{6.21}$$

where $l(e,u_e) = e^{\mathrm{T}}Qe + u_e^{\mathrm{T}}Ru_e$ is the utility function; Q and R are symmetric positive definite matrices with appropriate dimensions.

It is desirable to find the optimal feedback control u_e^* which stabilizes the system (6.19) and minimizes the cost functional (6.21). The kind of control is called admissible control.

Define the *Hamilton function* as

$$H(e,u_e,J_e) = J_e^{\mathrm{T}}(C^{\mathrm{T}}e + A^{\mathrm{T}}h_e + C_u^{\mathrm{T}}u_e) + e^{\mathrm{T}}Qe + u_e^{\mathrm{T}}Ru_e, \tag{6.22}$$

where $J_e = \partial J(e)/\partial e$.

The optimal value function $J^*(e)$ is defined as

$$J^*(e) = \min_{u_e \in \psi(\Omega)} \int_t^{\infty} l(e(\tau), u_e(e(\tau)))\mathrm{d}\tau, \tag{6.23}$$

and satisfies

$$0 = \min_{u_e \in \psi(\Omega)} (H(e, u_e, J_e^*)). \tag{6.24}$$

Further, we can obtain the optimal control u_e^* by solving $\partial H(e, u_e, J_e^*)/\partial u_e = 0$ as

$$u_e^* = -\frac{1}{2} R^{-1} C_u J_e^*, \tag{6.25}$$

where $J_e^* = \partial J^*(e)/\partial e$. Then, the overall optimal control input can be rewritten as $u^* = u_d + u_e^*$.

In the following, we will focus on the optimal feedback controller design using the ADP method, which is implemented by employing the critic NN and the action NN.

A neural network is utilized to approximate $J(e)$ as

$$J(e) = W_c^{\mathrm{T}} \phi_c(e) + \varepsilon_c, \tag{6.26}$$

where W_c is the unknown ideal constant weights and $\phi_c(e) : \mathbb{R}^n \to \mathbb{R}^{N_1}$ is called the critic NN activation function vector, N_1 is the number of neurons in the hidden layer, and ε_c is the critic NN approximation error.

The derivative of the cost function $J(e)$ with respect to e is

$$J_e = \nabla \phi_c^{\mathrm{T}} W_c + \nabla \varepsilon_c, \tag{6.27}$$

where $\nabla \phi_c \triangleq \partial \phi_c(e)/\partial e$ and $\nabla \varepsilon_c \triangleq \partial \varepsilon_c/\partial e$.

Let $\hat{W}_c$ be an estimate of W_c, then we have the estimate of $J(e)$ as follows:

$$\hat{J}(e) = \hat{W}_c^{\mathrm{T}} \phi_c(e). \tag{6.28}$$

Then, the approximate Hamilton function can be derived as follows:

$$\begin{aligned} H(e, u_e, \hat{W}_a) &= \hat{W}_c^{\mathrm{T}} \nabla \phi_c (C^{\mathrm{T}} e + A^{\mathrm{T}} h_e + C_u^{\mathrm{T}} u_e) + e^{\mathrm{T}} Q e + u_e^{\mathrm{T}} R u_e \\ &= e_c. \end{aligned} \tag{6.29}$$

Given any admissible control law u_e, we desire to select $\hat{W}_c$ to minimize the squared residual error $E_c(\hat{W}_c)$ as follows:

$$E_c(\hat{W}_c) = \frac{1}{2} e_c^{\mathrm{T}} e_c. \tag{6.30}$$

The weight update law for the critic NN is a gradient descent algorithm, which is given by

$$\dot{\hat{W}}_c = -\alpha_c \sigma_c (\phi_c^{\mathrm{T}} \hat{W}_c + e^{\mathrm{T}} Q e + u_e^{\mathrm{T}} R u_e), \tag{6.31}$$

where $\alpha_c > 0$ is the adaptive gain of the critic NN, $\sigma_c = \sigma/(\sigma^{\mathrm{T}}\sigma + 1)$, $\sigma = \nabla\phi_c(C^{\mathrm{T}}e + A^{\mathrm{T}}h_e + C_u^{\mathrm{T}}u_e)$. Therefore, there exists a positive constant $\sigma_{cM} > 1$ such that $\|\sigma_c\| \le \sigma_{cM}$. Define the weight estimation error of critic NN to be $\tilde{W}_c = \hat{W}_c - W_c$, and note that, for a fixed control law u_e, the Hamilton function (6.22) becomes

$$\begin{aligned} H(e, u_e, W_c) &= W_c^{\mathrm{T}}\nabla\phi_c(C^{\mathrm{T}}e + A^{\mathrm{T}}h_e + C_u^{\mathrm{T}}u_e) + e^{\mathrm{T}}Qe + u_e^{\mathrm{T}}Ru_e \\ &= \varepsilon_{\mathrm{HJB}}, \end{aligned} \tag{6.32}$$

where the residual error due to the NN approximation error is $\varepsilon_{\mathrm{HJB}} = -\nabla\varepsilon_c(C^{\mathrm{T}}e + A^{\mathrm{T}}h_e + C_u^{\mathrm{T}}u_e)$.

Rewriting (6.31) by using (6.32), we have

$$\dot{\tilde{W}}_c = -\alpha_c\sigma_c(\phi_c^{\mathrm{T}}\tilde{W}_c + \varepsilon_{\mathrm{HJB}}). \tag{6.33}$$

To begin the development of the feedback control law, u_e is approximated by the action NN as

$$u_e = W_a^{\mathrm{T}}\phi_a(e) + \varepsilon_a, \tag{6.34}$$

where W_a is the matrix of unknown ideal constant weights and $\phi_a(e) : \mathbb{R}^n \to \mathbb{R}^{N_2}$ is called the action NN activation function vector, N_2 is the number of neurons in the hidden layer, and ε_a is the action NN approximation error.

Let $\hat{W}_a$ be an estimate of W_a, the actual output can be expressed as

$$\hat{u}_e = \hat{W}_a^{\mathrm{T}}\phi_a(e). \tag{6.35}$$

The feedback error signal used for tuning action NN is defined to be the difference between the real feedback control input applied to the error system (6.19) and the desired control signal input minimizing (6.28) as

$$e_a = \hat{W}_a^{\mathrm{T}}\phi_a + \frac{1}{2}R^{-1}C_u\nabla\phi_c^{\mathrm{T}}\hat{W}_c. \tag{6.36}$$

The objective function to be minimized by the action NN is defined as

$$E_a(\hat{W}_a) = \frac{1}{2}e_a^{\mathrm{T}}e_a. \tag{6.37}$$

The weight update law for the action NN is a gradient descent algorithm, which is given by

$$\dot{\hat{W}}_a = -\alpha_a\phi_a\left(\hat{W}_a^{\mathrm{T}}\phi_a + \frac{1}{2}R^{-1}C_u\nabla\phi_c^{\mathrm{T}}\hat{W}_c\right)^{\mathrm{T}}, \tag{6.38}$$

where $\alpha_a > 0$ is the adaptive gain of the action NN.

Define the weight estimation error of action NN to be $\tilde{W}_a = \hat{W}_a - W_a$. Since the control law in (6.34) minimizes the infinite-horizon cost functional (6.26), from (6.25) we have

$$\varepsilon_a + W_a^{\mathrm{T}}\phi_a + \frac{1}{2}R^{-1}C_u\nabla\phi_c^{\mathrm{T}}W_c + \frac{1}{2}R^{-1}C_u\nabla\varepsilon_c = 0. \tag{6.39}$$

Combining (6.38) with (6.39), we have

$$\dot{\tilde{W}}_a = -\alpha_a\phi_a\left(\tilde{W}_a^{\mathrm{T}}\phi_a + \frac{1}{2}R^{-1}C_u\nabla\phi_c^{\mathrm{T}}\tilde{W}_c + \varepsilon_{12}\right)^{\mathrm{T}}, \tag{6.40}$$

where $\varepsilon_{12} = -(\varepsilon_a + R^{-1}C_u\nabla\varepsilon_c/2)$.

Remark 6.4 It is important to note that the tracking error must be persistently excited: enough for tuning the critic NN and action NN. In order to satisfy the *persistent excitation condition*, probing noise is added to the control input [17]. Further, the persistent excitation condition ensures $\|\sigma_c\| \geq \sigma_{cm}$ and $\|\phi_a\| \geq \phi_{am}$ with σ_{cm} and ϕ_{am} being positive constants.

Based on the above analysis, the optimal tracking controller is composed of the steady-state controller u_d and the optimal feedback controller u_e. As a result, the control input is written as

$$u = u_d + \hat{u}_e. \tag{6.41}$$

According to (6.35) and the error system (6.19), we have

$$\dot{e} = C^{\mathrm{T}}e + A^{\mathrm{T}}h_e + C_u^{\mathrm{T}}\hat{W}_a^{\mathrm{T}}\phi_a. \tag{6.42}$$

Subtracting and adding $C_u^{\mathrm{T}}W_a\phi_a$ to (6.42), and then recalling (6.34), (6.42) is rewritten as

$$\dot{e} = C^{\mathrm{T}}e + A^{\mathrm{T}}h_e + C_u^{\mathrm{T}}\tilde{W}_a^{\mathrm{T}}\phi_a + C_u^{\mathrm{T}}u_e - C_u^{\mathrm{T}}\varepsilon_a. \tag{6.43}$$

In the following, the stability analysis will be performed. First, the following assumption is made, which can reasonably be satisfied under the current problem settings.

Assumption 6.5

(a) The unknown ideal constant weights for the critic NN and the action NN, i.e., W_c and W_a, are upper bounded so that $\|W_c\| \leq W_{cM}$ and $\|W_a\| \leq W_{aM}$ with W_{cM} and W_{aM} being positive constants, respectively.
(b) The NN approximation errors ε_c and ε_a are upper bounded so that $\|\varepsilon_c\| \leq \varepsilon_{cM}$ and $\|\varepsilon_a\| \leq \varepsilon_{aM}$ with ε_{cM} and ε_{aM} being positive constants, respectively.
(c) The vectors of the activation functions of the critic NN and the action NN, i.e., ϕ_c and ϕ_a, are upper bounded so that $\|\phi_c(\cdot)\| \leq \phi_{cM}$ and $\|\phi_a(\cdot)\| \leq \phi_{aM}$ with ϕ_{cM} and ϕ_{aM} being positive constants, respectively.

(d) The gradients of the critic NN approximation error and the activation function vector are upper bounded so that $\|\nabla\varepsilon_c\| \le \varepsilon'_{cM}$ and $\|\nabla\phi_a\| \le \phi_{dM}$ with ε'_{cM} and ϕ_{dM} being positive constants. The residual error is upper bounded so that $\|\varepsilon_{\mathrm{HJB}}\| \le \varepsilon_{\mathrm{HJBM}}$ with $\varepsilon_{\mathrm{HJBM}}$ being positive constant.

Now we are ready to prove the following theorem.

Theorem 6.6 (cf. [25]) *Consider the system given by* (6.17) *and the desired trajectory* (6.18). *Let the control input be provided by* (6.41). *The weight updating laws of the critic NN and the action NN are given by* (6.31) *and* (6.38), *respectively. Let the initial action NN weights be chosen to generate an initial admissible control. Then, the tracking error e, the weight estimate errors $\tilde{W}_c$ and $\tilde{W}_a$ are uniformly ultimately bounded* (*UUB*) *with the bounds specifically given by* (6.51)–(6.53). *Moreover, the obtained control input u is close to the optimal control input u^* within a small bound ε_u, i.e., $\|u-u^*\| \le \varepsilon_u$ as $t \to \infty$ for a small positive constant ε_u.*

Proof Choose the following Lyapunov function candidate:

$$L(t) = L_1(t) + L_2(t) + L_3(t), \tag{6.44}$$

where $L_1(t) = \mathrm{tr}\{\tilde{W}_c^{\mathrm{T}}\tilde{W}_c\}/2\alpha_c$, $L_2(t) = \alpha_c\,\mathrm{tr}\{\tilde{W}_a^{\mathrm{T}}\tilde{W}_a\}/2\alpha_a$, and $L_3(t) = \alpha_c\alpha_a(e^{\mathrm{T}}e + \Gamma J(e))$ with $\Gamma > 0$.

According to Assumption 6.5 and using (6.21), (6.33), and (6.40), the time derivative of the Lyapunov function candidate (6.44) along the trajectories of the error system (6.43) is computed as follows:

$$\dot{L}(t) = \dot{L}_1(t) + \dot{L}_2(t) + \dot{L}_3(t), \tag{6.45}$$

where

$$\begin{aligned}
\dot{L}_1(t) &= \frac{1}{\alpha_c}\mathrm{tr}\{\tilde{W}_c^{\mathrm{T}}\dot{\tilde{W}}_c\} \\
&= \frac{1}{\alpha_c}\mathrm{tr}\{\tilde{W}_c^{\mathrm{T}}(-\alpha_c\sigma_c(\phi_c^{\mathrm{T}}\tilde{W}_c + \varepsilon_{\mathrm{HJB}}))\} \\
&\le -\left(\sigma_{cm}^2 - \frac{\alpha_c}{2}\sigma_{cM}^2\right)\|\tilde{W}_c\|^2 + \frac{1}{2\alpha_c}\varepsilon_{\mathrm{HJB}}^2, \\
\dot{L}_2(t) &= \frac{\alpha_c}{\alpha_a}\mathrm{tr}\{\tilde{W}_a^{\mathrm{T}}\dot{\tilde{W}}_a\} \\
&= \frac{\alpha_c}{\alpha_a}\mathrm{tr}\left\{\tilde{W}_a^{\mathrm{T}}\left(-\alpha_a\phi_a\left(\tilde{W}_a^{\mathrm{T}}\phi_a + \frac{1}{2}R^{-1}C_u\nabla\phi_c^{\mathrm{T}}\tilde{W}_c + \varepsilon_{12}\right)^{\mathrm{T}}\right)\right\} \\
&\le -\left(\alpha_c\phi_{am}^2 - \frac{3}{4}\alpha_c\alpha_a\phi_{aM}^2\right)\|\tilde{W}_a\|^2 \\
&\quad + \frac{\alpha_c}{4\alpha_a}\|R^{-1}\|^2\|C_u\|^2\phi_{dM}^2\|\tilde{W}_c\|^2 + \frac{\alpha_c}{2\alpha_a}\varepsilon_{12}^{\mathrm{T}}\varepsilon_{12},
\end{aligned}$$

$$\begin{aligned}
\dot{L}_3(t) &= 2\alpha_c\alpha_a e^{\mathrm{T}}\dot{e} + \alpha_c\alpha_a \Gamma(-e^{\mathrm{T}}Qe - u_e^{\mathrm{T}}Ru_e) \\
&= 2\alpha_c\alpha_a e^{\mathrm{T}}(C^{\mathrm{T}}e + A^T h_e + C_u^{\mathrm{T}}\tilde{W}_a^{\mathrm{T}}\phi_a + C_u^{\mathrm{T}}u_e \\
&\quad - C_u^{\mathrm{T}}\varepsilon_a) + \alpha_c\alpha_a\Gamma(-e^{\mathrm{T}}Qe - u_e^{\mathrm{T}}Ru_e) \\
&\le \alpha_c\alpha_a(2\|C\| + 3 + \|A\|^2 + k^2 - \Gamma\lambda_{\min}(Q))\|e\|^2 \\
&\quad + \alpha_c\alpha_a\phi_{aM}^2\|C_u\|^2\|\tilde{W}_a\|^2 + \alpha_c\alpha_a\|C_u\|^2\varepsilon_a^{\mathrm{T}}\varepsilon_a \\
&\quad + \alpha_c\alpha_a(\|C_u\|^2 - \Gamma\lambda_{\min}(R))\|u_e\|^2.
\end{aligned}$$

Then we obtain

$$\begin{aligned}
\dot{L}(t) \le &-\left(\sigma_{cm}^2 - \frac{\alpha_c}{2}\sigma_{cM}^2 - \frac{\alpha_c}{4\alpha_a}\|R^{-1}\|^2\|C_u\|^2\phi_{dM}^2\right)\|\tilde{W}_c\|^2 \\
&- \left(\alpha_c\phi_{am}^2 - \frac{3}{4}\alpha_c\alpha_a\phi_{aM}^2 - \alpha_c\alpha_a\phi_{aM}^2\|C_u\|^2\right)\|\tilde{W}_a\|^2 \\
&- \alpha_c\alpha_a(-\|C_u\|^2 + \Gamma\lambda_{\min}(R))\|u_e\|^2 \\
&- \alpha_c\alpha_a(-2\|C\| - 3 - \|A\|^2 - k^2 + \Gamma\lambda_{\min}(Q))\|e\|^2 \\
&+ \frac{1}{2\alpha_c}\varepsilon_{\mathrm{HJB}}^2 + \frac{\alpha_c}{2\alpha_a}\varepsilon_{12}^{\mathrm{T}}\varepsilon_{12} + \alpha_c\alpha_a\|C_u\|^2\varepsilon_a^{\mathrm{T}}\varepsilon_a. \qquad (6.46)
\end{aligned}$$

By using Assumption 6.5, we have $\|\varepsilon_{12}\| \le \varepsilon_{12M}$, where $\varepsilon_{12M} = \varepsilon_{aM} + R^{-1}C_u\varepsilon'_{cM}/2$. Then, we have

$$\frac{1}{2\alpha_c}\varepsilon_{\mathrm{HJB}}^2 + \frac{\alpha_c}{2\alpha_a}\varepsilon_{12}^{\mathrm{T}}\varepsilon_{12} + \alpha_c\alpha_a\|C_u\|^2\varepsilon_{12}^{\mathrm{T}}\varepsilon_{12} \le D_M, \qquad (6.47)$$

where $D_M = \varepsilon_{\mathrm{HJBM}}^2/(2\alpha_c) + \alpha_c\varepsilon_{12M}^2/(2\alpha_a) + \alpha_c\alpha_a\|C_u\|^2\varepsilon_{12M}^2$.

If Γ, α_c, and α_a are selected to satisfy

$$\Gamma > \max\left\{\frac{\|C_u\|^2}{\lambda_{\min}(R)}, \frac{2\|C\| + 3 + \|A\|^2 + k^2}{\lambda_{\min}(Q)}\right\}, \qquad (6.48)$$

$$\alpha_c < \frac{4\alpha_a\sigma_{cm}^2}{2\alpha_a\sigma_{cM}^2 + \|R^{-1}\|^2\|C_u\|^2\phi_{dM}^2}, \qquad (6.49)$$

$$\alpha_a < \frac{4\phi_{am}^2}{3\phi_{aM}^2 + 4\phi_{aM}^2\|C_u\|^2}, \qquad (6.50)$$

and given the following inequalities:

$$\begin{aligned}
\|e\| &> \sqrt{\frac{D_M}{\alpha_c\alpha_a(-2\|C\| - 3 - \|A\|^2 - k^2 + \Gamma\lambda_{\min}(Q))}} \\
&\triangleq b_e, \qquad (6.51)
\end{aligned}$$

or

$$\|\tilde{W}_c\| > \sqrt{\frac{4\alpha_a D_M}{4\alpha_a\sigma_{cm}^2 - 2\alpha_a\alpha_c\sigma_{cM}^2 - \alpha_c\|R^{-1}\|^2\|C_u\|^2\phi_{dM}^2}} \triangleq b_{\tilde{W}_c}, \tag{6.52}$$

or

$$\|\tilde{W}_a\| > \sqrt{\frac{4D_M}{4\alpha_c\phi_{am}^2 - 3\alpha_c\alpha_a\phi_{aM}^2 - 4\alpha_c\alpha_a\phi_{aM}^2\|C_u\|^2}} \triangleq b_{\tilde{W}_a}, \tag{6.53}$$

then we can conclude $\dot{L}(t) < 0$. Therefore, using Lyapunov theory [7], it can be concluded that the tracking error e, and the NN weight estimation errors $\tilde{W}_c$ and $\tilde{W}_a$ are UUB.

Next, we will prove $\|u - u^*\| \le \varepsilon_u$ as $t \to \infty$. Recalling the expression of u^* together with (6.34) and (6.41), we have

$$u - u^* = \tilde{W}_a^{\mathrm{T}}\phi_a + \varepsilon_a. \tag{6.54}$$

When $t \to \infty$, the upper bound of (6.54) is

$$\|u - u^*\| \le \varepsilon_u, \tag{6.55}$$

where $\varepsilon_u = b_{\tilde{W}_a}\phi_{aM} + \varepsilon_{aM}$.

This completes the proof. □

Remark 6.7 From (6.31) and (6.38), it is noted that the weights of critic NN and action NN are updated simultaneously in contrast with some standard ADP methods in which the weights of critic NN and action NN are updated sequentially.

Remark 6.8 If the NN approximation errors ε_c and ε_a are considered to be negligible, then from (6.47) we have $D_M = 0$, with $u \to u^*$. Otherwise, the obtained control input u is close to the optimal input u^* within a small bound ε_u.

Due to the presence of the NN approximation errors ε_c and ε_a, the tracking error is UUB instead of asymptotically convergent to zero. In the following, for improving the tracking performance, an additional robustifying term is developed to attenuate the NN approximation errors such that tracking error converges to zero asymptotically, which can be constructed in the form

$$u_r = \frac{K_r e}{e^{\mathrm{T}}e + \zeta}, \tag{6.56}$$

where $\zeta > 0$ is a constant, $K_r > K_{r\min}$ is a designed parameter. $K_{r\min}$ is selected to satisfy the following inequality:

$$K_{r\min} \ge \frac{D_M(e^{\mathrm{T}}e + \zeta)}{2\alpha_c\alpha_a\|C_u\|e^{\mathrm{T}}e}. \tag{6.57}$$

Then, the overall control input is given as

$$u_{ad} = u - u_r, \tag{6.58}$$

where u is the same as (6.41).

Applying (6.58) to the error system (6.17) and using (6.18), a new error system is obtained as follows:

$$\dot{e} = C^{\mathrm{T}} e + A^{\mathrm{T}} h_e + C_u^{\mathrm{T}} \tilde{W}_a^{\mathrm{T}} \phi_a + C_u^{\mathrm{T}} u_e - C_u^{\mathrm{T}} \varepsilon_a - C_u^{\mathrm{T}} u_r. \tag{6.59}$$

Theorem 6.9 (cf. [25]) *Consider the system given by* (6.17) *and the desired trajectory* (6.18). *Let the control input be provided by* (6.58). *The weight updating laws of the critic NN and the action NN are given by* (6.31) *and* (6.38), *respectively. Let the initial action NN weights be chosen to generate an initial admissible control. Then, the tracking error e and the weight estimation errors $\tilde{W}_c$ and $\tilde{W}_a$ will asymptotically converge to zero. Moreover, the obtained control input u_{ad} is close to the optimal control input u^* within a small bound δ_u, i.e., $\|u_{ad} - u^*\| \le \delta_u$ as $t \to \infty$ for a small positive constant δ_u.*

Proof Choose the same Lyapunov function candidate as that in Theorem 6.6. Differentiating the Lyapunov function candidate in (6.44) along the trajectories of the error system in (6.59), similar to the proof of Theorem 6.6, by using (6.56) and (6.57), we obtain

$$\begin{aligned}
\dot{L}(t) \le & -\left(\sigma_{cm}^2 - \frac{\alpha_c}{2}\sigma_{cM}^2 - \frac{\alpha_c}{4\alpha_a}\|R^{-1}\|^2\|C_u\|^2\phi_{dM}^2\right)\|\tilde{W}_c\|^2 \\
& - \left(\alpha_c\phi_{am}^2 - \frac{3}{4}\alpha_c\alpha_a\phi_{aM}^2 - \alpha_c\alpha_a\phi_{aM}^2\|C_u\|^2\right)\|\tilde{W}_a\|^2 \\
& - \alpha_c\alpha_a(-\|C_u\|^2 + \Gamma\lambda_{\min}(R))\|u_e\|^2 \\
& - \alpha_c\alpha_a(-2\|C\| - 3 - \|A\|^2 - k^2 + \Gamma\lambda_{\min}(Q))\|e\|^2.
\end{aligned} \tag{6.60}$$

Choosing Γ, α_c, and α_a as Theorem 6.6, we have $\dot{L}(t) \le 0$. Equations (6.44) and (6.60) guarantee that the tracking error e, NN weight estimation errors $\tilde{W}_c$ and $\tilde{W}_a$ are bounded, since L is nonincreasing. Because all the variables on the right-hand side of (6.59) are bounded, $\dot{e}$ is also bounded. From (6.60), we have

$$\dot{L}(t) \le -B_e\|e\|^2, \tag{6.61}$$

where $B_e = \alpha_c\alpha_a(-2\|C\| - 3 - \|A\|^2 - k^2 + \Gamma\lambda_{\min}(Q))$.

Integrating both sides of (6.61) and after some manipulations, we have

$$\int_0^\infty \|e\|^2 \mathrm{d}t \le B_e^{-1}(L(0) - L(\infty)). \tag{6.62}$$

Since the right side of (6.59) is bounded, $\|e\| \in \mathcal{L}_2$. Using Barbalat's lemma [7], we have $\lim_{t\to\infty}\|e\| = 0$. Similarly, we can prove that $\lim_{t\to\infty}\|\tilde{W}_c\| = 0$ and $\lim_{t\to\infty}\|\tilde{W}_a\| = 0$.

Next, we will prove $\|u_{ad} - u^*\| \le \delta_u$ as $t \to \infty$. From (6.34) and (6.58), we can have

$$u_{ad} - u^* = \tilde{W}_a^{\mathrm{T}} \phi_a + \varepsilon_a + u_r. \tag{6.63}$$

Since $\|e\| \to 0$ as $t \to \infty$, the robustifying control input $\|u_r\| \to 0$ as $t \to \infty$. Then, the upper bound of (6.63) is

$$\|u_{ad} - u^*\| \le \delta_u, \tag{6.64}$$

where $\delta_u = \varepsilon_{aM}$.

This completes the proof. □

Remark 6.10 From (6.55) and (6.64), it can be seen that δ_u is smaller than ε_u, which reveals the role of the robustifying term in making the obtained control input closer to the optimal control input.

6.2.3 Simulations

In this subsection, two examples are provided to demonstrate the effectiveness of the present approach.

Example 6.11 Consider the following affine nonlinear continuous-time system:

$$\begin{aligned} \dot{x}_1 &= -x_1 + x_2, \\ \dot{x}_2 &= -0.5x_1 - 0.5x_2(1 - (\cos(2x_1) + 2)^2) \\ &\quad + (\cos(2x_1) + 2)u. \end{aligned} \tag{6.65}$$

The cost functional is defined by (6.21) where Q and R are chosen as identity matrices of appropriate dimensions. The control object is to make x_1 follow the desired trajectory $x_{1d} = \sin(t)$. It is assumed that the system dynamics is unknown and input/output data are available.

First, a data-based model is established to estimate the nonlinear system dynamics. Let us select the RNN as (6.5) with $S = -30I_2$ and $\eta = 1.5$. The activation function $h(\hat{x})$ is selected as hyperbolic tangent function $\tanh(\hat{x})$. Select the design parameters in Theorem 6.2 as $\Gamma_1 = [1, 0.1; 0.1, 1]$, $\Gamma_2 = [1, 0.2; 0.2, 1]$, $\Gamma_3 = [1, 0.1; 0.1, 1]$, $\Gamma_4 = 0.2$, and $\Gamma_5 = 0.1$. Then, we can obtain the trajectories of the modeling error as shown in Fig. 6.1. It is observed that the obtained data-based model can reconstruct the nonlinear system dynamics successfully, as Theorem 6.2 predicted.

Then, based on the obtained data-based model, the approximate *optimal robust* controller is implemented for the unknown affine nonlinear continuous-time system (6.65). The activation function of the critic NN is selected as $\phi_c = [e_1^2 \; e_1 e_2 \; e_2^2]^{\mathrm{T}}$, the

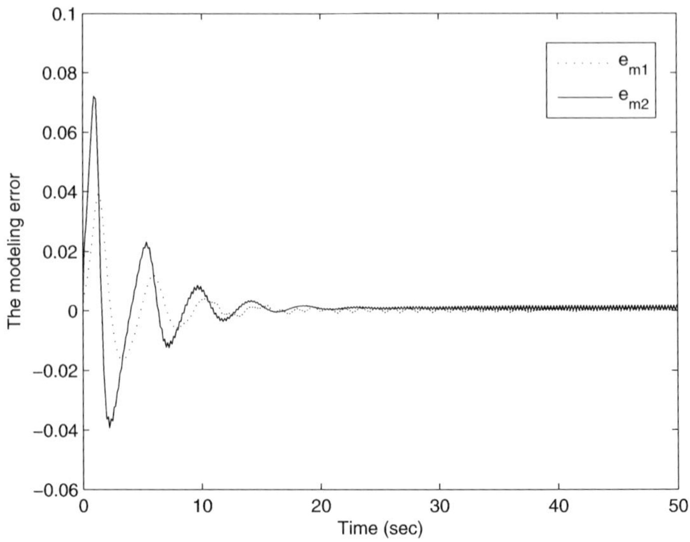

Fig. 6.1 The modeling error for the affine nonlinear system

critic NN weights are denoted $\hat{W}_c = [W_{c1}\ W_{c2}\ W_{c3}]^{\mathrm{T}}$. The activation function of the action NN ϕ_a is chosen as the gradient of the critic NN, the action NN weights are denoted $\hat{W}_a = [W_{a1}\ W_{a2}\ W_{a3}]^{\mathrm{T}}$. The adaptive gains for the critic NN and action NN are selected as $\alpha_c = 0.8$ and $\alpha_a = 0.5$, and the design parameters of the robustifying term are selected as $K_r = [20, 20]$, $\zeta = 1.2$. Additionally, the critic NN weights are set as $[1, 1, 1]^{\mathrm{T}}$ at the beginning of the simulation with the initial weights of the action NN chosen to reflect the initial admissible control. To maintain the excitation condition, probing noise is added to the control input for the first 5000 s.

After simulation, the trajectory of the tracking error is shown in Fig. 6.2. The convergence trajectories of the critic NN weights and action NN weights are shown in Figs. 6.3 and 6.4, respectively. For comparing the tracking performance, we apply the obtained optimal robust controller and the initial admissible controller to system (6.65) under the same initial state, and obtain the trajectories of tracking error as shown in Fig. 6.5, respectively. It can be seen from Fig. 6.5 that the present robust approximate optimal controller yields a better tracking performance than the initial admissible controller.

Example 6.12 Consider the following continuous-time nonaffine nonlinear system:

$$\begin{aligned}\dot{x}_1 &= x_2,\\ \dot{x}_2 &= x_1^2 + 0.15u^3 + 0.1(4 + x_2^2)u + \sin(0.1u).\end{aligned}\tag{6.66}$$

The cost functional is defined as Example 6.11. The control objective is to make x_1 follow the desired trajectory $x_{1d} = \sin(t)$. It is assumed that the system dynamics is unknown and input/output data are available.

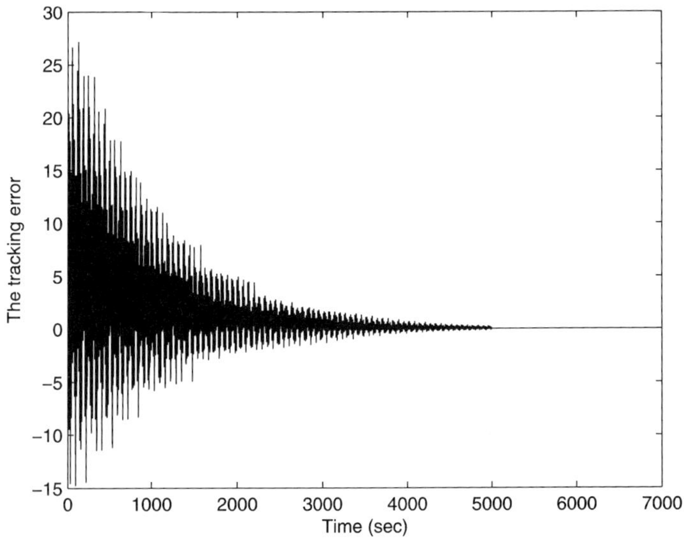

Fig. 6.2 The tracking error for the affine nonlinear system

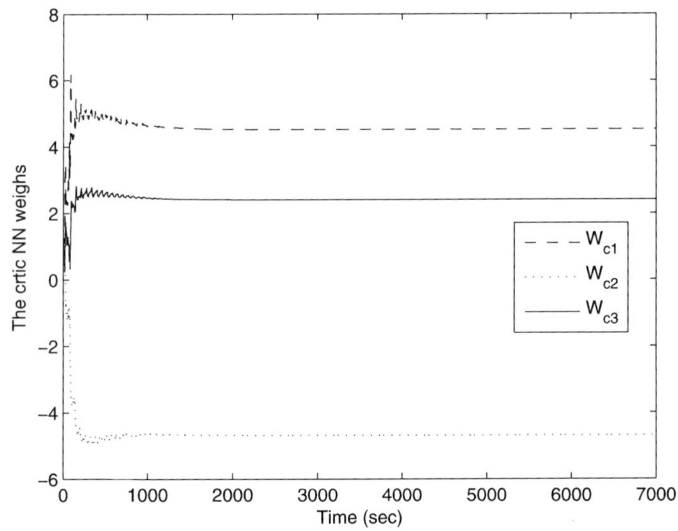

Fig. 6.3 The critic NN weights

Using a similar method as shown in Example 6.11, we can obtain the trajectories of modeling error as shown in Fig. 6.6. It is observed that the obtained data-based model learns the nonlinear system dynamics successfully, as Theorem 6.2 predicted. Then, based on the obtained data-based model we design the robust approximate optimal controller, which is then applied to the unknown nonaffine nonlinear system

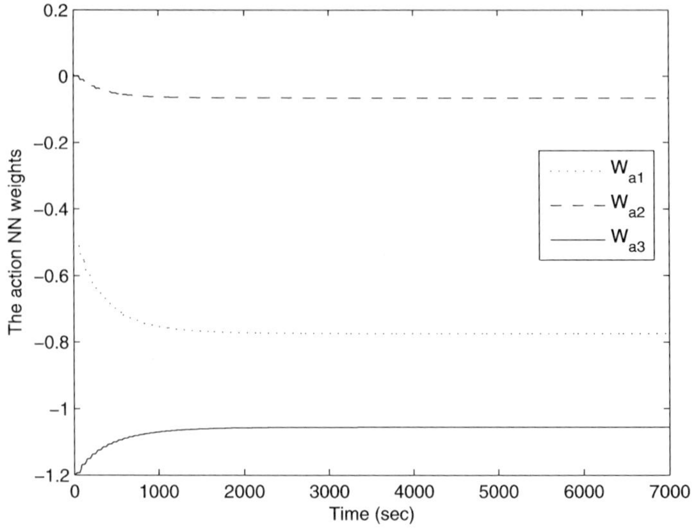

Fig. 6.4 The action NN weights

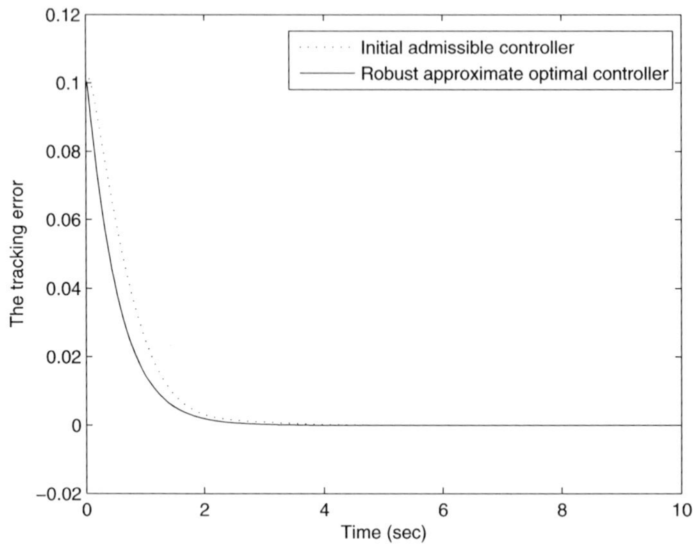

Fig. 6.5 The comparison result between initial admissible controller and robust approximate optimal controller

(6.66). The activation functions of the critic NN and action NN are the same as the ones in Example 6.11. The adaptive gains for critic NN and action NN are selected as $\alpha_c = 0.5$ and $\alpha_a = 0.2$ and the parameters of the robustifying term are selected as $K_r = [10, 10]$, $\zeta = 1.2$. Additionally, the critic NN weights are set as $[1, 1, 1]^{\mathrm{T}}$ at

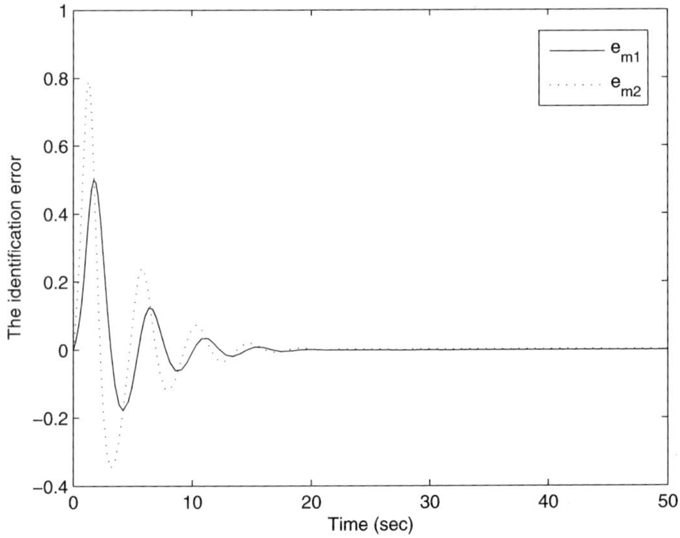

Fig. 6.6 The modeling error for the continuous-time nonaffine nonlinear system

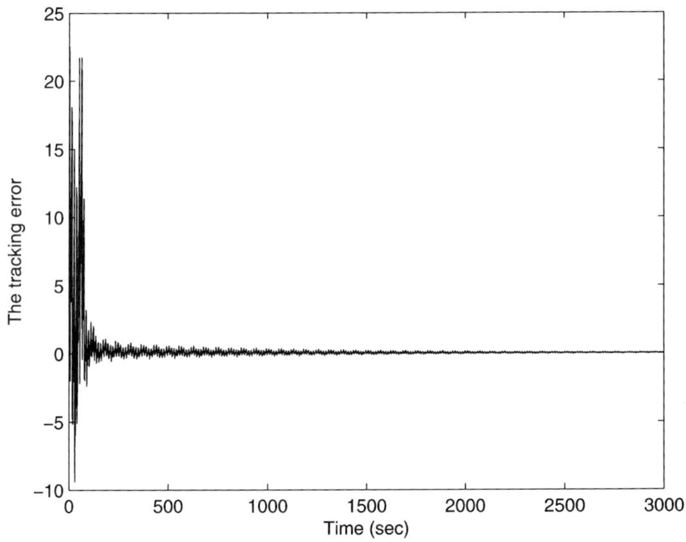

Fig. 6.7 The tracking error for the nonaffine nonlinear system

the beginning of the simulation with the initial weights of the action NN chosen to reflect the initial admissible control. Similarly, to maintain the excitation condition, probing noise is added to the control input for the first 1500 s.

After simulation, the trajectory of the tracking error is shown in Fig. 6.7. The convergence trajectories of the critic NN weights and action NN weights are shown

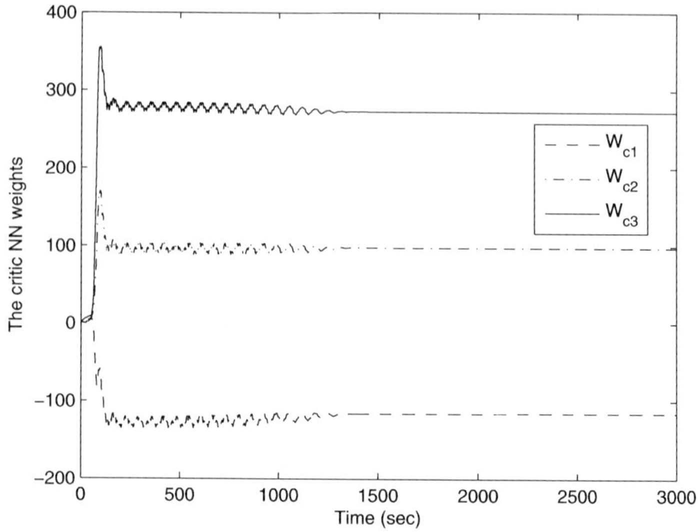

Fig. 6.8 The critic NN weights

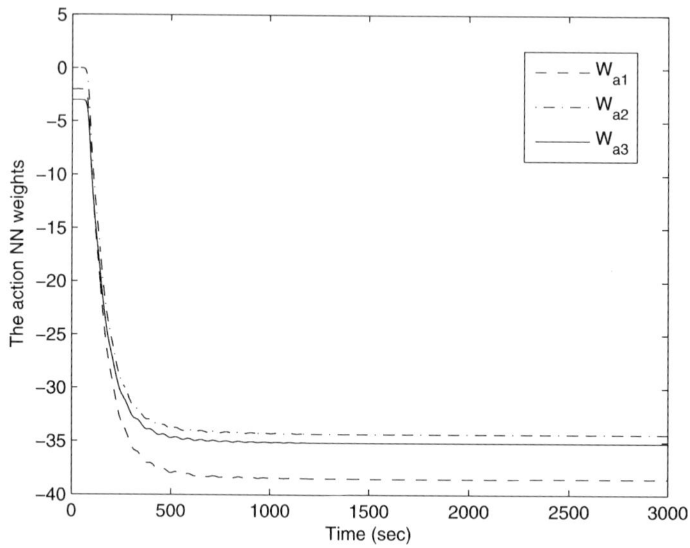

Fig. 6.9 The action NN weights

in Figs. 6.8 and 6.9, respectively. Similarly, for comparing the tracking performance, we apply the obtained robust optimal controller and the initial admissible controller to system (6.66) for the same initial state, and obtain the trajectories of tracking error as shown in Fig. 6.10, respectively. It can be seen from Fig. 6.10 that the present robust approximate optimal controller yields better tracking performance

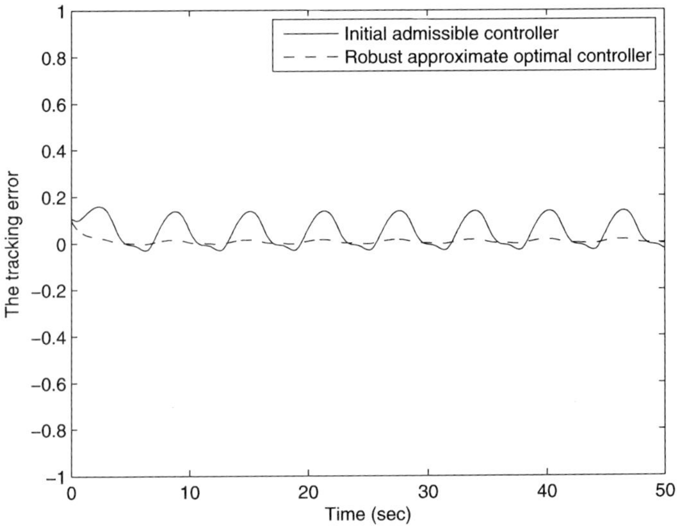

Fig. 6.10 The comparison result between initial admissible controller and robust approximate optimal controller

than the initial admissible controller. The simulation results reveal that the present controller can be applied to nonaffine nonlinear systems and we obtain satisfactory tracking performance even for the unknown system dynamics.

6.3 Optimal Feedback Control for Nonaffine Nonlinear Systems

In this section, we will study the optimal feedback control problem of a class of continuous-time nonaffine nonlinear systems via the ADP method.

6.3.1 Problem Formulation

Consider a class of continuous-time nonaffine nonlinear systems described as follows:

$$\begin{aligned}\dot{x}_i &= x_{i+1}, i = 1, \ldots, n-1, \\ \dot{x}_n &= f(x, u), \\ y &= x_1, \end{aligned} \tag{6.67}$$

where $x = [x_1, \ldots, x_n]^{\mathrm{T}} \in \mathbb{R}^n$ is the system state vector, $u \in \mathbb{R}$ is the control input, and $y \in \mathbb{R}$ is the output of the system; $f(x, u)$ is an unknown smooth function satisfying $f(0, 0) = 0$.

Assumption 6.13 The control effectiveness term $f_u(x,u) \triangleq \partial f(x,u)/\partial u$ has a known sign and is bounded away from zero, i.e., there exists $d_f > 0$ such that $|f_u(x,u)| \geq d_f$. Without loss of generality, we assume $f_u > 0$.

The control objective is to force the system output y to follow the desired trajectory y_d while ensuring that all the signals of the closed-loop system are bounded. Assume that y_d and its up to n times derivatives, namely $\dot{y}_d, y_d^{(2)}, \ldots,$ and $y_d^{(n)}$ are smooth, bounded, and available for design.

6.3.2 Robust Approximate Optimal Control Based on ADP Algorithm

Define the tracking error $\tilde{y} = y - y_d$, desired trajectory vector $\bar{y}_d = [y_d, \ldots, y_d^{(n-1)}]^{\mathrm{T}}$, tracking error vector $\tilde{x} = x - \bar{y}_d$, and the filtered tracking error as

$$r = \tilde{y}^{(n-1)} + \lambda_{n-1}\tilde{y}^{(n-2)} + \cdots + \lambda_2 \tilde{y}^{(1)} + \lambda_1 \tilde{y} = [\Lambda^{\mathrm{T}} \;\; 1]\tilde{x}, \tag{6.68}$$

where $\lambda_i, i = 1, \ldots, n-2$, are chosen such that the polynomial $H(s) = s^{(n-1)} + \lambda_{n-1}s^{(n-2)} + \cdots + \lambda_1$ is Hurwitz, and $\Lambda = [\lambda_1, \ldots, \lambda_{n-1}]^{\mathrm{T}}$.

With these definitions, the tracking error dynamics can be given as follows:

$$\dot{r} = -y_d^{(n)} + [0 \;\; \Lambda^{\mathrm{T}}]\tilde{x} + f(x,u). \tag{6.69}$$

Then, feedback linearization is performed by introducing a so-called pseudocontrol

$$\upsilon = \hat{f}(x,u), \tag{6.70}$$

where $\hat{f}(x,u)$ is an available approximation of $f(x,u)$ satisfying Assumption 6.13.

Assumption 6.14 $\hat{f}_u \triangleq \partial \hat{f}(x,u)/\partial u \geq d_{\hat{f}} > 0$ for some $d_{\hat{f}}$, and there exist some constants $b_l, b_u > 0$ such that $b_l \leq f_u/\hat{f}_u \leq b_u$.

By adding and subtracting υ on the right-hand side of (6.69), the tracking error dynamics can be rewritten as follows:

$$\dot{r} = -y_d^{(n)} + [0 \;\; \Lambda^{\mathrm{T}}]\tilde{x} + \hat{f}(x,u) + \Delta, \tag{6.71}$$

where $\Delta = f(x,u) - \hat{f}(x,u)$ is the modeling error.

The pseudocontrol υ is designed as

$$\upsilon = \upsilon_{rm} + \upsilon_{dc} - \upsilon_{ad} + \upsilon_r, \tag{6.72}$$

where $\upsilon_{rm} = y_d^{(n)} - [0\ \ \Lambda^{\mathrm{T}}]\tilde{x}$, $\upsilon_{dc} = -Kr$ is to stabilize the tracking error dynamics in the absence of a modeling error, υ_{ad} is used to approximately cancel the modeling error, and υ_r is a robustifying term.

By inverting (6.70), we have the following control law:

$$u = \hat{f}^{-1}(x, \upsilon). \tag{6.73}$$

Then, applying (6.73) to (6.71), we have

$$\dot{r} = -Kr + \Delta - \upsilon_{ad} + \upsilon_r. \tag{6.74}$$

The term $\Delta - \upsilon_{ad}$ can be expressed as

$$\Delta - \upsilon_{ad} = f(x, \hat{f}^{-1}(x, \upsilon_{rm} + \upsilon_{dc} - \upsilon_{ad} + \upsilon_r)) - \upsilon_{rm} - \upsilon_{dc} - \upsilon_r. \tag{6.75}$$

Reference [2] points out that Δ depends on υ_{ad} through (6.72) and (6.73), while υ_{ad} is designed to cancel Δ. A contraction mapping assumption on Δ with respect to υ_{ad} is required to guarantee the existence of a smooth function $\upsilon_{ad}^* = \Delta(\cdot, \upsilon_{ad}^*)$. To remove the contraction mapping assumption, we follow the work of [8]. Define $\upsilon_l \triangleq \upsilon_{rm} + \upsilon_{dc}$, $\upsilon^* \triangleq \hat{f}(x, f^{-1}(x, \upsilon_l))$. Then we have $\upsilon_l = f(x, \hat{f}^{-1}(x, \upsilon^*))$.

Using the mean value theorem [5], (6.75) can be expressed as

$$\begin{aligned}
\Delta - \upsilon_{ad} &= f(x, u) - \hat{f}(x, u) - \upsilon_{ad} \\
&= f(x, \hat{f}^{-1}(x, \upsilon)) - \upsilon_l + \upsilon_{ad} - \upsilon_r - \upsilon_{ad} \\
&= f(x, \hat{f}^{-1}(x, \upsilon)) - f(x, \hat{f}^{-1}(x, \upsilon^*)) - \upsilon_r \\
&= f_\upsilon(\bar{\upsilon})(\upsilon - \upsilon^*) - \upsilon_r \\
&= f_\upsilon(\bar{\upsilon})(\upsilon_l - \upsilon_{ad} + \upsilon_r - \hat{f}(x, f^{-1}(x, \upsilon_l))) - \upsilon_r \\
&= f_\upsilon(\bar{\upsilon})(-\upsilon_{ad} + \upsilon_r + \bar{\Delta}) - \upsilon_r,
\end{aligned} \tag{6.76}$$

where $\bar{\Delta} = \upsilon_l - \hat{f}(x, f^{-1}(x, \upsilon_l)) = f(x, \hat{f}^{-1}(x, \upsilon^*)) - f(x, f^{-1}(x, \upsilon_l))$ denotes the unknown uncertain term, $f_\upsilon(\bar{\upsilon}) \triangleq (\partial f/\partial u)(\partial u/\partial \upsilon)|_{\upsilon=\bar{\upsilon}} = (\partial f/\partial u)/(\partial \hat{f}/\partial u)|_{u=\hat{f}^{-1}(x,\bar{\upsilon})}$, $\bar{\upsilon} = \lambda\upsilon + (1-\lambda)\upsilon^*$ and $0 \le \lambda(\upsilon) \le 1$.

Then, substituting (6.76) into (6.74), we have

$$\dot{r} = -Kr + f_\upsilon(\bar{\upsilon})(-\upsilon_{ad} + \bar{\Delta}) + f_\upsilon(\bar{\upsilon})\upsilon_r. \tag{6.77}$$

If $\bar{\Delta}$ is known, υ_{ad} can be chosen as $\upsilon_{ad} = \bar{\Delta}$; then we let $\upsilon_r = 0$. Since $\bar{\Delta}$ is unknown, the desired υ_{ad} cannot be implemented directly. Instead, an action NN is employed to approximate $\bar{\Delta}$ as follows:

$$\bar{\Delta} = W_a^{\mathrm{T}}\phi_a(\bar{x}) + \varepsilon_a(\bar{x}), \tag{6.78}$$

where $\bar{x} = [1, x, \upsilon_l, \bar{y}_d]$, $W_a \in \mathbb{R}^N$ is the ideal weight of the action NN, $\phi_a(\bar{x}) \in \mathbb{R}^N$ is the basis function of the action NN, and $\varepsilon_a(\bar{x})$ is the reconstruction error of the action NN satisfying $\|\varepsilon_a(\bar{x})\| \le \varepsilon_a^*$.

Remark 6.15 Because the unknown nonlinear function $f(x,u)$ of (6.67) is an implicit function with respect to the control input u, traditional ADP methods can rarely be applied. To overcome this difficulty, the action NN is employed to approximate the derived unknown uncertain term $\bar{\Delta}$ instead of modeling the unknown system (6.67) directly.

The adaptive signal υ_{ad} is designed as

$$\upsilon_{ad} = \hat{W}_a^{\mathrm{T}} \phi_a(\bar{x}), \tag{6.79}$$

where $\hat{W}_a$ is the estimate of W_a.

Substituting (6.78) and (6.79) into (6.77), the following is immediate:

$$\begin{aligned}\dot{r} &= -Kr + (f_\upsilon(\bar{\upsilon}) - b_u)\tilde{W}_a^{\mathrm{T}} \phi_a(\bar{x}) + b_u \tilde{W}_a^{\mathrm{T}} \phi_a \\ &\quad + f_\upsilon(\bar{\upsilon})\upsilon_r + f_\upsilon(\bar{\upsilon})\varepsilon_a(\bar{x}),\end{aligned} \tag{6.80}$$

where $\tilde{W}_a = W_a - \hat{W}_a$.

Using adaptive bounding technique, the robustifying term υ_r is designed as

$$\upsilon_r = -\frac{b_r}{1-b_r} |\hat{\kappa}| \psi \tanh\left(\frac{\mathcal{R}\psi}{\alpha}\right), \tag{6.81}$$

where $b_r = 1 - (b_l/b_u) < 1$ and α is a design parameter, $\hat{\kappa}$ is the adaptive parameter, $\mathcal{R}$ and ψ will be defined later. Applying (6.81) to (6.80), the tracking error dynamics can be rewritten as

$$\begin{aligned}\dot{r} &= -Kr + b_u \tilde{W}_a^{\mathrm{T}} \phi_a + b_u \left[\upsilon_r + \left(\frac{f_\upsilon(\bar{\upsilon})}{b_u} - 1\right)\left(\upsilon_r + \tilde{W}_a^{\mathrm{T}} \phi_a\right)\right] \\ &\quad + f_\upsilon(\bar{\upsilon})\varepsilon_a(\bar{x}).\end{aligned} \tag{6.82}$$

Remark 6.16 Since $\tanh(\cdot)$ can maintain a continuous control signal while $\mathrm{sgn}(\cdot)$ will result in a chattering phenomenon due to its discontinuity, the robustifying term is designed based on $\tanh(\cdot)$ rather than $\mathrm{sgn}(\cdot)$.

Next, we choose the critic signal as [13]

$$\mathcal{R}_n = \mathcal{R} + |\mathcal{R}| W_c^{\mathrm{T}} \phi_c(r). \tag{6.83}$$

The first term $\mathcal{R}$ is called the primary critic signal which is defined in terms of the performance measure as

$$\mathcal{R} = \frac{\chi}{1+e^{-mr}} - \frac{\chi}{1+e^{mr}}, \tag{6.84}$$

where m is a positive constants and the value of $\mathcal{R}$ is bounded in the interval $[-\chi, \chi]$ with $\chi > 0$ being the critic slope gain. The second term $|\mathcal{R}| W_c^{\mathrm{T}} \phi_c(r)$ is called the second critic signal, where W_c is the ideal weight of critic NN and $\phi_c(r)$ is the basis function of critic NN. The actual output of the critic NN is $\hat{W}_c^{\mathrm{T}}\phi_c(r)$ where $\hat{W}_c$ is the estimate of W_c. Then, the actual critic signal can be expressed as $\hat{\mathcal{R}}_n = \mathcal{R} + |\mathcal{R}|\hat{W}_c^{\mathrm{T}}\phi_c(r)$. Define $\tilde{W}_c = W_c - \hat{W}_c$. It should be noted that $\mathcal{R}$ will approach zero when r approaches zero. Therefore we can conclude that $\hat{\mathcal{R}}_n$ will approach zero. As a learning signal, the critic signal $\hat{\mathcal{R}}_n$ is more informative than the filtered tracking error r. Consequently, a larger control input can be yielded and better tracking performance can be obtained.

Next, the uniformly ultimate boundedness of the closed-loop system is demonstrated by the Lyapunov method.

Assumption 6.17 The ideal weights of the action NN and the critic NN, i.e., W_a and W_c, are bounded above by unknown positive constants so that $\|W_a\| \le W_a^*$, $\|W_c\| \le W_c^*$.

Assumption 6.18 The activation functions of the action NN and critic NN, ϕ_a and ϕ_c, are bounded above by known positive constants, so that $\|\phi_a\| \le \phi_a^*$, $\|\phi_c\| \le \phi_c^*$.

Lemma 6.19 (cf. [4]) *The following inequality holds*:

$$\begin{aligned}\|\delta\| &\triangleq \|\mathcal{R}f_\upsilon(\bar{\upsilon})\varepsilon_a(\bar{x}) + b_u b_r|\mathcal{R}|\|\tilde{W}_a\|\|\phi_a\| \\ &\quad + b_u|\mathcal{R}|\mathrm{tr}\{\hat{W}_a^{\mathrm{T}}\phi_a\phi_c^{\mathrm{T}}W_a - W_a^{\mathrm{T}}\phi_a(\phi_c^{\mathrm{T}}\hat{W}_c)\}\| \\ &\le b_u b_r|\mathcal{R}|\kappa^*\psi,\end{aligned} \tag{6.85}$$

where κ^* is an unknown constant and $\psi = 1 + \|\hat{W}_a\| + \|\hat{W}_c\|$.

Proof Using Assumptions 6.13, 6.14, 6.17, and 6.18, the boundedness of $\varepsilon_a(\bar{x})$ and the inequality $\|\tilde{W}_a\| \le \|\hat{W}_a\| + \|W_a^*\|$, we have

$$\begin{aligned}\|\delta\| &\le |\mathcal{R}|b_u\varepsilon_a^* + b_u b_r|\mathcal{R}|\|\hat{W}_a\|\|\phi_a\| + b_u b_r|\mathcal{R}|W_a^*\|\phi_a\| \\ &\quad + b_u|\mathcal{R}|\phi_a^*\phi_c^*W_c^*\|\hat{W}_a\| + b_u|\mathcal{R}|\phi_a^*\phi_c^*W_a^*\|\hat{W}_c\| \\ &\le b_u b_r|\mathcal{R}|\kappa^*\psi,\end{aligned} \tag{6.86}$$

where $\kappa^* \triangleq \max\{\varepsilon_a^*/b_r + W_a^*\phi_a^*, \phi_a^*(1+\phi_c^*W_c^*/b_r), \phi_a^*\phi_c^*W_a^*/b_r\}$, $\psi = 1 + \|\hat{W}_a\| + \|\hat{W}_c\|$.

This completes the proof. □

Theorem 6.20 (cf. [4]) *Under Assumptions* 6.13, 6.14 *and* 6.17, *considering the closed-loop system consisting of system* (6.67) *and the control u provided by* (6.73), *the weights tuning laws of the action NN and critic NN are*

$$\dot{\hat{W}}_a = \alpha_1(\phi_a(\mathcal{R} + |\mathcal{R}|\hat{W}_c^{\mathrm{T}}\phi_c)^{\mathrm{T}} - K_{W_a}|\mathcal{R}|\hat{W}_a), \tag{6.87}$$

$$\dot{\hat{W}}_c = -\alpha_2(|\mathcal{R}|\phi_c(\hat{W}_a^{\mathrm{T}}\phi_a)^{\mathrm{T}} + K_{W_c}|\mathcal{R}|\hat{W}_c), \tag{6.88}$$

and the tuning law of the adaptive parameter is

$$\dot{\hat{\kappa}} = \alpha_3\left(\mathcal{R}\psi\tanh\left(\frac{\mathcal{R}\psi}{\alpha}\right) - K_\kappa\hat{\kappa}\right), \tag{6.89}$$

where α_1, α_2, α_3, K_{W_a}, K_{W_c}, and K_κ are the positive design parameters. If $K_\kappa > 1/(b_u b_r)$, then all the closed-loop signals are uniformly ultimately bounded.

Proof Choose a Lyapunov function candidate as

$$L = \rho(r) + \frac{b_u}{2\alpha_1}\mathrm{tr}\{\tilde{W}_a^{\mathrm{T}}\tilde{W}_a\} + \frac{b_u}{2\alpha_2}\mathrm{tr}\{\tilde{W}_c^{\mathrm{T}}\tilde{W}_c\} + \frac{b_u b_r}{2\alpha_3}\tilde{\kappa}^2, \tag{6.90}$$

where $\rho(r) = \chi(\ln(1+e^{mr}) + \ln(1+e^{-mr}))/m$, $\tilde{\kappa} = \kappa^* - \hat{\kappa}$. The time derivative of (6.90) can be expressed as

$$\dot{L} = \mathcal{R}\dot{r} + \frac{b_u}{\alpha_1}\mathrm{tr}\{\tilde{W}_a^{\mathrm{T}}\dot{\tilde{W}}_a\} + \frac{b_u}{\alpha_2}\mathrm{tr}\{\tilde{W}_c^{\mathrm{T}}\dot{\tilde{W}}_c\} + \frac{b_u b_r}{\alpha_3}\tilde{\kappa}\dot{\tilde{\kappa}}. \tag{6.91}$$

Substituting (6.82) into (6.91), we have

$$\begin{aligned}\dot{L} = \mathcal{R}\Bigg\{&-Kr + b_u\tilde{W}_a^{\mathrm{T}}\phi_a + b_u\left[\upsilon_r + \left(\frac{f_\upsilon(\bar{\upsilon})}{b_u} - 1\right)\left(f_\upsilon(\bar{\upsilon}) + \tilde{W}_a^{\mathrm{T}}\phi_a\right)\right]\Bigg\}\\ &+ \mathcal{R}f_\upsilon(\bar{\upsilon})\varepsilon_a(\bar{x}) + \frac{b_u}{\alpha_1}\mathrm{tr}\{\tilde{W}_a^{\mathrm{T}}\dot{\tilde{W}}_a\} + \frac{b_u}{\alpha_2}\mathrm{tr}\{\tilde{W}_c^{\mathrm{T}}\dot{\tilde{W}}_c\} + \frac{b_u b_r}{\alpha_3}\tilde{\kappa}\dot{\tilde{\kappa}}.\end{aligned} \tag{6.92}$$

Applying (6.87) to (6.92), we have

$$\begin{aligned}\dot{L} = \mathcal{R}\Bigg\{&-Kr + b_u\tilde{W}_a^{\mathrm{T}}\phi_a + b_u\left[\upsilon_r + \left(\frac{f_\upsilon(\bar{\upsilon})}{b_u} - 1\right)\left(f_\upsilon(\bar{\upsilon}) + \tilde{W}_a^{\mathrm{T}}\phi_a\right)\right]\Bigg\}\\ &+ \mathcal{R}f_\upsilon(\bar{\upsilon})\varepsilon_a(\bar{x}) - b_u\mathrm{tr}\left\{\tilde{W}_a^{\mathrm{T}}(\phi_a(\mathcal{R} + |\mathcal{R}|\hat{W}_c^{\mathrm{T}}\phi_c)^{\mathrm{T}} - K_{W_a}|\mathcal{R}|\hat{W}_a)\right\}\\ &+ \frac{b_u}{\alpha_2}\mathrm{tr}\{\tilde{W}_c^{\mathrm{T}}\dot{\tilde{W}}_c\} + \frac{b_u b_r}{\alpha_3}\tilde{\kappa}\dot{\tilde{\kappa}}.\end{aligned} \tag{6.93}$$

Due to the fact that

$$\begin{aligned} b_u\mathrm{tr}\{\tilde{W}_a^{\mathrm{T}}\phi_a\mathcal{R}\} &= \mathcal{R}b_u\tilde{W}_a^{\mathrm{T}}\phi_a,\\ \mathrm{tr}\{\tilde{W}_a^{\mathrm{T}}\phi_a|\mathcal{R}|(\hat{W}_c^{\mathrm{T}}\phi_c)^{\mathrm{T}}\} &= |\mathcal{R}|\tilde{W}_a^{\mathrm{T}}\phi_a\phi_c^{\mathrm{T}}\hat{W}_c\\ &= |\mathcal{R}|(-\hat{W}_a^{\mathrm{T}}\phi_a\phi_c^{\mathrm{T}}W_a + W_a^{\mathrm{T}}\phi_a\phi_c^{\mathrm{T}}\hat{W}_c + \hat{W}_a^{\mathrm{T}}\phi_a\phi_c^{\mathrm{T}}\tilde{W}_c),\end{aligned} \tag{6.94}$$

and using (6.88), (6.93) can be rewritten as

$$
\begin{aligned}
\dot{L} \leq & -\mathcal{R}Kr + \mathcal{R}b_u\left[\upsilon_r + \left(\frac{f_\upsilon(\bar{\upsilon})}{b_u} - 1\right)\left(f_\upsilon(\bar{\upsilon}) + \tilde{W}_a^{\mathrm{T}}\phi_a\right)\right] + \mathcal{R}f_\upsilon(\bar{\upsilon})\varepsilon_a(\bar{x}) \\
& + b_u K_{W_a}|\mathcal{R}|\mathrm{tr}\{\tilde{W}_a^{\mathrm{T}}\hat{W}_a\} + b_u K_{W_c}|\mathcal{R}|\mathrm{tr}\{\tilde{W}_c^{\mathrm{T}}\hat{W}_c\} \\
& + b_u|\mathcal{R}|(-\hat{W}_a^{\mathrm{T}}\phi_a(\phi_c^{\mathrm{T}}W_a) + W_a^{\mathrm{T}}\phi_a\phi_c^{\mathrm{T}}\hat{W}_c) + \frac{b_u b_r}{\alpha_3}\tilde{\kappa}\dot{\tilde{\kappa}}.
\end{aligned} \tag{6.95}
$$

With the robustifying term in (6.81), we have

$$
\mathcal{R}\upsilon_r = -\frac{b_r}{1-b_r}|\hat{\kappa}|\mathcal{R}\psi\,\mathrm{sgn}(\mathcal{R}\psi) + \frac{b_r}{1-b_r}|\hat{\kappa}|\left(|\mathcal{R}\psi| - \mathcal{R}\psi\tanh\left(\frac{\mathcal{R}\psi}{\alpha}\right)\right). \tag{6.96}
$$

According to Assumptions 6.13 and 6.14, we have

$$
\left|\left(\frac{f_\upsilon(\bar{\upsilon})}{b_u} - 1\right)\left(\upsilon_r + \tilde{W}_a^{\mathrm{T}}\phi_a\right)\right| \leq b_r|\upsilon_r| + b_r\|\tilde{W}_a\|\|\phi_a\|. \tag{6.97}
$$

Substituting (6.96) and (6.97) into (6.95), we have

$$
\begin{aligned}
\dot{L} \leq & -\mathcal{R}Kr + b_u\left[-\frac{b_r}{1-b_r}|\hat{\kappa}|\mathcal{R}\psi\,\mathrm{sgn}(\mathcal{R}\psi) + b_r\mathcal{R}|\upsilon_r|\right. \\
& \left. + \frac{b_r}{1-b_r}|\hat{\kappa}|\left(|\mathcal{R}\psi| - \mathcal{R}\psi\tanh\left(\frac{\mathcal{R}\psi}{\alpha}\right)\right)\right] \\
& + b_u K_{W_a}|\mathcal{R}|\mathrm{tr}\{\tilde{W}_a^{\mathrm{T}}\hat{W}_a\} + b_u K_{W_c}|\mathcal{R}|\mathrm{tr}\{\tilde{W}_c^{\mathrm{T}}\hat{W}_c\} \\
& + \delta + \frac{b_u b_r}{\alpha_3}\tilde{\kappa}\dot{\tilde{\kappa}}.
\end{aligned} \tag{6.98}
$$

According to Lemma 6.19 and the inequality $0 \leq |\mathcal{R}\psi| - \mathcal{R}\psi\tanh(\mathcal{R}\psi/\alpha) \leq \alpha$ for $\forall\alpha > 0$, $\mathcal{R}\psi \in \mathbb{R}$, we have

$$
\begin{aligned}
\dot{L} \leq & -\mathcal{R}Kr + b_u\left[-b_r|\hat{\kappa}|\mathcal{R}\psi\tanh\left(\frac{\mathcal{R}\psi}{\alpha}\right)\right. \\
& \left. + \frac{b_r}{1-b_r}|\hat{\kappa}|\left(|\mathcal{R}\psi| - \mathcal{R}\psi\tanh\left(\frac{\mathcal{R}\psi}{\alpha}\right)\right)\right] + b_u K_{W_a}|\mathcal{R}|tr\{\tilde{W}_a^{\mathrm{T}}\hat{W}_a\} \\
& + b_u K_{W_c}|\mathcal{R}|\mathrm{tr}\{\tilde{W}_c^{\mathrm{T}}\hat{W}_c\} + b_u b_r\kappa^*\left(\alpha + \mathcal{R}\psi\tanh\left(\frac{\mathcal{R}\psi}{\alpha}\right)\right) \\
& - b_u b_r\tilde{\kappa}\mathcal{R}\psi\tanh\left(\frac{\mathcal{R}\psi}{\alpha}\right) + b_u b_r K_\kappa\tilde{\kappa}\hat{\kappa}
\end{aligned}
$$

$$\leq -\mathcal{R}Kr + \frac{b_u b_r}{1-b_r}|\hat{\kappa}|\alpha + b_u b_r \kappa^* \alpha + b_u K_{W_a}|\mathcal{R}|\mathrm{tr}\{\tilde{W}_a^{\mathrm{T}}\hat{W}_a\}$$
$$+ b_u K_{W_c}|\mathcal{R}|\mathrm{tr}\{\tilde{W}_c^{\mathrm{T}}\hat{W}_c\} + b_u b_r K_\kappa \tilde{\kappa}\hat{\kappa}. \tag{6.99}$$

Let $b^* = b_u - b_l$. Using the relation $b_u b_r/(1-b_r) = (b_u/b_l)b^*$ and the inequality $|\hat{\kappa}| \leq |\tilde{\kappa}| + \kappa^*$, $\dot{L}$ is further derived as

$$\dot{L} \leq -\mathcal{R}Kr + \frac{b_u}{b_l}b^*|\tilde{\kappa}|\alpha + \frac{b_u}{b_l}b^*\kappa^*\alpha + b_u b_r \kappa^* \alpha$$
$$+ b_u K_{W_a}|\mathcal{R}|\mathrm{tr}\{\tilde{W}_a^{\mathrm{T}}\hat{W}_a\} + b_u K_{W_c}|\mathcal{R}|\mathrm{tr}\{\tilde{W}_c^{\mathrm{T}}\hat{W}_c\} + b_u b_r K_\kappa \tilde{\kappa}\hat{\kappa}. \tag{6.100}$$

Since $\mathrm{tr}\{\tilde{W}_a^{\mathrm{T}}\hat{W}_a\} \leq -\|\tilde{W}_a\|^2/2 + \|W_a\|^2/2$, $\mathrm{tr}\{\tilde{W}_c^{\mathrm{T}}\hat{W}_c\} \leq -\|\tilde{W}_2\|^2/2 + \|W_c\|^2/2$, $\tilde{\kappa}\hat{\kappa} \leq -|\tilde{\kappa}|^2/2 + \kappa^{*2}/2$, $(b_u/b_l)b^*|\tilde{\kappa}|\alpha \leq ((b_u/b_l)b^*\alpha)^2/2 + |\tilde{\kappa}|^2/2$, we have

$$\dot{L} \leq -\mathcal{R}Kr - \frac{b_u K_{W_a}|\mathcal{R}|}{2}\|\tilde{W}_a\|^2 - \frac{b_u K_{W_c}|\mathcal{R}|}{2}\|\tilde{W}_c\|^2$$
$$- \frac{b_u b_r}{2}\left(K_\kappa - \frac{1}{b_u b_r}\right)|\tilde{\kappa}|^2 + D, \tag{6.101}$$

where

$$D = \frac{b_u K_{W_a}|\mathcal{R}|}{2}\|W_a\|^2 + \frac{b_u K_{W_c}|\mathcal{R}|}{2}\|W_c\|^2 + \frac{b_u b_r K_\kappa}{2}\kappa^{*2}$$
$$+ \frac{1}{2}\left(\frac{b_u}{b_l}b^*\alpha\right)^2 + \frac{b_u}{b_l}b^*\kappa^*\alpha + b_u b_r \kappa^* \alpha. \tag{6.102}$$

Because $\mathcal{R}r > 0$ for any $r \neq 0$ and $\mathcal{R} \in [-\chi, \chi]$ and using Assumption 6.17, (6.101) becomes

$$\dot{L} \leq -\chi K|r| - \frac{b_u K_{W_a}|\mathcal{R}|}{2}\|\tilde{W}_a\|^2 - \frac{b_u K_{W_c}|\mathcal{R}|}{2}\|\tilde{W}_c\|^2$$
$$- \frac{b_u b_r}{2}\left(K_\kappa - \frac{1}{b_u b_r}\right)|\tilde{\kappa}|^2 + D_M, \tag{6.103}$$

where $\lambda_M \triangleq b_u K_{W_a}\chi W_a^{*2}/2 + b_u K_{W_c}\chi W_c^{*2}/2 + (b_u/b_l)b^*\kappa^*\alpha + b_u b_r \kappa^* \alpha + ((b_u/b_l)b^*\alpha)^2/2$.

With K_κ picked so that $K_\kappa > 1/(b_u b_r)$ is satisfied, the time derivative of L is guaranteed to be negative as long as the following hold:

$$|\tilde{r}| \geq \sqrt{\frac{D_M}{K\chi}}, \tag{6.104}$$

or

$$\|\tilde{W}_a\| \geq \sqrt{\frac{2D_M}{b_u K_{W_a}|\mathcal{R}|}} \geq \sqrt{\frac{2D_M}{b_u K_{W_a}\chi}}, \tag{6.105}$$

or

$$\|\tilde{W}_c\| \geq \sqrt{\frac{2D_M}{b_u K_{W_c}|\mathcal{R}|}} \geq \sqrt{\frac{2D_M}{b_u K_{W_c}\chi}}, \tag{6.106}$$

or

$$|\tilde{\kappa}| \geq \sqrt{\frac{2\lambda_M}{b_u b_r (K_\kappa - \frac{1}{b_u b_r})}}. \tag{6.107}$$

Therefore, according to the standard Lyapunov extensions [11], this demonstrates that the filtered tracking error, the weights estimation errors of the critic NN and action NN are uniformly ultimately bounded. □

Remark 6.21 It is interesting to note from (6.104)–(6.107) that arbitrarily small $|\tilde{r}|$, $\|\tilde{W}_a\|$, $\|\tilde{W}_c\|$ and $|\tilde{\kappa}|$ may be achieved by selecting the large fixed gain K, K_{W_a}, K_{W_c}, and K_κ or the critic slope gain χ, respectively.

6.3.3 Simulations

Example 6.22 Consider the following continuous-time nonaffine nonlinear system:

$$\begin{aligned}
\dot{x}_1 &= x_2, \\
\dot{x}_2 &= x_1 x_2^2 + x_2 e^{-1-x_1^2} - x_1 x_2 + (2 + 0.3\sin x_2^2)u + \cos(0.1u). \\
y &= x_1.
\end{aligned}$$

The control objective is to make the output y follow the desired trajectory y_d. The reference signal is selected as $y_d = \sin(t) + \cos(0.5t)$. We assume that the control design is performed by using the approximate model $\hat{f}(x, u) = 10u$. The controller parameters are chosen as $K = 30$, $\Lambda = 20$, $b_r = 1/3$, and $\chi = 20$. The critic NN and action NN consist of five and six hidden-layer nodes, respectively. The activation functions are selected as sigmoidal activation functions. The first-layer weights of both action NN and critic NN are selected randomly over an internal of $[-1, 1]$. The threshold weights for the first layer of both action NN and critic NN are uniformly randomly distributed between -10 and 10. The second-layer weights of action NN $\hat{W}_a$ is uniformly randomly initialized over an internal of $[-1, 1]$. The second-layer weights of the critic NN $\hat{W}_c$ is initialized at zero. The parameters of the critic signal are selected as $m = 2$. For weights and adaptive parameters updating, the design parameters are selected as $\alpha_1 = \alpha_2 = \alpha_3 = K_{W_a} = K_{W_c} = 0.1$, and $K_\kappa = 80$, which satisfies $K_\kappa > 1/(b_u b_r)$. Then, for the initial states $x(0) = [0, 0]^{\mathrm{T}}$, we apply the present controller to the system for 100 s. The simulation results are shown in Figs. 6.11–6.16. From Fig. 6.11, it is observed that the system output

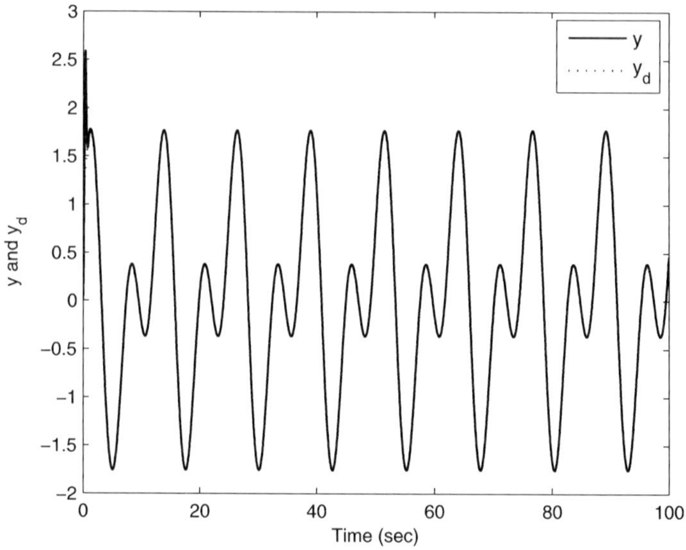

Fig. 6.11 The trajectories of y_d and y

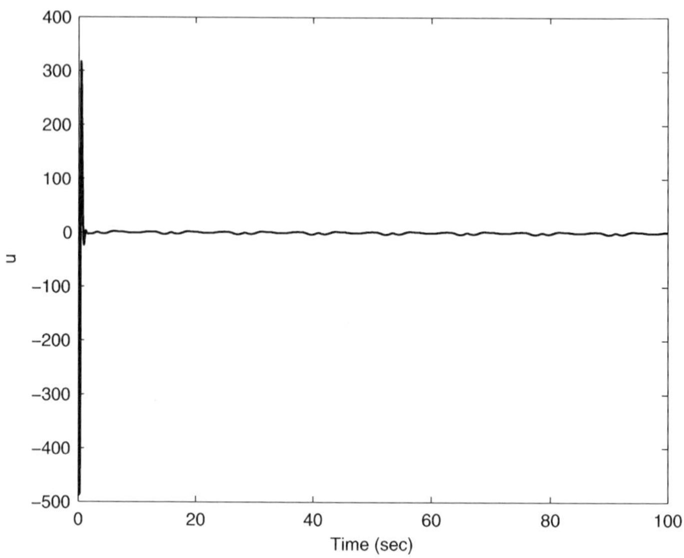

Fig. 6.12 The trajectory of control input u

y tracks the desired trajectory y_d fast and well. Figures 6.12, 6.13, 6.14 clearly show that the control input u is bounded. Figure 6.15 displays that the critic signal $\hat{\mathcal{R}}_n$ is bounded. From Fig. 6.16, it is observed that the adaptive parameter $\hat{\kappa}$ is

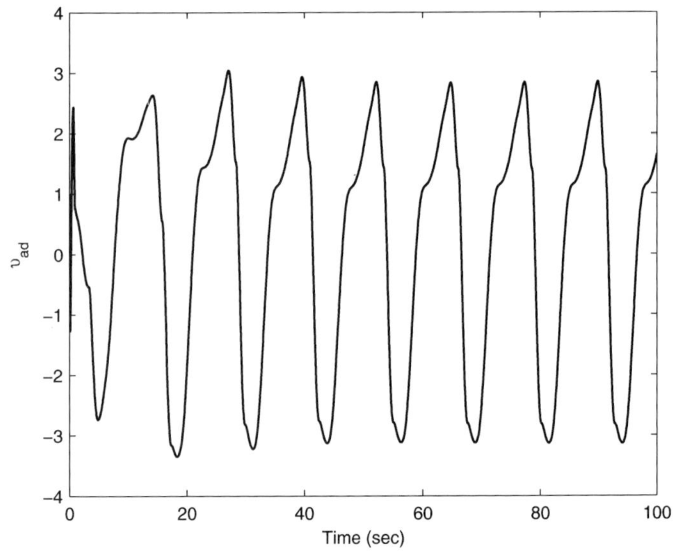

Fig. 6.13 The trajectory of υ_{ad}

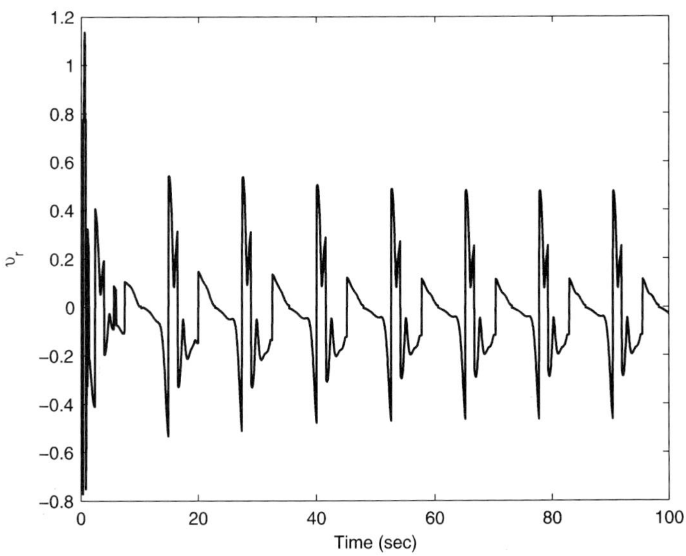

Fig. 6.14 The trajectory of υ_r

bounded. The simulation results show that the present ADP method can perform successful control and achieve the desired performance for the nonaffine nonlinear system.

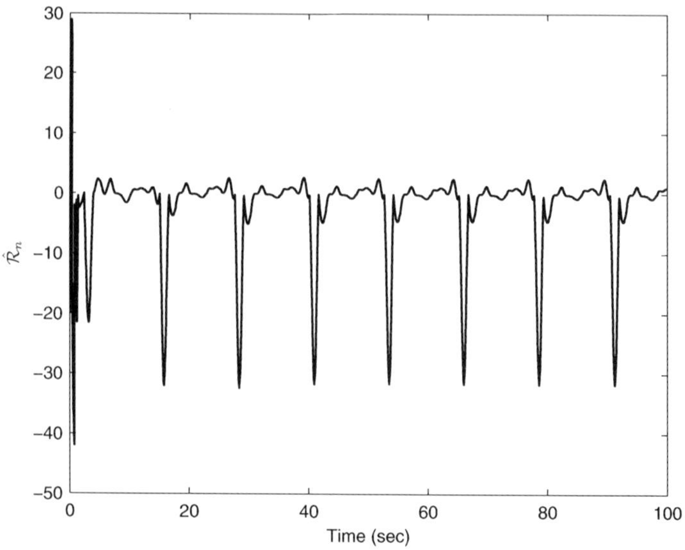

Fig. 6.15 The trajectory of $\hat{\mathcal{R}}_n$

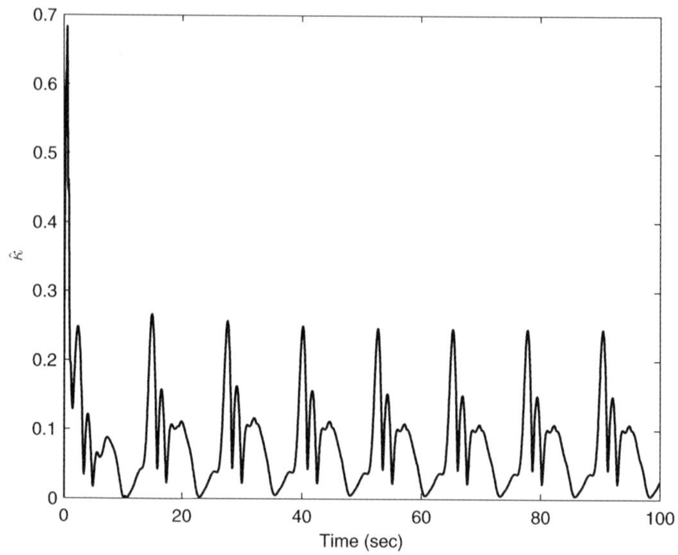

Fig. 6.16 The trajectory of $\hat{\kappa}$

6.4 Summary

In this chapter, we investigated the optimal feedback control problems of a class of unknown general continuous-time nonlinear systems and a class of continuous-time nonaffine nonlinear systems via the ADP approach. In Sect. 6.2, we developed an effective scheme to design the data-based optimal robust tracking controller for unknown general continuous-time nonlinear systems, in which only input/output data are required instead of the system dynamics. In Sect. 6.3, a novel ADP based robust neural network controller was developed for a class of continuous-time nonaffine nonlinear systems, which is the first attempt to extend the ADP approach to continuous-time nonaffine nonlinear systems. Numerical simulations have shown that the present methods are effective and can be used for a quite wide class of continuous-time nonlinear systems.

References

1. Al-Tamimi A, Lewis FL, Abu-Khalaf M (2007) Model-free Q-learning designs for linear discrete-time zero-sum games with application to H_∞ control. Automatica 43:473–481
2. Calise AJ, Hovakimyan N, Idan M (2001) Adaptive output feedback control of nonlinear systems using neural networks. Automatica 37:1201–1211
3. Chellaboina V, Haddad WM (2002) A unification between partial stability and stability theory for time-varying systems. IEEE Control Syst Mag 22:66–75
4. Cui LL, Luo YH, Zhang HG (2011) Adaptive critic design based robust neural network control for a class of continuous-time nonaffine nonlinear system. In: Proceedings of international conference on modelling, identification and control, Shanghai, pp 26–29
5. Ge S, Zhang J (2003) Neural-network control of nonaffine nonlinear system with zero dynamics by state and output feedback. IEEE Trans Neural Netw 14:900–918
6. Hanselmann T, Noakes L, Zaknich A (2007) Continuous-time adaptive critics. IEEE Trans Neural Netw 3:631–647
7. Khalil HK (2002) Nonlinear system. Prentice Hall, Englewood Cliffs
8. Kim N, Calise AJ (2007) Several extensions in methods for adaptive output feedback control. IEEE Trans Neural Netw 18:482–494
9. Kim YH, Lewis FL (2000) Reinforcement adaptive learning neural-net-based friction compensation control for high speed and precision. IEEE Trans Control Syst Technol 8:118–126
10. Kuljaca O, Lewis FL (2003) Adaptive critic design using non-linear network structure. Int J Adapt Control Signal Prog 17:431–445
11. Lewis FL, Jagannathan S, Yesildirek A (1999) Neural network control of robot manipulators and nonlinear systems. Taylor & Francis, London
12. Lin CK (2005) Adaptive critic autopilot design of bank-to-turn missiles using fuzzy basis function networks. IEEE Trans Syst Man Cybern, Part B, Cybern 35:197–206
13. Lin CK (2009) H_∞ reinforcement learning control of robot manipulators using fuzzy wavelet networks. Fuzzy Sets Syst 160:1765–1786
14. Liu ZW, Zhang HG, Zhang QL (2010) Novel stability analysis for recurrent neural networks with multiple delays via line integral-type L-K functional. IEEE Trans Neural Netw 21:1710–1718
15. Murray JJ, Cox CJ, Lendaris GG, Saeks R (2002) Adaptive dynamic programming. IEEE Trans Syst Man Cybern, Part B, Cybern 32:140–153
16. Rubio JDJ, Yu W (2007) Stability analysis of nonlinear system identification via delayed neural networks. IEEE Trans Circuits Syst II, Express Briefs 54:161–195
17. Vamvoudakis KG, Lewis FL (2010) Online actor-critic algorithm to solve the continuous-time infinite horizon optimal control problem. Automatica 46:878–888
18. Vrabie D, Lewis FL (2009) Neural network approach to continuous-time direct adaptive optimal control for partially unknown nonlinear system. Neural Netw 22:237–246

19. Vrabie D, Pastravanu O, Abu-Khalaf M, Lewis FL (2009) Adaptive optimal control for continuous-time linear systems based on policy iteration. Automatica 45:477–484
20. Wang ZS, Zhang HG, Yu W (2009) Robust stability of Cohen–Grossberg neural networks via state transmission matrix. IEEE Trans Neural Netw 20:169–174
21. Wang ZS, Zhang HG, Jiang B (2011) LMI-based approach for global asymptotic stability analysis of recurrent neural networks with various delays and structures. IEEE Trans Neural Netw 22:1032–1045
22. Zhang HG, Wang ZS, Liu DR (2008) Global asymptotic stability of recurrent neural networks with multiple time-varying delays. IEEE Trans Neural Netw 19:855–873
23. Zhang HG, Li M, Yang J (2009) Fuzzy model-based robust networked control for a class of nonlinear systems. IEEE Trans Syst Man Cybern, Part A, Syst Hum 39:437–447
24. Zhang HG, Liu ZW, Huang GB (2010) Novel weighting-delay-based stability criteria for recurrent neural networks with time-varying delay. IEEE Trans Neural Netw 21:91–106
25. Zhang HG, Cui LL, Zhang X, Luo YH (2011) Data-driven robust approximate optimal tracking control for unknown general nonlinear systems using adaptive dynamic programming method. IEEE Trans Neural Netw 22(12):2226–2236
26. Zhang HG, Xie XP, Tong SC (2011) Homogeneous polynomially parameter-dependent H_∞ filter designs of discrete-time fuzzy systems. IEEE Trans Syst Man Cybern, Part B, Cybern 41:1313–1322

Chapter 7
Several Special Optimal Feedback Control Designs Based on ADP

7.1 Introduction

In Chap. 6, we have discussed the optimal feedback control for continuous-time systems by ADP algorithms. In this chapter, several special optimal feedback control designs based on ADP are developed.

In this chapter, we first study the optimal control problem of affine nonlinear switched systems. Though there are some results on the control of switched systems, most existing results only deal with the state-feedback control problem for switched systems, such as [14, 16]. Hence, in order to seek the optimal solution, a novel two-stage adaptive dynamic programming (TSADP) is developed. The algorithm can be divided into two stages: first, for each possible mode, calculate associated value function, and then select the optimal mode for each state. It is shown that the value function at each iteration can be characterized pointwise by a set of smooth functions recursively generated from TSADP. Moreover, the optimal control policy, continuous control and switching control law included, is explicitly provided in a state-feedback form.

In the second part, the near-optimal control problem of nonlinear descriptor systems is solved by greedy iterative DHP algorithm. A nonquadratic cost functional is developed in order to deal with the actuator saturation problem. In this way, the greedy iterative DHP algorithm can be properly introduced to solve the optimal control problem of the original descriptor system.

In the third part, a novel near-optimal control design method for a class of nonlinear singularly perturbed systems is developed based on ADP algorithm. By the slow/fast HJB equations and starting from initial performances, the optimal performances are converged by neural network approximation and iteration between control laws and performance indices. It avoids solving the complex HJB equations directly.

In the fourth part, the near-optimal state-feedback control problem is solved by SNAC method for a class of nonlinear discrete-time systems with control constraints. Based on the costate function, the greedy iterative DHP algorithm is developed to solve the HJB equation of the system. At each step of the iterative algorithm,

H. Zhang et al., *Adaptive Dynamic Programming for Control*,
Communications and Control Engineering, DOI 10.1007/978-1-4471-4757-2_7,

a neural network is utilized to approximate the costate function, and then the optimal control policy of the system can be computed directly according to the costate function, which removes the action network appearing in the ordinary ADP method.

7.2 Optimal Feedback Control for a Class of Switched Systems

7.2.1 Problem Description

Consider the following affine nonlinear discrete-time switched system:

$$x(t+1) = f_{v(t)}(x(t)) + g_{v(t)}(x(t))u(t), \tag{7.1}$$

where $x(t) \in \mathbb{R}^n$ and $u(t) \in \mathbb{R}^m$ are the continuous state variable and control input, $v(t) \in \mathcal{M}$ is the switching control that determines which subsystem is switched on to operate. The sequence $\{\langle u(t), v(t)\rangle\}_0^\infty$ is called the hybrid control sequence for system (7.1). Generally, the control is a feedback form, thus we denote the hybrid feedback law as $\pi(x(t)) = \langle \mu(x(t)), \nu(x(t))\rangle : \mathbb{R}^n \to \mathbb{R}^m \times \mathcal{M}$, where $\mu(x(t))\colon \mathbb{R}^n \to \mathbb{R}^m$ and $\nu(x(t))\colon \mathbb{R}^n \to \mathcal{M}$. Then, the closed-loop system model with control policy π is given by

$$x(t+1) = f_{\nu(x(t))}(x(t)) + g_{\nu(x(t))}(x(t))\mu(x(t)). \tag{7.2}$$

The infinite horizon optimal control problem is to design a hybrid feedback control policy $\pi = \langle \mu, \nu\rangle$ that minimizes the following cost functional:

$$J_\pi(z) = \sum_{t=0}^{\infty} L(x(t), u(t), v(t)) \tag{7.3}$$

subject to (7.2), where $z \in \mathbb{R}^n$ is an arbitrary initial state, i.e., $x(0) = z$, $u(t) = \mu(x(t))$ and $v(t) = \nu(x(t))$. $L(x, u, v) = x^{\mathrm{T}} Q_v x + u^{\mathrm{T}} R_v u$, $Q_v > 0$ and $R_v > 0$ are the cost-to-go weighting matrices for the state and control, respectively.

Definition 7.1 (Admissible Control Policy) The control policy π is admissible if π stabilizes system (7.2) and $\forall z$, the total cost $J_\pi(z)$ is finite, denoted $\pi \in \Pi_A$.

The following assumption is necessary in the rest of this section.

Assumption 7.2 At least one subsystem is stabilizable.

Assumption 7.2 guarantees the existence of at least one admissible control policy, i.e., the set Π_A is not empty.

Thus, the discrete-time infinite horizon switched optimal control problem can be formulated below.

Problem 7.3 For a given initial state $z \in \mathbb{R}^n$, find the control policy $\pi \in \Pi_A$ that minimizes $J_\pi(z)$ subject to (7.2).

To solve this problem, define the optimal value function as

$$J^*(z) = \inf_{\pi \in \Pi_A} J_\pi(z), \tag{7.4}$$

which is characterized by the Bellman equation

$$J^*(z) = \inf_{u,v} \left\{ L(z,u,v) + J^*(f_v(z) + g_v(z)u) \right\}. \tag{7.5}$$

Then, based on traditional dynamic programming theory, the optimal value function can be computed by value iteration, i.e., to start from the initial value function V_0 and update iteratively according to

$$V_{k+1}(z) = \inf_{u,v} \{ L(z,u,v) + V_k(f_v(z) + g_v(z)u) \}. \tag{7.6}$$

Unfortunately, the value function for switched systems is in general nonsmooth, even for simple LQR problem (see [15]). Besides, there are two independent variables to minimize the function, thus the iteration with (7.6) is quite complicated and infeasible.

Notation $\mathcal{M} = 1, \ldots, M$ denotes the set of indices of subsystems. The set L_k is defined as $L_k = 1, \ldots, M^k$ where k is a positive integer or zero. $\mod(a,b)$ is an operator that seeks the remainder of a/b and if a is an exact multiple of b, the operator returns b. The operator $[a]$ takes the largest integer that is no greater than a. $\mathcal{N}$ represents the natural number set.

7.2.2 Optimal Feedback Control Based on Two-Stage ADP Algorithm

In this section, we develop a novel iterative algorithm, termed two-stage adaptive dynamic programming (TSADP), to seek the optimal value function J^*. The greatest advantage of our algorithm is that it transforms a multivariate optimal control problem to be a certain number of univariate optimal control problems at each iteration.

To start the iteration, set the initial condition as $V_0(z) = V_0^{(1)}(z) = 0$. For each $i \in \mathcal{M}$ and $l \in \mathcal{L}_0 = \{1\}$, solve

$$u_0^{(l,i)}(z) = \arg\min_u \left\{ L(z,u,i) + V_0^{(l)}(f_i(z) + g_i(z)u) \right\} \tag{7.7}$$

and then update the value function using $u_0^{(l,i)}$

$$V_1^{(\hat{l})}(z) = L(z, u_0^{(l,i)}, i) + V_0^{(l)}(f_i(z) + g_i(z)u_0^{(l,i)}), \tag{7.8}$$

where $\hat{l} = (l-1) \times M + i$.

Let

$$V_1(z) = \min_{\hat{l} \in \mathcal{L}_1} \left\{ V_1^{(\hat{l})}(z) \right\}, \tag{7.9}$$

where the set $\mathcal{L}_1$ has been defined in Sect. 7.2.1. Therefore, the value function of the first iteration is obtained.

Denote

$$\iota_0(z) = \arg\min_{\hat{l} \in \mathcal{L}_1} \left\{ V_1^{(\hat{l})}(z) \right\}. \tag{7.10}$$

Thus, we get the initial control policy $\pi_0 = \langle \mu_0, \nu_0 \rangle$, where

$$\nu_0(z) = \mod(\iota_0(z), M), \tag{7.11}$$

$$\mu_0(z) = u_0^{(\lfloor \iota_0(z)/M \rfloor + 1, \nu_0(z))}(z), \tag{7.12}$$

and the initial iteration finishes.

Suppose V_k is obtained from the $(k-1)$st iteration, i.e., $V_k(z) = \min_{l \in \mathcal{L}_k} \{V_k^{(l)}\}$. Then, for each $i \in \mathcal{M}$ and $l \in \mathcal{L}_k$, solve

$$u_k^{(l,i)}(z) = \arg\min_u \left\{ L(z,u,i) + V_k^{(l)}(f_i(z) + g_i(z)u) \right\}, \tag{7.13}$$

and update the value function $V_{k+1}^{(\hat{l})}$

$$V_{k+1}^{(\hat{l})}(z) = L(z, u_k^{(l,i)}, i) + V_k^{(l)}(f_i(z) + g_i(z)u_k^{(l,i)}), \tag{7.14}$$

where $\hat{l} = (l-1) \times M + i$. Then, we get

$$V_{k+1}(z) = \min_{\hat{l} \in \mathcal{L}_{k+1}} \left\{ V_{k+1}^{(\hat{l})}(z) \right\} \tag{7.15}$$

and

$$\iota_k(z) = \arg\min_{\hat{l} \in \mathcal{L}_{k+1}} \left\{ V_{k+1}^{(\hat{l})}(z) \right\}. \tag{7.16}$$

Therefore, the control policy at the kth iteration is $\pi_k = \langle \mu_k, \nu_k \rangle$, where

$$\nu_k(z) = \mod(\iota_k(z), M), \tag{7.17}$$

$$\mu_k(z) = u_k^{(\lfloor \iota_k(z)/M \rfloor + 1, \nu_k(z))}(z). \tag{7.18}$$

The iteration continues according to (7.13)–(7.18).

The core part of the two-stage adaptive dynamic programming algorithm is to solve (7.13) and (7.14) iteratively. Compared with (7.6), there is only one variable u to be determined, which is much easier to be dealt with. As a matter of fact, each iteration can be separated into two steps: one to calculate the optimal $u_k^{(l,i)}$ and $V_{k+1}^{(\hat{l})}$ with each fixed i and l; then to find the optimal i and l for each z, i.e., to formulate $\iota_k(z)$ and then the control policy $\pi_k(z)$. That is what the term "two-stage" implies.

Remark 7.4 Note that the subscripts of all variables used above, such as V_k, $V_k^{(\cdot)}$ and u_k demonstrate the kth iteration of our algorithm. Moreover, the superscript of $V_k^{(l)}$ represents the lth possible component of V_k, while the counterpart of $u_k^{(l,i)}$ indicates the lth component of V_{k-1} as well as the ith subsystem switched on.

Remark 7.5 Apparently, by induction, we conclude that all the functions $V_k^{(l)}$ are smooth because each of them is calculated by conventional value iteration with initial smooth function $V_0 = 0$.

Remark 7.6 Actually, unless the optimal value function is found, the controller law μ_k and ν_k are not used to update the value function V_{k+1}, then they can be omitted during the process.

Theorem 7.7 *Value function at the kth iteration can be expressed as* $V_k(z) = \min_{l\in\mathcal{L}_k}\{V_k^{(l)}\}$ *and the control policy* $\pi_k = \langle \mu_k, \nu_k \rangle$ *defined in* (7.18) *and* (7.17) *is optimal, i.e.,*

$$\langle \mu_k, \nu_k \rangle = \arg\min_{u,i} \{L(z,u,i) + V_k(f_i(z) + g_i(z)u)\}. \tag{7.19}$$

Proof The expression of the value function can be naturally concluded from previous derivation. Besides, according to the principle of dynamic programming, we have

$$\begin{aligned} V_{k+1}(z) &= \min_{u,i} \{L(z,u,i) + V_k(f_i(z) + g_i(z)u)\} \\ &= \min_{u,i} \left\{ L(z,u,i) + \min_{l\in\mathcal{L}_k} \left\{ V_k^{(l)}(f_i(z) + g_i(z)u) \right\} \right\} \\ &= \min_{l\in\mathcal{L}_k} \min_{u,i} \left\{ L(z,u,i) + V_k^{(l)}(f_i(z) + g_i(z)u) \right\} \\ &= \min_{l\in\mathcal{L}_k} \min_{i} \left\{ L(z,u_k^{(l,i)},i) + V_k^{(l)}(f_i(z) + g_i(z)u_k^{(l,i)}) \right\}. \end{aligned} \tag{7.20}$$

Denote

$$\langle \varsigma_k(z), \nu_k(z) \rangle = \arg\min_{l\in\mathcal{L}_k} \min_{i} \left\{ L(z,u_k^{(l,i)},i) + V_k^{(l)}(f_i(z) + g_i(z)u_k^{(l,i)}) \right\}. \tag{7.21}$$

Thus, $\langle u_k^{(\varsigma_k(z),\nu_k(z))}(z), \nu_k(z) \rangle$ is the optimal control law for the kth value iteration, i.e., minimizes the first equation in (7.20).

In addition, note that $\iota_k(z)$ in (7.16) can be expressed by $\varsigma_k(z)$, i.e., $\iota_k = (\varsigma_k - 1) \times M + \nu_k$. As a result, the control policy $\langle u_k^{(\varsigma_k(z),\nu_k(z))}(z), \nu_k(z)\rangle$ is equivalent to $\pi_k = \langle \mu_k, \nu_k\rangle$ defined in (7.18) and (7.17). Therefore, π_k satisfies (7.19). □

Remark 7.8 Different from normal optimal control problem, V_k for switched system is no longer a single smooth function, it becomes the pointwise minimum of a certain number of value functions obtained from two-stage adaptive dynamic programming. Similarly, the control policy π_k is also state-dependent instead of a single optimal feedback control law over the entire state space.

Remark 7.9 Related work has been published in some references, e.g., [11] proposed the master-slave procedure (MSP), which was based on some numerical search techniques such as MIQP to solve similar optimal control problems. Compared with the results in [5, 10–13], our contribution is that the two-stage adaptive dynamic programming explicitly presents the analytical expression of the optimal feedback control policy and its connection to conventional value iteration, cf. (7.17)–(7.19).

In this part, we will prove that the sequence of value functions $\{V_k\}$ obtained by TSADP is convergent, i.e., there exists a function V_∞ satisfying $V_\infty = \lim_{k\to\infty} V_k$, and derive the relation between V_∞ and the optimal value function.

Lemma 7.10 *Let* $\{\hat{\pi}_k = \langle \hat{\mu}_k, \hat{\nu}_k\rangle\}$ *be arbitrary sequence of control policy and* $\{\Lambda_k\}$ *be defined by*

$$\Lambda_{k+1}(z) = L(z, \hat{\mu}_k, \hat{\nu}_k) + \Lambda_k(f_{\hat{\nu}_k}(z) + g_{\hat{\nu}_k}(z)\hat{\mu}_k), \tag{7.22}$$

and $\{V_k\}$ *is obtained by TSADP. If* $V_0(z) = \Lambda_0(z) = 0$, *thus* $V_k(z) \le \Lambda_k(z)$, $\forall k \in \mathcal{N}, z \in \mathbb{R}^n$.

Proof According to Theorem 7.7, μ_k, ν_k minimize the right-hand side of (7.19) and because $V_0(z) = \Lambda_0(z) = 0$, then by induction it follows that $V_k(z) \le \Lambda_k(z)$, $\forall k, z$. □

Lemma 7.11 (cf. [4], Boundedness) *Under Assumption* 7.2, *there is a state-dependent upper bound* $B(z)$ *such that* $V_k(z) \le B(z)$, $\forall k \in \mathcal{N}, z \in \mathbb{R}^n$.

Proof Suppose the $\underline{i}$th subsystem $(f_{\underline{i}}, g_{\underline{i}})$ is stabilizable ($\underline{i} \in \mathcal{M}$). Denote $\{V_k^{(\underline{i})}\}$ as the sequence of value functions generated by the iteration only in the $\underline{i}$ th subsystem without switching, i.e., $V_{k+1}^{(\underline{i})}(z) = \min_u\{L(z, u, \underline{i}) + V_k^{(\underline{i})}(f_{\underline{i}}(z) + g_{\underline{i}}(z)u)\}$.

Because subsystem $\underline{i}$ is stabilizable, there must exist an admissible control law sequence $\{\hat{\pi}_k = \langle \hat{\mu}_k, \underline{i}\rangle\}$ where $\hat{\mu}_k$ can be obtained by

$$\hat{\mu}_k = \arg\min_u \left\{L(z, u, \underline{i}) + V_k^{(\underline{i})}(f_{\underline{i}}(z) + g_{\underline{i}}(z)u)\right\}, \tag{7.23}$$

and thus $V_k^{(i)}(z) < \infty$. Therefore, there exists a $B(z)$ such that $V_k^{(i)}(z) < B(z)$. According to Lemma 7.10, $V_k(z) \leq V_k^{(i)}(z)$. Consequently, we have $V_k(z) \leq V_k^{(i)}(z) < B(z)$. □

Lemma 7.12 (cf. [4], Monotonicity) *The sequence $\{V_k\}$ generated from TSADP is uniformly monotonically nondecreasing, i.e., $V_k(z) \leq V_{k+1}(z), \forall k \in \mathcal{N}, z \in \mathbb{R}^n$.*

Proof Define $\{\Lambda_k\}$ as

$$\Lambda_{k+1}(z) = L(z, \mu_{k+1}, \nu_{k+1}) + \Lambda_k(f_{\nu_{k+1}}(z) + g_{\nu_{k+1}}(z)\mu_{k+1}), \tag{7.24}$$

moreover,

$$V_{k+1}(z) = L(z, \mu_k, \nu_k) + V_k(f_{\nu_k}(z) + g_{\nu_k}(z)\mu_k), \tag{7.25}$$

with $\Lambda_0(z) = V_0(z) = 0$.

Next, we are ready to prove that $\Lambda_k(z) \leq V_{k+1}(z), \forall z$.

First, when $k = 0$, we have

$$V_1(z) - \Lambda_0(z) = L(z, \mu_0, \nu_0) \geq 0, \tag{7.26}$$

i.e., $\Lambda_0(z) \leq V_1(z)$, for $\forall z$.

Then, assuming $\Lambda_{k-1}(z) \leq V_k(z)$ holds for $\forall z$, we obtain

$$V_{k+1}(z) - \Lambda_k(z) = V_k(\hat{z}) - \Lambda_{k-1}(\hat{z}) \geq 0, \tag{7.27}$$

where $\hat{z} = f_{\nu_k}(z) + g_{\nu_k}(z)\mu_k$. This completes the proof that $\Lambda_k(z) \leq V_{k+1}(z)$.

Furthermore, because μ_k and ν_k are optimal for V_k, according to Lemma 7.10, we conclude that $V_k(z) \leq \Lambda_k(z)$. All in all, it follows that $V_k(z) \leq V_{k+1}(z), \forall k, z$. □

Finally, we study its convergence and optimality. Let $\tilde{\pi} = \{\langle \tilde{\mu}_k, \tilde{\nu}_k \rangle\}$ be an admissible control law sequence and construct the associated sequence $\{P_k(z)\}$ as follows

$$P_{k+1}(z) = L(z, \tilde{\mu}_k, \tilde{\nu}_k) + P_k(f_{\tilde{\nu}_k}(z) + g_{\tilde{\nu}_k}(z)\tilde{\mu}_k). \tag{7.28}$$

By induction, we get

$$\begin{aligned}
P_{k+1}(z) &= L(x(0), \tilde{\mu}_k, \tilde{\nu}_k) + P_k(f_{\tilde{\nu}_k}(x(0)) + g_{\tilde{\nu}_k}(x(0))\tilde{\mu}_k) \\
&= \sum_{j=0}^{1} L(x(j), \tilde{\mu}_{k-j}, \tilde{\nu}_{k-j}) + P_{k-1}(f_{\tilde{\nu}_{k-1}}(x(1)) + g_{\tilde{\nu}_{k-1}}(x(1))\tilde{\mu}_{k-1}) \\
&\ \ \vdots \\
&= \sum_{j=0}^{k} L(x(j), \tilde{\mu}_{k-j}, \tilde{\nu}_{k-j}),
\end{aligned} \tag{7.29}$$

where $x(j+1) = f_{\tilde{v}_{k-j}}(x(j)) + g_{\tilde{v}_{k-j}}(x(j))\tilde{\mu}_{k-j}$.

As $k \to \infty$, we obtain

$$P_\infty(z) = \lim_{k\to\infty} \sum_{j=0}^{k} L(x(j), \tilde{\mu}_{k-j}, \tilde{v}_{k-j}). \tag{7.30}$$

Define

$$J^*(z) = \inf_{\tilde{\pi}} P_\infty(z), \tag{7.31}$$

then we have the following theorem.

Theorem 7.13 (cf. [4], Convergence and Optimality) *Under Assumption* 7.2, *the sequence* $\{V_k\}$ *is convergent and as* $k \to \infty$, $V_k \to J^*$, *where* J^* *is the optimal value function satisfying*

$$J^*(z) = \inf_{u,v} \left\{ L(z, u, v) + J^*(f_v(z) + g_v(z)u) \right\}. \tag{7.32}$$

Proof From Lemmas 7.11 and 7.12, the sequence $\{V_k\}$ is bounded and monotonically nondecreasing, thus we easily conclude that the sequence gets convergent to V_∞.

Suppose the control policy is chosen to be $\tilde{\pi}$, Lemma 7.10 still holds, i.e., $\forall k, z$,

$$V_k(z) \le P_k(z), \tag{7.33}$$

then

$$\lim_{k\to\infty} V_k(z) \le \lim_{k\to\infty} P_k(z), \tag{7.34}$$

and

$$V_\infty(z) = \lim_{k\to\infty} V_k(z) \le \inf_{\tilde{\pi}} \lim_{k\to\infty} P_k(z) = J^*(z). \tag{7.35}$$

Moreover, from Lemma 7.11, we know that $\forall k$, V_k is bounded, so is V_∞. Thus, by definition of admissible control, the control sequence associated with the V_∞ must be admissible. Recalling the definition of J^* in (7.31), we obtain $J^* \le V_\infty$. Combining it with (7.35), we conclude that $J^* = V_\infty$, i.e., $J^* = \lim_{k\to\infty} V_k$.

Next, we commence to prove the J^* to be optimal.

On one hand, according to (7.6), for any k, z, u and v, we have

$$V_{k+1}(z) \le L(z, u, v) + V_k(f_v(z) + g_v(z)u). \tag{7.36}$$

From Lemma 7.12, the sequence $\{V_k\}$ is uniformly monotonically nondecreasing and gets convergent to J^*. Hence, $V_k \le J^*$ holds for any k, z. Thus, (7.36) becomes

$$V_{k+1}(z) \le L(z, u, v) + J^*(f_v(z) + g_v(z)u). \tag{7.37}$$

As $k \to \infty$, we get

$$J^*(z) \le L(z, u, v) + J^*(f_v(z) + g_v(z)u). \tag{7.38}$$

Since u and v are chosen arbitrarily, we have

$$J^*(z) \le \inf_{u,v} \left\{ L(z, u, v) + J^*(f_v(z) + g_v(z)u) \right\}. \tag{7.39}$$

On the other hand, V_{k+1} in (7.6) still satisfies $V_{k+1} \le J^*$. Therefore, (7.6) can be written as

$$J^*(z) \ge \inf_{u,v} \{ L(z, u, v) + V_k(f_v(z) + g_v(z)u) \}. \tag{7.40}$$

Let $k \to \infty$, we obtain

$$J^*(z) \ge \inf_{u,v} \left\{ L(z, u, v) + J^*(f_v(z) + g_v(z)u) \right\}. \tag{7.41}$$

Combining (7.41) with (7.39), we conclude (7.32), i.e., J^* is the optimal value function.

This completes the proof. □

From Theorem 7.13, we conclude that the sequence $\{V_k\}$ obtained by TSADP converges to the optimal value function.

It could be seen that the essential part of solving the switched optimal control problem is to compute (7.13) and (7.14) iteratively. However, considering the complexity and generality of nonlinear systems, there are still large difficulties in solving (7.13) and (7.14) analytically, though it is much easier compared to (7.6). Therefore, for implementation purposes, we apply two kinds of neural network structure, named the critic and action network, respectively, to approximate the value function and control law at each iteration.

Note that the neural network has a powerful ability to approximate any smooth nonlinear function to arbitrary degree of accuracy, but for discontinuous (or nonsmooth) functions, it is not efficient. Unfortunately, the value function for nonlinear switched systems at each iterative V_k is the pointwise minimum of all $V_k^{(l)}$ generated from TSADP, which means it is nonsmooth generally. Hence, neural network cannot be utilized directly to approximate the value function. In this section, a set of neural networks, instead of a single one, is employed to approximate V_k.

According to Remark 7.5, we realize that all $V_k^{(l)}(z)$ are smooth. Thus, we can use the following neural network structures to approximate value function $V_k^{(l)}(z)$ and control law $u_k^{(l,i)}(z)$, respectively:

$$V_k^{(l)}(z) = W_k^{(l)\mathrm{T}} \phi(z), \tag{7.42}$$

$$u_k^{(l,i)}(z) = U_k^{(l,i)\mathrm{T}} \sigma(z), \tag{7.43}$$

where $W_k^{(l)} = [W_{k,1}^{(l)}, \ldots, W_{k,N_V}^{(l)}]^{\mathrm{T}}$ and $U_k^{(l,i)} = [U_{k,1}^{(l,i)}, \ldots, U_{k,N_U}^{(l,i)}]^{\mathrm{T}}$ are NN weight vectors, N_V and N_U are the numbers of hidden-layer neurons for critic and action network, respectively. $\phi = [\phi_1, \ldots, \phi_{N_V}]^{\mathrm{T}}$ and $\sigma = [\sigma_1, \ldots, \sigma_{N_U}]^{\mathrm{T}}$, where $\phi_j(z)$ and $\sigma_j(z)$ are basis functions. Because it is required that $V_k^{(l)}(0) = 0$ and $u_k^{(l,i)}(0) = 0$, we select basis functions with $\phi_j(0) = 0$ and $\sigma_j(0) = 0$. Moreover, the value function is always positive definite and symmetric, thus it is convenient to require the ϕ_j to be symmetric, i.e., $\phi_j(z) = \phi_j(-z)$. Despite the requirements above, the choice of basis functions can be diverse. Many functions can be used, such as the hyperbolic tangent function $\tanh(x)$. Here, we make use of polynomials as basis functions whose terms are obtained from the expansion of the polynomial [2]

$$\sum_{j=1}^{\infty} \left(\sum_{k=1}^{n} x_k \right)^{2j}.$$

Particularly, the switching control $\nu_k(z)$ is not approximated by neural network because $\nu_k(z)$ is a piecewise function which take discrete values finitely within the set $\mathcal{M}$.

According to (7.42), the value function at kth iteration can be expressed as

$$V_k(z) = \min_{l \in \mathcal{L}_k} \left\{ V_k^{(l)} \right\} = \min_{W_k^{(l)} \in \mathcal{W}_k} \left\{ W_k^{(l)\mathrm{T}} \phi(z) \right\}, \tag{7.44}$$

where $\mathcal{W}_k = \{W_k^{(l)} \mid l \in \mathcal{L}_k\}$ is a set of weights for constructing the value function V_k. To meet the condition $V_0(z) = 0$, $\mathcal{W}_0$ is usually initialized with $\mathcal{W}_0 = \{W_0^{(1)} = [0, 0, \ldots, 0]^{\mathrm{T}}\}$.

Based on (7.44), the weight of action network is tuned to solve (7.13), which can be rewritten in the following form:

$$U_k^{(l,i)} = \arg\min_{U} \left\{ L(z, U^{\mathrm{T}}\sigma(z), i) + V_k^{(l)}(f_i(z) + g_i(z)U^{\mathrm{T}}\sigma(z)) \right\}. \tag{7.45}$$

Note that (7.45) is an implicit function for $U_k^{(l,i)}$, so it is difficult to solve the weight explicitly. Therefore, we adopt the gradient descent algorithm (GDA); then the updating formula for $U_k^{(l,i)}$ can be shown to be

$$\begin{aligned} U_{k,[j+1]}^{(l,i)} &= U_{k,[j]}^{(l,i)} - \alpha \frac{\partial (L(z, U_{k,[j]}^{(l,i)\mathrm{T}} \sigma(z), i) + V_k^{(l)}(f_i(z) + g_i(z) U_{k,[j]}^{(l,i)\mathrm{T}} \sigma(z)))}{\partial U_{k,[j]}^{(l,i)}} \\ &= U_{k,[j]}^{(l,i)} - \alpha \left(2\sigma(z) R_i U_{k,[j]}^{(l,i)\mathrm{T}} \sigma(z) + \sigma(z) g(z)^{\mathrm{T}} \frac{\partial \phi(\hat{z})}{\partial \hat{z}} W_k^{(l)\mathrm{T}} \right), \end{aligned} \tag{7.46}$$

where $\hat{z} = f_i(z) + g_i(z) U_{k,[j]}^{(l,i)\mathrm{T}} \sigma(z)$, α is a positive step size, j is the iteration step for GDA and as $j \to \infty$, $U_{k,[j]}^{(l,i)} \to U_k^{(l,i)}$.

Next, based on $U_k^{(l,i)}$ and $W_k^{(l)}$, we can compute the target cost function for the critic network according to

$$\mathrm{d}(z, W_k^{(l)}, U_k^{(l,i)}) = L(z, U_k^{(l,i)\mathrm{T}}\sigma(z), i) + W_k^{(l)\mathrm{T}}\phi(f_i(z) + g_i(z)U_k^{(l,i)\mathrm{T}}\sigma(z)). \tag{7.47}$$

Define the residual error as

$$e(z) = W_{k+1}^{(\hat{l})\mathrm{T}}\phi(z) - \mathrm{d}(z, W_k^{(l)}, U_k^{(l,i)}), \tag{7.48}$$

where $\hat{l} = (l-1) \times M + i$.

The task is to adjust the weight $W_{k+1}^{(\hat{l})}$, such that the error $e(z)$ is minimized on a training set $\Omega \in \mathbb{R}^n$. Note that the relation between $e(z)$ and $W_{k+1}^{(\hat{l})}$ is linear, for convenient, we employ the method of weighted residuals to realize the iteration. The weight $W_{k+1}^{(\hat{l})}$ is determined by projecting the residual error onto $\mathrm{d}e(z)/\mathrm{d}W_{k+1}^{(\hat{l})}$ and setting the result to zero for any $z \in \Omega$ using the inner product, i.e.,

$$\int_\Omega \frac{\partial e(z)}{\partial W_{k+1}^{(\hat{l})}} \cdot e^{\mathrm{T}}(z)\mathrm{d}z = 0. \tag{7.49}$$

Substitute $e(z)$ in (7.49) with (7.48), and then we have

$$\int_\Omega \phi(z)\left(W_{k+1}^{(\hat{l})\mathrm{T}}\phi(z) - \mathrm{d}(z, W_k^{(l)}, U_k^{(l,i)})\right)^{\mathrm{T}} \mathrm{d}z = 0. \tag{7.50}$$

Therefore, a unique solution for $W_{k+1}^{(\hat{l})}$ exists, i.e.,

$$W_{k+1}^{(\hat{l})} = \left(\int_\Omega \phi(z)\phi(z)^{\mathrm{T}}\mathrm{d}z\right)^{-1} \int_\Omega \phi(z)\mathrm{d}^{\mathrm{T}}(z, W_k^{(l)}, U_k^{(l,i)})\mathrm{d}z. \tag{7.51}$$

It is worth mentioning that the basis functions we choose, which are linearly independent, guarantee that the inverse in (7.51) exists, see [1] for details.

For each i and l, there will be a unique $W_{k+1}^{(\hat{l})}$ generated from (7.51). All the $W_{k+1}^{(\hat{l})}$ compose a new set $\mathcal{W}_{k+1} = \{W_{k+1}^{(\hat{l})} \mid \hat{l} \in \mathcal{L}_{k+1}\}$. Therefore, the value function at $(k+1)$th iteration can be expressed as

$$V_{k+1}(z) = \min_{W_{k+1}^{(\hat{l})} \in \mathcal{W}_{k+1}} \left\{W_{k+1}^{(\hat{l})\mathrm{T}}\phi(z)\right\}. \tag{7.52}$$

The iteration continues till $V_k(z)$ converges to the optimal value function $J^*(z)$ or $V_\infty(z)$. Based on $J^*(z)$, we obtain the function $\iota^*(z)$

$$\iota^*(z) = \arg\min_l \left\{V_\infty^{(l)}(z)\right\}, \tag{7.53}$$

and the optimal control policy $\pi^* = \langle \mu^*, \nu^* \rangle$, where

$$\nu^*(z) = \mod\left(\iota^*(z), M\right), \tag{7.54}$$

$$\mu^*(z) = u_\infty^{(\lfloor \iota^*(z)/M \rfloor + 1, \nu^*(z))}(z). \tag{7.55}$$

Note that at the kth iteration, the number of weights in $\mathcal{W}_k$ is M^k, which implies that the computation burden grows exponentially as k increases. However, as a matter of fact, not all the weights in $\mathcal{W}_k$ make contributes to characterize the value function V_k. The weights that are redundant can be directly removed to simplify the computation efficiently. To formalize this idea, firstly we introduce the following definition.

Definition 7.14 (Redundant Weight) A weight $\hat{W} \in \mathcal{W}$ is redundant if there exists a $W \in \mathcal{W} \setminus \{\hat{W}\}$ such that $W^{\mathrm{T}}\phi(z) \le \hat{W}^{\mathrm{T}}\phi(z)$ holds for all $z \in \mathbb{R}^n$.

Obviously, because the redundant weights are never utilized for the value function, there will be no influence on the optimality and convergence of our algorithm if they are ruled out.

By now, we can summarize the TSADP algorithm using neural networks as follows:

1. Initialization. Choose suitable basis functions $\phi(x)$ and $\sigma(x)$. Let $W_0^{(1)} = [0, 0, \ldots, 0]^{\mathrm{T}}$, $\mathcal{W}_0 = \{W_0^{(1)}\}$. $k = 0$.
2. For each $l = 1, 2, \ldots, M^k$, $i = 1, 2, \ldots, M$, update the weight $U_k^{(l,i)}$ according to (7.46) and adjust $W_{k+1}^{(\hat{l})}$ by (7.51) where $\hat{l} = (l-1) \times M + i$. All the weights $W_{k+1}^{(\hat{l})}$ compose the set $\mathcal{W}_{k+1}$.
3. Rule out redundant weights in $\mathcal{W}_{k+1}$ based on Definition 7.14.
4. If $V_{k+1} = V_k$, go to 5; else $k = k + 1$ and go to 2.
5. Compute the optimal control policy $\pi^* = \langle \mu^*, \nu^* \rangle$ by (7.55) and (7.54).

7.2.3 Simulations

In this section, to support the novel theoretical result, we offer a simulation example for the present TSADP algorithm.

Consider a general switched optimal control problem with the following two nonlinear subsystems.

Subsystem 1:

$$\begin{bmatrix} x_1(t+1) \\ x_2(t+1) \end{bmatrix} = \begin{bmatrix} 0.2\sin(x_1(t))x_2(t) \\ 0.5x_2(t) + 0.5u(t) \end{bmatrix}. \tag{7.56}$$

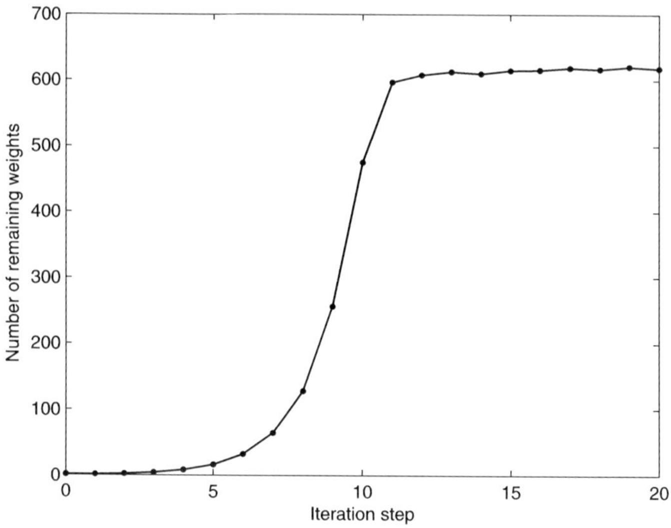

Fig. 7.1 Number of remaining weights in $\mathcal{W}_k$

Subsystem 2:

$$\begin{bmatrix} x_1(t+1) \\ x_2(t+1) \end{bmatrix} = \begin{bmatrix} 0.5x_1^3(t)x_2(t) \\ 0.3x_2(t)+0.5u(t) \end{bmatrix}. \tag{7.57}$$

The weighting matrices in cost functional are set to be $Q_1 = Q_2 = \left[\begin{smallmatrix} 1 & 0 \\ 0 & 1 \end{smallmatrix}\right]$, $R_1 = R_2 = 1$.

To initiate the algorithm, basis functions for critic and action networks are respectively chosen as

$$\begin{aligned} \phi(x) = {} & [x_1^2, x_1x_2, x_2^2, x_1^4, x_1^3x_2, x_1^2x_2^2, x_1x_2^3, x_2^4, x_1^6, x_1^5x_2, x_1^4x_2^2, \\ & x_1^3x_2^3, x_1^2x_2^4, x_1x_2^5, x_2^6]^{\mathrm{T}}, \\ \sigma(x) = {} & [x_1, x_2, x_1^3, x_1^2x_2, x_1x_2^2, x_2^3, x_1^5, x_1^4x_2, x_1^3x_2^2, x_1^2x_2^3, x_1x_2^4, x_2^5]^{\mathrm{T}}, \end{aligned}$$

with initial weight $W_0^{(1)} = [0, 0, \ldots, 0]^{\mathrm{T}}$. Training set is defined as

$$\Omega = \left\{[x_1, x_2]^{\mathrm{T}} \mid -2 \le x_1 \le 2, -2 \le x_2 \le 2\right\}.$$

After 20 steps iteration, the value function gets convergent. At each iteration, the number of remaining weights (after redundant weights are deleted) in $\mathcal{W}_k$ is shown in Fig. 7.1. At the beginning (when $k < 10$), the number increases exponentially, but eventually keeps almost constant. There is a slight perturbation, mainly because of the approximated errors incurred by neural networks.

Based on the final set $\mathcal{W}_{20}$, we can compute the optimal control law μ^* and ν^* by (7.55) and (7.54). According to ν^*, we can draw the figure of decision region of

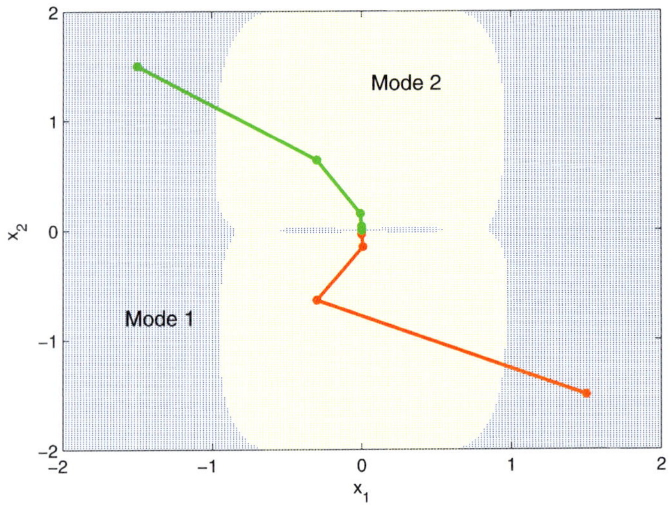

Fig. 7.2 State trajectories in the optimal decision region map. The *red one* started with initial condition $x_0 = [1.5, -1.5]^{\mathrm{T}}$, while the *green one* with $x_0 = [-1.5, 1.5]^{\mathrm{T}}$.

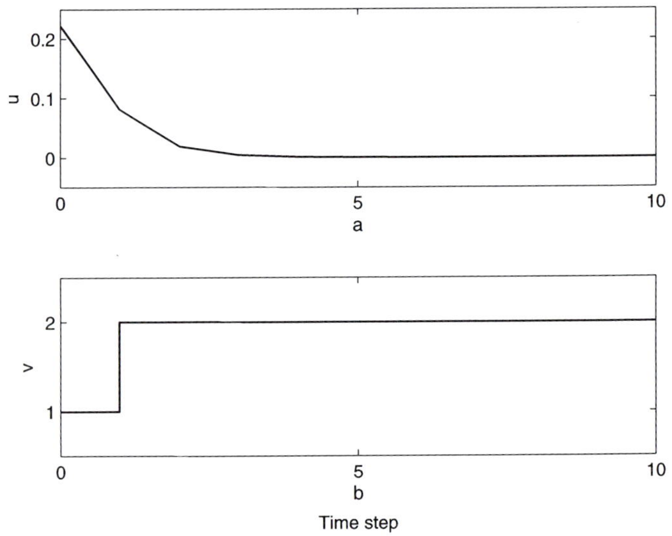

Fig. 7.3 Control inputs with initial condition $x_0 = [1.5, -1.5]^{\mathrm{T}}$

switching control, as illustrated in Fig. 7.2. States in the blue region choose mode 1, i.e., operate on the $1st$ subsystem, while states in the yellow one choose mode 2. To depict this better, we run the switched system in the optimal manner with two different initial states, $x_0 = [1.5, -1.5]^{\mathrm{T}}$ and $x_0 = [-1.5, 1.5]^{\mathrm{T}}$, respectively. The

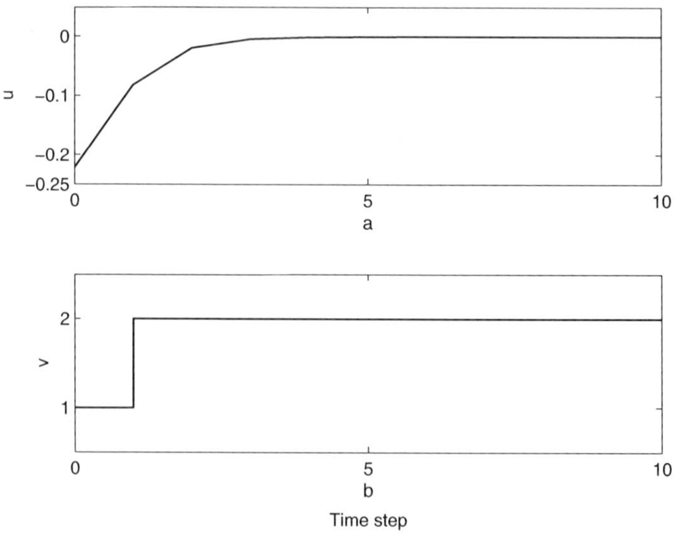

Fig. 7.4 Control inputs with initial condition $x_0 = [-1.5, 1.5]^{\mathrm{T}}$

two trajectories are shown in Fig. 7.2. The curves of the control input are shown in Figs. 7.3 and 7.4, respectively.

7.3 Optimal Feedback Control for a Class of Descriptor Systems

7.3.1 Problem Formulation

Consider a class of descriptor systems as follows:

$$x_1(k+1) = F_1(x_1(k), x_2(k), u(k)), \tag{7.58}$$

$$0 = F_2(x_1(k), x_2(k), u(k)), \tag{7.59}$$

where $x_1(k) \in \mathbb{R}^{n_1}$, $x_2(k) \in \mathbb{R}^{n_2}$, $F_1 \colon \mathbb{R}^{n_1} \times \mathbb{R}^{n_2} \times \mathbb{R}^m \to \mathbb{R}^{n_1}$, $F_2 \colon \mathbb{R}^{n_1} \times \mathbb{R}^{n_2} \times \mathbb{R}^m \to \mathbb{R}^{n_2}$, and the control $u(k) \in \Omega_u$, $\Omega_u = \{u(k) = [u_1(k), u_2(k), \ldots, u_m(k)]^{\mathrm{T}} \in \mathbb{R}^m \colon |u_i(k)| \le \bar{u}_i, i = 1, \ldots, m\}$, here $\bar{u}_i$ denotes the saturating bound for the ith actuator. Let $\bar{U}_i \in \mathbb{R}^{m \times m}$ be the constant diagonal matrix described as $\bar{U} = \operatorname{diag}\{\bar{u}_1, \bar{u}_2, \ldots, \bar{u}_m\}$.

For simplicity of analysis, we write (7.59) as

$$Ex(k+1) = F(x(k), u(k)), \tag{7.60}$$

where E is a singular matrix of rank n_1, $x(k) = \operatorname{col}\{x_1(k), x_2(k)\}$. Assume that F is Lipschitz continuous on a set Ω_x in $\mathbb{R}^n$ containing the origin, and that the system

(7.60) is controllable in the sense that there exists a continuous control law on Ω_x that asymptotically stabilizes the system.

In this section, we mainly discuss how to design an optimal controller for this class of descriptor systems. Therefore, it is desired to find $u(\cdot)$ which minimizes a generalized nonquadratic cost functional as follows:

$$J(Ex(k), u) = \sum_{i=k}^{\infty} \left\{x^{\mathrm{T}}(i)Qx(i) + W(u(i))\right\}, \tag{7.61}$$

where $W(u(i))$ is positive definite and Q is also positive definite.

For the optimal control problem, it is required to find a control $u(\cdot)$ which cannot only stabilize the system on Ω_x, obtaining a unique impulse-free solution, but also guarantee the cost functional $J(Ex(0))$ to be finite, i.e., admissible control. In the following part, we use $J^*(Ex(k))$ denote the minimal value of cost functional $J(Ex(k), u)$, which is also called optimal value function.

Definition 7.15 A control law $\eta(x) \in \Gamma$ is called as an admissible policy for (7.60) on Ω_x, if for any initial state Ex_0, not only the obtained closed-loop system has a unique impulse-free solution, $\eta(0) = 0$, but the cost functional $J(Ex(0))$ is guaranteed to be finite. Γ is called the admissible policy set. $\eta(x)$ is assumed to be continuously differentiable in its argument.

Assumption 7.16 There exists at least an admissible policy for the system (7.60). In a linear descriptor system, this is equivalent to the impulsive controllability of the system.

According to the Bellman optimality principle, we obtain

$$J^*(Ex(k)) = \min_{u(k)} \{x^{\mathrm{T}}(k)Qx(k) + W(u(k)) + J^*(Ex(k+1))\}. \tag{7.62}$$

For an unconstrained control problem, a common choice for $W(u(i))$ is $W(u(i)) = u^{\mathrm{T}}(i)Ru(i)$, where $R \in \mathbb{R}^{m\times m}$ is positive semi-definite. With the first order necessity condition, we compute the gradient of the right-hand side of (7.62) with respect to u as

$$\begin{aligned}\frac{\partial J^*(Ex(k))}{\partial u(k)} &= \frac{\partial (x^{\mathrm{T}}(k)Qx(k) + u^{\mathrm{T}}(k)Ru(k))}{\partial u(k)} \\ &\quad + \left(\frac{\partial x(k+1)}{\partial u(k)}\right)^{\mathrm{T}} \frac{\partial J^*(Ex(k+1))}{\partial x(k+1)} \\ &= 0.\end{aligned} \tag{7.63}$$

Therefore, we obtain

$$u^*(k) = -\frac{1}{2}R^{-1}\left(\frac{\partial x(k+1)}{\partial u(k)}\right)^{\mathrm{T}} \frac{\partial J^*(Ex(k+1))}{\partial x(k+1)}, \tag{7.64}$$

where J^* is the value function corresponding to the optimal control law u^*.

However, for constrained control problem, the above derivation is infeasible. To confront this bounded control problem, we introduce a nonquadratic functional derived from the idea by [8] as follows:

$$W(u(i)) = 2\int_0^{u(i)} \varphi^{-\mathrm{T}}(\bar{U}^{-1}s)\bar{U}R\mathrm{d}s,$$
$$\varphi^{-1}(u(i)) = [\psi^{-1}(u_1(i)),\dots,\psi^{-1}(u_m(i))], \tag{7.65}$$

where R is positive definite, $s \in \mathbb{R}^m$, $\varphi \in \mathbb{R}^m$, $\psi(\cdot)$ is a bounded one-to-one function satisfying $|\psi(\cdot)| \le 1$ and belonging to C^p $(p \ge 1)$ and $L_2(\Omega)$. Moreover, it is a monotonically increasing odd function with its first derivative bounded by a constant M. Such a function is easy to find; one example is the hyperbolic tangent function $\psi(\cdot) = \tanh(\cdot)$. It should be noticed that by the definition above, $W(u(i))$ is ensured to be positive definite because $\psi^{-1}(\cdot)$ is monotonic odd function and R is positive definite.

Substituting (7.65) into (7.62), we obtain

$$J^*(Ex(k)) = \min_{u(k)}\left\{x^{\mathrm{T}}(k)Qx(k) + 2\int_0^{u(k)} \varphi^{-\mathrm{T}}(\bar{U}^{-1}s)\bar{U}R\mathrm{d}s \right.$$
$$\left. + J^*(Ex(k+1))\right\}. \tag{7.66}$$

According to the necessary condition of optimal control, the following derivation can be given:

$$\frac{\partial J^*(Ex(k))}{\partial u(k)} = 2R\bar{U}\varphi^{-1}(\bar{U}^{-1}u(k)) + \left(\frac{\partial x(k+1)}{\partial u(k)}\right)^{\mathrm{T}} \frac{\partial J^*(Ex(k+1))}{\partial x(k+1)} = 0. \tag{7.67}$$

Therefore, the following formulation can be obtained:

$$u^*(k) = \bar{U}\varphi\left(-\frac{1}{2}(\bar{U}R)^{-1}\left(\frac{\partial x(k+1)}{\partial u(k)}\right)^{\mathrm{T}} \frac{\partial J^*(Ex(k+1))}{\partial x(k+1)}\right). \tag{7.68}$$

From (7.68), the optimal control $u^*(k)$ can be directly obtained if the value function $J^*(Ex(k+1))$ is known. However, there is currently no method for rigorously solving for the costate vector of this constrained optimal control problem. Therefore, in the next subsection we will discuss how to utilize an adaptive dynamic programming method, named GI-DHP, to solve the near-optimal control solution.

7.3.2 Optimal Controller Design for a Class of Descriptor Systems

In the following, we will give the derivation of the GI-DHP Algorithm.

Define the costate $E^{\mathrm{T}}\lambda(x(k)) = \frac{\partial V(Ex(k))}{\partial x(k)}$. Then, (7.68) can be written as

$$u^*(k) = \bar{U}\varphi\left(-\frac{1}{2}(\bar{U}R)^{-1}\left(\frac{\partial Ex(k+1)}{\partial u(k)}\right)^{\mathrm{T}}\lambda^*(x(k+1))\right). \tag{7.69}$$

However, there is currently no method for rigorously solving for the costate vector λ^* of this constrained optimal control problem. Therefore, in the next part, we develop an iterative algorithm for computing the costate vector λ^* and the optimal control u^*.

First, we start with initial value function $V^{[0]}(Ex(k)) = 0$ which is necessary to the value function. The initial costate vector is set to $\lambda_1^{[0]}(\cdot) = 0$ and then we compute the optimal control $u^{[0]}$ as follows:

$$v^{[0]}(x(k)) = \arg\min_{u(k)}\left\{x^{\mathrm{T}}(k)Qx(k) + W(v(k)) + V^{[0]}(Ex(k+1))\right\}. \tag{7.70}$$

In view of $E^{\mathrm{T}}\lambda^{[i]}(x(k)) = \frac{\partial V^{[i]}(Ex(k))}{\partial x(k)}$ and according to the first order necessity condition, we obtain

$$\frac{\partial(x^{\mathrm{T}}(k)Qx(k) + W(u(k)))}{\partial u(k)} + \left(\frac{\partial Ex(k+1)}{\partial u(k)}\right)^{\mathrm{T}}\lambda^{[0]}(x(k+1)) = 0, \tag{7.71}$$

i.e.,

$$\begin{aligned} v^{[0]}(k) &= \bar{U}\varphi\left(-\frac{1}{2}(\bar{U}R)^{-1}\left(\frac{\partial Ex(k+1)}{\partial u(k)}\right)^{\mathrm{T}}\lambda^{[0]}(x(k+1))\right) \\ &= \bar{U}\varphi\Bigg(-\frac{1}{2}(\bar{U}R)^{-1}\Bigg(\left(\frac{\partial F_1(\cdot)}{\partial u(k)}\right)^{\mathrm{T}}\lambda_1^{[0]}(x(k+1)) \\ &\quad + \left(\frac{\partial F_2(\cdot)}{\partial u(k)}\right)^{\mathrm{T}}\lambda_2^{[0]}(x(k+1))\Bigg)\Bigg), \end{aligned} \tag{7.72}$$

and then update the cost function by

$$V^{[1]}(Ex(k)) = x^{\mathrm{T}}(k)Qx(k) + W(v^{[0]}(k)) + V^{[0]}(Ex(k+1)). \tag{7.73}$$

For the costate vector, it can be updated by

$$\begin{aligned} E^{\mathrm{T}}\lambda^{[1]}(x(k)) = {}& \frac{\partial(x^{\mathrm{T}}(k)Qx(k) + W(v^{[0]}(k)))}{\partial x(k)} \\ & + \left(\frac{\partial Ex(k+1)}{\partial x(k)}\right)^{\mathrm{T}}\lambda^{[0]}(x(k+1)). \end{aligned} \tag{7.74}$$

Let $L^{[0]}(k) = x^{\mathrm{T}}(k)Qx(k) + W(v^{[0]}(k))$, and then the above equation can be expanded as

$$\lambda_1^{[1]}(x(k)) = \frac{\partial L^{[0]}(k)}{\partial x_1(k)} + \left(\frac{\partial F_1}{\partial x_1(k)}\right)^{\mathrm{T}} \lambda_1^{[0]}(x(k+1)) + \left(\frac{\partial F_2}{\partial x_1(k)}\right)^{\mathrm{T}} \lambda_2^{[0]}(x(k+1)) \tag{7.75}$$

and

$$0 = \frac{\partial L^{[0]}(k)}{\partial x_2(k)} + \left(\frac{\partial F_1}{\partial x_2(k)}\right)^{\mathrm{T}} \lambda_1^{[0]}(x(k+1)) + \left(\frac{\partial F_2}{\partial x_2(k)}\right)^{\mathrm{T}} \lambda_2^{[0]}(x(k+1)). \tag{7.76}$$

Therefore, for the ith iteration, we have

$$v^{[i]}(x(k)) = \arg \min_{u(k)} \left\{x^{\mathrm{T}}(k)Qx(k) + W(u(k)) + V^{[i]}(Ex(k+1))\right\}. \tag{7.77}$$

Also, according to the first order necessity condition, we obtain

$$\frac{\partial(x^{\mathrm{T}}(k)Qx(k) + W(u(k))}{\partial u(k)} + \left(\frac{\partial x(k+1)}{\partial u(k)}\right)^{\mathrm{T}} \frac{\partial V^{[i]}(Ex(k+1))}{\partial x(k+1)} = 0, \tag{7.78}$$

i.e.,

$$\begin{aligned} v^{[i]}(k) &= \bar{U}\varphi\left[-\frac{1}{2}(\bar{U}R)^{-1}\left(\frac{\partial Ex(k+1)}{\partial u(k)}\right)^{\mathrm{T}} \lambda^{[i]}(x(k+1))\right] \\ &= \bar{U}\varphi\Bigg[-\frac{1}{2}(\bar{U}R)^{-1}\Bigg(\left(\frac{\partial F_1(\cdot)}{\partial u(k)}\right)^{\mathrm{T}} \lambda_1^{[i]}(x(k+1)) \\ &\quad + \left(\frac{\partial F_2(\cdot)}{\partial u(k)}\right)^{\mathrm{T}} \lambda_2^{[i]}(x(k+1))\Bigg)\Bigg]. \end{aligned} \tag{7.79}$$

The updated value function is given as

$$V^{[i+1]}(Ex(k)) = x^{\mathrm{T}}(k)Qx(k) + W(v^{[i]}(k)) + V^{[i]}(Ex(k+1)). \tag{7.80}$$

In view of $E^{\mathrm{T}}\lambda^{[i+1]}(x(k)) = \frac{\partial V^{[i+1]}(Ex(k))}{\partial x(k)}$, we have

$$\frac{\partial V^{[i+1]}(Ex(k))}{\partial x(k)} = \frac{\partial(x^{\mathrm{T}}(k)Qx(k) + W(v^{[i]}(k)))}{\partial x(k)}$$

$$
\begin{aligned}
&+\left(\frac{\partial v^{[i]}(x(k))}{\partial x(k)}\right)^{\mathrm{T}} \frac{\partial(x^{\mathrm{T}}(k)Qx(k)+W(v^{[i]}(k)))}{\partial v^{[i]}(x(k))}\\
&+\left(\frac{\partial x(k+1)}{\partial x(k)}\right)^{\mathrm{T}} \frac{\partial V^{[i]}(Ex(k+1))}{\partial x(k+1)}\\
&+\left(\frac{\partial v^{[i]}(x(k))}{\partial x(k)}\right)^{\mathrm{T}}\left(\frac{\partial x(k+1)}{\partial v^{[i]}(x(k))}\right)^{\mathrm{T}} \frac{\partial V^{[i]}(Ex(k+1))}{\partial x(k+1)}\\
&=\frac{\partial(x(k)^{\mathrm{T}}Qx(k)+W(v^{[i]}(k)))}{\partial x(k)}+\left(\frac{\partial Ex(k+1)}{\partial x(k)}\right)^{\mathrm{T}}\lambda^{[i]}(x(k+1)).
\end{aligned}
\tag{7.81}
$$

Combining with (7.78), (7.81) can be changed into

$$
\begin{aligned}
E^{\mathrm{T}}\lambda^{[i+1]}(x(k))=&\frac{\partial(x^{\mathrm{T}}(k)Qx(k)+W(v^{[i]}(k)))}{\partial x(k)}\\
&+\left(\frac{\partial Ex(k+1)}{\partial x(k)}\right)^{\mathrm{T}}\lambda^{[i]}(x(k+1)).
\end{aligned}
\tag{7.82}
$$

Similarly, the above equation can be expanded into

$$
\begin{aligned}
\lambda_1^{[i+1]}(x(k))=&\frac{\partial L^{[i]}(k)}{\partial x_1(k)}+\left(\frac{\partial F_1}{\partial x_1(k)}\right)^{\mathrm{T}}\lambda_1^{[i]}(x(k+1))\\
&+\left(\frac{\partial F_2}{\partial x_1(k)}\right)^{\mathrm{T}}\lambda^{[i]}2(x(k+1))
\end{aligned}
\tag{7.83}
$$

and

$$
0=\frac{\partial L^{[i]}(k)}{\partial x_2(k)}+\left(\frac{\partial F_1}{\partial x_2(k)}\right)^{\mathrm{T}}\lambda_1^{[i]}(x(k+1))+\left(\frac{\partial F_2}{\partial x_2(k)}\right)^{\mathrm{T}}\lambda_2^{[i]}(x(k+1)). \tag{7.84}
$$

Therefore, the GI-DHP iterative scheme iterates between

$$
\begin{aligned}
v^{[i]}(k)=\bar{U}\varphi\Bigg[&-\frac{1}{2}(\bar{U}R)^{-1}\Bigg(\left(\frac{\partial F_1(\cdot)}{\partial u(k)}\right)^{\mathrm{T}}\lambda_1^{[i]}(x(k+1))\\
&+\left(\frac{\partial F_2(\cdot)}{\partial u(k)}\right)^{\mathrm{T}}\lambda_2^{[i]}(x(k+1))\Bigg)\Bigg],
\end{aligned}
\tag{7.85}
$$

$$
\begin{aligned}
\lambda_1^{[i+1]}(x(k))=&\frac{\partial L^{[i]}(k)}{\partial x_1(k)}+\left(\frac{\partial F_1}{\partial x_1(k)}\right)\lambda_1^{[i]}(x(k+1))\\
&+\left(\frac{\partial F_2}{\partial x_1(k)}\right)\lambda_2^{[i]}(x(k+1)),
\end{aligned}
\tag{7.86}
$$

and

$$0=\frac{\partial L^{[i]}(k)}{\partial x_2(k)}+\left(\frac{\partial F_1}{\partial x_2(k)}\right)\lambda_1^{[i]}(x(k+1))+\left(\frac{\partial F_2}{\partial x_2(k)}\right)\lambda_2^{[i]}(x(k+1)). \tag{7.87}$$

At each iteration step, the $\lambda_1^{[i]}(\cdot)$ is computed iteratively by (7.86) while the $\lambda_2^{[i]}(\cdot)$ is computed by the constraint equation (7.87). With known $\lambda_1^{[i]}(\cdot)$ and $\lambda_2^{[i]}(\cdot)$, the control $v_i(\cdot)$ can be obtained by (7.85).

In the following part, we will prove that by the above iterative process, the optimal control u^* and the costate λ^* can be obtained with $V^{[i]} \to J^*$, $\lambda^{[i]} \to \lambda^*$ and the control policy $v^{[i]} \to u^*$ as $i \to \infty$.

For convenience of analysis, we first present a lemma as follows.

Lemma 7.17 *Let $\mu^{[i]}$ be arbitrary sequence of control policies, and $v^{[i]}$ is the policies in* (7.77). *Let $V^{[i]}$ be* (7.80) *and $\Lambda^{[i]}$ as*

$$\Lambda^{[i+1]}(Ex(k))=x^{\mathrm{T}}(k)Qx(k)+W(\mu^{[i]}(k))+\Lambda^{[i]}(Ex(k+1)), \tag{7.88}$$

If $V^{[0]}=\Lambda^{[0]}=0$, then $V^{[i]}\le\Lambda^{[i]}$, $\forall i$.

Lemma 7.18 *Let the sequence $\left\{V^{[i]}\right\}$ be defined as* (7.80). *If the system is controllable, then there is an upper bound Y such that $0\le V^{[i]}\le Y\ \forall i$.*

Proof Let $\eta(x(k))$ be any stabilizing and admissible control input, and let $V^{[0]}=Z^{[0]}=0$ where $V^{[i]}$ is updated as (7.80) and $Z^{[i]}$ is updated as

$$Z^{[i+1]}(Ex(k))=x^{\mathrm{T}}(k)Qx(k)+W(\eta(k))+Z^{[i]}(Ex([k+1])). \tag{7.89}$$

It follows that the difference

$$\begin{aligned}
Z^{[i+1]}(Ex(k))-Z^{[i]}(Ex(k))&=Z^{[i]}(Ex(k+1))-Z^{[i-1]}(Ex(k+1))\\
&=Z^{[i-1]}(Ex(k+2))-Z^{[i-2]}(Ex(k+2))\\
&=Z^{[i-2]}(Ex(k+3))-Z^{[i-3]}(Ex(k+3))\\
&\ \ \vdots\\
&=Z^{[1]}(Ex(k+i))-Z^{[0]}(Ex(k+i)).
\end{aligned} \tag{7.90}$$

Then, the following relation can be obtained:

$$Z^{[i+1]}(Ex(k))-Z^{[i]}(Ex(k))=Z^{[1]}(Ex(k+i))-Z^{[0]}(Ex(k+i)). \tag{7.91}$$

Since $Z^{[0]}(Ex(\cdot))=0$, we have

$$Z^{[i+1]}(Ex(k))=Z^{[1]}(Ex(k+i))+Z^{[i]}(Ex(k))$$

$$
\begin{aligned}
&= Z^{[1]}(Ex(k+i)) + Z^{[1]}(Ex(k+i-1)) + Z^{[i-1]}(Ex(k)) \\
&= Z^{[1]}(Ex(k+i)) + Z^{[1]}(Ex(k+i-1)) \\
&\quad + Z^{[1]}(Ex(k+i-2)) + Z^{[i-2]}(Ex(k)) \\
&= Z^{[1]}(Ex(k+i)) + Z^{[1]}(Ex(k+i-1)) \\
&\quad + Z^{[1]}(Ex(k+i-2)) + \cdots + Z^{[1]}(Ex(k)).
\end{aligned} \tag{7.92}
$$

So (7.92) can be written as

$$
\begin{aligned}
Z^{[i+1]}(Ex(k)) &= \sum_{j=0}^{i} Z^{[1]}(Ex(k+j)) \\
&= \sum_{j=0}^{i} (x^{\mathrm{T}}(k+j)Qx(k+j) + W(\eta(x(k+j))) \\
&\leq \sum_{j=0}^{\infty} (x^{\mathrm{T}}(k+j)Qx(k+j) + W(\eta(x(k+j))).
\end{aligned} \tag{7.93}
$$

Noticing that the system is stable with the stabilizing and admissible control input $\eta(x(k))$, i.e., $x(k) \to 0$ as $k \to \infty$, we have

$$
\forall i: \ Z^{[i+1]}(Ex(k)) \leq \textstyle\sum_{j=0}^{\infty} Z^{[1]}(Ex(k+j)) \leq Y. \tag{7.94}
$$

From Lemma 7.17, we have

$$
\forall i: \ V^{[i+1]}(Ex(k)) \leq Z^{[i+1]}(Ex(k)) \leq Y. \tag{7.95}
$$

□

Now, Lemmas 7.17 and 7.18 will be used in the next main theorem.

Theorem 7.19 (cf. [7]) *Define the sequence $\{V^{[i]}\}$ as (7.80) with $V^{[0]} = 0$, the sequence $\{\lambda^{[i]}\}$ as (7.86) and (7.87) with $\lambda_1^{[0]}(\cdot) = 0$. Then, $\{V^{[i]}\}$ is a nondecreasing sequence in which $V^{[i+1]}(Ex(k)) \geq V^{[i]}(Ex(k))$, $\forall i$, and converge to the value function of the discrete-time HJB equation, i.e., $V^{[i]} \to J^*$ as $i \to \infty$, while the sequence $\{\lambda^{[i]}\}$ is also convergent with $\lambda^{[i]} \to \lambda^*$ as $i \to \infty$.*

Proof For the convenience of analysis, define a new sequence Φ^i as follows:

$$
\Phi^{[i+1]}(Ex(k)) = x^{\mathrm{T}}(k)Qx(k) + W(v^{[i+1]}(k)) + \Phi^{[i]}(Ex(k+1)), \tag{7.96}
$$

with $\Phi^{[0]} = V^{[0]} = 0$ and the policies $v^{[i]}$ defined as (7.77); the value function $V^{[i]}$ is updated by (7.80).

In the following, we prove $\Phi^{[i]}(Ex(k)) \le V^{[i+1]}(Ex(k))$ by mathematical induction.

First, we prove that it holds for $i = 0$. Noticing that

$$V^{[1]}(Ex(k)) - \Phi^{[0]}(Ex(k)) = x^{\mathrm{T}}(k)Qx(k) + W(v^{[0]}(k)) \ge 0, \tag{7.97}$$

for $i = 0$, we get

$$V^{[1]}(Ex(k)) \ge \Phi^{[0]}(Ex(k)). \tag{7.98}$$

Second, we assume that it holds for $i-1$, i.e., $V^{[i]}(Ex(k)) \ge \Phi^{[i-1]}(Ex(k))$, $\forall x(k)$. Then, for i, since

$$\Phi^{[i]}(Ex(k)) = x^{\mathrm{T}}(k)Qx(k) + W(v^{[i]}(k)) + \Phi^{[i-1]}(Ex(k+1)) \tag{7.99}$$

and

$$V^{[i+1]}(Ex(k)) = x^{\mathrm{T}}(k)Qx(k) + W(v^{[i]}(k)) + V^{[i]}(Ex(k+1)) \tag{7.100}$$

hold, we obtain

$$V^{[i+1]}(Ex(k)) - \Phi^{[i]}(Ex(k)) = V^{[i]}(Ex(k+1)) - \Phi^{[i-1]}(Ex(k+1)) \ge 0, \tag{7.101}$$

i.e., the following equation holds:

$$\Phi^{[i]}(Ex(k)) \le V^{[i+1]}(Ex(k)). \tag{7.102}$$

Therefore, the mathematical induction proof is completed.

Furthermore, from Lemma 7.17 we know that $V^{[i]}(Ex(k)) \le \Phi^{[i]}(Ex(k))$. Therefore, we have

$$V^{[i]}(Ex(k)) \le \Phi^{[i]}(Ex(k)) \le V^{[i+1]}(Ex(k)). \tag{7.103}$$

Hence, we can conclude that $\{V^{[i]}\}$ is a nondecreasing sequence in which $V^{[i+1]}(x(k)) \ge V^{[i]}(x(k))$ $\forall i$, and converge to the value function of the discrete-time HJB, i.e., $V^{[i]} \to J^*$ as $i \to \infty$, while the sequence $\{\lambda^{[i]}\}$ is also convergent with $\lambda^{[i]} \to \lambda^*$ as $i \to \infty$.

Since value function and the costate vector are convergent, according to (7.68) and (7.85), we can conclude that the corresponding control policy sequence $\{v^{[i]}\}$ converges to the optimal policy u^* as $i \to \infty$.

This completes the proof. □

7.3.3 Simulations

In this section, an example is given to show the effectiveness of the control scheme developed.

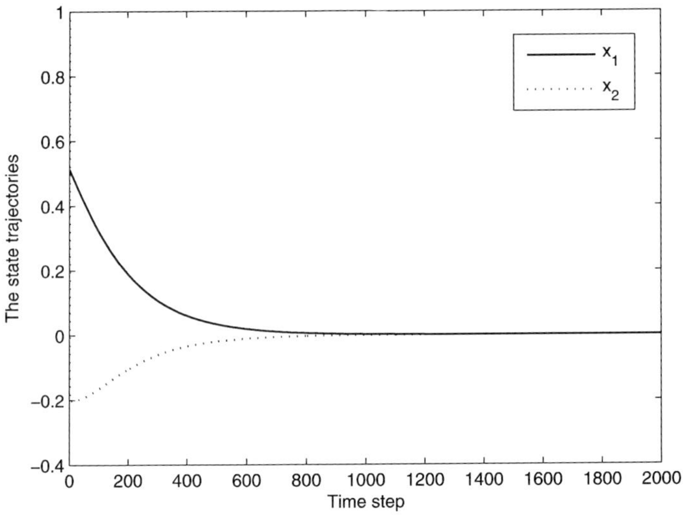

Fig. 7.5 The state variables curves

Consider the following linear system:

$$\begin{aligned}\dot{x}_1 &= x_2,\\ 0 &= x_2 + u,\end{aligned} \tag{7.104}$$

where the control constraint is set to $|u| \leq 1$. We first discretize this system with the Euler method with $t = kT$, T is the sampling period and k is the sample number.

Let $T = 0.001$ s, we obtain the discrete system as follows:

$$\begin{aligned}x_1(k+1) &= x_2(k)T + x_1(k),\\ 0 &= x_2(k) + u(k).\end{aligned} \tag{7.105}$$

Define the cost functional as

$$J(Ex(k), u) = \sum_{i=k}^{\infty} \left\{ x^{\mathrm{T}}(i)Qx(i) + 2\int_0^{u(i)} \tanh^{-T}(s/\bar{U})\bar{U}R\mathrm{d}s \right\}, \tag{7.106}$$

where the weight matrix is chosen as $Q = \left[\begin{smallmatrix}0.1 & 0\\ 0 & 0.1\end{smallmatrix}\right]$ and $R = 0.1$.

The critic network and the action network are chosen with the structure 1–8–1 and 2–8–1, respectively. We take 1000 groups of sampling data to train the network. After the iterative process, we apply the obtained optimal control policy to the system for 2000 time steps under the initial state vector $x_1(0) = 0.5$, $x_2(0) = 1$, and obtain the following results. The state trajectories are given as Fig. 7.5, the corresponding control curve is given as Fig. 7.6, and the convergence curve of the cost functional is given as Fig. 7.7.

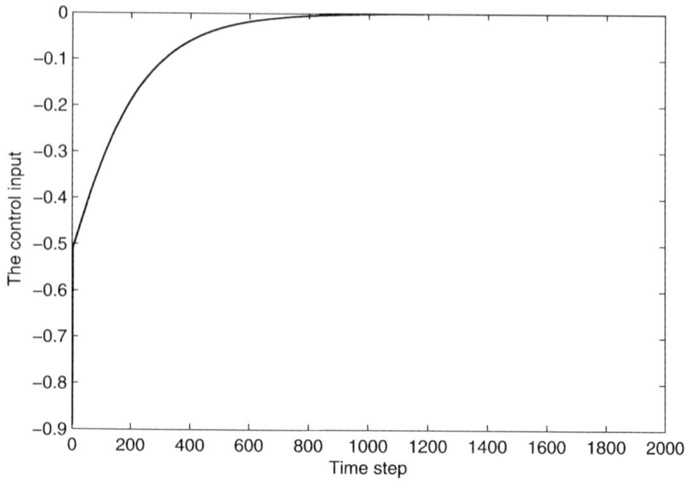

Fig. 7.6 The control curve

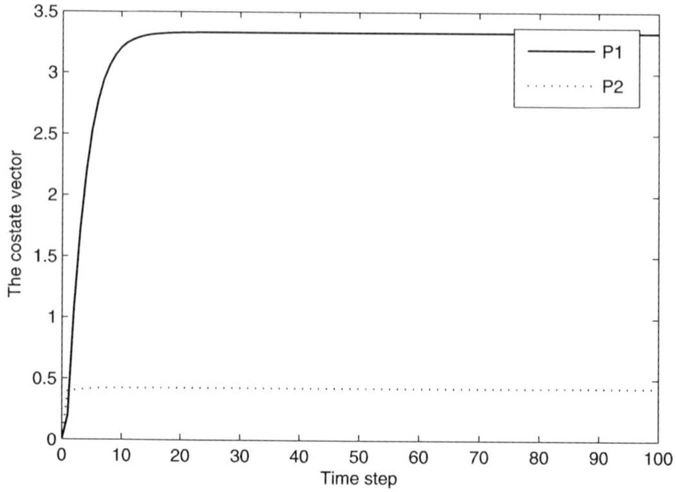

Fig. 7.7 The convergence curve of costate vector

7.4 Optimal Feedback Control for a Class of Singularly Perturbed Systems

7.4.1 Problem Formulation

Consider the following nonlinear singularly perturbed systems:

$$\dot{x}_1 = f_{11}(x_1) + f_{12}(x_1)x_2 + g_1(x_1)u, \tag{7.107}$$

$$\epsilon\dot{x}_2 = f_{21}(x_1) + f_{22}(x_1)x_2 + g_2(x_1)u, \tag{7.108}$$

where $x_1 \in \mathbb{R}^{n_1}$ and $x_2 \in \mathbb{R}^{n_2}$ are states, $u \in \mathbb{R}^m$ is control input. $0 < \epsilon \ll 1$ is perturbation parameter. f_{ij} and g_i are differentiable functions, where $f_{11}(0) = 0$ and $f_{21}(0) = 0$, i.e., guaranteeing that $(0, 0)$ is the equilibrium point. Suppose that f_{22} is invertible for all x_1, and system (7.107) and (7.108) is stabilizable in $\Omega \subset \mathbb{R}^{n_1+n_2}$.

The target is to design the optimal controller to minimize the following cost functional:

$$J = \int_0^\infty (x^{\mathrm{T}} Q x + u^{\mathrm{T}} R u)\mathrm{d}t, \tag{7.109}$$

where $x = [x_1 \;\; x_2]^{\mathrm{T}}$, $Q = [C_1 \;\; C_2]^{\mathrm{T}}[C_1 \;\; C_2] \geq 0$, $R > 0$.

Because of the existence of ϵ, the order of whole system will become much higher, which makes it almost impossible to solve the optimal control problem directly. Therefore, we will decompose system (7.107) and (7.108) as fast subsystem and slow subsystem in the next section.

First, let $\epsilon = 0$, and then (7.108) becomes

$$x_{2s} = -f_{22}^{-1}(x_{1s})f_{21}(x_{1s}) - f_{22}^{-1}(x_{1s})g_2(x_{1s})u_s. \tag{7.110}$$

Substituting (7.110) into (7.107), we get the slow subsystem

$$\dot{x}_{1s} = F_s(x_{1s}) + G_s(x_{1s})u_s, \tag{7.111}$$

and

$$\begin{bmatrix} C_1 & C_2 \end{bmatrix}\begin{bmatrix} x_{1s} \\ x_{2s} \end{bmatrix} = p(x_{1s}) + q(x_{1s})u_s, \tag{7.112}$$

where

$$\begin{aligned}
F_s(x_{1s}) &= f_{11}(x_{1s}) - f_{12}(x_{1s})f_{22}^{-1}(x_{1s})f_{21}(x_{1s}),\\
G_s(x_{1s}) &= g_1(x_{1s}) - f_{12}(x_{1s})f_{22}^{-1}(x_{1s})g_2(x_{1s}),\\
p(x_{1s}) &= C_1 x_{1s} - C_2 f_{22}^{-1}(x_{1s})f_{21}(x_{1s}),\\
q(x_{1s}) &= -C_2 f_{22}^{-1}(x_{1s})g_2(x_{1s}).
\end{aligned}$$

Thus, the cost functional for the slow subsystem is

$$J_s = \int_0^\infty l_s(x_{1s}, u_s)\mathrm{d}t, \tag{7.113}$$

where $l_s = p^{\mathrm{T}}p + 2p^{\mathrm{T}}qu_s + u_s^{\mathrm{T}}q^{\mathrm{T}}qu_s + u_s^{\mathrm{T}}Ru_s$.

From optimal control theory, the optimal controller u_s^* and optimal cost functional J_s^* of slow subsystem should satisfy the following HJB equation:

$$0 = \left(\frac{\partial J_s^*}{\partial x_{1s}}\right)^{\mathrm{T}} (F_s + G_s u_s^*) + l_s(x_{1s}, u_s^*), \tag{7.114}$$

$$u_s^* = -\frac{1}{2}\left(q^{\mathrm{T}}q + R\right)^{-1}\left(G_s^{\mathrm{T}}\frac{\partial J_s^*}{\partial x_{1s}} + 2q^{\mathrm{T}}p\right). \tag{7.115}$$

Set $x_{2f} = x_2 - x_{2s}$ and $u_f = u - u_s$. Suppose x_{2s} keeps constant, substituting them into (7.108), and then we obtain the fast subsystem

$$\epsilon\dot{x}_{2f} = F_f(x_1)x_{2f} + G_f(x_1)u_f, \tag{7.116}$$

where $F_f(x_1) = f_{22}(x_1)$, $G_f(x_1) = g_2(x_1)$. x_1 is a fixed parameter.

Now, the fast cost functional is

$$J_f = \int_0^\infty l_f(x_{2f}, u_f)\mathrm{d}t, \tag{7.117}$$

where $l_f = x_{2f}^{\mathrm{T}}C_2^{\mathrm{T}}C_2x_{2f} + u_f^{\mathrm{T}}Ru_f$.

Similarly, optimal controller u_f^* and cost functional J_f^* of the fast subsystem satisfy the following HJB equation:

$$0 = \left(\frac{\partial J_f^*}{\partial x_{2f}}\right)^{\mathrm{T}}(F_f + G_fu_f^*) + l_f(x_{2f}, u_f^*), \tag{7.118}$$

where

$$u_f^* = -\frac{1}{2}R^{-1}G_f^{\mathrm{T}}\frac{\partial J_f^*}{\partial x_{2f}}. \tag{7.119}$$

From u_s^* and u_f^*, we get the composite control for system (7.107) and (7.108):

$$u_c^* = u_s^* + u_f^*. \tag{7.120}$$

7.4.2 Optimal Controller Design for Singularly Perturbed Systems

7.4.2.1 Algorithm Design

Considering slow subsystem (7.111) and cost functional (7.113), choose one initial value function $V_s^{[0]}$, and then initial control law can be obtained as

$$v_s^{[0]} = -\frac{1}{2}\left(q^{\mathrm{T}}q + R\right)^{-1}\left(G_s^{\mathrm{T}}\frac{\partial V_s^{[0]}}{\partial x_{1s}} + 2q^{\mathrm{T}}p\right). \tag{7.121}$$

It is noted that the chosen $V_s^{[0]}$ should guarantee that the corresponding control law $v_s^{[0]}$ is stable for system (7.111).

Based on the initial control law $v_s^{[0]}$, we get one state trajectory of system (7.111) $x_{1s}^{[0]}(x_0, \cdot)$ for each initial state x_0.

According to $x_{1s}^{[0]}(x_0, \cdot)$ and $v_s^{[0]}$, we update the value function by

$$V_s^{[1]} = \int_0^\infty l_s(x_{1s}^{[0]}(x_0, t), u_s^{[0]})\mathrm{d}t. \tag{7.122}$$

Repeatedly, the iteration between control law $v_s^{[i]}$ and value function $V_s^{[i+1]}$ continues between

$$v_s^{[i]} = -\frac{1}{2}\left(q^{\mathrm{T}}q + R\right)^{-1}\left(G_s^{\mathrm{T}}\frac{\partial V_s^{[i]}}{\partial x_{1s}} + 2q^{\mathrm{T}}p\right), \tag{7.123}$$

and

$$V_s^{[i+1]} = \int_0^\infty l_s(x_{1s}^{[i]}(x_0, t), v_s^{[i]})\mathrm{d}t. \tag{7.124}$$

Similarly, for the fast subsystem (7.116) and value function (7.117), if the value function of fast subsystem at ith iteration is $V_f^{[i]}$, then the corresponding control law is

$$v_f^{[i]} = -\frac{1}{2}R^{-1}G_f^{\mathrm{T}}\frac{\partial V_f^{[i]}}{\partial x_{2f}}. \tag{7.125}$$

With the control law $v_f^{[i]}$, we can obtain the state trajectory of fast subsystem $x_{2f}^{[i]}(x_0, \cdot)$. Furthermore, we can update the value function by

$$V_f^{[i+1]} = \int_0^\infty l_f(x_{2f}^{[i]}(x_0, t), v_f^{[i]})\mathrm{d}t. \tag{7.126}$$

In the following, we are ready to prove the obtained $V_s^{(i)}$ and $V_f^{(i)}$ converge to the corresponding optimal value J_s^* and J_f^* eventually.

Theorem 7.20 (cf. [3]) *If initial control laws $v_s^{[0]}$ and $v_f^{[0]}$ are stable, then, for any $i = 1, 2, \ldots$, control laws $v_s^{[i]}$ and $v_f^{[i]}$ obtained according to* (7.123) *and* (7.125) *are also stable. Moreover, as $i \to \infty$, $V_s^{[i]} \to J_s^*$ and $V_f^{[i]} \to J_f^*$, $v_s^{[i]} \to u_s^*$ and $v_f^{[i]} \to u_f^*$.*

Proof We prove this theorem by induction.

First, for the slow subsystem, $v_s^{[0]}$ is obviously stable based on assumption. Then, assume that $v_s^{[i]}$ is stable. It is easy to see that (7.124) is equivalent to the iterative HJB equation (7.127):

$$\frac{\partial V_s^{(i+1)\,\mathrm{T}}}{\partial x_{1s}}(F_s + G_s v_s^{(i)}) = -l_s(x_{1s}, v_s^{(i)}). \tag{7.127}$$

Setting $x_{1s}=x_{1s}^{[i+1]}(x_0,\cdot)$, we get

$$\frac{\partial V_s^{[i+1]\mathrm{T}}}{\partial x_{1s}}F_s(x_{1s}^{[i+1]})=-\frac{\partial V_s^{[i+1]\mathrm{T}}}{\partial x_{1s}}G_s(x_{1s}^{[i+1]})v_s^{[i]}-l_s(x_{1s}^{[i+1]},v_s^{[i]}). \tag{7.128}$$

Therefore, we further obtain

$$\begin{aligned}\frac{\mathrm{d}V_s^{[i+1]}}{\mathrm{d}t}(x_{1s}^{[i+1]}(x_0,t))&=\frac{\partial V_s^{[i+1]\mathrm{T}}}{\partial x_{1s}}F_s(x_{1s}^{[i+1]})+\frac{\partial V_s^{[i+1]\mathrm{T}}}{\partial x_{1s}}G_s(x_{1s}^{[i+1]})v_s^{[i+1]}\\&=-l_s(x_{1s}^{[i+1]},v_s^{[i]})-\frac{\partial V_s^{[i+1]\mathrm{T}}}{\partial x_{1s}}G_s(x_{1s}^{[i+1]})v_s^{[i]}+\frac{\partial V_s^{[i+1]\mathrm{T}}}{\partial x_{1s}}G_s(x_{1s}^{[i+1]})v_s^{[i+1]}\\&=-p^{\mathrm{T}}\left[I-q(q^{\mathrm{T}}q+R)^{-1}q^{\mathrm{T}}\right]p-\frac{1}{4}\frac{\partial V_s^{[i+1]\mathrm{T}}}{\partial x_{1s}}G_s(q^{\mathrm{T}}q+R)^{-1}G_s^{\mathrm{T}}\frac{\partial V_s^{[i+1]}}{\partial x_{1s}}\\&\quad-\frac{1}{4}\left[\frac{\partial V_s^{[i+1]}}{\partial x_{1s}}-\frac{\partial V_s^{[i]}}{\partial x_{1s}}\right]^{\mathrm{T}}G_s(q^{\mathrm{T}}q+R)^{-1}G_s^{\mathrm{T}}\times\left[\frac{\partial V_s^{[i+1]}}{\partial x_{1s}}-\frac{\partial V_s^{[i]}}{\partial x_{1s}}\right].\end{aligned} \tag{7.129}$$

Because $R>0$, by Schur supplement, we have

$$q^{\mathrm{T}}q+R-q^{\mathrm{T}}q>0\Leftrightarrow\begin{bmatrix}I & q\\ q^{\mathrm{T}} & q^{\mathrm{T}}q+R\end{bmatrix}>0\Leftrightarrow I-q(q^{\mathrm{T}}q+R)^{-1}q^{\mathrm{T}}>0.$$

Apparently, the first term in the last equation of (7.129) is less than 0.

Thus, for $x_{1s}^{[i+1]}(x_0,t)\neq 0$, $\mathrm{d}V_s^{[i+1]}(x_{1s}^{[i+1]}(x_0,t))/\mathrm{d}t<0$. Then, $V_s^{[i+1]}$ can be seen as the Lyapunov function for the slow subsystem, i.e., $v_s^{[i+1]}$ is stable.

As a result, for any $i=1,2,\dots$, control law $v_s^{[i]}$ obtained by (7.123) is stable.

Besides, we obtain

$$\begin{aligned}\frac{\mathrm{d}V_s^{[i+1]}}{\mathrm{d}t}(x_{1s}^{[i]}(x_0,t))&=p^{\mathrm{T}}q(q^{\mathrm{T}}q+R)^{-1}\left(G_s^{\mathrm{T}}\frac{\partial V_s^{[i]}}{\partial x}+2q^{\mathrm{T}}p\right)-p^{\mathrm{T}}p\\&\quad-\frac{1}{4}\left(G_s^{\mathrm{T}}\frac{\partial V_s^{[i]}}{\partial x}+2q^{\mathrm{T}}p\right)^{\mathrm{T}}(q^{\mathrm{T}}q+R)^{-1}\\&\quad\times\left(G_s^{\mathrm{T}}\frac{\partial V_s^{[i]}}{\partial x}+2q^{\mathrm{T}}p\right).\end{aligned} \tag{7.130}$$

Based on (7.129), we have

$$\frac{\mathrm{d}}{\mathrm{d}t}[V_s^{(i+1)}-V_s^{[i]}](x_{1s}^{[i]}(x_0,t))=\frac{1}{4}\left(\frac{\partial V_s^{[i]}}{\partial x}-\frac{\partial V_s^{[i-1]}}{\partial x}\right)^{\mathrm{T}}$$

$$\times G_s(q^{\mathrm{T}}q+R)^{-1}G_s^{\mathrm{T}}\left(\frac{\partial V_s^{[i]}}{\partial x}-\frac{\partial V_s^{[i-1]}}{\partial x}\right)$$
$$>0. \tag{7.131}$$

Since $x_{1s}^{[i]}(x_0,t)$ will converge to 0 eventually, $V_s^{(i+1)}-V_s^{(i)}$ converges to 0 as well. Thus, based on (7.131), we deduce $V_s^{[i+1]}<V_s^{[i]}$. Therefore, sequence $\{V_s^{(i)}\}_0^\infty$ is decreasing monotonously with lower bound J_s^*. So, $\{V_s^{[i]}\}_0^\infty$ is convergent. Suppose the convergent value is $V_s^{[\infty]}$, and then $V_s^{[\infty]}$ satisfies the iterative HJB equation (7.127), i.e.,

$$\frac{\partial V_s^{[\infty]\mathrm{T}}}{\partial x_{1s}}(F_s+G_s v_s^{[\infty]})=-l_s(x_{1s},v_s^{[\infty]}), \tag{7.132}$$

which is the HJB equation of the slow subsystem (7.128). According to the uniqueness of optimal control, we have $V_s^{[\infty]}=J_s^*$. Thus, we obtain $v_s^{[\infty]}=u_s^*$.

On the other hand, the result of fast subsystem can be deduced similarly, using different iteration expressions of control law and value function, i.e., (7.125) and (7.126), respectively. □

7.4.2.2 Neural Network Approximation

The ith iterative value function $V_s^{[i]}$ of the slow subsystem is approximated by the following NN:

$$V_s^{[i]}=W_s^{[i]\mathrm{T}}\Phi_s(x_{1s}), \tag{7.133}$$

where $W_s^{[i]}=[w_{s1}^{[i]},w_{s2}^{[i]},\dots,w_{sN_s}^{[i]\mathrm{T}}]$ are the matrix of weight. $\Phi_s(x_{1s})=[\phi_{s1},\phi_{s2},\dots,\phi_{sN_s}]^{\mathrm{T}}$ is the vector of basis function. Each $\{\phi_{si}(x_{1s})\}_1^{N_s}$ is linear independence. N_s is the number of basis function. The fast subsystem is approximated by the same NN, $V_f^{[i]}=W_f^{[i]\mathrm{T}}\Phi_f(x_1,x_{2f})$.

Select $x_j^{[i]}=x_{1s}^{[i]}(x_0,t_j)(j=1,2,\dots,r_s)$ from $x_{1s}^{[i]}(x_0,\cdot)$, $r_s\ge N_s$ as the initial state. The value function of slow subsystem becomes

$$V_s^{[i+1]}(x_j^{[i]})=\int_{t_j}^{\infty}l_s\left(x_{1s}^{[i]}(x_0,t),v_s^{[i]}\right)\mathrm{d}t. \tag{7.134}$$

For each $x_j^{[i]}$, $V_s^{[i+1]}(x_j^{[i]})=W_s^{[i+1]\mathrm{T}}\Phi_s(x_j^{[i]})$ must hold, and satisfies

$$\Gamma_s=\Psi_s W_s^{[i+1]}, \tag{7.135}$$

$\Gamma_s=[V_s^{[i+1]}(x_1^{[i]}),\dots,V_s^{[i+1]}(x_{r_s}^{[i]})]^{\mathrm{T}}$, $\Psi_s=[\Phi_s(x_1^{[i]}),\dots,\Phi_s(x_{r_s}^{[i]})]^{\mathrm{T}}$.

The weight-update rule is based on least square algorithm, which is given by

$$W_s^{[i+1]} = \left(\Psi_s^{\mathrm{T}}\Psi_s\right)^{-1} \Psi_s^{\mathrm{T}} \Gamma_s, \tag{7.136}$$

and $v_s^{[i+1]}$ becomes

$$v_s^{[i+1]} = -\frac{1}{2}\left(q^{\mathrm{T}}q + R\right)^{-1}\left(G_s^{\mathrm{T}}\frac{\partial \Phi_s^{\mathrm{T}}}{\partial x_{1s}} W_s^{[i+1]} + 2q^{\mathrm{T}}p\right). \tag{7.137}$$

Select $x_k^{[i]} = x_{2f}^{[i]}(x_0, t_k)$ $(k = 1, 2, \ldots, r_f)$ from $x_{2f}^{[i]}(x_0, \cdot)$, $r_f \geq N_f$, as the initial state. The value function of fast subsystem becomes

$$V_f^{[i+1]}(x_k^{[i]}) = \int_{t_k}^{\infty} l_f\left(x_{2f}^{[i]}(x_0, t), v_f^{[i]}\right) \mathrm{d}t. \tag{7.138}$$

Update $W_f^{[i+1]}$ according to (7.139). $V_f^{[i+1]}$ satisfies

$$W_f^{[i+1]} = \left(\Psi_f^{\mathrm{T}}\Psi_f\right)^{-1} \Psi_f^{\mathrm{T}} \Gamma_f, \tag{7.139}$$

where $\Gamma_f = [V_f^{[i+1]}(x_1^{[i]}), \ldots, V_f^{[i+1]}(x_{r_f}^{[i]})]^{\mathrm{T}}$, $\Psi_f = [\Phi_f(x_1^{[i]}), \ldots, \Phi_f(x_{r_f}^{[i]})]^{\mathrm{T}}$.

Hence, we get

$$v_f^{[i+1]} = -\frac{1}{2}R^{-1}G_f^{\mathrm{T}}\frac{\partial \Phi_f^{\mathrm{T}}}{\partial x_{2f}} W_f^{[i+1]}. \tag{7.140}$$

The design procedure of the approximate optimal controller for singularly perturbed systems is summarized as follows:

1. Select $V_s^{[0]}$ and $V_f^{[0]}$, and let $v_s^{[0]}$ and $v_f^{[0]}$ be admissible control.
2. Select random state as x_0. Get the trajectories of states $x_{1s}^{[i]}(x_0, \cdot)$, $x_{2f}^{[i]}(x_0, \cdot)$ under $v_s^{[i]}$, $v_f^{[i]}$ $(i = 1, 2, \ldots)$.
3. Update value functions $V_s^{[i+1]}$ and $V_f^{[i+1]}$. Then, compute $V_s^{[i+1]}(x_j^{[i]})$ and $V_f^{[i+1]}(x_k^{[i]})$, $j \neq k$. Update $W_s^{[i+1]}$ and $W_f^{[i+1]}$ according to (7.136) and (7.139). Update $v_s^{[i+1]}$ and $v_f^{[i+1]}$ according to (7.137) and (7.140).
4. If $\|W_s^{[i+1]} - W_s^{[i]}\| \leq \beta_s$ and $\|W_f^{[i+1]} - W_f^{[i]}\| \leq \beta_f$, stop; else, $i = i + 1$, go to Step 2.

The final controller is composed by u_s and u_f, i.e., $v_c^{[\infty]} = v_s^{[\infty]} + v_f^{[\infty]}$. Next, we are ready to prove that $v_c^{[\infty]}$ is the approximate optimal controller.

Theorem 7.21 (cf. [3]) *For any small constant α, there exist N_s and N_f such that $\|v_c^{[\infty]} - u^*\| \leq \alpha$ holds.*

Proof From Theorem 7.20, we know that we always get $v_s^{[i]} \to u_s^*$ and $v_f^{[i]} \to u_f^*$ when $i \to \infty$ if N_s and N_f are chosen appropriately.

It means that for $\forall \alpha_s$ and $\forall \alpha_f$, there exists i such that $\|v_s^{[i]} - u_s^*\| \le \alpha_s$ and $\|v_f^{[i]} - u_f^*\| \le \alpha_f$ hold.

Hence, if ϵ is small enough such that $\alpha - |O(\epsilon)| > 0$ holds, we get $\|v_c^{[\infty]} - u_c^*\| \le \alpha - |O(\epsilon)|, \forall \alpha$.

Thus, we further obtain

$$\begin{aligned}\|v_c^{[\infty]} - u^*\| &= \|v_c^{[\infty]} - u_c^* + u_c^* - u^*\| \\ &\le \|v_c^{[\infty]} - u_c^*\| + \|u_c^* - u^*\| \\ &\le \alpha - |O(\epsilon)| + |O(\epsilon)| \\ &= \alpha.\end{aligned}$$

□

7.4.3 Simulations

In this section, an example is provided to demonstrate the effectiveness of the present method.

Consider the following singularly perturbed system:

$$\begin{bmatrix}\dot{x}_1 \\ \epsilon\dot{x}_2\end{bmatrix} = \begin{bmatrix}-x_1 + x_2 \\ -\sin x_1 - x_2\end{bmatrix} + \begin{bmatrix}0 \\ 1\end{bmatrix} u, \tag{7.141}$$

where $\epsilon = 0.01$. The cost functional is chosen as (7.109), where $Q = 2I_2$ and $R = 1$.

First, two NNs are used to approximate fast subsystem and slow subsystem. Select $\Phi_s = [x_{1s}^2 \;\; x_{1s}^4 \;\; x_{1s}^6 \;\; x_{1s}^8]$, $\Phi_f = x_{2f}^2$. Initial weights are chosen as $W_s = [0 \;\; 0 \;\; 0 \;\; 0]^{\mathrm{T}}$, $W_f = 0$. According to (7.134), (7.136), (7.138), and (7.139), the weights will converge to

$$\Phi_s = [0.8917 \;\; -0.0181 \;\; 0 \;\; 0], \quad \Phi_f = 0.7258.$$

The state and input trajectories of slow subsystem and fast subsystem at different iteration number are shown in Figs. 7.8 and 7.9, respectively. The states and input trajectories with the composite control is shown in Fig. 7.10.

7.5 Optimal Feedback Control for a Class of Constrained Systems Via SNAC

7.5.1 Problem Formulation

Consider the following nonlinear system:

$$x(k+1) = f(x(k)) + g(x(k))u(k), \tag{7.142}$$

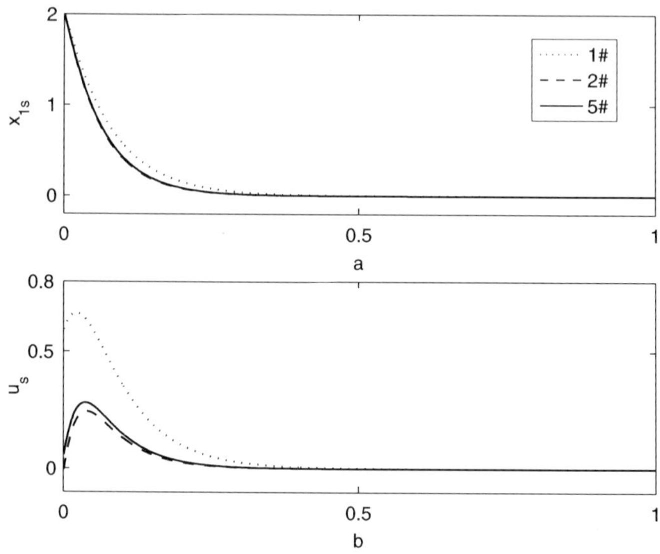

Fig. 7.8 State and input trajectories of the slow system at different iteration number

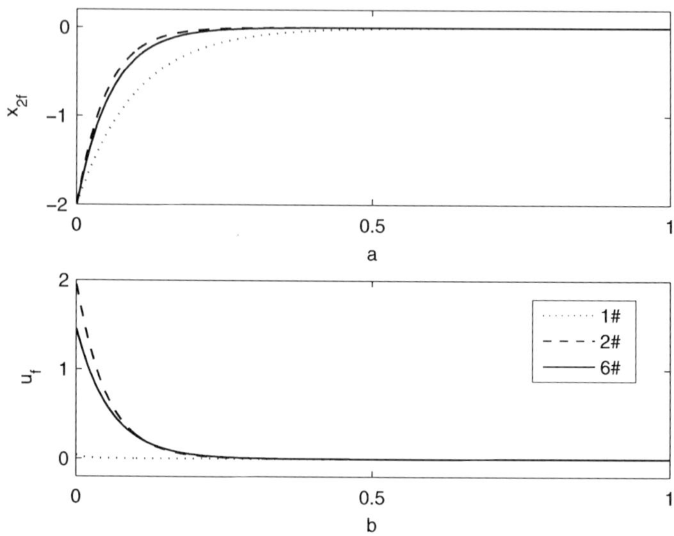

Fig. 7.9 State and input trajectories of the fast system at different iteration number

where $x(k) \in \mathbb{R}^n$ is the state vector, and $f: \mathbb{R}^n \to \mathbb{R}^n$ and $g: \mathbb{R}^n \to \mathbb{R}^{n\times m}$ are differentiable with respect to their arguments with $f(0) = 0$. Assume that $f + gu$ is Lipschitz continuous on a set Ω in $\mathbb{R}^n$ containing the origin, and that the system (7.142) is controllable in the sense that there exists a continuous control law on Ω_x that asymptotically stabilizes the system. We denote $u(k) \in \Omega_u$, $\Omega_u = \{u(k) =$

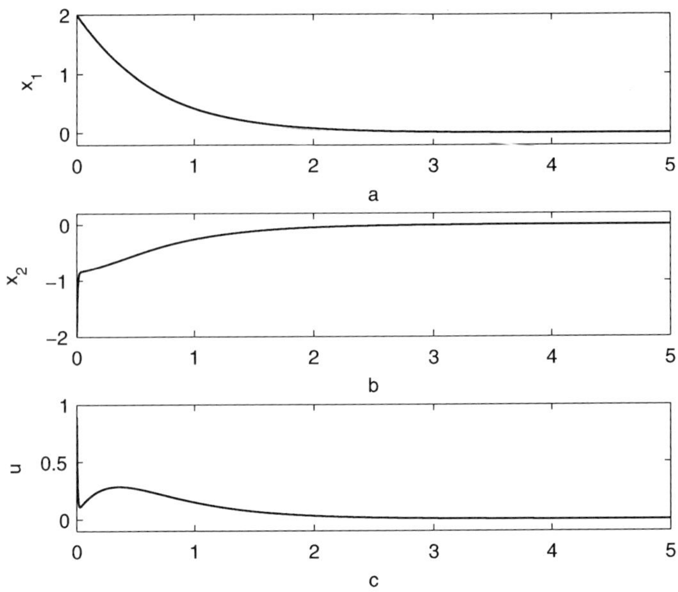

Fig. 7.10 State and input trajectories with composite control

$[u_1(k), u_2(k), \ldots, u_m(k)]^{\mathrm{T}} \in \mathbb{R}^m \colon |u_i(k)| \le \bar{u}_i, i = 1, \ldots, m\}$, where $\bar{u}_i$ is the saturating bound for the ith actuator. Let $\bar{U} \in \mathbb{R}^{m\times m}$ be the constant diagonal matrix given by $\bar{U} = \mathrm{diag}\{\bar{u}_1, \bar{u}_2, \ldots, \bar{u}_m\}$.

In this section, we mainly discuss how to design an optimal state-feedback controller for the system (7.142). It is desired to find the optimal control law $u(\cdot)$ to minimize the generalized cost functional as follows:

$$J(x(k), u) = \sum_{i=k}^{\infty} \big\{x^{\mathrm{T}}(i) Q x(i) + W(u(i))\big\}, \tag{7.143}$$

where $W(u(i))$ is positive definite, and the weight matrix Q is also positive definite.

For the optimal control problem, the state-feedback control law $u(\cdot)$ must not only stabilize the system on Ω_x, but also guarantee that (7.143) is finite. Such a control law is said to be admissible.

Let J^* denotes the optimal value function. According to Bellman's optimality principle, we have

$$J^*(x(k)) = \min_{u(k)} \Big\{x^{\mathrm{T}}(k) Q x(k) + W(u(k)) + J^*(x(k+1))\Big\}. \tag{7.144}$$

For the unconstrained control problem, $W(u(i))$ is commonly chosen as the quadratic form of the control input $u(i)$, i.e., $W(u(i)) = u^{\mathrm{T}}(i) R u(i)$, where $R \in \mathbb{R}^{m\times m}$ is semi-positive definite. Here, we assume that the value function is smooth. According to the necessary condition for optimality, we obtain

$$u^*(k) = -\frac{1}{2}R^{-1}g^{\mathrm{T}}(x(k))\frac{\partial J^*(x(k+1))}{\partial x(k+1)}, \tag{7.145}$$

where J^* is the optimal value function with respect to the optimal control law u^*.

However, the above derivation is not suitable for constrained optimal control problem. To guarantee bounded controls, inspired by [8], a nonquadratic functional is introduced as follows:

$$\begin{aligned} W(u(i)) &= 2\int_0^{u(i)} \varphi^{-\mathrm{T}}(\bar{U}^{-1}s)\bar{U}R\mathrm{d}s, \\ \varphi^{-1}(u(i)) &= [\varphi^{-1}(u_1(i)), \ldots, \varphi^{-1}(u_m(i))], \end{aligned} \tag{7.146}$$

where R is positive definite and assumed to be diagonal for simplicity of analysis, $s \in \mathbb{R}^m$, $\varphi \in \mathbb{R}^m$, $\varphi(\cdot)$ is a bounded one-to-one function satisfying $|\varphi(\cdot)| \le 1$ and belonging to $C^p(p \ge 1)$ and $L_2(\Omega_x)$. Moreover, it is a monotonically increasing odd function with its first derivative bounded by M. Such function is easy to find, and one example is the hyperbolic tangent function $\varphi(\cdot) = \tanh(\cdot)$. It should be noticed that by the definition above, $W(u(i))$ is ensured to be positive definite because $\varphi^{-1}(\cdot)$ is a monotonic odd function and R is positive definite.

Substituting (7.146) into (7.144), the HJB equation can be derived as follows:

$$J^*(x(k)) = \min_{u(k)}\left\{x^{\mathrm{T}}(k)Qx(k) + 2\int_0^{u(k)} \varphi^{-\mathrm{T}}(\bar{U}^{-1}s)\bar{U}R\mathrm{d}s + J^*(x(k+1))\right\}. \tag{7.147}$$

Assuming that the value function is smooth, and according to the necessary condition for optimality, we obtain

$$\frac{\partial J^*(x(k))}{\partial u(k)} = 2R\bar{U}\varphi^{-1}(\bar{U}^{-1}u(k)) + \left(\frac{\partial x(k+1)}{\partial u(k)}\right)^{\mathrm{T}}\frac{\partial J^*(x(k+1))}{\partial x(k+1)} = 0. \tag{7.148}$$

Therefore, letting $J_x^*(x(k+1)) = \partial J^*(x(k+1))/\partial x(k+1)$, the optimal control law can be computed as follows:

$$u^*(k) = \bar{U}\varphi\left(-\frac{1}{2}(\bar{U}R)^{-1}g^{\mathrm{T}}(x(k))J_x^*(x(k+1))\right). \tag{7.149}$$

If the optimal value function J^* is known, the optimal control law $u^*(k)$ can be obtained from (7.149). However, there is currently no method for solving this value function of the constrained optimal control problem. Therefore, in the next section we will discuss how to utilize the greedy iterative DHP algorithm to seek the near-optimal control solution.

7.5.2 Optimal Controller Design for Constrained Systems via SNAC

For convenience, in the sequel, $W(u(k))$ is used to denote the nonquadratic functional $2\int_0^{u(k)} \varphi^{-\mathrm{T}}(\bar{U}^{-1}s)\bar{U}R\mathrm{d}s$.

Define the costate function $\lambda(x(k)) = \frac{\partial J(x(k))}{\partial x(k)}$. Then, (7.149) can be rewritten as

$$u^*(k) = \bar{U}\varphi\Big(-\frac{1}{2}(\bar{U}R)^{-1}g^{\mathrm{T}}(x(k))\lambda^*(x(k+1))\Big). \tag{7.150}$$

By definition, the costate function $\lambda^*(x(k))$ satisfies

$$\begin{aligned}
\lambda^*(x(k)) &= \frac{\partial J^*(x(k))}{\partial x(k)} \\
&= \frac{\partial(x^{\mathrm{T}}(k)Qx(k) + W(u(k)))}{\partial x(k)} \\
&\quad + \left(\frac{\partial u(x(k))}{\partial x(k)}\right)^{\mathrm{T}} \frac{\partial(x^{\mathrm{T}}(k)Qx(k) + W(u(k)))}{\partial u(x(k))} \\
&\quad + \left(\frac{\partial x(k+1)}{\partial x(k)}\right)^{\mathrm{T}} \frac{\partial J^*(x(k+1))}{\partial x(k+1)} \\
&\quad + \left(\frac{\partial u(x(k))}{\partial x(k)}\right)^{\mathrm{T}} \left(\frac{\partial x(k+1)}{\partial u(x(k))}\right)^{\mathrm{T}} \frac{\partial J^*(x(k+1))}{\partial x(k+1)} \\
&= \left(\frac{\partial u(x(k))}{\partial x(k)}\right)^{\mathrm{T}} \left[\frac{\partial(x^{\mathrm{T}}(k)Qx(k) + W(u(k)))}{\partial u(x(k))} + \left(\frac{\partial x(k+1)}{\partial u(x(k))}\right)^{\mathrm{T}}\right. \\
&\quad \left.\times \frac{\partial J^*(x(k+1))}{\partial x(k+1)}\right] + \frac{\partial(x^{\mathrm{T}}(k)Qx(k) + W(u(k)))}{\partial x(k)} \\
&\quad + \left(\frac{\partial x(k+1)}{\partial x(k)}\right)^{\mathrm{T}} \frac{\partial J^*(x(k+1))}{\partial x(k+1)} \\
&= 2Qx(k) + \left(\frac{\partial x(k+1)}{\partial x(k)}\right)^{\mathrm{T}} \lambda^*(x(k+1)).
\end{aligned} \tag{7.151}$$

The optimal control law can be solved from (7.150) if the costate function λ^* can be obtained from (7.151). However, it is very difficult to solve (7.151) analytically due to the two-point boundary value problems of partial difference equation. Therefore, a greedy iterative DHP algorithm is developed to obtain the costate function λ^* and the optimal control law u^* in the sequel.

First, we start with initial value function $V_0(\cdot) = 0$ and initial costate function $\lambda_0(\cdot) = 0$, Then, we find the law of single control vector v_0 as follows:

$$v_0(x(k)) = \arg\min_{u(k)} \big\{x^{\mathrm{T}}(k)Qx(k) + W(u(k)) + V_0(x(k+1))\big\}. \tag{7.152}$$

According to the optimal principle, $v_0(x)$ can be derived as follows:

$$v_0(x(k)) = \bar{U}\varphi\Big(-\frac{1}{2}(\bar{U}R)^{-1}g^{\mathrm{T}}(x(k))\lambda_0(x(k+1))\Big). \tag{7.153}$$

Thus, we have

$$x(k+1) = f(x(k)) + g(x(k))v_0(x(k)), \tag{7.154}$$

and

$$\begin{aligned} &v_0(x(k+1)) \\ &\quad = \arg\min_{u(k+1)} \big\{x^{\mathrm{T}}(k+1)Qx(k+1) + W(u(k+1)) + V_0(x(k+2))\big\}. \end{aligned} \tag{7.155}$$

According to $\lambda_0(x(k+2)) = \partial V_0(x(k+2))/\partial x(k+2)$, and using (7.155), we have

$$v_0(x(k+1)) = \bar{U}\varphi\Big(-\frac{1}{2}(\bar{U}R)^{-1}g^{\mathrm{T}}(x(k+1))\lambda_0(x(k+2))\Big). \tag{7.156}$$

Then, the value function can be updated as

$$V_1(x(k+1)) = x^{\mathrm{T}}(k+1)Qx(k+1) + W(v_0(k+1)) + V_0(x(k+2)). \tag{7.157}$$

Therefore, for $i = 0, 1, \ldots$, the iterative ADP algorithm is realized by implementing the iterations between

$$\begin{aligned} &v_i(x(k+1)) \\ &\quad = \arg\min_{u(k+1)} \big\{x^{\mathrm{T}}(k+1)Qx(k+1) + W(u(k+1)) + V_i(x(k+2))\big\} \end{aligned} \tag{7.158}$$

and

$$\begin{aligned} V_{i+1}(x(k+1)) &= \min_{u(k+1)} \big\{x^{\mathrm{T}}(k+1)Qx(k+1) + W(u(k+1)) + V_i(x(k+2))\big\} \\ &= x^{\mathrm{T}}(k+1)Qx(k+1) + W(v_i(k+1)) + V_i(x(k+2)) \end{aligned} \tag{7.159}$$

until convergence.

Obviously, if we assume that the value function V_i is smooth, $v_i(k+1)$ in (7.158) can further be solved as

$$\begin{aligned} \frac{\partial V_{i+1}(x(k+1))}{\partial v(k+1)} &= \frac{\partial(x^{\mathrm{T}}(k+1)Qx(k+1) + W(v_i(k+1))}{\partial v(k+1)} \\ &\quad + g^{\mathrm{T}}(x(k+1))\frac{\partial V_i(x(k+2))}{\partial x(k+2)} \\ &= 0. \end{aligned} \tag{7.160}$$

That is,

$$v_i(k+1)=\bar{U}\varphi\Big(-\frac{1}{2}(\bar{U}R)^{-1}g^{\mathrm{T}}(x(k+1))\lambda_i(x(k+2))\Big). \tag{7.161}$$

Because $\lambda_{i+1}(x(k+1))=\partial V_{i+1}(x(k+1))/\partial x(k+1)$, similarly to the derivative process of (7.151), we have

$$\begin{aligned}
&\lambda_{i+1}(x(k+1))\\
&\quad=\frac{\partial(x^{\mathrm{T}}(k+1)Qx(k+1)+W(v_i(k+1)))}{\partial x(k+1)}\\
&\qquad+\left(\frac{\partial v_i(x(k+1))}{\partial x(k+1)}\right)^{\mathrm{T}}\frac{\partial(x^{\mathrm{T}}(k+1)Qx(k+1)+W(v_i(k+1)))}{\partial v_i(x(k+1))}\\
&\qquad+\left(\frac{\partial x(k+2)}{\partial x(k+1)}\right)^{\mathrm{T}}\frac{\partial V_i(x(k+2))}{\partial x(k+2)}\\
&\qquad+\left(\frac{\partial v_i(x(k+1))}{\partial x(k+1)}\right)^{\mathrm{T}}\left(\frac{\partial x(k+2)}{\partial v_i(x(k+1))}\right)^{\mathrm{T}}\frac{\partial V_i(x(k+2))}{\partial x(k+2)}\\
&\quad=\frac{\partial(x^{\mathrm{T}}(k+1)Qx(k+1)+W(v_i(k+1)))}{\partial x(k+1)}+g^{\mathrm{T}}(x(k+1))\frac{\partial V_i(x(k+2))}{\partial x(k+2)}.
\end{aligned} \tag{7.162}$$

That is,

$$\lambda_{i+1}(x(k+1))=2Qx(k+1)+g^{\mathrm{T}}(x(k+1))\lambda_i(x(k+2)). \tag{7.163}$$

Therefore, we obtain the optimal control as follows:

$$v_i(x(k))=\bar{U}\varphi\left(-\frac{1}{2}(\bar{U}R)^{-1}g^{\mathrm{T}}(x(k))\lambda_i(x(k+1))\right). \tag{7.164}$$

Hence, the iteration between (7.158) and (7.159) is an implementation of the iteration between (7.163) and (7.164).

In the following, we present the convergence analysis of the greedy iterative DHP algorithm. We first present two lemmas before presenting our theorems.

Lemma 7.22 *Let μ_i be an arbitrary sequence of control laws, and v_i be the control law sequence as in* (7.158). *Let V_i be as in* (7.159) *and Λ_i be*

$$\Lambda_{i+1}(x(k+1))=x^{\mathrm{T}}(k+1)Qx(k+1)+W(\mu_i(k+1))+\Lambda_i(x(k+2)). \tag{7.165}$$

If $V_0=\Lambda_0=0$, then $V_i\le\Lambda_i$, $\forall i$.

Lemma 7.23 *Let $\{V_i\}$ be defined as in*(7.159). *If the system is controllable, then there is an upper bound Y such that $0\le V_i\le Y$, $\forall i$.*

Proof Let $\eta(x(k+1))$ be a sequence of stabilizing and admissible control laws, and let $V_0(\cdot) = Z_0(\cdot) = 0$, where V_i is updated by (7.159) and Z_i is updated by

$$Z_{i+1}(x(k+1)) = x^{\mathrm{T}}(k+1)Qx(k+1) + W(\eta(x(k+1))) + Z_i(x(k+2)). \tag{7.166}$$

Then, we obtain

$$\begin{aligned} Z_{i+1}(x(k+1)) - Z_i(x(k+1)) &= Z_i(x(k+2)) - Z_{i-1}(x(k+2)) \\ &= Z_{i-1}(x(k+3)) - Z_{i-2}(x(k+3)) \\ &= Z_{i-2}(x(k+4)) - Z_{i-3}(x(k+4)) \\ &\ \ \vdots \\ &= Z_1(x(k+i+1)) - Z_0(x(k+i+1)). \end{aligned} \tag{7.167}$$

Thus, the following relation can be obtained:

$$Z_{i+1}(x(k+1)) = Z_1(x(k+i+1)) + Z_i(x(k+1)) - Z_0(x(k+i+1)). \tag{7.168}$$

Since $Z_0(\cdot) = 0$, it follows that

$$\begin{aligned} Z_{i+1}(x(k+1)) &= Z_1(x(k+i+1)) + Z_i(x(k+1)) \\ &= Z_1(x(k+i+1)) + Z_1(x(k+i)) + Z_{i-1}(x(k+1)) \\ &= Z_1(x(k+i+1)) + Z_1(x(k+i)) \\ &\quad + Z_1(x(k+i-1)) + \cdots + Z_1(x(k+1)). \end{aligned} \tag{7.169}$$

Then, (7.169) can further be rewritten as

$$\begin{aligned} Z_{i+1}(x(k+1)) &= \sum_{j=0}^{i} Z_1(x(k+j+1)) \\ &= \sum_{j=0}^{i} \left\{x^{\mathrm{T}}(k+j+1)Qx(k+j+1) + W(\eta(x(k+j+1)))\right\} \\ &\le \sum_{j=0}^{\infty} \left\{x^{\mathrm{T}}(k+j+1)Qx(k+j+1) + W(\eta(x(k+j+1)))\right\}. \end{aligned} \tag{7.170}$$

Note that $\eta(x(k+1))$ is an admissible control law sequence, i.e., $x(k+1) \to 0$ as $k \to \infty$, Therefore, there exists an upper bound Y such that

$$\forall i:\ Z_{i+1}(x(k+1)) \le \sum_{j=0}^{\infty} Z_1(x(k+j+1)) \le Y. \tag{7.171}$$

Combining with Lemma 1, we obtain

$$\forall i: \; V_{i+1}(x(k+1)) \le Z_{i+1}(x(k+1)) \le Y. \tag{7.172}$$

The proof is completed. □

Next, Lemmas 7.22 and 7.23 will be used in the proof of our main theorems.

Theorem 7.24 (cf. [6]) *Define the value function sequence* $\{V_i\}$ *as in* (7.159) *with* $V_0 = 0$, *and the control law sequence* $\{v_i\}$ *as in* (7.163). *Then, we can conclude that* $\{V_i\}$ *is a nondecreasing sequence satisfying* $V_{i+1}(x(k+1)) \ge V_i(x(k+1)), \forall i$, *and is convergent to the value function of the discrete-time HJB equation, i.e.,* $V_i \to J^*$ *as* $i \to \infty$. *Meanwhile, we conclude that the costate function sequence* $\{\lambda_i\}$ *and the control law sequence* $\{v_i\}$ *are also convergent, i.e.,* $\lambda_i \to \lambda^*$ *and* $v_i \to u^*$ *as* $i \to \infty$.

Proof For convenience of analysis, define a new sequence Φ as follows:

$$\Phi_{i+1}(x(k+1)) = x^{\mathrm{T}}(k+1)Qx(k+1) + W(v_{i+1}(k+1)) + \Phi_i(x(k+2)), \tag{7.173}$$

with $\Phi_0 = V_0 = 0$. The control law sequence v_i is updated by (7.158) and the value function sequence V_i is updated by (7.159).

In the following, we prove that $\Phi_i(x(k+1)) \le V_{i+1}(x(k+1))$ by mathematical induction.

First, we prove that it holds for $i = 0$. Noticing that

$$V_1(x(k+1)) - \Phi_0(x(k+1)) = x^{\mathrm{T}}(k+1)Qx(k+1) + W(v_0(k+1)) \ge 0, \tag{7.174}$$

we have

$$V_1(x(k+1)) \ge \Phi_0(x(k+1)). \tag{7.175}$$

Second, we assume that it holds for $i-1$, i.e., for any $x(k+1)$, $V_i(x(k+1)) \ge \Phi_{i-1}(x(k+1))$. Then, for i, since

$$\Phi_i(x(k+1)) = x^{\mathrm{T}}(k+1)Qx(k+1) + W(v_i(k+1)) + \Phi_{i-1}(x(k+2)) \tag{7.176}$$

and

$$V_{i+1}(x(k+1)) = x^{\mathrm{T}}(k+1)Qx(k+1) + W(v_i(k+1)) + V_i(x(k+2)) \tag{7.177}$$

hold, we obtain

$$V_{i+1}(x(k+1)) - \Phi_i(x(k+1)) = V_i(x(k+2)) - \Phi_{i-1}(x(k+2)) \ge 0, \tag{7.178}$$

i.e., the following equation holds:

$$\Phi_i(x(k+1)) \le V_{i+1}(x(k+1)). \tag{7.179}$$

Therefore, for $\forall i$, (7.179) is proved by mathematical induction.

Furthermore, from Lemma 7.22, we know that $V_i(x(k+1)) \le \Phi_i(x(k+1))$. Therefore, we have

$$V_i(x(k+1)) \le \Phi_i(x(k+1)) \le V_{i+1}(x(k+1)). \qquad (7.180)$$

Hence, we can conclude that the value function sequence $\{V_i\}$ is a nondecreasing sequence satisfying $V_{i+1}(x(k+1)) \ge V_i(x(k+1))$, $\forall i$, and convergent to the optimal value function of the discrete-time HJB equation, i.e., $V_i \to J^*$ as $i \to \infty$. The costate sequence $\{\lambda_i\}$ is also convergent, i.e., $\lambda_i \to \lambda^*$ as $i \to \infty$.

Since the value function sequence and the costate function sequence are convergent, according to (7.163) and (7.164), we conclude that $\{v_i\}$ converges to the optimal control law u^* as $i \to \infty$. □

For linear systems, if the cost functional is quadratic, then the optimal control law is linear. For nonlinear system, this is not true. So we use neural network to approximate costate function, and get the optimal control law.

In this part, we use single network GI-DHP algorithm to obtain the optimal costate function and optimal control law. SNAC can remove the action network appearing in the ordinary ADP method. It simplifies the structure of the system, saves storage space and reduces computation. In addition, it eliminates approximation error of neural network, and improves the calculation accuracy. We have known that the control law is implicit function which is difficult to solve. So the costate function should be transformed. As $\lambda(x(k+1))$ can be expressed by $x(k)$, so we let $\bar{\lambda}(x(k)) = \lambda(x(k+1))$. Then, the critic neural network can be used to approximate $\bar{\lambda}(x(k))$.

Equation (7.163) can be written as

$$\bar{\lambda}_{i+1}(x(k)) = 2Qx(k+1) + g^{\mathrm{T}}(x(k+1))\bar{\lambda}_i(x(k+1)). \qquad (7.181)$$

The optimal control law can be obtained as follows:

$$v_i(x(k)) = \bar{U}\varphi\left(-\frac{1}{2}(\bar{U}R)^{-1}g^{\mathrm{T}}(x(k))\bar{\lambda}_i(x(k))\right). \qquad (7.182)$$

Once we get the costate function $\bar{\lambda}(x(k))$, then according to (7.182), we can get the optimal control law. Let $\bar{\lambda}_0(\cdot) = 0$, and then the iteration between (7.163) and (7.164) is changed into the iteration between (7.181) and (7.182).

Based on the above analysis, we give the structure in Fig. 7.11.

For each iteration, we use three layer forward neural network to approximate $\lambda_i(x(k+1))$

$$\hat{\lambda}_i(x(k+1)) = \hat{\bar{\lambda}}_i(x(k)) = \bar{w}_i^{\mathrm{T}}\sigma(\bar{v}_i^{\mathrm{T}}x(k)), \qquad (7.183)$$

where $\bar{w}_i$ and $\bar{v}_i$ is the weights of output and hidden layers, $\sigma(\cdot) \in \mathbb{R}^l$, $[\sigma(z)]_p = \frac{e^{z_p} - e^{-z_p}}{e^{z_p} + e^{-z_p}}$, $p = 1, \ldots, l$, is the activation function of hidden layer, and l is the number as hidden node.

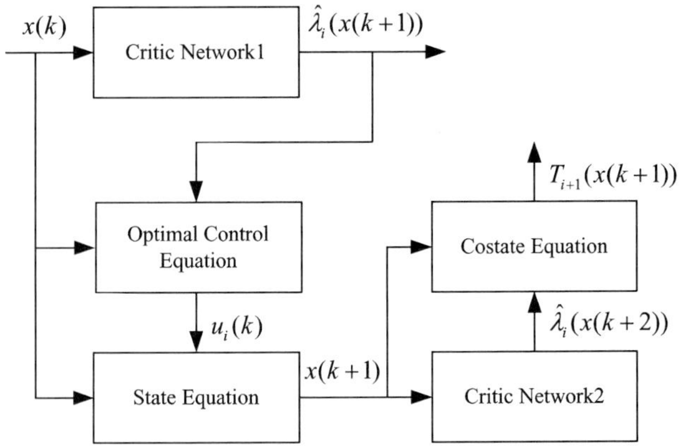

Fig. 7.11 The structure diagram of the single network GI-DHP algorithm

According to (7.163), the costate function can be expressed as

$$\begin{aligned}\hat{\lambda}_{i+1}(x(k+1)) &= 2Qx(k+1) + g^{\mathrm{T}}(x(k+1))\hat{\lambda}_i(x(k+2)) \\ &= 2Qx(k+1) + g^{\mathrm{T}}(x(k+1))\bar{w}_i^{\mathrm{T}}\sigma(\bar{v}_i^{\mathrm{T}}x(k+1)),\end{aligned} \tag{7.184}$$

where $x(k+1)$ can be obtained by $x(k)$ and $u_i(x(k))$.

Define the error function of critic network

$$e_j(k+1) = \hat{\lambda}_{i(j)}(x(k), \bar{w}_{i(j)}, \bar{v}_{i(j)}) - \lambda_{i+1}(x(k+1)). \tag{7.185}$$

The weights in the critic network are updated to minimize the following performance measure:

$$E_j(k+1) = \frac{1}{2}e_j^{\mathrm{T}}(k+1)e_j(k+1). \tag{7.186}$$

The weight updating rule for critic network is chosen as a gradient-based adaptation rule

$$\bar{w}_{i(j+1)}(k+1) = \bar{w}_{i(j)}(k+1) - \alpha\left[\frac{\partial E_j(k+1)}{\partial \bar{w}_{i(j)}(k+1)}\right], \tag{7.187}$$

$$\bar{v}_{i(j+1)}(k+1) = \bar{v}_{i(j)}(k+1) - \alpha\left[\frac{\partial E_j(k+1)}{\partial \bar{v}_{i(j)}(k+1)}\right], \tag{7.188}$$

where $\alpha > 0$ is the learning rate and j is the inner-loop iterative step for updating the weight parameters.

After the approximation of $\lambda_i(x(k+1))$, i.e., $\bar{\lambda}_i(x(k))$, we get the optimal control law from (7.182).

7.5.3 Simulations

In this section, two examples are provided to demonstrate the effectiveness of the control scheme.

Example 7.25 Consider the following Van der Pol oscillator system in [9]:

$$\begin{aligned}\dot{x}_1 &= x_2,\\ \dot{x}_2 &= \alpha(1-x_1^2)x_2 - x_1 + (1+x_1^2+x_2^2)u,\end{aligned} \tag{7.189}$$

where $\alpha = 0.05$, control constraint is set to $|u| \le 0.2$.

We first use the Euler method to discretize the system, where $t = kT$, T is sampling period, k is sampling step. If T is small enough compared with the system time constant, then the discretization method is precise enough.

So, let $T = 0.1$; we have

$$\begin{aligned}x_1(k+1) &= x_1(k) + 0.1x_2(k),\\ x_2(k+1) &= -x_1(k) + 0.005(1-x_1^2(k))x_2(k) + x_2(k)\\ &\quad + (1+x_1^2(k)+x_2^2(k))u(k).\end{aligned} \tag{7.190}$$

Define the cost functional as

$$J(x(k),u) = \sum_{i=k}^{\infty}\left\{x^{\mathrm{T}}(i)Qx(i) + 2\bar{U}\int_0^{u(i)} \tanh^{-\mathrm{T}}(\bar{U}s)R\mathrm{d}s\right\}, \tag{7.191}$$

where $\bar{U} = 0.2$ and $Q = \left[\begin{smallmatrix}3 & 0\\ 0 & 3\end{smallmatrix}\right]$, $R = [\,2\,]$.

We choose neural networks as the critic network with the structures 2–9–2. The initial weights of the output layer are zero to guarantee the initial output of critic network is zero, i.e., $\lambda_0 = 0$. The initial weights of the hidden layer are chosen randomly in $[-1\ 1]$. We train the critic network for 1200 iteration steps and 2000 time steps. The critic network is trained until the given accuracy 10^{-10}. The convergence curves of the costate function are shown in Fig. 7.12. Furthermore, for $x_1(0) = 0.1$ and $x_2(0) = 0.3$, we run the system for 200 time steps, we get the state trajectory in Fig. 7.13 and control trajectory in Fig. 7.14.

Moreover, in order to make comparison with the controller designed without considering the actuator saturation, we also present the system responses obtained by the controller designed regardless of the actuator saturation. After simulation, the state curves is shown in Fig. 7.15 and the control curve is shown in Fig. 7.16.

From the simulation results, we can see that the iterative costate function sequences do converge to the optimal ones with very fast speed, which also indicates the validity of the iterative ADP algorithm for dealing with constrained nonlinear systems.

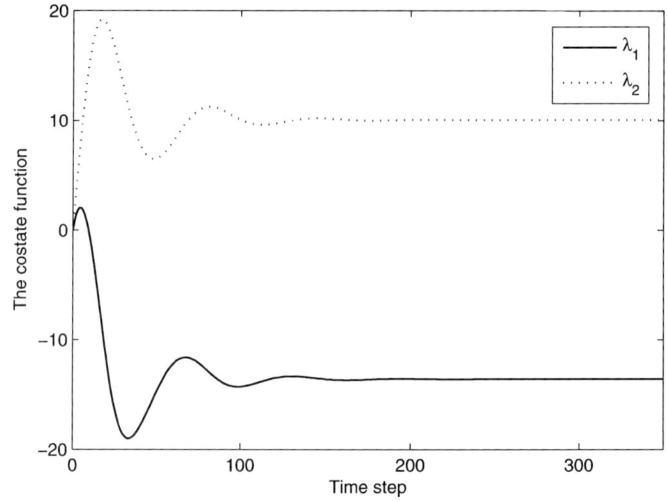

Fig. 7.12 The convergence process of the costate function

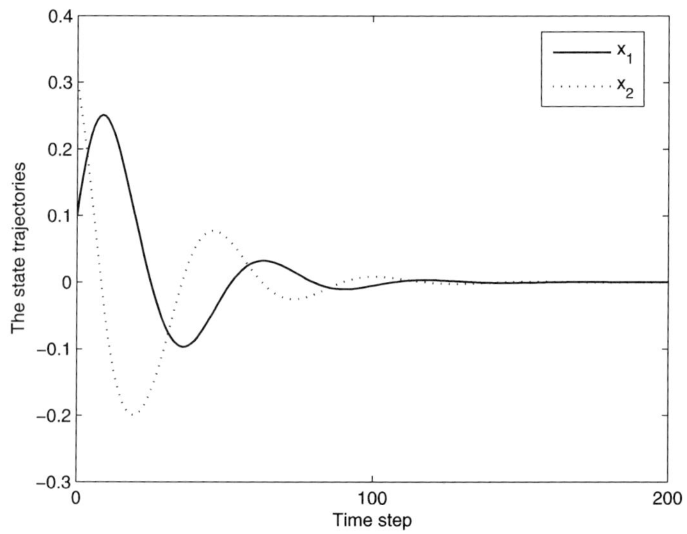

Fig. 7.13 The state variables curves

Example 7.26 CSTR system is the first order process of heat release. According to the quality of conservation and the principle of conservation of energy, the relationship between quality of materials C and reactor temperature T is

$$V_{ol}\frac{\mathrm{d}\mathbf{C}}{\mathrm{d}t} = \zeta(\mathbf{C_0} - \mathbf{C}) - V_{ol}R_a,$$

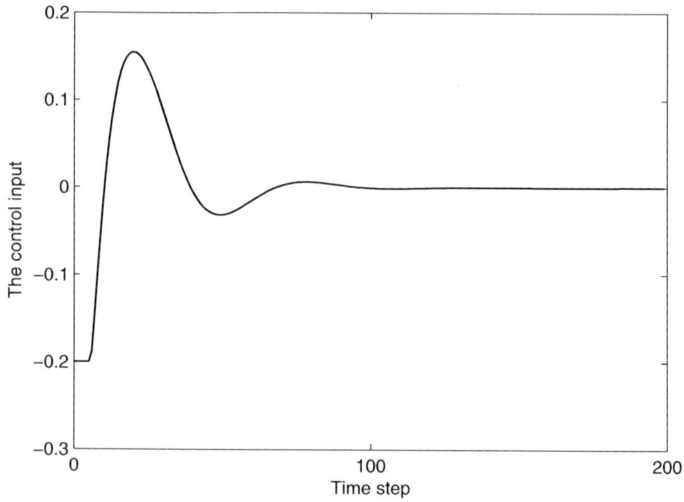

Fig. 7.14 The control variables curves

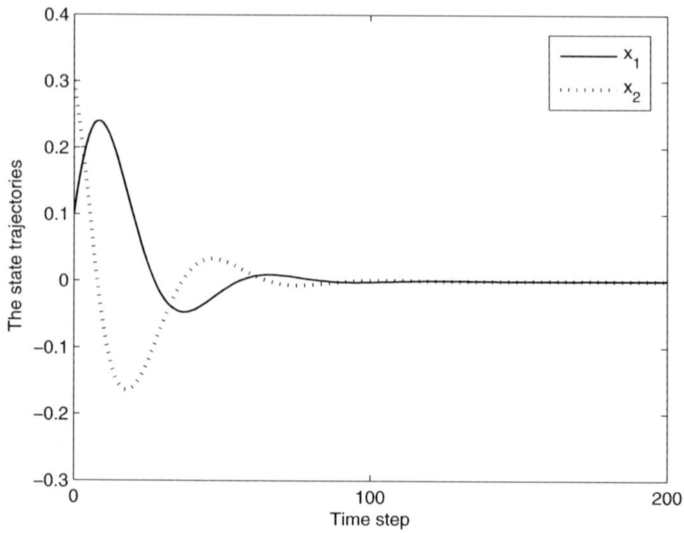

Fig. 7.15 The state variables curves

$$V_{ol}C_p\frac{\mathrm{d}\mathbf{T}}{\mathrm{d}t}=\zeta(\mathbf{T_0}-\mathbf{T})+(\Delta H)V_{ol}R_a-U(\mathbf{T}-\mathbf{T_w}), \tag{7.192}$$

where V_{ol} is the reactor volume, ζ is the raw materials in and out of the flow, $\mathbf{C_0}$ is the expectation of raw material of the supply, R_a is the unit volume response rate, C_p is the raw materials of the unit volume produced by the heat energy, $\mathbf{T_0}$ is the expected temperature, ΔH is the reaction heat; if it is an exothermic reaction

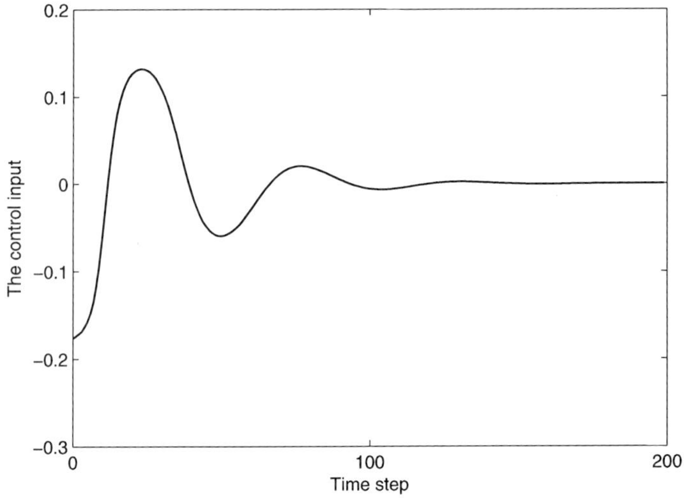

Fig. 7.16 The control variables curves

system, $\varDelta H$ should be taken positive, U is the heat transfer coefficient of the surface of the reactor, and $\mathbf{T_w}$ is the coolant average temperature.

Suppose that the first order reaction rate for the dynamic equation is

$$R_a = \kappa_0 \mathbf{C} \exp\left(\frac{-\mathbf{Q}}{\mathbf{T}}\right),$$

where κ_0 is the rate constant, $\mathbf{Q}$ is the Arrhenius activation energy and we have the ratio of the gas constant. By transformation, we have

$$x_1 = \frac{\mathbf{C_0} - \mathbf{C}}{\mathbf{C_0}}, \quad x_2 = \frac{\mathbf{T} - \mathbf{T_0}}{\mathbf{Q}},$$
$$\nu = \frac{t}{\frac{V_{ol}}{\zeta}}, \quad u = \frac{\mathbf{T_w} - \mathbf{T_0}}{\mathbf{Q}},$$

and then (7.192) can be written as

$$\begin{aligned} \frac{\mathrm{d}x_1}{\mathrm{d}\nu} &= -x_1 + D_a(1 - x_1)\exp\left(-\frac{1}{x_2 + \rho}\right) \frac{\mathrm{d}x_2}{\mathrm{d}\nu} \\ &= -(1 + \varrho)x_2 + H D_a(1 - x_1)\exp\left(-\frac{1}{x_2 + \rho}\right) + \varrho u, \end{aligned} \tag{7.193}$$

where

$$D_a = \frac{\kappa_0 V_{ol}}{\zeta}, \quad H = \frac{(\varDelta H)\mathbf{C_0}}{\mathbf{Q} C_p}, \quad \rho = \frac{\mathbf{T_0}}{\mathbf{Q}}, \quad \varrho = \frac{U}{\zeta C_p}.$$

For D_a, H, ρ, and ϱ, (7.193) has one, two and three equilibrium points. Select the corresponding control $u = u_e$. If

$$D_a = 0.072, \quad H = 8, \quad \rho = 20, \quad \varrho = 0.3,$$

and $u_e = 0$, we see that (7.193) has one equilibrium point, i.e.,

$$x_e = \begin{bmatrix} 0.0642 \\ 0.3948 \end{bmatrix}.$$

Defining the following new state variables:

$$\begin{aligned} \iota x_1 &= x_1 - x_{e1}, \\ \iota x_2 &= x_2 - x_{e2}, \end{aligned}$$

and then (7.193) can be written as

$$\begin{aligned} \frac{d\iota x_1}{d\nu} &= -\iota x_1 - x_{e1} + D_a(1 - \iota x_1 - x_{e1}) \exp\left(-\frac{1}{\iota x_2 + x_{e2} + \rho}\right), \\ \frac{d\iota x_2}{d\nu} &= -(1+\varrho)(\iota x_2 + x_{e2}) + H D_a(1 - \iota x_1 - x_{e1}) \exp\left(-\frac{1}{\iota x_2 + x_{e2} + \rho}\right) + \varrho u. \end{aligned} \tag{7.194}$$

Supposing $|u| \le 0.05$ and $T = 0.01$, we discretize (7.194) as

$$\begin{aligned} \iota x_1(k+1) &= -\iota x_1(k)T - x_{e1}T \\ &\quad + D_a(1 - \iota x_1(k) - x_{e1}) \exp\left(-\frac{1}{\iota x_2(k) + x_{e2} + \rho}\right) T + \iota x_1(k), \\ \iota x_2(k+1) &= -(1+\varrho)(\iota x_2(k) + x_{e2})T \\ &\quad + H D_a(1 - \iota x_1(k) - x_{e1}) \exp\left(-\frac{1}{\iota x_2(k) + x_{e2} + \rho}\right) T \\ &\quad + \iota x_2(k) + \varrho T u. \end{aligned} \tag{7.195}$$

The cost functional is selected as in Example 7.25, where $\bar{U} = 0.05$, $Q = \left[\begin{smallmatrix} 2 & 0 \\ 0 & 2 \end{smallmatrix}\right]$, and $R = [0.5]$.

We choose a neural network as the critic network with the structures 2–11–2.

The initial weight vector of the output layer is set to zero vector, i.e., $\lambda_0 = 0$. The initial weights of the hidden layer are chosen randomly in $[-1\ 1]$. We train the critic network for 1600 iteration steps and 2000 time steps. The critic network is trained until the given accuracy 10^{-10}. The convergence curves of the costate function are shown in Fig. 7.17. Furthermore, for $\iota x_1(0) = 1$ and $\iota x_2(0) = -1$, we run the system

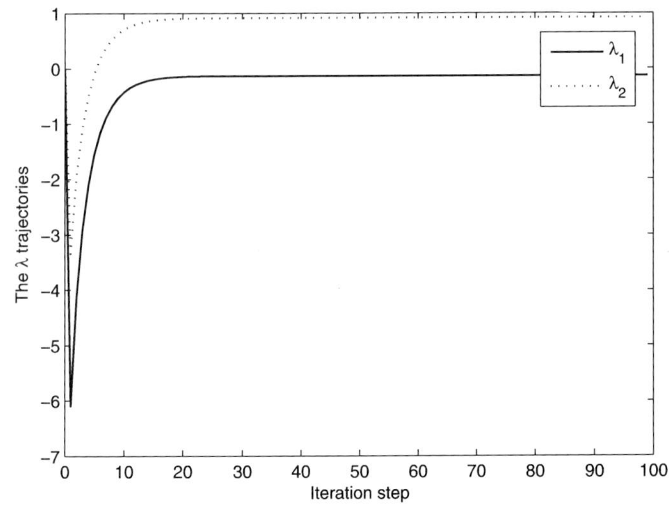

Fig. 7.17 The convergence process of the costate function

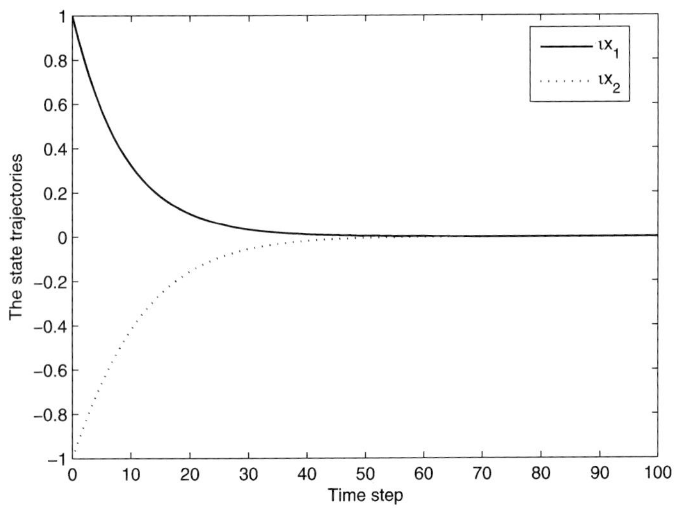

Fig. 7.18 The state variables curves

for 100 time steps, we get the state trajectory in Fig. 7.18 and control trajectory in Fig. 7.19.

As Example 7.25, in order to make a comparison with the controller designed without considering the actuator saturation, we also present the system responses obtained by the controller designed regardless of the actuator saturation. After simulation, the state curves is shown in Fig. 7.20 and the control curve is shown in Fig. 7.21.

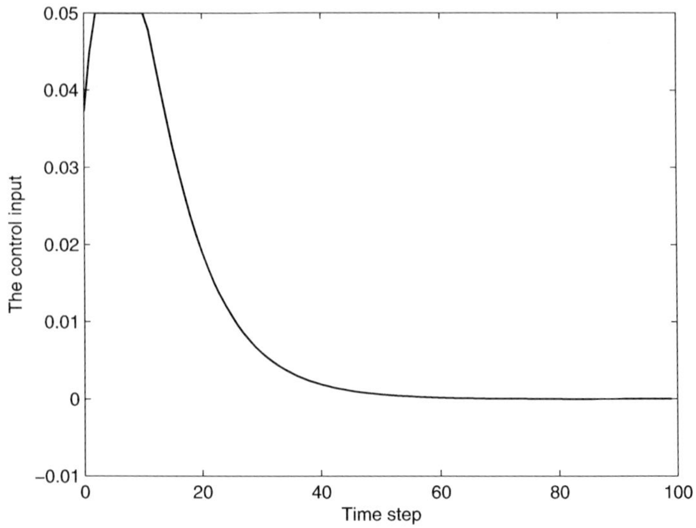

Fig. 7.19 The control variables curves

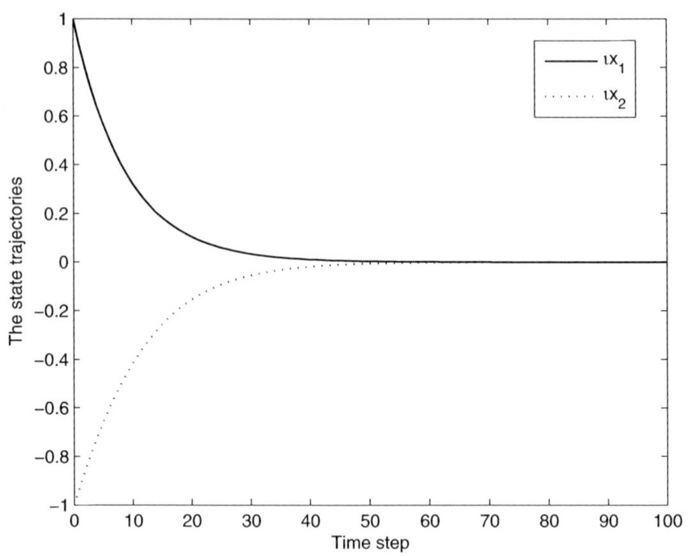

Fig. 7.20 The state variables curves

From the simulation results, we see that the smooth optimal control law is obtained and the actuators saturation phenomenon is avoided by a nonquadratic functional, which also indicates the validity of the iterative ADP algorithm for dealing with constrained nonlinear systems.

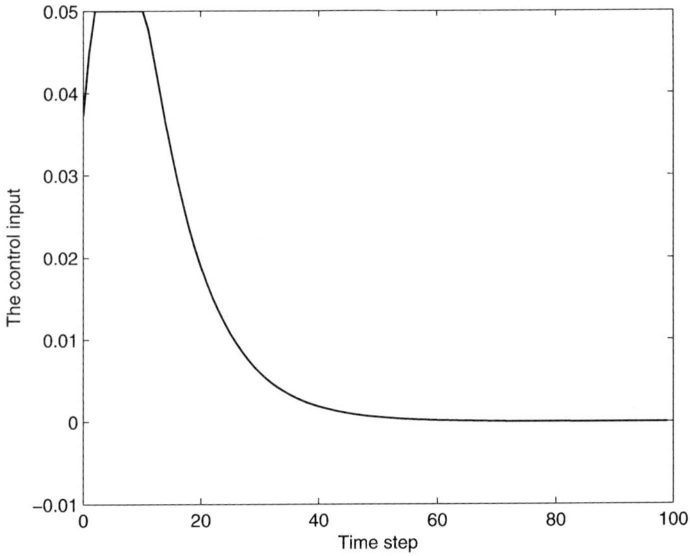

Fig. 7.21 The control variables curves

7.6 Summary

In this chapter, we studied several special optimal feedback control problems based on ADP approach. In Sect. 7.2, the infinite horizon optimal control of affine nonlinear discrete switched systems was investigated by using a two-stage ADP method. In Sect. 7.3, the near-optimal control problem of nonlinear descriptor systems was solved. Then, in Sect. 7.4, a novel near-optimal control design method for a class of nonlinear singularly perturbed systems was developed. In Sect. 7.5, based on the single network GI-DHP algorithm, the near-optimal state-feedback problem of nonlinear constrained discrete-time systems was studied.

References

1. Al-Tamimi A, Lewis F, Abu-Khalaf M (2008) Discrete-time nonlinear HJB solution using approximate dynamic programming: convergence proof. IEEE Trans Syst Man Cybern, Part B, Cybern 38:943–949
2. Beard R (1995) Improving the closed-loop performance of nonlinear systems. PhD dissertation, Rensselaer Polytechnic Institute, Troy, NY
3. Cao N, Zhang HG, Luo YH, Feng DZ, Liu Y (2011) Suboptimal control of a class of nonlinear singularly perturbed systems. Control Theory Appl 28(5):688–692
4. Cao N, Zhang HG, Luo YH, Feng DZ (2012) Infinite horizon optimal control of affine nonlinear discrete switched systems using two-stage approximate dynamic programming. Int J Syst Sci 43(9):1673–1682
5. Lincoln B, Rantzer A (2006) Relaxing dynamic programming. IEEE Trans Autom Control 51:1249–1260

6. Luo YH, Zhang HG, Cao N, Chen B (2009) Near-optimal stabilization for a class of nonlinear systems with control constraint based on single network greedy iterative DHP algorithm. Acta Autom Sin 35(11):1436–1445
7. Luo YH, Liu Z, Yang D (2010) Greedy iterative DHP algorithm-based near-optimal control for a class of nonlinear descriptor systems with actuator saturating. In: Proceedings of the 9th IEEE international conference on cognitive informatics, pp 788–793
8. Lyshevski SE (1998) Nonlinear discrete-time systems: constrained optimization and application of nonquadratic costs. In: Proceedings of the American control conference, Philadelphia, USA, pp 3699–3703
9. Padhi R, Unnikrishnan N, Wang X, Balakrishnan SN (2006) A single network adaptive critic (SNAC) architecture for optimal control synthesis for a class of nonlinear systems. Neural Netw 19(10):1648–1660
10. Rantzer A (2005) On approximate dynamic programming in switching systems. In: Proceeding of the IEEE conference on decision and control and the European control conference, Seville, Spain, pp 1391–1396
11. Seatzu C, Corona D, Giua A, Bempoard A (2006) Optimal control of continuous time switched affine systems. IEEE Trans Autom Control 51:726–741
12. Xu XP, Antsaklis PJ (2000) Optimal control of switched systems: new results and open problems. In: Proceeding of the American control conference, Chicago, Illinois, pp 2683–2687
13. Xu XP, Antsaklis PJ (2003) Results and perspectives on computational methods for optimal control of switched systems. Hybrid systems: computation and control (HSCC). Springer, Berlin, pp 540–555
14. Yang H, Jiang B, Cocquempot V, Zhang HG (2011) Stabilization of switched nonlinear systems with all unstable modes: application to multi-agent systems. IEEE Trans Autom Control 56(9):2230–2235
15. Zhang W, Hu J, Abate A (2009) On the value functions of the discrete-time switched LQR problem. IEEE Trans Autom Control 54:2669–2674
16. Zhang HG, Liu Z, Huang GB (2010) Novel delay-dependent robust stability analysis for switched neutral-type neural network with time-varying delays via SC technique. IEEE Trans Syst Man Cybern, Part B, Cybern 40(6):1480–1491

Chapter 8
Zero-Sum Games for Discrete-Time Systems Based on Model-Free ADP

8.1 Introduction

A large class of complicated practical systems are controlled by more than one controller or decision maker, each using an individual strategy. These controllers often operate in a group with a general cost functional as a game [8, 11]. Zero-sum game theory has been widely applied to decision making and control engineering problems [7]. Recently, much work has been done in the area of dealing with zero-sum games based on ADP methods [2–6], most of which are only for the one-dimensional systems. In the real world, many complicated control systems are described by 2-dimensional (2-D) structures [12, 15], such as thermal processes, image processing, signal filtering, etc. However, the optimal recurrent equation, the so called Hamilton–Jacobi–Isaacs (HJI) equation, is invalid in the 2-D structure. Moreover, the exact models of many 2-D systems cannot be obtained inherently. To solve the aforementioned problems, in Sect. 8.2, we will study the zero-sum games for a class of discrete-time Roesser types 2-D systems based on model-free ADP method. First, we will propose the optimality principle for 2-D systems and obtain the expressions of optimal control pair for the zero-sum games. Then, a data-based iterative ADP algorithm is developed without the requirement of the exact system model.

It is well known that some states of the system may not be measurable in many practical situations. So it is desirable to use available input–output data, rather than the states of system in the design of controller. In Sect. 8.3, we will develop a novel data-based optimal output feedback control scheme via the ADP algorithm, in which neither the exact system model nor the full states of the system are needed.

8.2 Zero-Sum Differential Games for a Class of Discrete-Time 2-D Systems

In this section, we will use the model-free ADP method to obtain the optimal control pair iteratively which makes the performance index function reach the saddle point of the zero-sum games for a class of discrete-time 2-D systems.

H. Zhang et al., *Adaptive Dynamic Programming for Control*,
Communications and Control Engineering, DOI 10.1007/978-1-4471-4757-2_8,

8.2.1 Problem Formulation

Basically, we consider the following discrete-time linear Roesser type *2-D systems*:

$$x^+(k,l) = Ax(k,l) + Bu(k,l) + Cw(k,l) \tag{8.1}$$

$$x^h(0,l) = f(l), x^v(k,0) = g(k) \tag{8.2}$$

with

$$x(k,l) = \begin{bmatrix} x^h(k,l) \\ x^v(k,l) \end{bmatrix}, \ x^+(k,l) = \begin{bmatrix} x^h(k+1,l) \\ x^v(k,l+1) \end{bmatrix}, \tag{8.3}$$

where $x^h(k,l)$ is the horizontal state in $\mathbb{R}^{n_1}$, $x^v(k,l)$ is the vertical state in $\mathbb{R}^{n_2}$, $u(k,l)$ and $w(k,l)$ are the control input in $\mathbb{R}^{m_1}$ and $\mathbb{R}^{m_2}$. Let the system matrices be $A \in \mathbb{R}^{(n_1+n_2)\times(n_1+n_2)}$, $B \in \mathbb{R}^{(n_1+n_2)\times n_1}$, and $C \in \mathbb{R}^{(n_1+n_2)\times m_2}$. It is assumed that all the system matrices are nonsingular, and that the system matrices can be expressed by

$$A = \begin{bmatrix} A_1 & A_2 \\ A_3 & A_4 \end{bmatrix}, \ B = \begin{bmatrix} B_1 \\ B_2 \end{bmatrix}, \ C = \begin{bmatrix} C_1 \\ C_2 \end{bmatrix}. \tag{8.4}$$

The functions $f(l)$ and $g(k)$ are the corresponding boundary conditions along two independent directions.

We define the following notation:

$$\begin{aligned} &(k,l) \le (m,n) && \text{if and only if } k \le m \text{ and } l \le n, \\ &(k,l) = (m,n) && \text{if and only if } k = m \text{ and } l = n, \\ &(k,l) < (m,n) && \text{if and only if } (k,l) \le (m,n) \text{ and} \\ & && (k,l) \ne (m,n). \end{aligned} \tag{8.5}$$

Then, the infinite-time cost functional for the 2-D system (8.1) is given by

$$\begin{aligned} J(x(0,0),u,w) = & \sum_{(0,0)\le(k,l)<(\infty,\infty)}\sum \big(x^{\mathrm{T}}(k,l)Qx(k,l) \\ & + u^{\mathrm{T}}(k,l)Ru(k,l) + w^{\mathrm{T}}(k,l)Sw(k,l)\big), \end{aligned} \tag{8.6}$$

where $Q \ge 0$, $R > 0$, and $S < 0$ have suitable dimensions, and $L(x(k,l),u(k,l)) = x^{\mathrm{T}}(k,l)Qx(k,l) + u^{\mathrm{T}}(k,l)Ru(k,l) + w^{\mathrm{T}}(k,l)Sw(k,l)$ is the utility function. For the above *zero-sum game*, the two control variables u and w are chosen, respectively, by player I and player II, where player I tries to minimize the cost functional $J(x(0,0),u,w)$, while player II attempts to maximize it. The following assumptions are needed; they are in effect in the remaining sections.

Assumption 8.1 The 2-D system (8.1) is controllable under the control variables u and w.

Assumption 8.2 For the boundary conditions of the 2-D system (8.1), the terms $\sum_{k=0}^{\infty} x^{v\mathrm{T}}(k,0)x^v(k,0)$, $\sum_{l=0}^{\infty} x^{h\mathrm{T}}(0,l)x^h(0,l)$ and

$$\sum\sum_{(0,0)\le(k,l)<(\infty,\infty)} x^{v\mathrm{T}}(k,0)x^h(0,l)$$

are all bounded.

Assumption 8.3 There exists a unique *saddle point* of the zero-sum games for the 2-D system (8.1).

There are some important characteristics that must be pointed out. First, for the 1-D control system, the boundary condition is just an initial point of state, while the boundary conditions of 2-D system are two given state trajectories along two different directions. Second, for the zero-sum games of 2-D system under the infinite time horizon, the boundary state trajectories are uncontrollable, and therefore the terms $\sum_{l=0}^{\infty} x^{\mathrm{T}}(k,0)Qx(k,0)$, $\sum_{l=0}^{\infty} x^{\mathrm{T}}(0,l)Qx(0,l)$ and $\sum\sum_{(0,0)\le(k,l)<(\infty,\infty)} x^{\mathrm{T}}(k,0)Qx(0,l)$ may be infinite, which means that the cost functional (8.6) is infinite. Therefore, Assumption 8.2 is necessary. Third, the boundary conditions $f(l)$ and $g(k)$ in (8.2) should be boundary, but not necessary smooth or continuous functions. For example, let

$$f(l) = \begin{cases} c & l \le \mathbb{T}, \\ 0 & l > \mathbb{T}, \end{cases} \tag{8.7}$$

where c is any real constant number and $\mathbb{T}$ is given real number. So Assumption 8.2 is not very strong.

According to Assumption 8.3, the optimal value function can be expressed as

$$\begin{aligned} J^*(x(k,l)) &= \min_u \max_w \sum\sum_{(k,l)\le(i,j)<(\infty,\infty)} \big\{x^{\mathrm{T}}(i,j)Qx(i,j) \\ &\quad + u^{\mathrm{T}}(i,j)Ru(i,j) + w^{\mathrm{T}}(i,j)Sw(i,j)\big\} \\ &= \max_w \min_u \sum\sum_{(k,l)\le(i,j)<(\infty,\infty)} \big\{x^{\mathrm{T}}(i,j)Qx(i,j) \\ &\quad + u^{\mathrm{T}}(i,j)Ru(i,j) + w^{\mathrm{T}}(i,j)Sw(i,j)\big\}. \end{aligned} \tag{8.8}$$

For the zero-sum games for 1-D systems, the optimal value function can be written by a recurrent formulation according to the dynamic programming principle. However, for the zero-sum games for 2-D systems, the dynamic programming principle may not be true. The main difficulty lies in the state of the 2-D system in the next stage coupling with the states of two different directions in the current stage; then the dynamic programming equation of the zero-sum games for 2-D systems does not exist. So in this section, we propose an optimality principle for the 2-D system and obtain the expressions of an optimal control pair for the zero-sum games.

In the following, we will propose the optimality principle for the zero-sum games for 2-D systems, and discuss the properties of the optimal control pair derived by the optimality principle.

Theorem 8.4 (cf. [16]) *If the cost functional is defined as* (8.6), *let* $u(k,l)$ *minimize the cost functional* (8.6) *and* $w(k,l)$ *maximize it subject to the system equation* (8.1); *then there are* $(n+m)$*-dimensional vector sequences* $\lambda(k,l)$ *and* $\lambda^+(k,l)$ *defined as*

$$\lambda^+(k,l)=\begin{bmatrix}\lambda^h(k+1,l)\\ \lambda^v(k,l+1)\end{bmatrix},\quad \lambda(k,l)=\begin{bmatrix}\lambda^h(k,l)\\ \lambda^v(k,l)\end{bmatrix}, \tag{8.9}$$

where $\lambda^h(k,l)\in\mathbb{R}^{n_1}$, $\lambda^v(k,l)\in\mathbb{R}^{n_2}$, *such that for all* $(0,0)\le(k,l)<(\infty,\infty)$ *we have the*

1. *State equation*:

$$x^+(k,l)=\frac{\partial H(k,l)}{\partial\lambda^+(k,l)}, \tag{8.10}$$

2. *Costate equation*:

$$\lambda(k,l)=\frac{\partial H(k,l)}{\partial x(k,l)}, \tag{8.11}$$

3. *Stationarity equation*:

$$\begin{aligned}0&=\frac{\partial H(k,l)}{\partial u(k,l)},\\ 0&=\frac{\partial H(k,l)}{\partial w(k,l)},\end{aligned} \tag{8.12}$$

where $H(k,l)$ *is a Hamilton function defined as*

$$\begin{aligned}H(k,l)&=x^{\mathrm{T}}(k,l)Qx(k,l)+u^{\mathrm{T}}(k,l)Ru(k,l)\\&\quad+w^{\mathrm{T}}(k,l)Sw(k,l)+\lambda^{+\mathrm{T}}(k,l)(Ax(k,l)\\&\quad+Bu(k,l)+Cw(k,l)).\end{aligned} \tag{8.13}$$

Proof By examining the gradients of each of the state equations, i.e., the vector of partial derivatives with respect to all the variables $x(k,l)$, $u(k,l)$, and $w(k,l)$ appearing in (8.6), they are easily seen to be linearly independent. The optimum of the cost functional (8.6) is therefore a regular point of the system (8.1) (see [13], pp. 187). The existence of linear independent $\lambda(k,l)$, $(0,0) \le (k,l) < (\infty,\infty)$ immediately follows from Lagrange multiplier theory (see [13], pp. 187–198).

Let

$$J'(x(k,l),u,w) = \sum_{(k,l)\le(i,j)<(\infty,\infty)}\sum \Big\{x^{\mathrm{T}}(i,j)Qx(i,j) + u^{\mathrm{T}}(i,j)Ru(i,j) + w^{\mathrm{T}}(i,j)Sw(i,j) + \lambda^{+\mathrm{T}}(i,j)(Ax(i,j)+Bu(i,j)+Cw(i,j) - x^{+}(i,j))\Big\}. \tag{8.14}$$

We introduce the Hamilton function of (8.13) and rewrite (8.14) as

$$J'(x(k,l),u,w) = \sum_{(k,l)\le(i,j)<(\infty,\infty)}\sum \big\{H(i,j) - \lambda^{+\mathrm{T}}(i,j)x^{+}(i,j)\big\}. \tag{8.15}$$

The last term in the previous double summation can be expanded as follows:

$$\sum_{(k,l)\le(i,j)<(\infty,\infty)}\sum \lambda^{+\mathrm{T}}(i,j)x^{+}(i,j) = \sum_{(k,l)\le(i,j)<(\infty,\infty)}\sum \Big[\lambda^{\mathrm{T}}(i,j)x(i,j) - \sum_{j=l}^{\infty}\lambda^{h\mathrm{T}}(0,j)x^{h}(0,j) - \sum_{i=k}^{\infty}\lambda^{v\mathrm{T}}(i,0)x^{h}(i,0)\Big]. \tag{8.16}$$

According to the Lagrange multiplier theory, the increment J' due to increments in all $x(k,l)$, $u(k,l)$, $w(k,l)$, and $\lambda(k,l)$ must be zero at a constrained minimum. Hence

$$\mathrm{d}J'(x(k,l),u,w) = \sum_{(k,l)\le(i,j)<(\infty,\infty)}\sum \Big\{\Big[\frac{\partial H(i,j)}{\partial u(i,j)}\Big]\mathrm{d}u(i,j) + \Big[\frac{\partial H(i,j)}{\partial w(i,j)}\Big]\mathrm{d}w(i,j) + \Big[\frac{\partial H(i,j)}{\partial \lambda^{+}(i,j)} - x^{+}(i,j)\Big]\mathrm{d}\lambda^{+}(i,j) + \Big[\frac{\partial H(i,j)}{\partial x(i,j)} - \lambda(i,j)\Big]\mathrm{d}x(i,j)\Big\}. \tag{8.17}$$

Equation (8.17) yields (8.10)–(8.12) by the following remarks:

1. Increments $du(i,j)$, $dw(i,j)$, $dx(i,j)$, and $d\lambda(i,j)$, with the exception of $dx^{h}(0,j)$ and $dx^{v}(i,0)$, are independent arbitrary vectors.

2. $dx^h(0, j) = 0$, $dx^v(i, 0) = 0$, since $x^h(i, j)$ and $x^v(i, j)$ are fixed boundary conditions. □

According to (8.12), the optimal control $u^*(k,l)$ and $w^*(k,l)$ can be expressed as

$$u^*(k,l) = -\frac{1}{2}R^{-1}B^{\mathrm{T}}\lambda^+(k,l), \tag{8.18}$$

and

$$w^*(k,l) = -\frac{1}{2}S^{-1}C^{\mathrm{T}}\lambda^+(k,l). \tag{8.19}$$

Theorem 8.5 (cf. [16]) *For the system* (8.1) *with respect to the cost functional* (8.6), *if the controls* $u(k,l)$ *and* $w(k,l)$ *are expressed as* (8.18) *and* (8.19) *respectively, then the optimal Hamilton function* (8.13) *satisfies a certain Riccati function.*

Proof For the zero-sum games of 2-D linear system, the optimal state feedback control should also be linearly dependent on the system state. As the system function is time-invariant, there exists a matrix P that satisfies

$$\lambda(k,l) = 2Px(k,l). \tag{8.20}$$

Then (8.18) and (8.19) can be rewritten as

$$u^*(k,l) = -B^{\mathrm{T}}P(Ax(k,l) + Bu(k,l) + Cw(k,l)), \tag{8.21}$$

and

$$w^*(k,l) = -C^{\mathrm{T}}P(Ax(k,l) + Bu(k,l) + Cw(k,l)). \tag{8.22}$$

So the optimal state feedback control $u(k,l)$ and $w(k,l)$ can be expressed as

$$\begin{aligned} u^*(k,l) &= -(R + B^{\mathrm{T}}PB - B^{\mathrm{T}}PC(S + C^{\mathrm{T}}PC)^{-1}C^{\mathrm{T}}PB)^{-1} \\ &\quad \times (B^{\mathrm{T}}PA - B^{\mathrm{T}}PC(S + C^{\mathrm{T}}PC)^{-1}C^{\mathrm{T}}PA)x(k,l), \end{aligned} \tag{8.23}$$

and

$$\begin{aligned} w^*(k,l) &= -(S + C^{\mathrm{T}}PC - C^{\mathrm{T}}PB(R + B^{\mathrm{T}}PB)^{-1}B^{\mathrm{T}}PC)^{-1} \\ &\quad \times (C^{\mathrm{T}}PA - C^{\mathrm{T}}PB(R + B^{\mathrm{T}}PB)^{-1}B^{\mathrm{T}}PA)x(k,l). \end{aligned} \tag{8.24}$$

According to (8.11) we have

$$\begin{aligned} 2Px(k,l) &= 2Qx(k,l) + 2A^{\mathrm{T}}Px^+(k,l) \\ &= 2Qx(k,l) + 2A^{\mathrm{T}}P(Ax(k,l) + Bu^*(k,l) \\ &\quad + Cw^*(k,l)). \end{aligned} \tag{8.25}$$

Substituting (8.23) and (8.24) into (8.25), we have the following Riccati function:

$$\begin{aligned}P &= Q + A^{\mathrm{T}}PA - A^{\mathrm{T}}PB(R + B^{\mathrm{T}}PB - B^{\mathrm{T}}PC(S \\ &\quad + C^{\mathrm{T}}PC)^{-1}C^{\mathrm{T}}PB)^{-1}B^{\mathrm{T}}PA + A^{\mathrm{T}}PB(R + B^{\mathrm{T}}PB \\ &\quad - B^{\mathrm{T}}PC(S + C^{\mathrm{T}}PC)^{-1}C^{\mathrm{T}}PB)^{-1}B^{\mathrm{T}}PC(S \\ &\quad + C^{\mathrm{T}}PC)^{-1}C^{\mathrm{T}}PA - A^{\mathrm{T}}PC(S + C^{\mathrm{T}}PC - C^{\mathrm{T}}PB \\ &\quad \times (R + B^{\mathrm{T}}PB)^{-1}B^{\mathrm{T}}PC)^{-1}C^{\mathrm{T}}PA + A^{\mathrm{T}}PC(S \\ &\quad + C^{\mathrm{T}}PC - C^{\mathrm{T}}PB(R + B^{\mathrm{T}}PB)^{-1}B^{\mathrm{T}}PC)^{-1}C^{\mathrm{T}}PB \\ &\quad \times (R + B^{\mathrm{T}}PB)^{-1}B^{\mathrm{T}}PA.\end{aligned} \tag{8.26}$$

The proof is completed. □

As the zero-sum game has a saddle point and is solvable, in order to obtain the unique feedback saddle point in the strictly feedback stabilizing control policy class, the following inequalities should be satisfied (see [8] for details):

$$P > 0, \tag{8.27}$$

$$S + C^{\mathrm{T}}PC < 0, \tag{8.28}$$

and

$$R + B^{\mathrm{T}}PB > 0. \tag{8.29}$$

Theorem 8.6 (cf. [16]) *For the system* (8.1) *with respect to the cost functional* (8.6), *if the optimal control* $u^*(k,l)$ *and* $w^*(k,l)$ *are expressed as* (8.18) *and* (8.19), *respectively, then the optimal value function* $J^*(x(k,l))$ *is a quadratic function dependent on the state* $x(k,l)$.

Proof Substituting (8.18) and (8.19) into the Hamilton function (8.13), we have

$$\begin{aligned}H(k,l) &= x^{\mathrm{T}}(k,l)Qx(k,l) + \frac{1}{4}\lambda^{+\mathrm{T}}(k,l)BR^{-1}B^{\mathrm{T}}\lambda^{+}(k,l) \\ &\quad + \frac{1}{4}\lambda^{+\mathrm{T}}(k,l)CS^{-1}C^{\mathrm{T}}\lambda^{+}(k,l) + \lambda^{+}(k,l)(Ax(k,l) \\ &\quad - \frac{1}{2}BR^{-1}B^{\mathrm{T}}\lambda^{+}(k,l) - \frac{1}{2}CS^{-1}C^{\mathrm{T}}\lambda^{+}(k,l)).\end{aligned} \tag{8.30}$$

Then, according to (8.20), (8.23), and (8.24), we have

$$\begin{aligned}H(k,l) &= x^{\mathrm{T}}(k,l)\big(Q + A^{\mathrm{T}}PA - A^{\mathrm{T}}PB(R + B^{\mathrm{T}}PB \\ &\quad - B^{\mathrm{T}}PC(S + C^{\mathrm{T}}PC)^{-1}C^{\mathrm{T}}PB)^{-1}B^{\mathrm{T}}PA \\ &\quad + A^{\mathrm{T}}PB(R + B^{\mathrm{T}}PB - B^{\mathrm{T}}PC(S + C^{\mathrm{T}}PC)^{-1}\end{aligned}$$

$$
\begin{aligned}
&\times C^{\mathrm{T}}PB)^{-1}B^{\mathrm{T}}PC(S+C^{\mathrm{T}}PC)^{-1}C^{\mathrm{T}}PA \\
&- A^{\mathrm{T}}PC(S+C^{\mathrm{T}}PC-C^{\mathrm{T}}PB(R+B^{\mathrm{T}}PB)^{-1} \\
&\times B^{\mathrm{T}}PC)^{-1}C^{\mathrm{T}}PA + A^{\mathrm{T}}PC(S+C^{\mathrm{T}}PC \\
&- C^{\mathrm{T}}PB(R+B^{\mathrm{T}}PB)^{-1}B^{\mathrm{T}}PC)^{-1}C^{\mathrm{T}}PB \\
&\times (R+B^{\mathrm{T}}PB)^{-1}B^{\mathrm{T}}PA)\big)x(k,l).
\end{aligned}
\tag{8.31}
$$

According to (8.26) and the optimality principle, we immediately have

$$
H(k,l) = x^{\mathrm{T}}(k,l)Px(k,l) = J^*(x(k,l)). \tag{8.32}
$$

So, $V^*(x(k,l))$ is a quadratic function dependent on the state $x(k,l)$.

The proof is completed. □

According to Theorem 8.6, we have the following corollary.

Corollary 8.7 *For the system* (8.1) *with respect to the cost functional* (8.6), *if the control $u(k,l)$ and $w(k,l)$ are expressed as* (8.18) *and* (8.19), *respectively, then the system is stable.*

Proof According to the definition of the cost functional in (8.8), let $(k,l) \to (\infty,\infty)$; we have

$$
\begin{aligned}
J^*(x(\infty,\infty)) &= H(\infty,\infty) \\
&= x^{\mathrm{T}}(\infty,\infty)Qx(\infty,\infty) + u^{*\mathrm{T}}(\infty,\infty) \\
&\quad \times Ru^*(\infty,\infty) + w^{*\mathrm{T}}(\infty,\infty)Sw^*(\infty,\infty).
\end{aligned}
\tag{8.33}
$$

On the other hand, according to (8.13), let $(k,l) \to (\infty,\infty)$; we have

$$
\begin{aligned}
H(\infty,\infty) = {}& x^{\mathrm{T}}(\infty,\infty)Qx(\infty,\infty) + u^{*\mathrm{T}}(\infty,\infty) \\
&\times Ru^*(\infty,\infty) + w^{*\mathrm{T}}(\infty,\infty)Sw^*(\infty,\infty) \\
&+ \lambda^{+\mathrm{T}}(\infty,\infty)(Ax(\infty,\infty) + Bu(\infty,\infty) \\
&+ Cw(\infty,\infty)).
\end{aligned}
\tag{8.34}
$$

According to (8.20), we have

$$
\begin{aligned}
H(\infty,\infty) = {}& x^{\mathrm{T}}(\infty,\infty)Qx(\infty,\infty) + u^{*\mathrm{T}}(\infty,\infty) \\
&\times Ru^*(\infty,\infty) + w^{*\mathrm{T}}(\infty,\infty)Sw^*(\infty,\infty) \\
&+ (Ax(\infty,\infty) + Bu(\infty,\infty) + Cw(\infty,\infty))^{\mathrm{T}} \\
&\times 2P(Ax(\infty,\infty) + Bu(\infty,\infty) + Cw(\infty,\infty)).
\end{aligned}
\tag{8.35}
$$

Then, we have

$$\begin{aligned}&(Ax(\infty,\infty)+Bu(\infty,\infty)+Cw(\infty,\infty))^{\mathrm{T}}\\&\quad\times 2P(Ax(\infty,\infty)+Bu(\infty,\infty)+Cw(\infty,\infty))=0.\end{aligned}\tag{8.36}$$

As the optimal control $u^*(k,l)$ and $w^*(k,l)$ are the state feedback control expressed in (8.23) and (8.24), respectively, and $P>0$, we obtain

$$\lim_{(k,l)\to\infty} x(k,l)=0.\tag{8.37}$$

The proof is completed. □

8.2.2 Data-Based Optimal Control via Iterative ADP Algorithm

In this subsection, based on the optimality principle, we will expand the ADP method to 2-D systems.

As the optimal control u^* and w^* are both linear feedback dependent on the state, let

$$\begin{aligned}u^*(k,l)&=K^*x(k,l),\\w^*(k,l)&=L^*x(k,l).\end{aligned}\tag{8.38}$$

Let

$$H(x(k,l),u(k,l),w(k,l))=[x^{\mathrm{T}}(k,l)\ u^{\mathrm{T}}(k,l)\ w^{\mathrm{T}}(k,l)]H\begin{bmatrix}x(k,l)\\u(k,l)\\w(k,l)\end{bmatrix}.\tag{8.39}$$

Then, according to (8.13), we have

$$\begin{aligned}\begin{bmatrix}H_{xx}&H_{xu}&H_{xw}\\H_{ux}&H_{uu}&H_{uw}\\H_{wx}&H_{wu}&H_{ww}\end{bmatrix}&=\begin{bmatrix}Q&0&0\\0&R&0\\0&0&S\end{bmatrix}+\begin{bmatrix}A&B&C\\K^*A&K^*B&K^*C\\L^*A&L^*B&L^*C\end{bmatrix}^{\mathrm{T}}H\\&\quad\times\begin{bmatrix}A&B&C\\K^*A&K^*B&K^*C\\L^*A&L^*B&L^*C\end{bmatrix}\\&=\begin{bmatrix}A^{\mathrm{T}}PA+Q&A^{\mathrm{T}}PB&A^{\mathrm{T}}PC\\B^{\mathrm{T}}PA&B^{\mathrm{T}}PB+R&B^{\mathrm{T}}PC\\C^{\mathrm{T}}PA&C^{\mathrm{T}}PB&C^{\mathrm{T}}PC+S\end{bmatrix},\end{aligned}\tag{8.40}$$

where $P=[I\ K^{*\mathrm{T}}\ L^{*\mathrm{T}}]H\begin{bmatrix}I\\K^*\\L^*\end{bmatrix}$.

According to (8.12) and (8.40), we have

$$0 = \frac{\partial H(k,l)}{\partial u(k,l)},$$
$$0 = 2H_{ux}x(k,l) + 2H_{uw}w(k,l) + 2H_{uu}u(k,l), \tag{8.41}$$

and

$$0 = \frac{\partial H(k,l)}{\partial w(k,l)},$$
$$0 = 2H_{wx}x(k,l) + 2H_{wu}u(k,l) + 2H_{ww}w(k,l). \tag{8.42}$$

Substituting (8.42) into (8.41), we get

$$u^*(k,l) = (H_{uu} - H_{uw}H_{ww}^{-1}H_{wu})^{-1}(H_{uw}H_{ww}^{-1}H_{wx} - H_{ux})x(k,l). \tag{8.43}$$

Taking (8.43) into (8.42), we get

$$w^*(k,l) = (H_{ww} - H_{wu}H_{uu}^{-1}H_{uw})^{-1}(H_{wu}H_{uu}^{-1}H_{ux} - H_{wx})x(k,l). \tag{8.44}$$

So we have

$$K^* = (H_{uu} - H_{uw}H_{ww}^{-1}H_{wu})^{-1}(H_{uw}H_{ww}^{-1}H_{wx} - H_{ux}), \tag{8.45}$$

and

$$L^* = (H_{ww} - H_{wu}H_{uu}^{-1}H_{uw})^{-1}(H_{wu}H_{uu}^{-1}H_{ux} - H_{wx}). \tag{8.46}$$

Although we have obtained the optimal control pair expressed in (8.45) and (8.46) with the information of the matrix H, we see that the matrix H in (8.40) is also unsolvable directly. Therefore, an iterative ADP algorithm is developed and the convergence property is also discussed.

8.2.2.1 The Derivation of Data-Based Iterative ADP Algorithm

We start with an initial Hamilton function $H_0(k,l) = 0$ which is not necessarily optimal, and then we obtain the function $H_i(k,l)$ for solving the following equation with the iteration performance $i \geq 0$:

$$\begin{aligned}
&\begin{bmatrix} x^{\mathrm{T}}(k,l) & u^{\mathrm{T}}(k,l) & w^{\mathrm{T}}(k,l) \end{bmatrix} H_{i+1} \begin{bmatrix} x(k,l) \\ u(k,l) \\ w(k,l) \end{bmatrix} \\
&\quad = \min_u \max_w \left\{ \begin{bmatrix} x^{\mathrm{T}(k,l)} & u^{\mathrm{T}}(k,l) & w^{\mathrm{T}}(k,l) \end{bmatrix} \begin{bmatrix} Q & 0 & 0 \\ 0 & R & 0 \\ 0 & 0 & S \end{bmatrix} \begin{bmatrix} x(k,l) \\ u(k,l) \\ w(k,l) \end{bmatrix} \right.
\end{aligned}$$

$$
\begin{aligned}
&+ \begin{bmatrix} x^{\mathrm{T}}(k,l) & u^{\mathrm{T}}(k,l) & w^{\mathrm{T}}(k,l) \end{bmatrix} \begin{bmatrix} A & B & C \\ KA & KB & KC \\ LA & LB & LC \end{bmatrix}^{\mathrm{T}} H_i \\
&\times \begin{bmatrix} A & B & C \\ KA & KB & KC \\ LA & L^*B & LC \end{bmatrix} \begin{bmatrix} x(k,l) \\ u(k,l) \\ w(k,l) \end{bmatrix} \Bigg\} \\
&= \begin{bmatrix} x^{\mathrm{T}}(k,l) & u_i^{\mathrm{T}}(k,l) & w_i^{\mathrm{T}}(k,l) \end{bmatrix} \begin{bmatrix} Q & 0 & 0 \\ 0 & R & 0 \\ 0 & 0 & S \end{bmatrix} \begin{bmatrix} x(k,l) \\ u_i(k,l) \\ w_i(k,l) \end{bmatrix} \\
&+ \begin{bmatrix} x^{\mathrm{T}}(k,l) \, u_i^{\mathrm{T}}(k,l) \, w_i^{\mathrm{T}}(k,l) \end{bmatrix} \begin{bmatrix} A & B & C \\ K_iA & K_iB & K_iC \\ L_iA & L_iB & L_iC \end{bmatrix}^{\mathrm{T}} H_i \\
&\times \begin{bmatrix} A & B & C \\ K_iA & K_iB & K_iC \\ L_iA & L_iB & L_iC \end{bmatrix} \begin{bmatrix} x(k,l) \\ u_i(k,l) \\ w_i(k,l) \end{bmatrix}.
\end{aligned} \tag{8.47}
$$

Then, according to (8.43) and (8.44), the iterative control laws can be expressed as

$$
K_i = (H_{uu}^{[i]} - H_{uw}^{[i]}(H_{ww}^{[i]})^{-1}H_{wu}^{[i]})^{-1}(H_{uw}^{[i]}(H_{ww}^{[i]})^{-1}H_{wx}^{[i]} - H_{ux}^{[i]}), \tag{8.48}
$$

and

$$
L_i = (H_{ww}^{[i]} - H_{wu}^{[i]}(H_{uu}^{[i]})^{-1}H_{uw}^{[i]})^{-1}(H_{wu}^{[i]}(H_{uu}^{[i]})^{-1}H_{ux}^{[i]} - H_{wx}^{[i]}). \tag{8.49}
$$

Therefore, the iterative controls can be given as

$$
u_i(k,l) = K_i x(k,l), \tag{8.50}
$$

and

$$
w_i(k,l) = L_i x(k,l). \tag{8.51}
$$

As we see, the iterative control law K_i and L_i can be updated by the H matrix without the system information. While the iteration of P_i changes into the iteration of H_i, the property of the iterative H_i should be discussed. So in the following part, the convergence and optimal properties are established. Also, we will show that the iteration of P_i is the same as the iteration of H_i.

8.2.2.2 Properties of Data-Based Iterative ADP Algorithm

First, we give the following lemmas, which are necessary for the proof.

Lemma 8.8 *The matrices H_{i+1}, K_i and L_{i+1} can be expressed as*

$$H_{i+1}=\begin{bmatrix} A^{\mathrm{T}}P_iA+Q & A^{\mathrm{T}}P_iB & A^{\mathrm{T}}P_iC \\ B^{\mathrm{T}}P_iA & B^{\mathrm{T}}P_iB+R & B^{\mathrm{T}}P_iC \\ C^{\mathrm{T}}P_iA & C^{\mathrm{T}}P_iB & C^{\mathrm{T}}P_iC+S \end{bmatrix}, \tag{8.52}$$

$$\begin{aligned} K_{i+1} &= -(R+B^{\mathrm{T}}P_iB-B^{\mathrm{T}}P_iC(S+C^{\mathrm{T}}P_iC)^{-1}C^{\mathrm{T}}P_iB)^{-1} \\ &\quad \times (B^{\mathrm{T}}P_iA-B^{\mathrm{T}}P_iC(S+C^{\mathrm{T}}P_iC)^{-1}C^{\mathrm{T}}P_iA), \end{aligned} \tag{8.53}$$

and

$$\begin{aligned} L_{i+1} &= -(S+C^{\mathrm{T}}P_iC-C^{\mathrm{T}}P_iB(R+B^{\mathrm{T}}P_iB)^{-1}B^{\mathrm{T}}P_iC)^{-1} \\ &\quad \times (C^{\mathrm{T}}P_iA-C^{\mathrm{T}}P_iB(R+B^{\mathrm{T}}P_iB)^{-1}B^{\mathrm{T}}P_iA), \end{aligned} \tag{8.54}$$

where P_i is given as

$$P_i=\begin{bmatrix} I & K_i^{\mathrm{T}} & L_i^{\mathrm{T}} \end{bmatrix} H_i \begin{bmatrix} I \\ K_i \\ L_i \end{bmatrix}. \tag{8.55}$$

Proof According to (8.47), we have

$$\begin{aligned} H_{i+1} &= \begin{bmatrix} Q & 0 & 0 \\ 0 & R & 0 \\ 0 & 0 & S \end{bmatrix} + \begin{bmatrix} A & B & C \\ K_iA & K_iB & K_iC \\ L_iA & L_iB & L_iC \end{bmatrix}^{\mathrm{T}} H_i \begin{bmatrix} A & B & C \\ K_iA & K_iB & K_iC \\ L_iA & L_iB & L_iC \end{bmatrix} \\ &= \begin{bmatrix} Q & 0 & 0 \\ 0 & R & 0 \\ 0 & 0 & S \end{bmatrix} + \begin{bmatrix} A^{\mathrm{T}} \\ B^{\mathrm{T}} \\ C^{\mathrm{T}} \end{bmatrix} \begin{bmatrix} I & K_i^{\mathrm{T}} & L_i^{\mathrm{T}} \end{bmatrix} H_i \begin{bmatrix} I \\ K_i^{\mathrm{T}} \\ L_i^{\mathrm{T}} \end{bmatrix} \begin{bmatrix} A & B & C \end{bmatrix}. \end{aligned} \tag{8.56}$$

Substituting (8.55) into (8.56), it follows that

$$H_{i+1}=\begin{bmatrix} A^{\mathrm{T}}P_iA+Q & A^{\mathrm{T}}P_iB & A^{\mathrm{T}}P_iC \\ B^{\mathrm{T}}P_iA & B^{\mathrm{T}}P_iB+R & B^{\mathrm{T}}P_iC \\ C^{\mathrm{T}}P_iA & C^{\mathrm{T}}P_iB & C^{\mathrm{T}}P_iC+S \end{bmatrix}. \tag{8.57}$$

According to (8.23) and (8.45), we have

$$\begin{aligned} K_i &= -(R+B^{\mathrm{T}}P_iB-B^{\mathrm{T}}P_iC(S+C^{\mathrm{T}}P_iC)^{-1}C^{\mathrm{T}}P_iB)^{-1} \\ &\quad \times (B^{\mathrm{T}}P_iA-B^{\mathrm{T}}P_iC(S+C^{\mathrm{T}}P_iC)^{-1}C^{\mathrm{T}}P_iA) \\ &= (H_{uu}^{[i]}-H_{uw}^{[i]}(H_{ww}^{[i]})^{-1}H_{wu}^{[i]})^{-1}(H_{uw}^{[i]}(H_{ww}^{[i]})^{-1}H_{wx}^{[i]}-H_{ux}^{[i]}). \end{aligned} \tag{8.58}$$

According to (8.24) and (8.46), we have

$$\begin{aligned} L_i &= -(S + C^{\mathrm{T}} P_i C - C^{\mathrm{T}} P_i B(R + B^{\mathrm{T}} P_i B)^{-1} B^{\mathrm{T}} P_i C)^{-1} \\ &\quad \times (C^{\mathrm{T}} P_i A - C^{\mathrm{T}} P_i B(R + B^{\mathrm{T}} P_i B)^{-1} B^{\mathrm{T}} P_i A) \\ &= (H_{ww}^{[i]} - H_{wu}^{[i]}(H_{uu}^{[i]})^{-1} H_{uw}^{[i]})^{-1}(H_{wu}^{[i]}(H_{uu}^{[i]})^{-1} H_{ux}^{[i]} - H_{wx}^{[i]}). \end{aligned} \tag{8.59}$$

The proof is completed. □

Lemma 8.9 *Iterating on H_i is similar to iterating on P_i by*

$$\begin{aligned} P_{i+1} &= Q + A^{\mathrm{T}} P_i A - A^{\mathrm{T}} P_i B(R + B^{\mathrm{T}} P_i B - B^{\mathrm{T}} P_i C(S \\ &\quad + C^{\mathrm{T}} P_i C)^{-1} C^{\mathrm{T}} P_i B)^{-1} B^{\mathrm{T}} P_i A + A^{\mathrm{T}} P_i B(R \\ &\quad + B^{\mathrm{T}} P_i B - B^{\mathrm{T}} P_i C(S + C^{\mathrm{T}} P_i C)^{-1} C^{\mathrm{T}} P_i B)^{-1} B^{\mathrm{T}} \\ &\quad \times P_i C(S + C^{\mathrm{T}} P_i C)^{-1} C^{\mathrm{T}} P_i A - A^{\mathrm{T}} P_i C(S + C^{\mathrm{T}} P_i C \\ &\quad - C^{\mathrm{T}} P_i B(R + B^{\mathrm{T}} P_i B)^{-1} B^{\mathrm{T}} P_i C)^{-1} C^{\mathrm{T}} P_i A \\ &\quad + A^{\mathrm{T}} P_i C(S + C^{\mathrm{T}} P_i C - C^{\mathrm{T}} P_i B(R + B^{\mathrm{T}} P_i B)^{-1} B^{\mathrm{T}} \\ &\quad \times P_i C)^{-1} C^{\mathrm{T}} P_i B(R + B^{\mathrm{T}} P_i B)^{-1} B^{\mathrm{T}} P_i A, \end{aligned} \tag{8.60}$$

where P_i is defined in (8.55).

Proof From (8.55), we have

$$P_{i+1} = \begin{bmatrix} I & K_{i+1}^{\mathrm{T}} & L_{i+1}^{\mathrm{T}} \end{bmatrix} H_{i+1} \begin{bmatrix} I \\ K_{i+1} \\ L_{i+1} \end{bmatrix}. \tag{8.61}$$

Taking (8.52), we obtain

$$\begin{aligned} P_{i+1} &= \begin{bmatrix} I & K_{i+1}^{\mathrm{T}} & L_{i+1}^{\mathrm{T}} \end{bmatrix} \\ &\quad \times \begin{bmatrix} A^{\mathrm{T}} P_i A + Q & A^{\mathrm{T}} P_i B & A^{\mathrm{T}} P_i C \\ B^{\mathrm{T}} P_i A & B^{\mathrm{T}} P_i B + R & B^{\mathrm{T}} P_i C \\ C^{\mathrm{T}} P_i A & C^{\mathrm{T}} P_i B & C^{\mathrm{T}} T P_i C + S \end{bmatrix} \begin{bmatrix} I \\ K_{i+1} \\ L_{i+1} \end{bmatrix}. \end{aligned} \tag{8.62}$$

Substituting (8.53) and (8.54) into (8.62), we have

$$\begin{aligned} P_{i+1} &= Q + A^{\mathrm{T}} P_i A - A^{\mathrm{T}} P_i B(R + B^{\mathrm{T}} P_i B - B^{\mathrm{T}} P_i C(S \\ &\quad + C^{\mathrm{T}} P_i C)^{-1} C^{\mathrm{T}} P_i B)^{-1} B^{\mathrm{T}} P_i A + A^{\mathrm{T}} P_i B(R \\ &\quad + B^{\mathrm{T}} P_i B - B^{\mathrm{T}} P_i C(S + C^{\mathrm{T}} P_i C)^{-1} C^{\mathrm{T}} P_i B)^{-1} B^{\mathrm{T}} \\ &\quad \times P_i C(S + C^{\mathrm{T}} P_i C)^{-1} C^{\mathrm{T}} P_i A - A^{\mathrm{T}} P_i C(S \end{aligned}$$

$$
\begin{aligned}
&+C^{\mathrm{T}}P_iC-C^{\mathrm{T}}P_iB(R+B^{\mathrm{T}}P_iB)^{-1}B^{\mathrm{T}}P_iC)^{-1}C^{\mathrm{T}}\\
&\times P_iA+A^{\mathrm{T}}P_iC(S+C^{\mathrm{T}}P_iC-C^{\mathrm{T}}P_iB(R+B^{\mathrm{T}}P_i\\
&\times B)^{-1}B^{\mathrm{T}}P_iC)^{-1}C^{\mathrm{T}}P_iB(R+B^{T}P_iB)^{-1}B^{\mathrm{T}}P_iA. \qquad (8.63)
\end{aligned}
$$

The proof is completed. □

From Lemma 8.9, we have

$$
x^{\mathrm{T}}(k,l)P_ix(k,l)=\left[x^{\mathrm{T}}(k,l)\; u_i^{\mathrm{T}}(k,l)\; w_i^{\mathrm{T}}(k,l)\right]H_i\begin{bmatrix}x(k,l)\\ u_i(k,l)\\ w_i(k,l)\end{bmatrix}. \qquad (8.64)
$$

Then (8.47) can be expressed as

$$
\begin{aligned}
H_{i+1}(k,l)&=\left[x^{\mathrm{T}}(k,l)\; u_i^{\mathrm{T}}(k,l)\; w_i^{\mathrm{T}}(k,l)\right]\begin{bmatrix}Q&0&0\\0&R&0\\0&0&S\end{bmatrix}\begin{bmatrix}x(k,l)\\ u_i(k,l)\\ w_i(k,l)\end{bmatrix}+H_i^{+}(k,l)\\
&=\left[x^{\mathrm{T}}(k,l)\; u_i^{\mathrm{T}}(k,l)\; w_i^{\mathrm{T}}(k,l)\right]\begin{bmatrix}Q&0&0\\0&R&0\\0&0&S\end{bmatrix}\begin{bmatrix}x(k,l)\\ u_i(k,l)\\ w_i(k,l)\end{bmatrix}\\
&\quad+x^{+\mathrm{T}}(k,l)P_ixx^{+}(k,l), \qquad (8.65)
\end{aligned}
$$

where

$$
\begin{aligned}
H_i^{+}(k,l)&=\left[x^{\mathrm{T}}(k,l)\; u_i^{\mathrm{T}}(k,l)\; w_i^{\mathrm{T}}(k,l)\right]\begin{bmatrix}A&B&C\\K_iA&K_iB&K_iC\\L_iA&L_iB&L_iC\end{bmatrix}^{\mathrm{T}}H_i\\
&\quad\times\begin{bmatrix}A&B&C\\K_iA&K_iB&K_iC\\L_iA&L_iB&L_iC\end{bmatrix}\begin{bmatrix}x(k,l)\\ u_i(k,l)\\ w_i(k,l)\end{bmatrix}, \qquad (8.66)
\end{aligned}
$$

and

$$
H_{i+1}(k,l)=\left[x^{\mathrm{T}}(k,l)\; u_i^{\mathrm{T}}(k,l)\; w_i^{\mathrm{T}}(k,l)\right]H_{i+1}\begin{bmatrix}x(k,l)\\ u_i(k,l)\\ w_i(k,l)\end{bmatrix}. \qquad (8.67)
$$

Theorem 8.10 (cf. [16]) *For the 2-D system* (8.1) *with respect to the cost functional* (8.6), *if the saddle point exists under the state feedback control* $u(k,l)$ *and* $w(k,l)$, *respectively, then the iteration on* (8.47) *will converge to the optimal value function.*

Proof In [9], it is shown that the iterating algebraic Riccati equation (8.63) is convergent, for $i\to\infty$ with $P_0=0$. So let $\lim_{i\to\infty}P_i=P^*$. Then, for $i\to\infty$, we have

$$u(k,l) = -(R + B^{\mathrm{T}}P^*B - B^{\mathrm{T}}P^*C(S + C^{\mathrm{T}}P^*C)^{-1}C^{\mathrm{T}}P^*B)^{-1}$$
$$\times (B^{\mathrm{T}}P^*A - B^{\mathrm{T}}P^*C(S + C^{\mathrm{T}}P^*C)^{-1}C^{\mathrm{T}}P^*A)x(k,l), \tag{8.68}$$

and

$$w(k,l) = -(S + C^{\mathrm{T}}P^*C - C^{\mathrm{T}}P^*B(R + B^{\mathrm{T}}P^*B)^{-1}B^{\mathrm{T}}P^*C)^{-1}$$
$$\times (C^{\mathrm{T}}P^*A - C^{\mathrm{T}}P^*B(R + B^{\mathrm{T}}P^*B)^{-1}B^{\mathrm{T}}P^*A)x(k,l), \tag{8.69}$$

where P^* satisfies the algebraic Riccati equation (8.26).

According to Theorem 8.4, the controls $u(k,l)$ and $w(k,l)$ in (8.68) and (8.69) are both optimal. So we have

$$J^*(x(k,l)) = x^{\mathrm{T}}(k,l)P^*x(k,l). \tag{8.70}$$

On the other hand, according to Lemma 8.8, we have

$$\begin{aligned} \lim_{i\to\infty} H_{i+1} &= \lim_{i\to\infty} \begin{bmatrix} A^{\mathrm{T}}P_iA + Q & A^{\mathrm{T}}P_iB & A^{\mathrm{T}}P_iC \\ B^{\mathrm{T}}P_iA & B^{\mathrm{T}}P_iB + R & B^{\mathrm{T}}P_iC \\ C^{\mathrm{T}}P_iA & C^{\mathrm{T}}P_iB & C^{\mathrm{T}}P_iC + S \end{bmatrix} \\ &= \begin{bmatrix} A^{\mathrm{T}}P^*A + Q & A^{\mathrm{T}}P^*B & A^{\mathrm{T}}P^*C \\ B^{\mathrm{T}}P^*A & B^{\mathrm{T}}P^*B + R & B^{\mathrm{T}}P^*C \\ C^{\mathrm{T}}P^*A & C^{\mathrm{T}}P^*B & C^{\mathrm{T}}P^*C + S \end{bmatrix}. \end{aligned} \tag{8.71}$$

So we obtain

$$H_i \to \begin{bmatrix} A^{\mathrm{T}}P^*A + Q & A^{\mathrm{T}}P^*B & A^{\mathrm{T}}P^*C \\ B^{\mathrm{T}}P^*A & B^{\mathrm{T}}P^*B + R & B^{\mathrm{T}}P^*C \\ C^{\mathrm{T}}P^*A & C^{\mathrm{T}}P^*B & C^{\mathrm{T}}P^*C + S \end{bmatrix} \tag{8.72}$$

as $i \to \infty$.

The proof is completed. □

In the iterative ADP algorithm, the Hamilton function $H_i^+(k,l)$ is generally difficult to obtain. Therefor, a parameter structure is necessary to approximate the actual $H_i^+(k,l)$. Next, a neural network, called a critic network, is adopted to approximate $H_i^+(k,l)$. Similarly, we adopt two neural networks (called action networks) to approximate the controls $u(k,l)$ and $w(k,l)$, respectively. Let the output of the action networks be expressed by

$$\hat{u}_i(k,l) = K_i x(k,l), \tag{8.73}$$

and

$$\hat{w}_i(k,l) = L_i x(k,l). \tag{8.74}$$

The output of critic network is expressed by

$$z_i^{\mathrm{T}}(k,l)H_i z_i(k,l) = h_i^{\mathrm{T}}\bar{z}_i, \tag{8.75}$$

where $z_i(k,l) = [x^{\mathrm{T}}(k,l)\ u_i^{\mathrm{T}}(k,l)\ w_i^{\mathrm{T}}(k,l)]^{\mathrm{T}}$, $z_i(k,l) \in \mathbb{R}^{n_1+n_2+m_1+m_2=q}$, $\bar{z}_i = (z_1^2, z_1 z_2, \dots, z_1 z_q, z_2^2, z_2 z_3, \dots, z_2 z_q, \dots, z_{q-1} z_q, z_q^2)$ is the Kronecker product quadratic polynomial basis vector. Furthermore $h = bv(H)$ with $v(\cdot)$ a vector function that acts on $q \times q$ matrix, and this gives a $q(q+1)/2 \times 1$ column vector.

To solve $H_{i+1}(k,l)$, the right-hand side of (8.65) can be written as

$$\begin{aligned} d(z_i(k,l), H_i) = {}& \left[x^{\mathrm{T}}(k,l)\ u_i^{\mathrm{T}}(k,l)\ w_i^{\mathrm{T}}(k,l)\right] \\ & \times \begin{bmatrix} Q & 0 & 0 \\ 0 & R & 0 \\ 0 & 0 & S \end{bmatrix} \begin{bmatrix} x(k,l) \\ u_i(k,l) \\ w_i(k,l) \end{bmatrix} + H_i^{+}(k,l), \end{aligned} \tag{8.76}$$

which can be regarded as the desired target function satisfying

$$h_{i+1}^{\mathrm{T}}\bar{z}_i(k,l) = d(z_i(k,l), H_i). \tag{8.77}$$

So we obtain

$$h_{i+1} = (\bar{z}_i(k,l)\bar{z}_i^{\mathrm{T}}(k,l))^{-1}\bar{z}_i(k,l)d(z_i(k,l), H_i). \tag{8.78}$$

Remark 8.11 In (8.78), we see that the matrix $\bar{z}_i(k,l)\bar{z}_i^{\mathrm{T}}(k,l)$ is generally invertible. To overcome this problem, two methods are developed. First, we can compute $(b\bar{z}_i(k,l)\bar{z}_i^{T}(k,l))^{-1}$ by the Moore–Penrose pseudoinverse technique, where $\bar{z}_i(k,l)\bar{z}_i^{\mathrm{T}}(k,l) \neq 0$, for $\forall k, l$. Second, we can use the least-square technique to obtain the inverse of the matrix $\bar{z}_i(k,l)\bar{z}_i^{\mathrm{T}}(k,l)$. We adopt the second method here.

To implement the least-square method, white noise is added into the controls (8.50) and (8.51), respectively. Then we have

$$\tilde{u}_i(k,l) = K_i x(k,l) + \xi_1, \tag{8.79}$$

and

$$\tilde{w}_i(k,l) = L_i x(k,l) + \xi_2, \tag{8.80}$$

where $\xi_1(0,\sigma_1^2)$ and $\xi_2(0,\sigma_2^2)$ are both zero-mean white noises with variances σ_1^2 and σ_2^2, respectively. So $z_i(k,l)$ in (8.75) can be written as

$$\tilde{z}_{\mathrm{i}}(k,l) = \begin{bmatrix} x(k,l) \\ \tilde{u}_i(k,l) \\ \tilde{w}_i(k,l) \end{bmatrix} = \begin{bmatrix} x(k,l) \\ u_i(k,l) + \xi_1 \\ w_i(k,l) + \xi_2 \end{bmatrix} = \begin{bmatrix} x(k,l) \\ u_{\mathrm{i}}(k,l) \\ w_{\mathrm{i}}(k,l) \end{bmatrix} + \begin{bmatrix} 0 \\ \xi_1 \\ \xi_2 \end{bmatrix}. \tag{8.81}$$

Evaluating h_{i+1} at the N points $p_1, p_2, \dots, p_N$, we have

$$h_{i+1} = (Z_N Z_N^{\mathrm{T}})^{-1} Z_N \hat{Y}_N, \tag{8.82}$$

where

$$Z_N = \left[\bar{z}_N(p_1)\ \bar{z}(p_2)\ \ldots\ \bar{z}(p_N)\right], \tag{8.83}$$

and

$$\hat{Y}_N = \left[d(z_N(p_1), H_i)\ d(z(p_2), H_i)\ \ldots\ d(z(p_N), H_i)]^{\mathrm{T}}\right]. \tag{8.84}$$

Then, we obtain

$$H_{i+1} = \bar{g}(h_{i+1}) \tag{8.85}$$

through the Kronecker method, and the feedback control law K_{i+1} and L_{i+1} can be obtained according to (8.48) and (8.49), respectively. According to the condition of the least-square solution, the number of sampling points N should satisfy the inequality

$$N \geq \frac{1}{2}\left(2q \times (2q+1)\right). \tag{8.86}$$

The least-square method in (8.82) can be solved in real time by collecting enough data points generated from $d(z_i(k,l), H_i)$ in (8.76). What we are required to know is the state and control data $x(k,l), u_i(k,l), w_i(k,l)$, and $H_i^+(k,l)$. Therefore, in the present iterative ADP method, the model of the system is not required to update the critic and the action networks.

Given the above preparation, now the present data-based iterative ADP algorithm is summarized as follows:

1. Give the boundary condition $x^h(0,l) = f(l)$ and $x^v(k,0) = g(k)$. Let $P_0 = 0$, $K_0 = 0$ and $L_0 = 0$. Give the computation accuracy ε.
2. According to the N sampling points, compute Z_N and $\hat{Y}_N$ based on (8.83) and (8.84).
3. Compute h_i according to (8.82) and compute H_i according to (8.85) through the Kronecker method.
4. Compute the feedback control laws by

$$K_{i+1} = (H_{uu}^{[i]} - H_{uw}^{[i]}(H_{ww}^{[i]})^{-1}H_{wu}^{[i]})^{-1}(H_{uw}^{[i]}(H_{ww}^{[i]})^{-1}H_{wx}^{[i]} - H_{ux}^{[i]}), \tag{8.87}$$

and

$$L_{i+1} = (H_{ww}^{[i]} - H_{wu}^{[i]}(H_{uu}^{[i]})^{-1}H_{uw}^{[i]})^{-1}(H_{wu}^{[i]}(H_{uu}^{[i]})^{-1}H_{ux}^{[i]} - H_{wx}^{[i]}). \tag{8.88}$$

5. If

$$\|h_{i+1} - h_i\| \leqslant \varepsilon, \tag{8.89}$$

go to Step 7; otherwise, go to Step 6.
6. Set $i = i+1$, go to Step 2.
7. Stop.

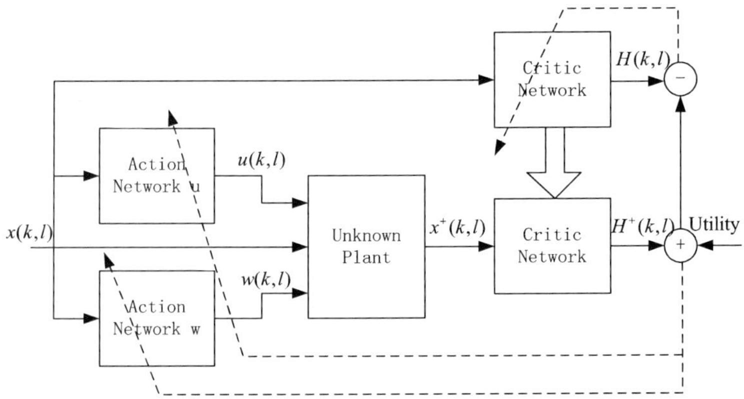

Fig. 8.1 The structure diagram of the algorithm

8.2.2.3 Neural Network Implementation

In the following, neural networks are constructed to implement the iterative ADP algorithm. There are several ADP structures that can be chosen. As the HDP structure is basic and convenient to realize, we will use HDP to implement the iterative ADP algorithm.

Assume that the number of hidden layer neurons is denoted by l, the weight matrix between the input layer and hidden layer is denoted by V, and the weight matrix between the hidden layer and output layer is denoted by W. Then, the output of the three-layer NN is represented by

$$\hat{F}(X, V, W) = W^{\mathrm{T}}\sigma(V^{\mathrm{T}}X), \tag{8.90}$$

where we have the activation function $\sigma(V^{\mathrm{T}}X) \in \mathbb{R}^l, [\sigma(z)]_i = (e^{z_i} - e^{-z_i})/(e^{z_i} + e^{-z_i}), i = 1, \dots, l$.

The NN estimation error can be expressed by

$$F(X) = F(X, V^*, W^*) + \varepsilon(X), \tag{8.91}$$

where V^*, and W^* are the ideal weight parameters, $\varepsilon(X)$ is the reconstruction error.

Here, there are three neural networks, which are critic network and action network u and action network w. All the neural networks are chosen as three-layer feedforward networks. The whole structure diagram is shown in Fig. 8.1. The utility term in the figure denotes $x^{\mathrm{T}}(k,l)Qx(k,l) + u^{\mathrm{T}}(k,l)Ru(k,l) + w^{\mathrm{T}}(k,l)Sw(k,l)$.

8.2.2.4 Critic Network

The critic network is used to approximate the Hamilton function $H(k,l)$. The output of the critic network is denoted

$$\hat{H}_i(k,l) = W_{ci}^{\mathrm{T}}\sigma(V_{ci}^{\mathrm{T}}x(k,l)). \tag{8.92}$$

The target function can be written as

$$H_{i+1}(k,l) = \left[x^{\mathrm{T}}(k,l)\; u_i^{\mathrm{T}}(k,l)\; w_i^{\mathrm{T}}(k,l)\right] \begin{bmatrix} Q & 0 & 0 \\ 0 & R & 0 \\ 0 & 0 & S \end{bmatrix} \begin{bmatrix} x(k,l) \\ u_i(k,l) \\ w_i(k,l) \end{bmatrix} + H_i^{+}(k,l). \tag{8.93}$$

Then, we define the error function for the critic network as

$$e_{ci}(k,l) = \hat{H}_{i+1}(k,l) - H_{i+1}(k,l). \tag{8.94}$$

The objective function to be minimized in the critic network is

$$E_{ci}(k,l) = \frac{1}{2} e_{ci}^{\mathrm{T}}(k,l) e_{ci}(k,l). \tag{8.95}$$

Therefore, the gradient-based weight updating rule for the critic network is given by

$$w_{c(i+1)}(k,l) = w_{ci}(k,l) + \Delta w_{ci}(k,l), \tag{8.96}$$

$$\Delta w_{ci}(k,l) = \alpha_c \left[-\frac{\partial E_{ci}(k,l)}{\partial w_{ci}(k,l)} \right], \tag{8.97}$$

$$\frac{\partial E_{ci}(k,l)}{\partial w_{ci}(k,l)} = \frac{\partial E_{ci}(k,l)}{\partial \hat{H}_i(k,l)} \frac{\partial \hat{H}_i(k,l)}{\partial w_{ci}(k,l)}, \tag{8.98}$$

where $\alpha_c > 0$ is the learning rate of critic network and $w_c(k,l)$ is the weight vector in the critic network.

8.2.2.5 Action Networks

Action networks are used to approximate the iterative optimal controls. There are two action networks, which are used to approximate the optimal control u and w, respectively.

For the action network that approximates the control $u(k,l)$, the state $x(k,l)$ is used as input to create the optimal control $u(k,l)$. The output can be formulated as

$$\hat{u}_i(k,l) = W_{ai}^{\mathrm{T}} \sigma(V_{ai}^{\mathrm{T}} x(k,l)). \tag{8.99}$$

Therefore, we define the output error of the action network as

$$e_{ai}(k,l) = \hat{u}_i(k,l) - u_i(k,l), \tag{8.100}$$

where $u_i(k,l)$ is the target function, which can be described by

$$\begin{aligned} u_i(k,l) = {} & (H_{ww}^{[i]} - H_{wu}^{[i]} (H_{uu}^{[i]})^{-1} H_{uw}^{[i]})^{-1} \\ & \times (H_{wu}^{[i]} (H_{uu}^{[i]})^{-1} H_{ux}^{[i]} - H_{wx}^{[i]}) x(k,l), \end{aligned} \tag{8.101}$$

where H_i can be obtained according to the Kronecker product in (8.85).

The weights of the action network are updated to minimize the following performance error measure:

$$E_{ai}(k,l) = \frac{1}{2} e_{ai}^{\mathrm{T}} e_{ai}. \tag{8.102}$$

The weight updating algorithm is similar to the one for the critic network. By the gradient descent rule, we obtain

$$w_{a(i+1)}(k,l) = w_{ai}(k,l) + \Delta w_{ai}(k,l), \tag{8.103}$$

$$\Delta w_{ai}(k,l) = \beta_a \left[-\frac{\partial E_{ai}(k,l)}{\partial w_{ai}(k,l)} \right], \tag{8.104}$$

$$\frac{\partial E_{ai}(k,l)}{\partial w_{ai}(k,l)} = \frac{\partial E_{ai}(k,l)}{\partial e_{ai}(k,l)} \frac{\partial e_{ai}(k,l)}{\partial u_i(k,l)} \frac{\partial u_i(k,l)}{\partial w_{ai}(k,l)}, \tag{8.105}$$

where $\beta_{\mathrm{a}} > 0$ is the learning rate of the action network.

For the action network w, the state $x(k,l)$ is used as input to create the optimal control $w(k,l)$. The target of the action network w can be expressed as

$$\begin{aligned} w_i(k,l) = {} & (H_{ww}^{[i]} - H_{wu}^{[i]} (H_{uu}^{[i]})^{-1} H_{uw}^{[i]})^{-1} \\ & \times (H_{wu}^{[i]} (H_{uu}^{[i]})^{-1} H_{ux}^{[i]} - H_{wx}^{[i]}) x(k,l). \end{aligned} \tag{8.106}$$

All the update rules of action network w are the same as the update rules of action network u, and they are omitted here.

8.2.3 Simulations

In this subsection, the present method is applied to an air drying process control.

Example 8.12 The dynamical processes can be described by the following Darboux equation:

$$\frac{\partial^2 x(s,t)}{\partial s \partial t} = a_1 \frac{\partial x(s,t)}{\partial t} + a_2 \frac{\partial x(s,t)}{\partial x} + a_0 x(s,t) + bu(s,t) + cw(s,t), \tag{8.107}$$

with the initial and boundary conditions

$$x^h(0,t) = \begin{cases} 0.5 & t \le 4, \\ 0 & t > 4, \end{cases} \quad \text{and} \quad x^v(s,0) = \begin{cases} 1 & s \le 4, \\ 0 & s > 4, \end{cases} \tag{8.108}$$

where $x(s,t)$ is an unknown function, a_0, a_1, a_2, b, and c are real coefficients, and $u(s,t)$ and $w(s,t)$ are the input functions. The variable x is for the humidity which is the system state, s is for the location of the air, and t is the processing time.

Let $a_0 = 0.2$, $a_1 = 0.3$, $a_2 = 0.1$, $b = 0.3$, and $c = 0.25$. The quadratic cost functional is formulated as

$$J = \int_{t=0}^{\infty}\int_{s=0}^{\infty} \{x^{\mathrm{T}}(s,t)Qx(s,t) + u^{\mathrm{T}}(s,t)Ru(s,t) + w^{\mathrm{T}}(s,t)Sw(s,t)\}\,\mathrm{d}t\mathrm{d}x. \tag{8.109}$$

The discretization method for the system (8.107) is similar to the method in [14]. Suppose that the sampling periods of the digital control system are chosen as $X = 0.1\,\mathrm{cm}$ and $T = 0.1\,\mathrm{s}$. Following the methodology presented in [14], we compute the discretized system equation as

$$\begin{bmatrix} x^h((k+1)X, lT) \\ x^v(kX, (l+1)T) \end{bmatrix} = \begin{bmatrix} 0.7408 & 0.2765 \\ 0.0952 & 0.9048 \end{bmatrix} \begin{bmatrix} x^h(kX, lT) \\ x^v(kX, lT) \end{bmatrix} + \begin{bmatrix} 0.0259 \\ 0 \end{bmatrix} u(kX, lT) + \begin{bmatrix} 0 \\ 0.0564 \end{bmatrix} w(kX, lT), \tag{8.110}$$

with the boundary conditions

$$x^h(0, lT) = \begin{cases} 0.5 & l \le 40, \\ 0 & l > 40, \end{cases} \quad \text{and} \quad x^v(kX, 0) = \begin{cases} 1 & k \le 40, \\ 0 & k > 40, \end{cases} \tag{8.111}$$

and the discredited cost functional

$$J(x(0,0), u, w) = \sum_{k=0}^{\infty}\sum_{l=0}^{\infty} \{x^{\mathrm{T}}(Xk, Tl)Qx(Xk, Tl) + u^{\mathrm{T}}(Xk, Tl)Ru(Xk, Tl) + w^{\mathrm{T}}(Xk, Tl)Sw(Xk, Tl)\}. \tag{8.112}$$

We implement the iterative algorithm at $(k, l) = (0, 0)$. We choose three-layer neural networks as the critic network, the action network u, and the action network w with the structure 2–8–1, 2–8–1 and 2–8–1, respectively. The initial weights of action networks and critic network are all set to be random in $[-0.5, 0.5]$. Then, the critic network and the action networks are trained for $i = 50$ times so that the given accuracy $\varepsilon = 10^{-6}$ is reached. In the training process, the learning rate is $\beta_a = \alpha_c = 0.05$. The evaluating point number is set as $N = 40$ for every iteration. Choose the small white noise as $\xi_1(0, 0.01)$, $\xi_2(0, 0.01)$. The convergence trajectory of the value function is shown in Fig. 8.2. Then, we apply the optimal control to the system for $k = 40$, $l = 40$ time steps and obtain the following results. The state trajectories are given as Figs. 8.3 and 8.4. The control trajectories are given as Figs. 8.5 and 8.6, respectively.

From the simulation results, we see that the present iterative ADP algorithm obtains good effects. In [14], Tsai just studied the model-based optimal control in case

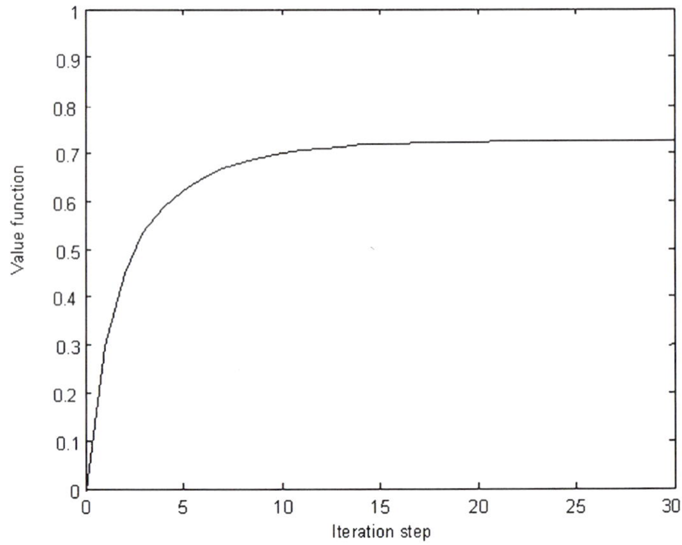

Fig. 8.2 The convergence of value function

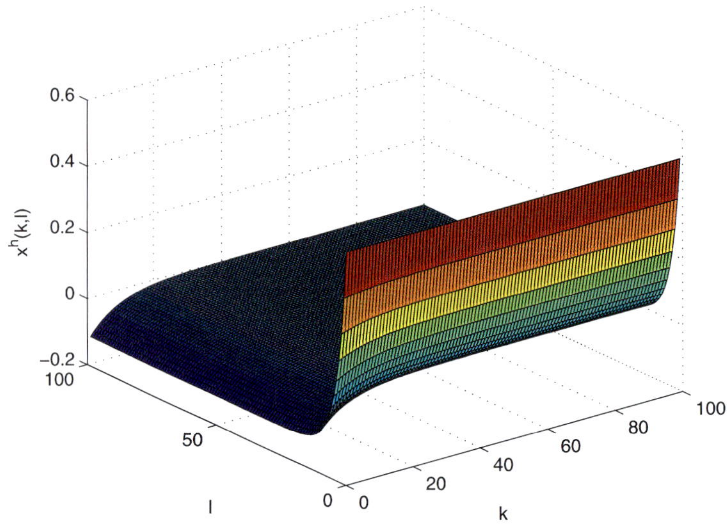

Fig. 8.3 The trajectories of the state variables x^h

of a finite horizon. In this section, using the iterative ADP algorithm, the optimal control scheme for 2-D system in infinite horizon can also be obtained without the system model. So the present algorithm in this section is more effective than the method in [14] for industrial process control.

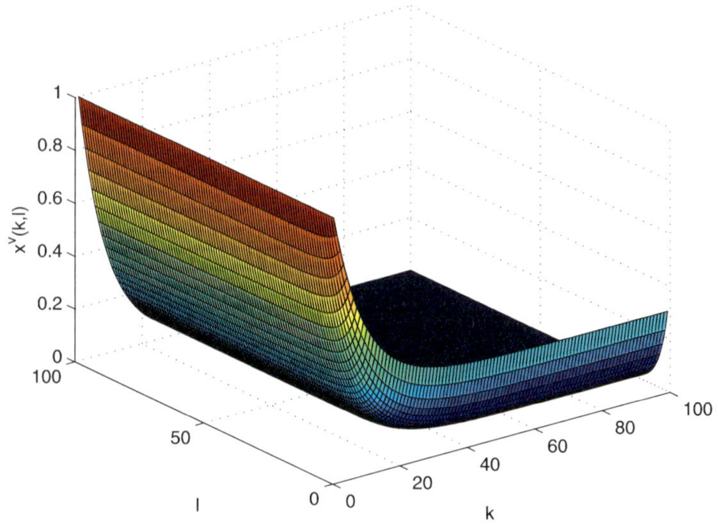

Fig. 8.4 The trajectories of the state variables x^v

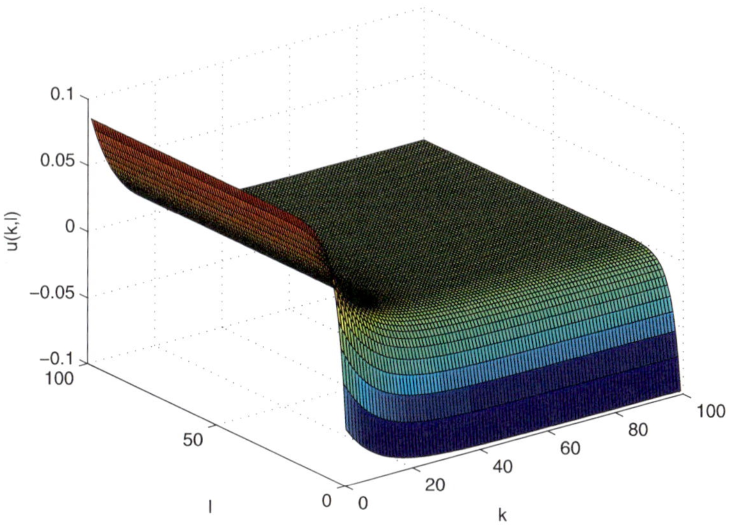

Fig. 8.5 The trajectories of the optimal control u

8.3 Zero-Sum Games for a Class of Discrete-Time Systems via Model-Free ADP

A novel model-free ADP method using output feedback is developed for zero-sum games of a class of discrete-time systems in this section. It is worthy to note that not only the model of the exact system is not required, but neither is the information

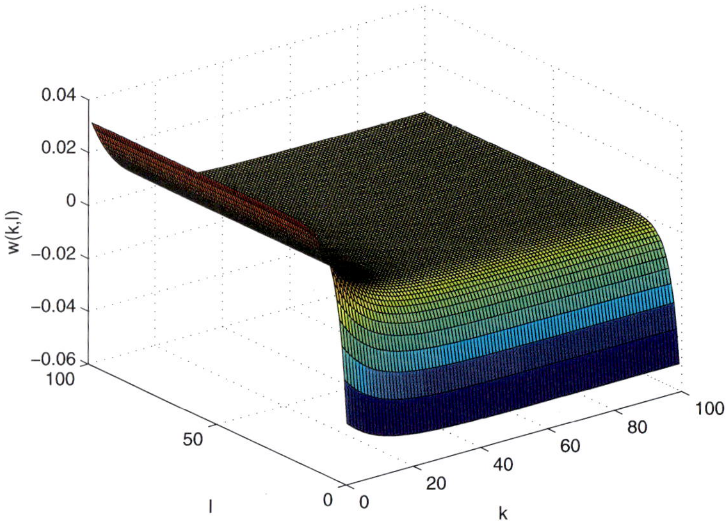

Fig. 8.6 The trajectories of the optimal control w

of the system states. Only the data measured from input and output are required for reaching the saddle point of the zero-sum games by using the present data-based iterative ADP algorithm.

8.3.1 Problem Formulation

Consider the following discrete-time linear systems:

$$\begin{aligned} x(k+1) &= Ax(k) + Bu(k) + Ew(k), \\ y(k) &= Cx(k), \end{aligned} \tag{8.113}$$

where $x(k) \in \mathbb{R}^n$ is the state, $u(k) \in \mathbb{R}^{m_1}$ is the control input, $w(k) \in \mathbb{R}^{m_2}$ is the disturbance input, and $y(k) \in \mathbb{R}^p$ is the measurable output.

The cost functional for (8.113) is given as follows:

$$\begin{aligned} J(x(k), u, w) &= \sum_{i=k}^{\infty} (y_i^{\mathrm{T}} Q y_i + u_i^{\mathrm{T}} u_i - \gamma^2 w_i^{\mathrm{T}} w_i) \\ &= l(x(k), u(k), w(k)) + J(x(k+1), u, w), \end{aligned} \tag{8.114}$$

where $Q \geq 0$ has suitable dimension, γ is a prescribed fixed value, and $l(x(k), u(k), w(k)) = y^{\mathrm{T}}(k)Qy(k) + u^{\mathrm{T}}(k)u(k) - \gamma^2 w^{\mathrm{T}}(k)w(k)$ is the utility function. For the above zero-sum games, two control variables u and w are chosen, respectively, by player I and player II, where player I tries to minimize the cost functional

$V(x(k), u, w)$, while player II tries to maximize it. The following assumptions are needed; these are in effect in the remaining sections.

Assumption 8.13 The system (8.113) is controllable and observable under the control variables u and w.

Assumption 8.14 The order of the system and an upper bound on its "observable index" are known.

Assumption 8.15 There exists a unique saddle point of the zero-sum games for the system (8.113).

According to Assumption 8.15 and using the dynamic programming principle, the optimal value function can be expressed as

$$\begin{aligned} J^*(x(k)) &= \min_u \max_w (l(x(k), u(k), w(k)) + J^*(x(k+1), u, w)) \\ &= \max_w \min_u (l(x(k), u(k), w(k)) + J^*(x(k+1), u, w)). \end{aligned} \tag{8.115}$$

Because the system (8.113) is linear, there exists a matrix P that satisfies

$$J^*(x(k)) = x^{\mathrm{T}}(k) P x(k), \tag{8.116}$$

where $P \geq 0$ is the solution of following generalized algebraic Riccati equation (GARE):

$$\begin{aligned} P = {} & A^{\mathrm{T}} P A + C^{\mathrm{T}} Q C - [A^{\mathrm{T}} P B \;\; A^{\mathrm{T}} P E] \\ & \times \begin{bmatrix} I + B^{\mathrm{T}} P B & B^{\mathrm{T}} P E \\ E^{\mathrm{T}} P B & E^{\mathrm{T}} P E - \gamma^2 I \end{bmatrix}^{-1} \begin{bmatrix} B^{\mathrm{T}} P A \\ E^{\mathrm{T}} P A \end{bmatrix}. \end{aligned} \tag{8.117}$$

The optimal control $u^*(k)$ and $w^*(k)$ can be derived as

$$\begin{aligned} u^*(k) = {} & -(I + B^{\mathrm{T}} P B - B^{\mathrm{T}} P E (E^{\mathrm{T}} P E - \gamma^2 I)^{-1} E^{\mathrm{T}} P B)^{-1} \\ & \times (B^{\mathrm{T}} P A - B^{\mathrm{T}} P E (E^{\mathrm{T}} P E - \gamma^2 I)^{-1} E^{\mathrm{T}} P A) x(k), \end{aligned} \tag{8.118}$$

$$\begin{aligned} w^*(k) = {} & -(E^{\mathrm{T}} P E - \gamma^2 I - E^{\mathrm{T}} P B (I + B^{\mathrm{T}} P B)^{-1} B^{\mathrm{T}} P E)^{-1} \\ & \times (E^{\mathrm{T}} P A - E^{\mathrm{T}} P B (I + B^{\mathrm{T}} P B)^{-1} B^{\mathrm{T}} P A) x(k). \end{aligned} \tag{8.119}$$

For obtaining the unique feedback saddle point in the strictly feedback stabilizing control policy class, the following inequalities should be satisfied [7]:

$$I + B^{\mathrm{T}} T P B > 0, \tag{8.120}$$

$$E^{\mathrm{T}} P E - \gamma^2 I > 0. \tag{8.121}$$

From (8.117)–(8.119), we see that the system model is needed not only for solving the GARE equation, but also for computing the optimal controls $u^*(k)$ and $w^*(k)$. To circumvent this requirement, [5] proposed the optimal strategies for discrete-time linear system quadratic zero-sum games without knowing the system dynamical matrices by using the Q-learning method under the assumption that all the states of the plant are available for feedback. Next, we propose a novel data-based optimal output feedback control scheme via the ADP algorithm, in which neither the system model nor the system states are needed.

8.3.2 Data-Based Optimal Output Feedback Control via ADP Algorithm

Inspired by the works in [1], according to Assumptions 8.13 and 8.14, the dynamics can be rewritten on a time horizon $[k-N,k]$ as the expanded state equation

$$x(k)=A^N x(k-N)+U_N\bar{u}_{k-1,k-N}+F_N\bar{w}_{k-1,k-N}, \tag{8.122}$$

$$\bar{y}_{k-1,k-N}=V_N x(k-N)+T_N\bar{u}_{k-1,k-N}+G_N\bar{w}_{k-1,k-N}, \tag{8.123}$$

where

$$\begin{aligned}
\bar{u}_{k-1,k-N}&=[u(k-1)\ \ u(k-2)\ \ \ldots\ \ u(k-N)]^{\mathrm{T}}\in\mathbb{R}^{m_1N},\\
\bar{w}_{k-1,k-N}&=[w(k-1)\ \ w(k-2)\ \ \ldots\ \ w(k-N)]^{\mathrm{T}}\in\mathbb{R}^{m_2N},\\
\bar{y}_{k-1,k-N}&=\left[y(k-1)\ \ y(k-2)\ \ \ldots\ \ y(k-N)\right]^{\mathrm{T}}\in\mathbb{R}^{pN},\\
V_N&=\left[CA^{N-1}\ \ \ldots\ \ CA\ \ C\right]^{\mathrm{T}},\\
U_N&=\left[B\ \ AB\ \ A^2B\ \ \ldots\ \ A^{N-1}B\right],\\
F_N&=\left[E\ \ AE\ \ A^2E\ \ \ldots\ \ A^{N-1}E\right],\\
T_N&=\begin{bmatrix}0 & CB & CAB & \ldots & CA^{N-2}B\\ 0 & 0 & CB & \ldots & CA^{N-3}B\\ \vdots & \vdots & \ddots & \ddots & \vdots\\ 0 & \ldots & \ldots & 0 & CB\\ 0 & 0 & 0 & 0 & 0\end{bmatrix},\\
G_N&=\begin{bmatrix}0 & CE & CAE & \ldots & CA^{N-2}E\\ 0 & 0 & CE & \ldots & CA^{N-3}E\\ \vdots & \vdots & \ddots & \ddots & \vdots\\ 0 & \ldots & \ldots & 0 & CE\\ 0 & 0 & 0 & 0 & 0\end{bmatrix}.
\end{aligned} \tag{8.124}$$

Since the system (8.113) is observable under the control variables u and w, there exists a upper bound on the observability index K, such that $\text{rank}(V_N) < n$ for $N < K$ and $\text{rank}(V_N) = n$ for $N > K$. Choose $N > K$; we have $\text{rank}(V_N) = n$. Then, there exists a matrix $M \in \mathbb{R}^{n\times pN}$ such that

$$A^N = MV_N. \tag{8.125}$$

Since $\text{rank}(V_N) = n$, its left inverse is given as

$$V_N^+ = (V_N^{\mathrm{T}} V_N)^{-1} V_N^{\mathrm{T}}. \tag{8.126}$$

Then M can be rearranged as the sum of two parts,

$$M = A^N V_N^+ + Z(I - V_N V_N^+) \equiv M_0 + M_1, \tag{8.127}$$

for any matrix Z, with M_0 denoting the minimum norm operator and $P(R^{\perp})(V_N) = I - V_N V_N^+$ being the projection onto a range perpendicular to V_N.

Theorem 8.16 (cf. [10]) *Using Assumption* 8.14, *the system state* $x(k)$ *is given uniquely in terms of the measured input–output data sequences* $\bar{y}_{k-1,k-N}$, $\bar{u}_{k-1,k-N}$, *and* $\bar{w}_{k-1,k-N}$ *by*

$$\begin{aligned} x(k) &= M_0\bar{y}_{k-1,k-N} + (U_N - M_0T_N)\bar{u}_{k-1,k-N} \\ &\quad + (F_N - M_0G_N)\bar{w}_{k-1,k-N} \\ &= M_y\bar{y}_{k-1,k-N} + M_u\bar{u}_{k-1,k-N} + M_w\bar{w}_{k-1,k-N}, \end{aligned} \tag{8.128}$$

where $M_y = M_0$, $M_u = U_N - M_0T_N$, *and* $M_w = F_N - M_0G_N$.

Proof Multiplying $x(k-N)$ by both sides of (8.123), and using (8.125), we have

$$\begin{aligned} A^N x(k-N) &= MV_N x(k-N) \\ &= M\bar{y}_{k-1,k-N} - MT_N\bar{u}_{k-1,k-N} - MG_N\bar{w}_{k-1,k-N}. \end{aligned} \tag{8.129}$$

Then, using (8.127), we have

$$\begin{aligned} (M_0 + M_1)V_N x(k-N) &= (M_0 + M_1)\bar{y}_{k-1,k-N} - (M_0 + M_1)T_N\bar{u}_{k-1,k-N} \\ &\quad - (M_0 + M_1)G_N\bar{w}_{k-1,k-N}. \end{aligned} \tag{8.130}$$

Since $M_1V_N = 0$, we have

$$MV_N x(k-N) = M_0V_N x(k-N). \tag{8.131}$$

Multiplying both sides of (8.123), we have

$$0 = M_1\bar{y}_{k-1,k-N} - M_1T_N\bar{u}_{k-1,k-N} - M_1G_N w_{k-1,k-N} \qquad \forall M_1 \text{ s.t. } M_1V_N = 0. \tag{8.132}$$

Then, using (8.129)–(8.132), we have

$$\begin{aligned}A^N x(k-N) &= M_0 V_N x(k-N)\\ &= M_0 \bar{y}_{k-1,k-N} - M_0 T_N \bar{u}_{k-1,k-N} - M_0 G_N \bar{w}_{k-1,k-N}.\end{aligned} \tag{8.133}$$

From (8.122), it follows that

$$\begin{aligned}x(k) &= M_0 \bar{y}_{k-1,k-N} + (U_N - M_0 T_N)\bar{u}_{k-1,k-N}\\ &\quad + (F_N - M_0 G_N)\bar{w}_{k-1,k-N}\\ &= M_y \bar{y}_{k-1,k-N} + M_u \bar{u}_{k-1,k-N} + M_w \bar{w}_{k-1,k-N}.\end{aligned} \tag{8.134}$$

This completes the proof. □

Remark 8.17 From (8.134), we see that the state variable is completely eliminated from the right-hand side of (8.113). That is, $x(k)$ can be reconstructed by the available measured input–output data sequence $\bar{y}_{k-1,k-N}$, $\bar{u}_{k-1,k-N}$, and $\bar{w}_{k-1,k-N}$.

Next, we will derive the optimal controls u^* and w^* by using the available measurable input–output data.

Define

$$\bar{z}_{k-1,k-N} = \begin{bmatrix} \bar{u}_{k-1,k-N}\\ \bar{w}_{k-1,k-N}\\ \bar{y}_{k-1,k-N}\end{bmatrix}. \tag{8.135}$$

We have

$$\begin{aligned}V(x(k),u,w) &= x(k)^{\mathrm{T}} P x(k)\\ &= \bar{z}^{\mathrm{T}}_{k-1,k-N}\begin{bmatrix} M_u^{\mathrm{T}}\\ M_w^{\mathrm{T}}\\ M_y^{\mathrm{T}}\end{bmatrix} P[M_u \ \ M_w \ \ M_y]\bar{z}_{k-1,k-N}\\ &= \bar{z}^{\mathrm{T}}_{k-1,k-N}\begin{bmatrix} M_u^{\mathrm{T}}PM_u & M_u^{\mathrm{T}}PM_w & M_u^{\mathrm{T}}PM_y\\ M_w^{\mathrm{T}}PM_u & M_w^{\mathrm{T}}PM_w & M_w^{\mathrm{T}}PM_y\\ M_y^{\mathrm{T}}PM_u & M_y^{\mathrm{T}}PM_w & M_y^{\mathrm{T}}PM_y\end{bmatrix}\bar{z}_{k-1,k-N}\\ &= \bar{z}^{\mathrm{T}}_{k-1,k-N}\bar{P}\bar{z}_{k-1,k-N},\end{aligned} \tag{8.136}$$

where $\bar{z} \in \mathbb{R}^{(m_1+m_2+p)N}$ and $\bar{P} \in \mathbb{R}^{(m_1+m_2+p)N\times(m_1+m_2+p)N}$.

By rewriting (8.115) and using (8.134), we have

$$\begin{aligned}(u^*(k), w^*(k)) &= \arg\min_u \max_w \big\{ y^{\mathrm{T}}(k)Qy(k) + u^{\mathrm{T}}(k)u(k) - \gamma^2 w^{\mathrm{T}}(k)w(k)\\ &\quad + \bar{z}^{\mathrm{T}}_{k,k-N+1}\bar{P}\bar{z}_{k,k-N+1}\big\}.\end{aligned} \tag{8.137}$$

Partition $\bar{z}^{\mathrm{T}}_{k,k-N+1}\bar{P}\bar{z}_{k,k-N+1}$ as

$$\bar{z}^{\mathrm{T}}_{k,k-N+1}\bar{P}\bar{z}_{k,k-N+1}$$
$$=\begin{bmatrix} u(k) \\ \bar{u}_{k-1,k-N+1} \\ w(k) \\ \bar{w}_{k-1,k-N+1} \\ \bar{y}_{k,k-N+1} \end{bmatrix}^{\mathrm{T}} \begin{bmatrix} P_{uu} & P_{u\bar{u}} & P_{uw} & P_{u\bar{w}} & P_{u\bar{y}} \\ P_{\bar{u}u} & P_{\bar{u}\bar{u}} & P_{\bar{u}w} & P_{\bar{u}\bar{w}} & P_{\bar{u}\bar{y}} \\ P_{wu} & P_{w\bar{u}} & P_{ww} & P_{w\bar{w}} & P_{w\bar{y}} \\ P_{\bar{w}u} & P_{\bar{w}\bar{u}} & P_{\bar{w}w} & P_{\bar{w}\bar{w}} & P_{\bar{w}\bar{y}} \\ P_{\bar{y}u} & P_{\bar{y}\bar{u}} & P_{\bar{y}w} & P_{\bar{y}\bar{w}} & P_{\bar{y}\bar{y}} \end{bmatrix} \begin{bmatrix} u(k) \\ \bar{u}_{k-1,k-N+1} \\ w(k) \\ \bar{w}_{k-1,k-N+1} \\ \bar{y}_{k,k-N+1} \end{bmatrix}. \tag{8.138}$$

Then differentiating with respect to $u(k)$ to perform the minimization in (8.137) yields

$$\begin{aligned} u(k) = {} & -(I+P_{uu})^{-1}(P_{u\bar{u}}\bar{u}_{k-1,k-N+1}+P_{uw}w(k) \\ & + P_{u\bar{w}}\bar{w}_{k-1,k-N+1}+P_{u\bar{y}}\bar{y}_{k,k-N+1}). \end{aligned} \tag{8.139}$$

Similarly, differentiating with respect to $w(k)$ to perform the maximization in (8.137) yields

$$\begin{aligned} w(k) = {} & -(P_{ww}-\gamma^2 I)(P_{wu}u(k)+P_{w\bar{u}}\bar{u}_{k-1,k-N+1} \\ & + P_{w\bar{w}}\bar{w}_{k-1,k-N+1}+P_{w\bar{y}}\bar{y}_{k,k-N+1}). \end{aligned} \tag{8.140}$$

Thus, substitute (8.140) into (8.139), and $u^*(k)$ can be derived as

$$\begin{aligned} u^*(k) = {} & -(I+P_{uu}-P_{uw}(P_{ww}-\gamma^2 I)P_{wu})^{-1}[-P_{uw}(P_{ww}-\gamma^2 I)^{-1} \\ & \times (P_{w\bar{u}}\bar{u}_{k-1,k-N+1}+P_{w\bar{w}}\bar{w}_{k-1,k-N+1}+P_{w\bar{y}}\bar{y}_{k,k-N+1}) \\ & + P_{u\bar{u}}\bar{u}_{k-1,k-N+1}+P_{u\bar{w}}\bar{w}_{k-1,k-N+1}+P_{u\bar{y}}\bar{y}_{k,k-N+1}]. \end{aligned} \tag{8.141}$$

Similarly, substitute (8.139) into (8.140), and $w^*(k)$ can be derived as

$$\begin{aligned} w^*(k) = {} & -(P_{ww}-\gamma^2 I-P_{wu}(I+P_{uu})^{-1}P_{uw})^{-1}[-P_{wu}(R+P_{uu})^{-1} \\ & \times (P_{u\bar{u}}\bar{u}_{k-1,k-N+1}+P_{u\bar{w}}\bar{w}_{k-1,k-N+1}+P_{u\bar{y}}\bar{y}_{k,k-N+1}) \\ & + P_{w\bar{u}}\bar{u}_{k-1,k-N+1}+P_{w\bar{w}}\bar{w}_{k-1,k-N+1}+P_{w\bar{y}}\bar{y}_{k,k-N+1}]. \end{aligned} \tag{8.142}$$

Comparing (8.118)–(8.119) with (8.141)–(8.142), it is important to note that the optimal control u^* and w^* expressed by (8.141)–(8.142) can be obtained if $\bar{P}$ can be solved. Neither knowledge of the system dynamics A, B, E, and C, nor information of the measurable state $x(k)$ is needed.

In the following, the uniqueness of the optimal control is discussed.

Theorem 8.18 (cf. [10]) *The optimal controls generated by* (8.141) *and* (8.142) *are independent of* M_1 *and depend only on* M_0. *Moreover,* (8.141) *and* (8.142) *are equivalent to*

$$
\begin{aligned}
u^*(k) = &-(I + B^{\mathrm{T}}PB - B^{\mathrm{T}}PE(E^{\mathrm{T}}PE - \gamma^2 I)^{-1}E^{\mathrm{T}}PB)^{-1} \\
&\times \big[- P_{uw}(P_{ww} - \gamma^2 I)^{-1} \\
&\times (P_{w\bar{u}}\bar{u}_{k-1,k-N+1} + P_{w\bar{w}}\bar{w}_{k-1,k-N+1} + P_{w\bar{y}}\bar{y}_{k,k-N+1}) \\
&+ P_{u\bar{u}}\bar{u}_{k-1,k-N+1} + P_{u\bar{w}}\bar{w}_{k-1,k-N+1} + P_{u\bar{y}}\bar{y}_{k,k-N+1}\big],
\end{aligned} \tag{8.143}
$$

$$
\begin{aligned}
w^*(k) = &-[E^{\mathrm{T}}PE - \gamma^2 I - E^{\mathrm{T}}PB(I + B^{\mathrm{T}}PB)^{-1}B^{\mathrm{T}}PE]^{-1} \\
&\times \big[- P_{wu}(I + P_{uu})^{-1} \\
&\times (P_{u\bar{u}}\bar{u}_{k-1,k-N+1} + P_{u\bar{w}}\bar{w}_{k-1,k-N+1} + P_{u\bar{y}}\bar{y}_{k,k-N+1}) \\
&+ P_{w\bar{u}}\bar{u}_{k-1,k-N+1} + P_{w\bar{w}}\bar{w}_{k-1,k-N+1} + P_{w\bar{y}}\bar{y}_{k,k-N+1}\big],
\end{aligned} \tag{8.144}
$$

where $P_{w\bar{u}}$, $P_{w\bar{w}}$, $P_{w\bar{y}}$, $P_{u\bar{u}}$, $P_{u\bar{w}}$, and $P_{u\bar{y}}$ depend only on M_0.

Proof Rewrite (8.139) as

$$
\begin{aligned}
0 &= u(k) + P_{uu}u(k) + P_{u\bar{u}}\bar{u}_{k-1,k-N+1} + P_{uw}w(k) \\
&\quad + P_{u\bar{w}}\bar{w}_{k-1,k-N+1} + P_{u\bar{y}}\bar{y}_{k,k-N+1} \\
&= u(k) + [P_{uu} \;\; P_{u\bar{u}} \;\; | P_{uw} \;\; P_{u\bar{w}} \;\; | P_{u\bar{y}}] \begin{bmatrix} \bar{u}_{k,k-N+1} \\ \bar{w}_{k,k-N+1} \\ \bar{y}_{k,k-N+1} \end{bmatrix}.
\end{aligned} \tag{8.145}
$$

According to (8.136) and (8.138), we have

$$
\begin{aligned}
[P_{uu} \;\; P_{u\bar{u}}] &= [I_{m_1} \;\; 0]M_u^{\mathrm{T}}PM_u, \\
[P_{uw} \;\; P_{u\bar{w}}] &= [I_{m_1} \;\; 0]M_u^{\mathrm{T}}PM_w, \\
P_{u\bar{y}} &= [I_{m_1} \;\; 0]M_u^{\mathrm{T}}PM_y.
\end{aligned} \tag{8.146}
$$

After substituting (8.146) into (8.145), we have

$$
\begin{aligned}
0 &= u(k) + [P_{uu} \;\; P_{u\bar{u}} \;\; | P_{uw} \;\; P_{u\bar{w}} \;\; | P_{u\bar{y}}] \begin{bmatrix} \bar{u}_{k,k-N+1} \\ \bar{w}_{k,k-N+1} \\ \bar{y}_{k,k-N+1} \end{bmatrix} \\
&= u(k) + [I_{m_1} \;\; 0]M_u^{\mathrm{T}}P \\
&\quad \times [M_u\bar{u}_{k,k-N+1} + M_w\bar{w}_{k,k-N+1} + M_y\bar{y}_{k,k-N+1}] \\
&= u(k) + [I_{m_1} \;\; 0]M_u^{\mathrm{T}}P \\
&\quad \times [M_0\bar{y}_{k-1,k-N} + (U_N - M_0T_N)\bar{u}_{k-1,k-N} \\
&\quad + (F_N - M_0G_N)\bar{w}_{k-1,k-N}].
\end{aligned} \tag{8.147}
$$

According to Theorem 8.16, (8.147) is unique and independent of M_1. Using the structure of U_N, T_N, F_N, and G_N, it follows that

$$M_u \begin{bmatrix} I_{m_1} \\ 0 \end{bmatrix} = (U_N - M_0 T_N) \begin{bmatrix} I_{m_1} \\ 0 \end{bmatrix} = B, \tag{8.148}$$

$$M_w \begin{bmatrix} I_{m_2} \\ 0 \end{bmatrix} = (F_N - M_0 G_N) \begin{bmatrix} I_{m_2} \\ 0 \end{bmatrix} = E. \tag{8.149}$$

Then,

$$P_{uu} = [I_{m_1} \ \ 0] M_u^{\mathrm{T}} P M_u \begin{bmatrix} I_{m_1} \\ 0 \end{bmatrix} = B^{\mathrm{T}} P B, \tag{8.150}$$

$$P_{uw} = [I_{m_1} \ \ 0] M_u^{\mathrm{T}} P M_w \begin{bmatrix} I_{m_2} \\ 0 \end{bmatrix} = B^{\mathrm{T}} P E, \tag{8.151}$$

$$P_{wu} = [I_{m_2} \ \ 0] M_w^{\mathrm{T}} P M_u \begin{bmatrix} I_{m_1} \\ 0 \end{bmatrix} = E^{\mathrm{T}} P B. \tag{8.152}$$

Substituting (8.150)–(8.152) to (8.139), we obtain

$$\begin{aligned} u(k) = &-(I + B^{\mathrm{T}} P B)^{-1} (P_{u\bar{u}} \bar{u}_{k-1,k-N+1} + B^{\mathrm{T}} P E w(k) \\ &+ P_{u\bar{w}} \bar{w}_{k-1,k-N+1} + P_{u\bar{y}} \bar{y}_{k,k-N+1}). \end{aligned} \tag{8.153}$$

Similarly, we obtain

$$\begin{aligned} w(k) = &-(E^{\mathrm{T}} P E - \gamma^2 I)(E^{\mathrm{T}} P B u(k) + P_{w\bar{u}} \bar{u}_{k-1,k-N+1} \\ &+ P_{w\bar{w}} \bar{w}_{k-1,k-N+1} + P_{w\bar{y}} \bar{y}_{k,k-N+1}). \end{aligned} \tag{8.154}$$

Then, substituting (8.154) into (8.153), we have (8.143). By a similar method, (8.144) can also be obtained.

This completes the proof. □

Although we have obtained the optimal control expressed in (8.141)–(8.142) with the information of the matrix $\bar{P}$, it is not straightforward to solve $\bar{P}$ in most cases. Therefore, next we propose the data-based iterative ADP algorithm for zero-sum games of (8.113).

First, the initial value of $\bar{P}$ is set as $\bar{P}_0 = 0$, which is not necessarily optimal.

Then, for $i = 0, 1, \ldots$, the matrix $\bar{P}_i$ associated with the cost function $V_i(x(k), u, w)$ and control policy $u_i(k)$ and $w_i(k)$ can be updated by implementing the iteration between

$$\begin{aligned} \bar{z}_{k-1,k-N}^{\mathrm{T}} \bar{P}_{i+1} \bar{z}_{k-1,k-N} = &\, y^{\mathrm{T}}(k) Q y(k) + u_i^{\mathrm{T}}(k) u_i(k) - \gamma^2 w_i^{\mathrm{T}}(k) w_i(k) \\ &+ \bar{z}_{k,k-N}^{\mathrm{T}} \bar{P}_i \bar{z}_{k,k-N}, \end{aligned} \tag{8.155}$$

and

$$\begin{aligned} u_{i+1}(k) = &-(I + P_{uu}^{[i+1]} + P_{uw}^{[i+1]}(P_{ww}^{[i+1]} - \gamma^2 I) P_{wu}^{[i+1]})^{-1} \\ &\times \big[- P_{uw}^{[i+1]}(P_{ww}^{[i+1]} - \gamma^2 I)^{-1} (P_{w\bar{u}}^{[i+1]} \bar{u}_{k-1,k-N+1} \\ &+ P_{w\bar{w}}^{[i+1]} \bar{w}_{k-1,k-N+1} + P_{w\bar{y}}^{[i+1]} \bar{y}_{k,k-N+1}) \\ &+ P_{u\bar{u}}^{[i+1]} \bar{u}_{k-1,k-N+1} + P_{u\bar{w}}^{[i+1]} \bar{w}_{k-1,k-N+1} \\ &+ P_{u\bar{y}}^{[i+1]} \bar{y}_{k,k-N+1} \big], \end{aligned} \tag{8.156}$$

$$\begin{aligned} w_{i+1}(k) = &-(P_{wu}^{[i+1]} - \gamma^2 I - P_{wu}^{[i+1]}(I + P_{uu}^{[i+1]})^{-1} P_{uw}^{[i+1]})^{-1} \\ &\times \big[- P_{wu}^{[i+1]}(I + P_{uu}^{[i+1]})^{-1} (P_{u\bar{u}}^{[i+1]} \bar{u}_{k-1,k-N+1} \\ &+ P_{u\bar{w}}^{[i+1]} \bar{w}_{k-1,k-N+1} + P_{u\bar{y}}^{[i+1]} \bar{y}_{k,k-N+1}) \\ &+ P_{w\bar{u}}^{[i+1]} \bar{u}_{k-1,k-N+1} + P_{w\bar{w}}^{[i+1]} \bar{w}_{k-1,k-N+1} \\ &+ P_{w\bar{y}}^{[i+1]} \bar{y}_{k,k-N+1} \big] \end{aligned} \tag{8.157}$$

until convergence.

In the following, we discuss the implementation of the present data-based iterative ADP algorithm. Rewrite (8.155) as

$$\begin{aligned} \operatorname{stk}(\bar{P}_{i+1})[\bar{z}_{k-1,k-N} \otimes \bar{z}_{k-1,k-N}] = &\, y^{\mathrm{T}}(k) Q y(k) + u_i^{\mathrm{T}}(k) u_i(k) - \gamma^2 w_i^{\mathrm{T}}(k) w_i(k) \\ &+ \bar{z}_{k,k-N+1} \bar{P}_i \bar{z}_{k,k-N+1} \end{aligned} \tag{8.158}$$

with $\otimes$ being the Kronecker product and $\operatorname{stk}(\cdot)$ being the column stacking operator. Thus we obtain

$$\begin{aligned} \operatorname{stk}(\bar{P}_{i+1}) = &\, ((\bar{z}_{k-1,k-N} \otimes \bar{z}_{k-1,k-N})(\bar{z}_{k-1,k-N} \otimes \bar{z}_{k-1,k-N})^{\mathrm{T}})^{-1} \\ &\times (\bar{z}_{k-1,k-N} \otimes \bar{z}_{k-1,k-N}) \big[y^{\mathrm{T}}(k) Q y(k) + u_i^{\mathrm{T}}(k) u_i(k) \\ &- \gamma^2 w_i^{\mathrm{T}}(k) w_i(k) + \bar{z}_{k,k-N+1} \bar{P}_i \bar{z}_{k,k-N+1} \big]. \end{aligned} \tag{8.159}$$

Remark 8.19 It should be noted that the matrix $(\bar{z}_{k-1,k-N} \otimes \bar{z}_{k-1,k-N})(\bar{z}_{k-1,k-N} \otimes \bar{z}_{k-1,k-N})^{\mathrm{T}})$ is generally invertible. To address this problem, we use the least-square technique to obtain the inverse of the matrix $(\bar{z}_{k-1,k-N} \otimes \bar{z}_{k-1,k-N})(\bar{z}_{k-1,k-N} \otimes \bar{z}_{k-1,k-N})^{\mathrm{T}})$.

To implement the least-square technique, probing noise is added into the $u(k)$ and $w(k)$, i.e., $\hat{u}(k) = u(k) + \xi_1$, $\hat{w}(k) = w(k) + \xi_2$, respectively. According to the condition of the least-square solution, the number of sampling points should be not less than $[(m_1 + m_2 + p)N][(m_1 + m_2 + p)N + 1]/2$. The least-square problem can be solved in real time by collecting enough data points.

Remark 8.20 It should be noted that what we need to know is the measurements of the current utility,

$$r(x(k),u,w) = y^{\mathrm{T}}(k)Qy(k) + u^{\mathrm{T}}(k)u(k) - \gamma^2 w^{\mathrm{T}}(k)w(k),$$

and the available measured input–output data sequence $\bar{u}_{k-1,k-N}$, $\bar{w}_{k-1,k-N}$, and $\bar{y}_{k,k-N}$. The exact system model and full information regarding the system state $x(k)$ are not needed. The control policy given by (8.141)–(8.142) is actually a dynamic autoregressive moving-average (ARMA) regulator in terms of the past input and the current and past output.

8.3.3 Simulations

Example 8.21 Consider the discrete-time system as follows:

$$\begin{aligned} x(k+1) &= \begin{bmatrix} 1.1 & -0.3 \\ 1 & 0 \end{bmatrix} x(k) + \begin{bmatrix} 1 \\ 0 \end{bmatrix} u(k) + \begin{bmatrix} 0 \\ 1 \end{bmatrix} w(k), \\ y(k) &= \begin{bmatrix} 1 \\ -0.8 \end{bmatrix} x(k), \end{aligned} \tag{8.160}$$

where Q in the cost functional is the identity matrix of appropriate dimension and γ is chosen as 0.2. The solution of the GARE equation (8.117) is

$$P = \begin{bmatrix} 0.9925 & -0.7932 \\ -0.7932 & 0.6335 \end{bmatrix}. \tag{8.161}$$

From (8.136), we have

$$\bar{P} = \begin{bmatrix} M_u^{\mathrm{T}} P M_u & M_u^{\mathrm{T}} P M_w & M_u^{\mathrm{T}} P M_y \\ M_w^{\mathrm{T}} P M_u & M_w^{\mathrm{T}} P M_w & M_w^{\mathrm{T}} P M_y \\ M_y^{\mathrm{T}} P M_u & M_y^{\mathrm{T}} P M_w & M_y^{\mathrm{T}} P M_y \end{bmatrix}. \tag{8.162}$$

Then $\bar{P}$ can be calculated using (8.134) as follows:

$$\bar{P} = \begin{bmatrix} 1.5598 & -1.4724 & -0.2538 & 0.2135 & 1.9193 & -0.5522 \\ -1.4724 & 0.8120 & 0.2287 & -0.1177 & -1.2881 & 0.3045 \\ -0.2538 & 0.2287 & 0.0411 & -0.0332 & -0.3025 & 0.0858 \\ 0.2135 & -0.1177 & -0.0332 & 0.0171 & 0.1868 & -0.0442 \\ 1.9193 & -1.2881 & -0.3025 & 0.1868 & 1.8871 & -0.4830 \\ -0.5522 & 0.3045 & 0.0858 & -0.0442 & -0.4830 & 0.1142 \end{bmatrix}. \tag{8.163}$$

Next, the present data-based iterative ADP algorithm is applied for the system (8.160). In order to implement the least-square technique, probing noise is added to

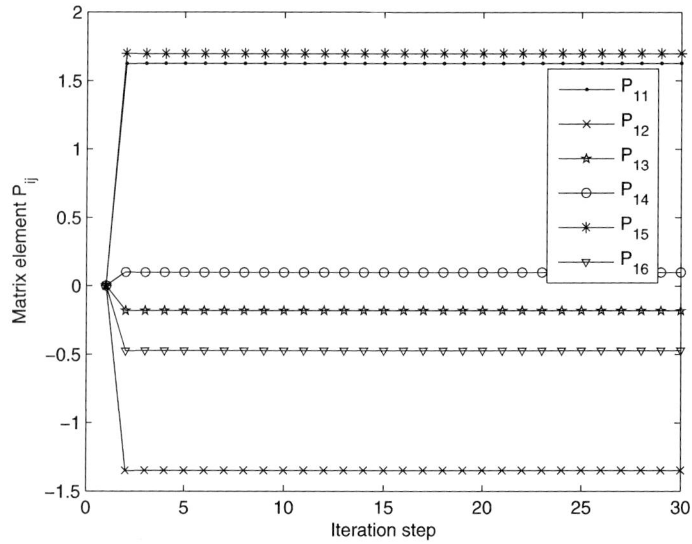

Fig. 8.7 The convergence of P_{ij} where P_{ij} denotes the element on col i, row j of the estimated $\bar{P}$, $i = 1$, $j = 1, \ldots, 6$

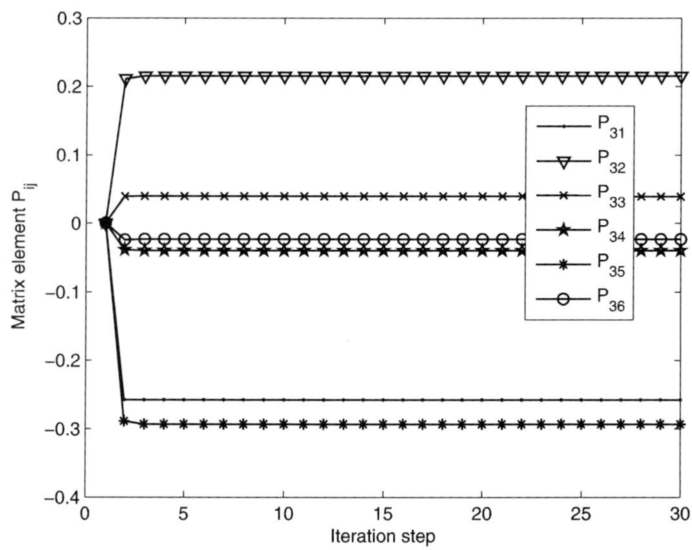

Fig. 8.8 The convergence of P_{ij} where P_{ij} denotes the element on col i, row j of the estimated $\bar{P}$, $i = 3$, $j = 1, \ldots, 6$

the control input u and w. In Figs. 8.7 and 8.8, the convergence of the estimated $\bar{P}$ to the optimal one is shown. The system state trajectories are shown in Fig. 8.9. By the simulation results, the effectiveness of the present method is validated.

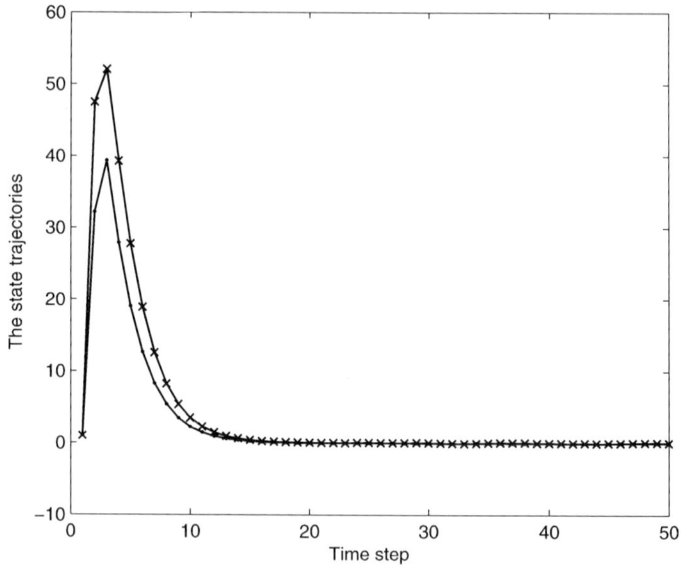

Fig. 8.9 The trajectories of system states

8.4 Summary

We studied the zero-sum games for discrete-time systems based on the model-free ADP method in this chapter. In Sect. 8.2, an effective data-based optimal control scheme was developed via the iterative ADP algorithm to find the optimal controller of a class of discrete-time zero-sum games for Roesser type 2-D systems. The present method allows implementation without the system model. In Sect. 8.2, a data-based optimal output feedback controller was developed for solving the zero-sum games of a class of discrete-time systems, whose merit is that not only knowledge of system model is not required, but also information of the system states is not. The theoretical analysis and simulation study showed the validity of the present methods.

References

1. Aangenent W, Kostic D, de Jager B, Van de Molengraft R, Steinbuch M (2005) Data-based optimal control. In: Proceedings of American control conference, Portland, pp 1460–1465
2. Abu-Khalaf M, Lewis FL (2008) Neurodynamic programming and zero-sum games for constrained control systems. IEEE Trans Neural Netw 19:1243–1252
3. Abu-Khalaf M, Lewis FL, Huang J (2006) Policy iterations on the Hamilton–Jacobi–Isaacs equation for H_∞ state feedback control with input saturation. IEEE Trans Autom Control 51:1989–1995
4. Al-Tamimi A, Abu-Khalaf M, Lewis FL (2007) Adaptive critic designs for discrete-time zero-sum games with application to H_∞ control. IEEE Trans Syst Man Cybern, Part B, Cybern 37:240–247

5. Al-Tamimi A, Lewis FL, Abu-Khalaf M (2007) Model-free q-learning designs for linear discrete-time zero-sum games with application to H_∞ control. Automatica 43:473–481
6. Al-Tamimi A, Lewis FL, Abu-Khalaf M (2007) Model-free Q-learning designs for linear discrete-time zero-sum games with application to H-infinity control. Automatica 43:473–481
7. Basar T, Bernhard P (1995) H_∞ optimal control and related minimax design problems. Birkhauser, Basel
8. Basar T, Olsder GJ (1982) Dynamic noncooperative game theory. Academic Press, New York
9. Bertsekas DP (2003) Convex analysis and optimization. Athena Scientific, Boston
10. Cui LL, Zhang HG, Zhang X, Luo YH (2011) Adaptive critic design based output feedback control for discrete-time zero-sum games. In: Proceedings of IEEE symposium on adaptive dynamic programming and reinforcement learning, France, pp 190–195
11. Hua X, Mizukami K (1994) Linear-quadratic zero-sum differential games for generalized state space systems. IEEE Trans Autom Control 39:143–147
12. Li CJ, Fadali MS (1991) Optimal control of 2-D systems. IEEE Trans Autom Control 36:223–228
13. Luenberger DG (1969) Optimization by vector space methods. Wiley, New York
14. Tsai JS, Li JS, Shieh LS (2002) Discretized quadratic optimal control for continuous-time two-dimensional systems. IEEE Trans Circuits Syst I, Fundam Theory Appl 49:116–125
15. Uetake Y (1992) Optimal smoothing for noncausal 2-D systems based on a descriptor model. IEEE Trans Autom Control 37:1840–1845
16. Wei QL, Zhang HG, Cui LL (2009) Data-based optimal control for discrete-time zero-sum games of 2-D systems using adaptive critic designs. Acta Autom Sin 35:682–692

Chapter 9
Nonlinear Games for a Class of Continuous-Time Systems Based on ADP

9.1 Introduction

Game theory is concerned with the study of decision making in situations where two or more rational opponents are involved under conditions of conflicting interests. This has been widely investigated by many authors [5, 7, 8, 12, 13]. Though the nonlinear optimal solution in term of Hamilton–Jacobi–Bellman equation is hard to obtain directly [4], it is still fortunate that there is only one controller or decision maker. In the previous chapter, we have studied discrete-time zero-sum games based on the ADP method. In this chapter, we will consider continuous-time games.

For zero-sum differential games, the existence of the saddle point is proposed before obtaining the saddle point in much of the literature [1, 6, 11]. In many real world applications, however, the saddle point of a game may not exist, which means that we can only obtain the mixed optimal solution of the game. In Sect. 9.2, we will study how to obtain the saddle point without complex existence conditions of the saddle point and how to obtain the mixed optimal solution when the saddle point does not exist based on the ADP method for a class of affine nonlinear zero-sum games. Note that many applications of practical zero-sum games have nonaffine control input. In Sect. 9.3, we will focus on finite horizon zero-sum games for a class of nonaffine nonlinear systems.

The non-zero-sum differential games theory also has a number of potential applications in control engineering, economics and the military field [9]. For zero-sum differential games, two players work on a cost functional together and minimax it. However, for non-zero-sum games, the control objective is to find a set of policies that guarantee the stability of the system and minimize the individual performance function to yield a Nash equilibrium. In Sect. 9.4, non-zero-sum differential games will be studied using a single network ADP.

H. Zhang et al., *Adaptive Dynamic Programming for Control*,
Communications and Control Engineering, DOI 10.1007/978-1-4471-4757-2_9,

9.2 Infinite Horizon Zero-Sum Games for a Class of Affine Nonlinear Systems

In this section, the nonlinear infinite horizon zero-sum differential games is studied. We propose a new iterative ADP method which is effective for both the situation that the saddle point does and does not exist. For the situation that the saddle point exists, the existence conditions of the saddle point are avoided. The value function can reach the saddle point using the present iterative ADP method. For the situation that the saddle point does not exist, the mixed optimal value function is obtained under a deterministic mixed optimal control scheme, using the present iterative ADP algorithm.

9.2.1 Problem Formulation

Consider the following two-person zero-sum differential games. The system is described by the continuous-time affine nonlinear equation

$$\dot{x}(t) = f(x(t), u(t), w(t)) = f(x(t)) + g(x(t))u(t) + k(x(t))w(t), \tag{9.1}$$

where $x(t) \in \mathbb{R}^n$, $u(t) \in \mathbb{R}^k$, $w(t) \in \mathbb{R}^m$, and the initial condition $x(0) = x_0$ is given.

The cost functional is the generalized quadratic form given by

$$J(x(0), u, w) = \int_0^\infty l(x, u, w)\mathrm{d}t, \tag{9.2}$$

where $l(x, u, w) = x^{\mathrm{T}}Ax + u^{\mathrm{T}}Bu + w^{\mathrm{T}}Cw + 2u^{\mathrm{T}}Dw + 2x^{\mathrm{T}}Eu + 2x^{\mathrm{T}}Fw$. The matrices A, B, C, D, E, and F have suitable dimensions and $A \geq 0$, $B > 0$, and $C < 0$. According to the situation of two players we have the following definitions. Let $\overline{J}(x) := \inf_u \sup_w J(x, u, w)$ be the *upper value function* and $\underline{J}(x) := \sup_w \inf_u J(x, u, w)$ be the *lower value function* with the obvious inequality $\overline{J}(x) \geq \underline{J}(x)$. Define the optimal control pairs to be $(\overline{u}, \overline{w})$ and $(\underline{u}, \underline{w})$ for upper and lower value functions, respectively. Then, we have $\overline{J}(x) = J(x, \overline{u}, \overline{w})$ and $\underline{J}(x) = J(x, \underline{u}, \underline{w})$.

If both $\overline{J}(x)$ and $\underline{J}(x)$ exist and

$$\overline{J}(x) = \underline{J}(x) = J^*(x) \tag{9.3}$$

holds, we say that the saddle point exists and the corresponding optimal control pair is denoted by (u^*, w^*).

We have the following lemma.

Lemma 9.1 *If the nonlinear system* (9.1) *is controllable and both the upper value function and lower value function exist, then* $\overline{J}(x)$ *is a solution of the following* *upper Hamilton–Jacobi–Isaacs (HJI) equation*:

$$\inf_u \sup_w \{\overline{J}_t + \overline{J}_x^{\mathrm{T}} f(x,u,w) + l(x,u,w)\} = 0, \tag{9.4}$$

which is denoted by $\mathrm{HJI}(\overline{J}(x), \overline{u}, \overline{w}) = 0$ *and* $\underline{J}(x)$ *is a solution of the following lower HJI equation*:

$$\sup_w \inf_u \{\underline{J}_t + \underline{J}_x^{\mathrm{T}} f(x,u,w) + l(x,u,w)\} = 0, \tag{9.5}$$

which is denoted by $\mathrm{HJI}(\underline{J}(x), \underline{u}, \underline{w}) = 0$.

9.2.2 Zero-Sum Differential Games Based on Iterative ADP Algorithm

As the HJI equations (9.4) and (9.5) cannot be solved in general, in the following, a new iterative ADP method for zero-sum differential games is developed.

9.2.2.1 Derivation of the Iterative ADP Method

The goal of the present iterative ADP method is to obtain the saddle point. As the saddle point may not exist, this motivates us to obtain the mixed optimal value function $J^o(x)$ where $\underline{J}(x) \le J^o(x) \le \overline{J}(x)$.

Theorem 9.2 (cf. [15]) *Let* $(\overline{u}, \overline{w})$ *be the optimal control pair for* $\overline{J}(x)$ *and* $(\underline{u}, \underline{w})$ *be the optimal control pair for* $\underline{J}(x)$. *Then, there exist control pairs* $(\overline{u}, w)$ *and* $(u, \underline{w})$ *which lead to* $J^o(x) = J(x, \overline{u}, w) = J(x, u, \underline{w})$. *Furthermore, if the saddle point exists, then* $J^o(x) = J^*(x)$.

Proof According to the definition of $\overline{J}(x)$, we have $J(x, \overline{u}, w) \le J(x, \overline{u}, \overline{w})$. As $J^o(x)$ is a mixed optimal value function, we also have $J^o(x) \le J(x, \overline{u}, \overline{w})$. As the system (9.1) is controllable and w is continuous on $\mathbb{R}^m$, there exists a control pair $(\overline{u}, w)$ which makes $J^o(x) = J(x, \overline{u}, w)$. On the other hand, we have $J^o(x) \ge J(x, \underline{u}, \underline{w})$. We also have $J(x, u, \underline{w}) \ge J(x, \underline{u}, \underline{w})$. As u is continuous on $\mathbb{R}^k$, there exists a control pair $(u, \underline{w})$ which makes $J^o(x) = J(x, u, \underline{w})$. If the saddle point exists, we have (9.3). On the other hand, $\underline{J}(x) \le J^o(x) \le \overline{J}(x)$. Then, clearly $J^o(x) = J^*(x)$. □

If (9.3) holds, we have a saddle point; if not, we adopt a mixed trajectory to obtain the mixed optimal solution of the game. To apply the mixed trajectory method, the game matrix is necessary under the trajectory sets of the control pair (u, w). Small Gaussian noises $\gamma_u \in R^k$ and $\gamma_w \in R^m$ are introduced that are added to the optimal control $\underline{u}$ and $\overline{w}$, respectively, where $\gamma_u^i(0,\sigma_i^2)$, $i=1,\dots,k$, and $\gamma_w^j(0,\sigma_j^2)$, $j=1,\dots,m$, are zero-mean Gaussian noises with variances σ_i^2 and σ_j^2, respectively.

We define the expected value function as $E(J(x)) = \min_{P_{Ii}} \max_{P_{IIj}} \sum_{i=1}^{2}\sum_{j=1}^{2} P_{Ii}L_{ij}P_{IIj}$, where we let $L_{11} = J(x,\overline{u},\overline{w})$, $L_{12} = J(x,(\underline{u}+\gamma_u),\underline{w})$, $L_{21} = J(x,\underline{u},\underline{w})$ and $L_{22} = J(x,\overline{u},(\overline{w}+\gamma_w))$. Let $\sum_{i=1}^{2} P_{Ii}=1$ and $P_{Ii}>0$. Let $\sum_{j=1}^{2} P_{IIj}=1$ and $P_{IIj}>0$. Next, let N be a large enough positive integer. Calculating the expected value function N times, we can obtain $E_1(J(x)), E_2(J(x)), \dots, E_N(J(x))$. Then, the mixed optimal value function can be written as

$$J^o(x) = E(E_i(J(x))) = \frac{1}{N}\sum_{i=1}^{N} E_i(J(x)).$$

Remark 9.3 In the classical mixed trajectory method, the whole control sets $\mathbb{R}^k$ and $\mathbb{R}^m$ should be searched under some distribution functions. As there are no constraints for both controls, we see that there exist controls that cause the system to be unstable. This is not permitted for real-world control systems. Thus, it is impossible to search the whole control sets and we can only search the local area around the stable controls which guarantees stability of the system. This is the reason why the small Gaussian noises γ_u and γ_w are introduced. So the meaning of the Gaussian noises can be seen in terms of the local stable area of the control pairs. A proposition will be given to show that the control pair chosen in the local area is stable (see Proposition 9.14). Similar work can also be found in [3, 14].

We can see that the mixed optimal solution is a mathematically expected value which means that it cannot be obtained in reality once the trajectories are determined. For most practical optimal control problems, however, the expected optimal solution (or mixed optimal solution) has to be achieved. To overcome this difficulty, a new method is developed in this section. Let $\alpha = (J^o(x)-\underline{J}(x))/(\overline{J}(x)-\underline{J}(x))$. Then, $J^o(x)$ can be written as $J^o(x) = \alpha\overline{J}(x) + (1-\alpha)\underline{J}(x)$. Let $l^o(x,\overline{u},\overline{w},\underline{u},\underline{w}) = \alpha l(x,\overline{u},\overline{w}) + (1-\alpha)l(x,\underline{u},\underline{w})$. We have $J^o(x(0)) = \int_0^\infty l^o\mathrm{d}t$. According to Theorem 9.2, the mixed optimal control pair can be obtained by regulating the control w in the control pair $(\overline{u},\overline{w})$ that minimizes the error between $\mathcal{J}(x)$ and $J^o(x)$ where the value function $\mathcal{J}(x)$ is defined as $\mathcal{J}(x(0)) = J(x(0),\overline{u},w) = \int_0^\infty l(x,\overline{u},w)\mathrm{d}t$ and $\underline{J}(x(0)) \le \mathcal{J}(x(0)) \le \overline{J}(x(0))$.

Define $\widetilde{J}(x(0)) = \int_0^\infty \widetilde{l}(x,w)dx$, where $\widetilde{l}(x,w) = l(x,\overline{u},w) - l^o(x,\overline{u},\overline{w},\underline{u},\underline{w})$. Then, the problem can be described as $\min_w(\widetilde{J}(x))^2$.

According to the principle of optimality, when $\widetilde{J}(x) \ge 0$ we have the following HJB equation:

$$\mathrm{HJB}(\widetilde{J}(x),w) := \min_{w}\{\widetilde{J}_t(x) + \widetilde{J}_x f(x,u,w) + \widetilde{l}(x,w)\} = 0. \tag{9.6}$$

For $\widetilde{J}(x) < 0$, we have $-\widetilde{J}(x) = -(\mathcal{J}(x) - J^o(x)) > 0$, and we can obtain the same HJB equation as (9.6).

9.2.2.2 The Iterative ADP Algorithm

Given the above preparation, we now formulate the iterative ADP algorithm for zero-sum differential games as follows:

1. Initialize the algorithm with a stabilizing control pair $(u^{[0]}, w^{[0]})$, and the value function is $V^{[0]}$. Choose the computation precision $\zeta > 0$. Set $i = 0$.
2. For the upper value function, let

$$\overline{V}^{[i]}(x(0)) = \int_0^\infty l(x, \overline{u}^{[i+1]}, \overline{w}^{[i+1]})\mathrm{d}t, \tag{9.7}$$

where the iterative optimal control pair is formulated as

$$\begin{aligned}\overline{u}^{[i+1]} = &-\frac{1}{2}(B - DC^{-1}D^{\mathrm{T}})^{-1}\big(2(k^{\mathrm{T}} - DC^{-1}F^{\mathrm{T}})x \\ &+ (g^{\mathrm{T}}(x) - DC^{-1}k^{\mathrm{T}}(x))\overline{V}_x^{[i]}\big),\end{aligned} \tag{9.8}$$

and

$$\overline{w}^{[i+1]} = -\frac{1}{2}C^{-1}\big(2D^{\mathrm{T}}\overline{u}^{[i+1]} + 2F^{\mathrm{T}}x + k^{\mathrm{T}}(x)\overline{V}_x^{[i]}\big). \tag{9.9}$$

$(\overline{u}^{[i]}, \overline{w}^{[i]})$ satisfies the HJI equation $\mathrm{HJI}(\overline{V}^{[i]}(x), \overline{u}^{[i]}, \overline{w}^{[i]}) = 0$, and $\overline{V}_x^{[i]} = d\overline{V}^{[i]}(x)/dx$.
3. If $|\overline{V}^{[i+1]}(x(0)) - \overline{V}^{[i]}(x(0))| < \zeta$, let $\overline{u} = \overline{u}^{[i]}$, $\overline{w} = \overline{w}^{[i]}$ and $\overline{J}(x) = \overline{V}^{[i+1]}(x)$. Set $i = 0$ and go to Step 4. Else, set $i = i + 1$ and go to Step 2.
4. For the lower value function, let

$$\underline{V}^{[i]}(x(0)) = \int_0^\infty l(x, \underline{u}^{[i+1]}, \underline{w}^{[i+1]})\mathrm{d}t, \tag{9.10}$$

where the iterative optimal control pair is formulated as

$$\underline{u}^{[i+1]} = -\frac{1}{2}g^{-1}(2D\underline{w}^{[i+1]} + 2k^{\mathrm{T}}x + g^{\mathrm{T}}(x)\underline{V}_x^{[i]}), \tag{9.11}$$

and

$$\begin{aligned}\underline{w}^{[i+1]} = &-\frac{1}{2}(C - D^{\mathrm{T}}BD)^{-1}(2(F^{\mathrm{T}} - D^{\mathrm{T}}g^{-1}E)x \\ &+ (k^{\mathrm{T}}(x) - D^{\mathrm{T}}g^{-1}g^{\mathrm{T}}(x))\underline{V}_x^{[i]}).\end{aligned} \tag{9.12}$$

$(\underline{u}^{[i]}, \underline{w}^{[i]})$ satisfies the HJI equation $\mathrm{HJI}(\underline{V}^{[i]}(x), \underline{u}^{[i]}, \underline{w}^{[i]}) = 0$, and $\underline{V}_x^{[i]} = d\underline{V}^{[i]}(x)/dx$.
5. If $|\underline{V}^{[i+1]}(x(0)) - \underline{V}^{[i]}(x(0))| < \zeta$, let $\underline{u} = \underline{u}^{[i]}$, $\underline{w} = \underline{w}^{[i]}$ and $\underline{J}(x) = \underline{V}^{[i+1]}(x)$. Set $i = 0$ and go to Step 6. Else, set $i = i + 1$ and go to Step 4.
6. If $|\overline{J}(x(0)) - \underline{J}(x(0))| < \zeta$, stop, and the saddle point is achieved. Else set $i = 0$ and go to the next step.

7. Regulate the control w for the upper value function and let

$$\begin{aligned}\widetilde{J}^{[i+1]}(x(0)) &= \mathcal{V}^{[i+1]}(x(0)) - J^{o}(x(0)) \\ &= \int_0^\infty \widetilde{l}(x, \overline{u}, w^{[i]})\mathrm{d}t.\end{aligned} \tag{9.13}$$

The iterative optimal control is formulated as

$$w^{[i]} = -\frac{1}{2}C^{-1}(2D^{\mathrm{T}}\overline{u} + 2F^{\mathrm{T}}x + k^{\mathrm{T}}(x)\widetilde{V}_x^{[i+1]}), \tag{9.14}$$

where $\widetilde{V}_x^{[i]} = d\widetilde{V}^{[i]}(x)/dx$.

8. If $|\mathcal{V}^{[i+1]}(x(0)) - J^{o}(x(0))| < \zeta$, stop. Else, set $i = i + 1$ and go to Step 7.

9.2.2.3 Properties of the Iterative ADP Algorithm

In this part, some results are presented to show the stability and convergence of the present iterative ADP algorithm.

Theorem 9.4 (cf. [15]) *If for* $\forall i \geq 0$, $\mathrm{HJI}(\overline{V}^{[i]}(x), \overline{u}^{[i]}, \overline{w}^{[i]}) = 0$ *holds, and for* $\forall t$, $l(x, \overline{u}^{[i]}, \overline{w}^{[i]}) \geq 0$, *then the control pairs* $(\overline{u}^{[i]}, \overline{w}^{[i]})$ *make system* (9.1) *asymptotically stable.*

Proof According to (9.7), for $\forall t$, taking the derivative of $\overline{V}^{[i]}(x)$, we have

$$\frac{\mathrm{d}\overline{V}^{[i]}(x)}{\mathrm{d}t} = \overline{V}_x^{[i]\mathrm{T}}\left(f(x) + g(x)\overline{u}^{[i+1]} + k(x)\overline{w}^{[i+1]}\right). \tag{9.15}$$

From the HJI equation we have

$$0 = \overline{V}_x^{[i]\mathrm{T}} f(x, \overline{u}^{[i]}, \overline{w}^{[i]}) + l(x, \overline{u}^{[i]}, \overline{w}^{[i]}). \tag{9.16}$$

Combining (9.15) and (9.16), we get

$$\begin{aligned}\frac{\mathrm{d}\overline{V}^{[i]}(x)}{\mathrm{d}t} &= \overline{V}_x^{[i]\mathrm{T}}(g(x) - k(x)C^{-1}D^{\mathrm{T}})(\overline{u}^{[i+1]} - \overline{u}^{[i]}) \\ &\quad - x^{\mathrm{T}}Ax - \overline{u}^{[i]\mathrm{T}}(B - DC^{-1}D^{\mathrm{T}})\overline{u}^{[i]} - \frac{1}{4}\overline{V}_x^{[i]\mathrm{T}}k(x)C^{-1}k^{\mathrm{T}}(x)\overline{V}_x^{[i]} \\ &\quad - 2x^{\mathrm{T}}(E - FC^{-1}D^{\mathrm{T}})\overline{u}^{[i+1]} + x^{\mathrm{T}}FC^{-1}F^{\mathrm{T}}x.\end{aligned} \tag{9.17}$$

According to (9.8) we have

$$\frac{\mathrm{d}\overline{V}^{[i]}(x)}{\mathrm{d}t} = -\,(\overline{u}^{[i+1]} - \overline{u}^{[i]})^{\mathrm{T}}(B - DC^{-1}D^{\mathrm{T}})$$

$$\times(\overline{u}^{[i+1]}-\overline{u}^{[i]})-l(x,\overline{u}^{[i+1]},\overline{w}^{(i+1)})$$
$$\leq 0. \tag{9.18}$$

So, $\overline{V}^{[i]}(x)$ is a Lyapunov function. Let $\varepsilon>0$ and $\|x(t_0)\|<\delta(\varepsilon)$. Then, there exist two functions $\alpha(\|x\|)$ and $\beta(\|x\|)$ which belong to class $\mathcal{K}$ and satisfy

$$\alpha(\varepsilon)\geq\beta(\delta)\geq\overline{V}^{[i]}(x(t_0))\geq\overline{V}^{[i]}(x(t))\geq\alpha(\|x\|). \tag{9.19}$$

Therefore, system (9.1) is asymptotically stable. □

Theorem 9.5 (cf. [15]) *If for* $\forall i\geq 0$, $\mathrm{HJI}(\underline{V}^{[i]}(x),\underline{u}^{[i]},\underline{w}^{[i]})=0$ *holds, and for* $\forall t$, $l(x,\underline{u}^{[i]},\underline{w}^{[i]})<0$, *then the control pairs* $(\underline{u}^{[i]},\underline{w}^{[i]})$ *make system* (9.1) *asymptotically stable.*

Corollary 9.6 *If for* $\forall i\geq 0$, $\mathrm{HJI}(\underline{V}^{[i]}(x),\underline{u}^{[i]},\underline{w}^{[i]})=0$ *holds, and for* $\forall t$, $l(x,\underline{u}^{[i]},\underline{w}^{[i]})\geq 0$, *then the control pairs* $(\underline{u}^{[i]},\underline{w}^{[i]})$ *make system* (9.1) *asymptotically stable.*

Proof As $\underline{V}^{[i]}(x)\leq\overline{V}^{[i]}(x)$ and $l(x,\underline{u}^{[i]},\underline{w}^{[i]})\geq 0$, we have $0\leq\underline{V}^{[i]}(x)\leq\overline{V}^{[i]}(x)$.

From Theorem 9.4, we know that for $\forall t_0$, there exist two functions $\alpha(\|x\|)$ and $\beta(\|x\|)$ which belong to class $\mathcal{K}$ and satisfy (9.19).

As $\overline{V}^{[i]}(x)\to 0$, there exist time instants t_1 and t_2 (without loss of generality, let $t_0<t_1<t_2$) that satisfy

$$\overline{V}^{[i]}(x(t_0))\geq\overline{V}^{[i]}(x(t_1))\geq\underline{V}^{[i]}(x(t_0))\geq\overline{V}^{[i]}(x(t_2)). \tag{9.20}$$

Choose $\varepsilon_1>0$ that satisfies $\underline{V}^{[i]}(x(t_0))\geq\alpha(\varepsilon_1)\geq\overline{V}^{[i]}(x(t_2))$. Then, there exists $\delta_1(\varepsilon_1)>0$ that makes $\alpha(\varepsilon_1)\geq\beta(\delta_1)\geq\overline{V}^{[i]}(x(t_2))$. Then, we obtain

$$\underline{V}^{[i]}(x(t_0))\geq\alpha(\varepsilon_1)\geq\beta(\delta_1)\geq\overline{V}^{[i]}(x(t_2))\geq\overline{V}^{[i]}(x(t))\geq\underline{V}^{[i]}(x(t))\geq\alpha(\|x\|). \tag{9.21}$$

According to (9.19), we have

$$\alpha(\varepsilon)\geq\beta(\delta)\geq\underline{V}^{[i]}(x(t_0))\geq\alpha(\varepsilon_1)\geq\beta(\delta_1)\geq\underline{V}^{[i]}(x(t))\geq\alpha(\|x\|). \tag{9.22}$$

Since $\alpha(\|x\|)$ belongs to class $\mathcal{K}$, we obtain $\|x\|\leq\varepsilon$.

Therefore, we conclude that the system (9.1) is asymptotically stable. □

Corollary 9.7 *If for* $\forall i\geq 0$, $\mathrm{HJI}(\overline{V}^{[i]}(x),\overline{u}^{[i]},\overline{w}^{[i]})=0$ *holds, and for* $\forall t$, $l(x,\overline{u}^{[i]},\overline{w}^{[i]})<0$, *then the control pairs* $(\overline{u}^{[i]},\overline{w}^{[i]})$ *make system* (9.1) *asymptotically stable.*

Theorem 9.8 (cf. [15]) *If for* $\forall i \geq 0$, $\mathrm{HJI}(\overline{V}^{[i]}(x), \overline{u}^{[i]}, \overline{w}^{[i]}) = 0$ *holds, and* $l(x, \overline{u}^{[i]}, \overline{w}^{[i]})$ *is the utility function, then the control pairs* $(\overline{u}^{[i]}, \overline{w}^{[i]})$ *make system* (9.1) *asymptotically stable.*

Proof For the time sequence $t_0 < t_1 < t_2 < \cdots < t_m < t_{m+1} < \cdots$, without loss of generality, we assume $l(x, \overline{u}^{[i]}, \overline{w}^{[i]}) \geq 0$ in $[t_{2n}, t_{(2n+1)})$ and $l(x, \overline{u}^{[i]}, \overline{w}^{[i]}) < 0$ in $[t_{2n+1}, t_{(2(n+1))})$ where $n = 0, 1, \ldots$.

Then, for $t \in [t_0, t_1)$ we have $l(x, \overline{u}^{[i]}, \overline{w}^{[i]}) \geq 0$ and $\int_{t_0}^{t_1} l(x, \overline{u}^{[i]}, \overline{w}^{[i]}) \mathrm{d}t \geq 0$. According to Theorem 9.4, we have $\|x(t_0)\| \geq \|x(t)\| \geq \|x(t_1)\|$.

For $t \in [t_1, t_2)$ we have $l(x, \overline{u}^{[i]}, \overline{w}^{[i]}) < 0$ and $\int_{t_1}^{t_2} l(x, \overline{u}^{[i]}, \overline{w}^{[i]}) \mathrm{d}t < 0$. According to Corollary 9.7, we have $\|x(t_1)\| > \|x(t)\| > \|x(t_2)\|$. So we obtain $\|x(t_0)\| \geq \|x(t)\| > \|x(t_2)\|$, for $\forall t \in [t_0, t_2)$.

Using mathematical induction, for $\forall t$, we have $\|x(t')\| \leq \|x(t)\|$ where $t' \in [t, \infty)$. So we conclude that the system (9.1) is asymptotically stable, and the proof is completed. □

Theorem 9.9 (cf. [15]) *If for* $\forall i \geq 0$, $\mathrm{HJI}(\underline{V}^{[i]}(x), \underline{u}^{[i]}, \underline{w}^{[i]}) = 0$ *holds, and* $l(x, \underline{u}^{[i]}, \underline{w}^{[i]})$ *is the utility function, then the control pairs* $(\underline{u}^{[i]}, \underline{w}^{[i]})$ *make system* (9.1) *asymptotically stable.*

Next, we will give the convergence proof of the iterative ADP algorithm.

Proposition 9.10 *If for* $\forall i \geq 0$, $\mathrm{HJI}(\overline{V}^{[i]}(x), \overline{u}^{[i]}, \overline{w}^{[i]}) = 0$ *holds, then the control pairs* $(\overline{u}^{[i]}, \overline{w}^{[i]})$ *make the upper value function* $\overline{V}^{[i]}(x) \to \bar{J}(x)$ *as* $i \to \infty$.

Proof According to $HJI(\overline{V}^{[i]}(x), \overline{u}^{[i]}, \overline{w}^{[i]}) = 0$, we obtain $\mathrm{d}\overline{V}^{[i+1]}(x)/\mathrm{d}t$ by replacing the index "i" by the index "$i+1$":

$$\begin{aligned}\frac{\mathrm{d}\overline{V}^{[i+1]}(x)}{\mathrm{d}t} &= -(x^{\mathrm{T}}Ax + \overline{u}^{(i+1)\mathrm{T}}(B - DC^{-1}D^{\mathrm{T}})\overline{u}^{[i+1]} \\ &\quad + \frac{1}{4}\overline{V}_x^{[i]T} k(x)C^{-1}k^{\mathrm{T}}(x)\overline{V}_x^{[i]} + 2x^{\mathrm{T}}(E - FC^{-1}D^{\mathrm{T}})\overline{u}^{[i+1]} \\ &\quad - x^{\mathrm{T}}FC^{-1}F^{\mathrm{T}}x).\end{aligned} \tag{9.23}$$

According to (9.18), we obtain

$$\begin{aligned}\frac{\mathrm{d}(\overline{V}^{[i+1]}(x) - \overline{V}^{[i]}(x))}{\mathrm{d}t} &= \frac{\mathrm{d}\overline{V}^{[i+1]}(x)}{\mathrm{d}t} - \frac{\mathrm{d}\overline{V}^{[i]}(x)}{\mathrm{d}t} \\ &= (\overline{u}^{[i+1]} - \overline{u}^{[i]})^{\mathrm{T}}(B - DC^{-1}D^{\mathrm{T}}(\overline{u}^{[i+1]} - \overline{u}^{[i]}) \\ &> 0.\end{aligned} \tag{9.24}$$

Since the system (9.1) is asymptotically stable, its state trajectories x converge to zero, and so does $\overline{V}^{[i+1]}(x) - \overline{V}^{[i]}(x)$. Since $d(\overline{V}^{[i+1]}(x) - \overline{V}^{[i]}(x))/\mathrm{d}t \geq 0$

on these trajectories, it implies that $\overline{V}^{[i+1]}(x) - \overline{V}^{[i]}(x) \le 0$; that is $\overline{V}^{[i+1]}(x) \le \overline{V}^{[i]}(x)$. Thus, $\overline{V}^{[i]}(x)$ is convergent as $i \to \infty$.

Next, we define $\lim_{i\to\infty} \overline{V}^{[i]}(x) = \overline{V}^{[\infty]}(x)$.

For $\forall i$, let $\overline{w}^* = \arg\max_w \{\int_t^{\hat{t}} l(x,u,w)\mathrm{d}\tau + \overline{V}^{[i]}(x(\hat{t}))\}$. Then, according to the principle of optimality, we have

$$\begin{aligned}\overline{V}^{[i]}(x) &\le \sup_w \left\{\int_t^{\hat{t}} l(x,u,w)\mathrm{d}\tau + \overline{V}^{[i]}(x(\hat{t}))\right\}\\ &= \int_t^{\hat{t}} l(x,u,\overline{w}^*)\mathrm{d}\tau + \overline{V}^{[i]}(x(\hat{t})).\end{aligned} \tag{9.25}$$

Since $\overline{V}^{[i+1]}(x) \le \overline{V}^{[i]}(x)$, we have $\overline{V}^{[\infty]}(x) \le \int_t^{\hat{t}} l(x,u,\overline{w}^*)\mathrm{d}\tau + \overline{V}^{[i]}(x(\hat{t}))$.

Letting $i \to \infty$, we obtain $\overline{V}^{[\infty]}(x) \le \int_t^{\hat{t}} l(x,u,\overline{w}^*)\mathrm{d}\tau + \overline{V}^{[\infty]}(x(\hat{t}))$. So, we have $\overline{V}^{[\infty]}(x) \le \inf_u \sup_w \{\int_t^{\hat{t}} l(x,u,w)\mathrm{d}t + \overline{V}^{[i]}(x(\hat{t}))\}$.

Let $\epsilon > 0$ be an arbitrary positive number. Since the upper value function is nonincreasing and convergent, there exists a positive integer i such that $\overline{V}^{[i]}(x) - \epsilon \le \overline{V}^{[\infty]}(x) \le \overline{V}^{[i]}(x)$.

Let $\overline{u}^* = \arg\min_u \{\int_t^{\hat{t}} l(x,u,\overline{w}^*)\mathrm{d}\tau + \overline{V}^{[i]}(x(\hat{t}))\}$. Then we get $\overline{V}^{[i]}(x) = \int_t^{\hat{t}} l(x,\overline{u}^*,\overline{w}^*)\mathrm{d}\tau + \overline{V}^{[i]}(x(\hat{t}))$.

Thus, we have

$$\begin{aligned}\overline{V}^{[\infty]}(x) &\ge \int_t^{\hat{t}} l(x,\overline{u}^*,\overline{w}^*)\mathrm{d}\tau + \overline{V}^{[i]}(x(\hat{t})) - \epsilon\\ &\ge \int_t^{\hat{t}} l(x,\overline{u}^*,\overline{w}^*)\mathrm{d}\tau + \overline{V}^{[\infty]}(x(\hat{t})) - \epsilon\\ &= \inf_u \sup_w \left\{\int_t^{\hat{t}} l(x,u,w)\mathrm{d}\tau + \overline{V}^{[\infty]}(x(\hat{t}))\right\} - \epsilon.\end{aligned} \tag{9.26}$$

Since ϵ is arbitrary, we have

$$\overline{V}^{[\infty]}(x) \ge \inf_u \sup_w \left\{\int_t^{\hat{t}} l(x,u,w)\mathrm{d}\tau + \overline{V}^{[\infty]}(x(\hat{t}))\right\}.$$

Therefore, we obtain

$$\overline{V}^{[\infty]}(x) = \inf_u \sup_w \left\{\int_t^{\hat{t}} l(x,u,w)\mathrm{d}\tau + \overline{V}^{[\infty]}(x(\hat{t}))\right\}.$$

Let $\hat{t} \to \infty$, we have

$$\overline{V}^{[\infty]}(x) = \inf_u \sup_w J(x,u,w) = \overline{J}(x).$$

□

Proposition 9.11 *If for* $\forall i \geq 0$, $\mathrm{HJI}(\underline{V}^{[i]}(x), \underline{u}^{[i]}, \underline{w}^{[i]}) = 0$ *holds, then the control pairs* $(\underline{u}^{[i]}, \underline{w}^{[i]})$ *make the lower value function* $\underline{V}^{[i]}(x) \to \underline{J}(x)$ *as* $i \to \infty$.

Theorem 9.12 (cf. [15]) *If the saddle point of the zero-sum differential game exists, then the control pairs* $(\overline{u}^{[i]}, \overline{w}^{[i]})$ *and* $(\underline{u}^{[i]}, \underline{w}^{[i]})$ *make* $\overline{V}^{[i]}(x) \to J^*(x)$ *and* $\underline{V}^{[i]}(x) \to J^*(x)$, *respectively, as* $i \to \infty$.

Proof For the upper value function, according to Proposition 9.10, we have $\overline{V}^{[i]}(x) \to \overline{J}(x)$ under the control pairs $(\overline{u}^{[i]}, \overline{w}^{[i]})$ as $i \to \infty$. So the optimal control pair for the upper value function satisfies $\overline{J}(x) = J(x, \overline{u}, \overline{w}) = \inf_u \sup_w J(x, u, w)$.

On the other hand, there exists an optimal control pair (u^*, w^*) making the value reach the saddle point. According to the property of the saddle point, the optimal control pair (u^*, w^*) satisfies $J^*(x) = J(x, u^*, w^*) = \inf_u \sup_w J(x, u, w)$.

So, we have $\overline{V}^{[i]}(x) \to J^*(x)$ under the control pair $(\overline{u}^{[i]}, \overline{w}^{[i]})$ as $i \to \infty$. Similarly, we can derive $\underline{V}^{[i]}(x) \to J^*(x)$ under the control pairs $(\underline{u}^{[i]}, \underline{w}^{[i]})$ as $i \to \infty$. □

Remark 9.13 From the proofs we see that the complex existence conditions of the saddle point in [1, 2] are not necessary. If the saddle point exists, the iterative value functions can converge to the saddle point using the present iterative ADP algorithm.

In the following part, we emphasize that when the saddle point does not exist, the mixed optimal solution can be obtained effectively using the iterative ADP algorithm.

Proposition 9.14 *If* $\overline{u} \in \mathbb{R}^k$, $w^{[i]} \in \mathbb{R}^m$ *and the utility function is* $\tilde{l}(x, w^{[i]}) = l(x, \overline{u}, w^{[i]}) - l^o(x, \overline{u}, \overline{w}, \underline{u}, \underline{w})$, *and* $w^{[i]}$ *is expressed in* (9.14), *then the control pairs* $(\overline{u}, w^{[i]})$ *make the system* (9.1) *asymptotically stable.*

Proposition 9.15 *If* $\overline{u} \in \mathbb{R}^k$, $w^{[i]} \in \mathbb{R}^m$ *and for* $\forall t$, *the utility function* $\tilde{l}(x, w^{[i]}) \geq 0$, *then the control pairs* $(\overline{u}, w^{[i]})$ *make* $\widetilde{V}^{[i]}(x)$ *a nonincreasing convergent sequence as* $i \to \infty$.

Proposition 9.16 *If* $\overline{u} \in \mathbb{R}^k$, $w^{[i]} \in \mathbb{R}^m$ *and for* $\forall t$, *the utility function* $\tilde{l}(x, w^{[i]}) < 0$, *then the control pairs* $(\overline{u}, w^{[i]})$ *make* $\tilde{V}^{[i]}(x)$ *a nondecreasing convergent sequence as* $i \to \infty$.

Theorem 9.17 (cf. [15]) *If* $\overline{u} \in \mathbb{R}^k$, $w^{[i]} \in \mathbb{R}^m$, *and* $\tilde{l}(x, w^{[i]})$ *is the utility function, then the control pairs* $(\overline{u}, w^{[i]})$ *make* $\widetilde{V}^{[i]}(x)$ *convergent as* $i \to \infty$.

Proof For the time sequence $t_0 < t_1 < t_2 < \cdots < t_m < t_{m+1} < \cdots$, without loss of generality, we suppose $\tilde{l}(x, w^{[i]}) \geq 0$ in $[t_{2n}, t_{2n+1})$ and $\tilde{l}(x, w^{[i]}) < 0$ in $[t_{2n+1}, t_{2(n+1)})$, where $n = 0, 1, \ldots$.

For $t \in [t_{2n}, t_{2n+1})$ we have $\tilde{l}(x, w^{[i]}) \geq 0$ and $\int_{t_0}^{t_1} \tilde{l}(x, w^{[i]}) dt \geq 0$. According to Proposition 9.15, we have $\widetilde{V}^{[i+1]}(x) \leq \widetilde{V}^{[i]}(x)$. For $t \in [t_{2n+1}, t_{2(n+1)})$ we have $\tilde{l}(x, w^{[i]}) < 0$ and $\int_{t_1}^{t_2} \tilde{l}(x, w^{[i]}) dt < 0$. According to Proposition 9.16 we have $\widetilde{V}^{[i+1]}(x) > \widetilde{V}^{[i]}(x)$. Then, for $\forall t_0$, we have

$$\begin{aligned} \left\| \widetilde{V}^{[i+1]}(x(t_0)) \right\| &= \left\| \int_{t_0}^{t_1} \tilde{l}(x, w^{[i]}) dt \right\| + \left\| \int_{t_1}^{t_2} \tilde{l}(x, w^{[i]}) dt \right\| \\ &\quad + \ldots + \left\| \int_{t_m}^{t_{(m+1)}} \tilde{l}(x, w^{[i]}) dt \right\| + \ldots \\ &< \left\| \widetilde{V}^{[i]}(x(t_0)) \right\|. \end{aligned} \tag{9.27}$$

So, $\widetilde{V}^{[i]}(x)$ is convergent as $i \to \infty$. □

Theorem 9.18 (cf. [15]) *If $\overline{u} \in R^k$, $w^{[i]} \in R^m$, and $\tilde{l}(x, w^{[i]})$ is the utility function, then the control pairs $(\overline{u}, w^{[i]})$ make $\mathcal{V}^{[i]}(x) \to J^o(x)$ as $i \to \infty$.*

Proof It is proved by contradiction. Suppose that the control pair $(\overline{u}, w^{[i]})$ makes the value function $\mathcal{V}^{[i]}(x)$ converge to $\mathcal{J}'(x)$ and $\mathcal{J}'(x) \neq J^o(x)$.

According to Theorem 9.17, based on the principle of optimality, as $i \to \infty$ we have the HJB equation $\text{HJB}(\widetilde{J}(x), w) = 0$.

From the assumptions we know that $|\mathcal{V}^{[i]}(x) - J^o(x)| \neq 0$ as $i \to \infty$. From Theorem 9.5, we know that there exists a control pair $(\overline{u}, w')$ that makes $J(x, \overline{u}, w') = J^o(x)$, which minimizes the performance index function $\widetilde{J}(x)$. According to the principle of optimality, we also have the HJB equation $\text{HJB}(\widetilde{J}(x), w') = 0$.

It is a contradiction. So the assumption does not hold. Thus, we have $\mathcal{V}^{[i]}(x) \to J^o(x)$ as $i \to \infty$. □

Remark 9.19 For the situation where the saddle point does not exist, the methods in [1, 2] are all invalid. Using our iterative ADP method, the iterative value function reaches the mixed optimal value function $J^o(x)$ under the deterministic control pair. Therefore, we emphasize that the present iterative ADP method is more effective.

9.2.3 Simulations

Example 9.20 The dynamics of the benchmark nonlinear plant can be expressed by system (9.1) where

$$f(x) = \begin{bmatrix} x_2 & \dfrac{-x_1 + \varepsilon x_4^2 \sin x_3}{1 - \varepsilon^2 \cos^2 x_3} & x_4 & \dfrac{\varepsilon \cos x_3 (x_1 - \varepsilon x_4^2 \sin x_3)}{1 - \varepsilon^2 \cos^2 x_3} \end{bmatrix}^{\mathrm{T}},$$

$$g(x) = \begin{bmatrix} 0 & \dfrac{-\varepsilon \cos x_3}{1 - \varepsilon^2 \cos^2 x_3} & 0 & \dfrac{1}{1 - \varepsilon^2 \cos^2 x_3} \end{bmatrix}^{\mathrm{T}},$$

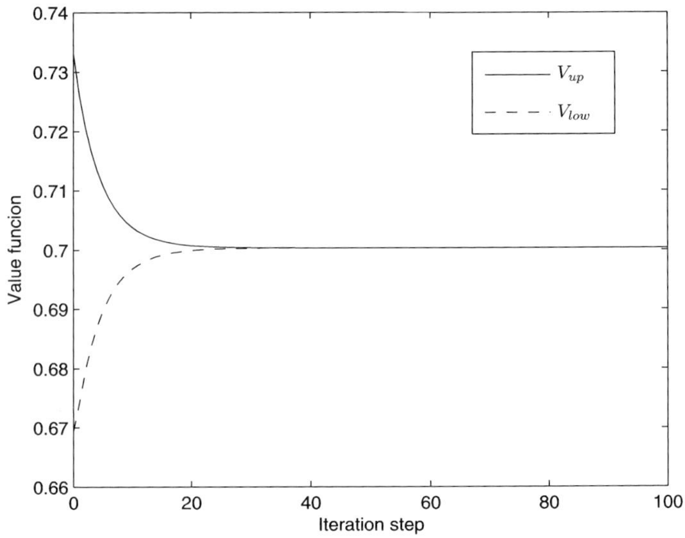

Fig. 9.1 Trajectories of upper and lower value function

$$k(x)=\begin{bmatrix} 0 & \dfrac{1}{1-\varepsilon^2\cos^2 x_3} & 0 & \dfrac{-\varepsilon\cos x_3}{1-\varepsilon^2\cos^2 x_3} \end{bmatrix}^{\mathrm{T}}, \tag{9.28}$$

and $\varepsilon = 0.2$. The initial state is given as $x(0) = [1, 1, 1, 1]^{\mathrm{T}}$. The cost functional is defined by (9.2) where the utility function is expressed as $l(x, u, w) = x_1^2 + 0.1x_2^2 + 0.1x_3^2 + 0.1x_4^2 + \|u\|^2 - \gamma^2\|w\|^2$ and $\gamma^2 = 10$.

Any differential structure can be used to implement the iterative ADP method. For facilitating the implementation of the algorithm, we choose three-layer neural networks as the critic networks with the structure of 4–8–1. The structures of the u and w for the upper value function are 4–8–1 and 5–8–1; while they are 5–8–1 and 4–8–1 for the lower one. The initial weights are all randomly chosen in $[-0.1,\ 0.1]$. Then, for each i, the critic network and the action networks are trained for 1000 time steps so that the given accuracy $\zeta = 10^{-6}$ is reached. Let the learning rate $\eta = 0.01$. The iterative ADP method runs for $i = 70$ times and the convergence trajectory of the value function is shown in Fig. 9.1. We can see that the saddle point of the game exists. Then, we apply the controller to the benchmark system and run for $T_f = 60$ seconds. The optimal control trajectories are shown in Fig. 9.2. The corresponding state trajectories are shown in Figs. 9.3 and 9.4, respectively.

Remark 9.21 The simulation results illustrate the effectiveness of the present iterative ADP algorithm. If the saddle point exists, the iterative control pairs $(\overline{u}^{[i]}, \overline{w}^{[i]})$ and $(\underline{u}^{[i]}, \underline{w}^{[i]})$ can make the iterative value functions reach the saddle point, while the existence conditions of the saddle point are avoided.

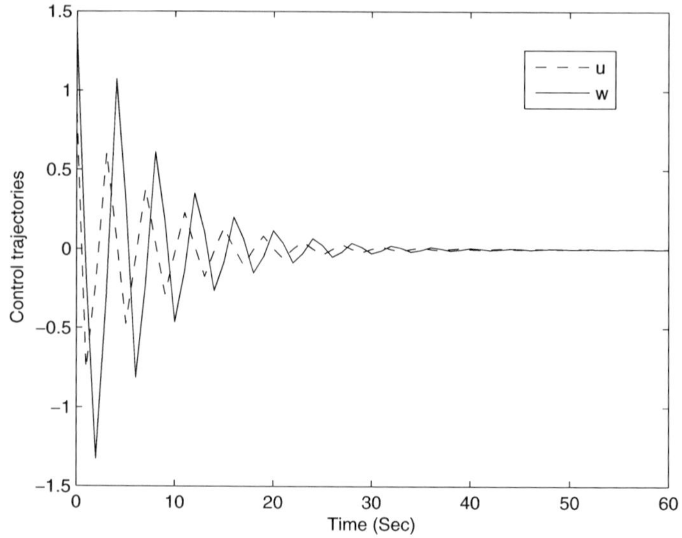

Fig. 9.2 Trajectories of the controls

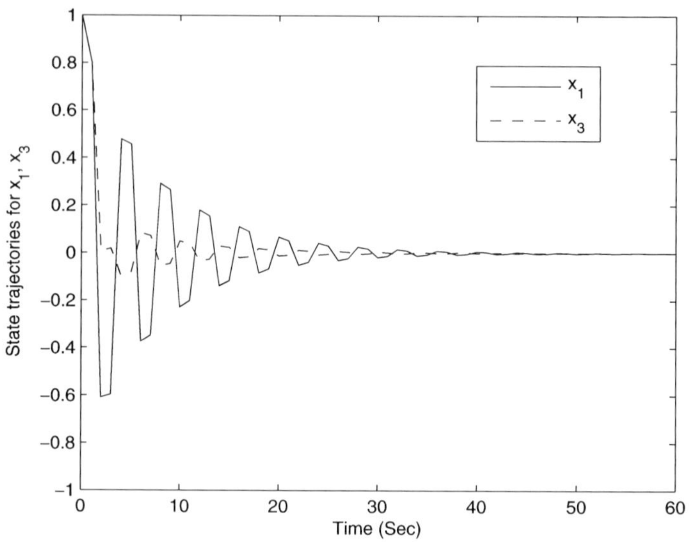

Fig. 9.3 Trajectories of state x_1 and x_3

Example 9.22 In this example, we just change the utility function to

$$\begin{aligned} l(x,u,w) = {} & x_1^2 + 0.1x_2^2 + 0.1x_3^2 + 0.1x_4^2 + \|u\|^2 - \gamma^2\|w\|^2 - 0.1uw \\ & + 0.1x^{\mathrm{T}}u + 0.1x^{\mathrm{T}}w, \end{aligned}$$

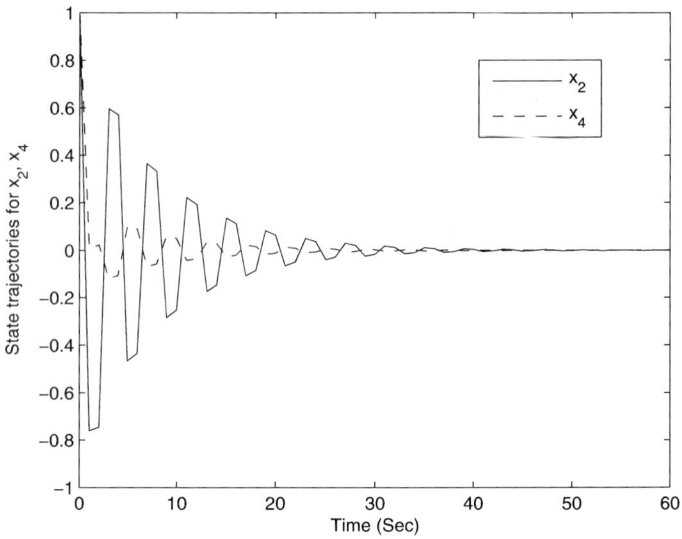

Fig. 9.4 Trajectories of state x_2 and x_4

and all other conditions are the same as the ones in Example 9.20. We obtain $\overline{J}(x(0)) = 0.65297$ and $\underline{J}(x(0)) = 0.44713$, with trajectories shown in Figs. 9.5(a) and (b), respectively. Obviously, the saddle point does not exist. Thus, the method in [1] is invalid. Using the present mixed trajectory method, we choose the Gaussian noises $\gamma_u(0, 0.05^2)$ and $\gamma_w(0, 0.05^2)$. Let $N = 5000$ times. The value function trajectories are shown in Fig. 9.5(c). Then, we obtain the value of the mixed optimal value function $J^o(x(0)) = 0.55235$ and then $\alpha = 0.5936$. Regulating the control w to obtain the trajectory of the mixed optimal value function displayed in Fig. 9.5. The state trajectories are shown in Figs. 9.6(a) and 9.7, respectively. The corresponding control trajectories are shown in Figs. 9.8 and 9.9, respectively.

9.3 Finite Horizon Zero-Sum Games for a Class of Nonlinear Systems

In this section, a new iterative approach is derived to solve optimal policies of finite horizon quadratic zero-sum games for a class of continuous-time nonaffine nonlinear system. Through the iterative algorithm between two sequences, which are a sequence of state trajectories of linear quadratic zero-sum games and a sequence of corresponding Riccati differential equations, the optimal policies for nonaffine nonlinear zero-sum games are given. Under very mild conditions of local Lipschitz continuity, the convergence of approximating linear time-varying sequences is proved.

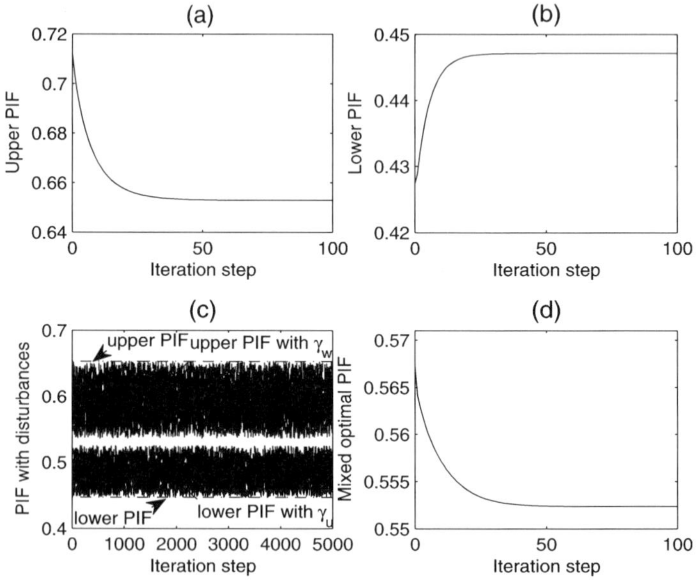

Fig. 9.5 Performance index function trajectories. (**a**) Trajectory of upper value function. (**b**) Trajectory of lower value function. (**c**) Performance index functions with disturbances. (**d**) Trajectory of the mixed optimal performance index function

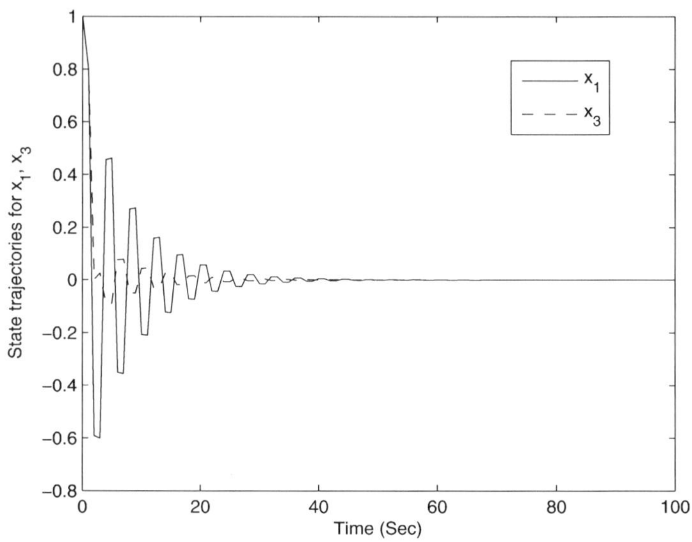

Fig. 9.6 Trajectories of state x_1 and x_3

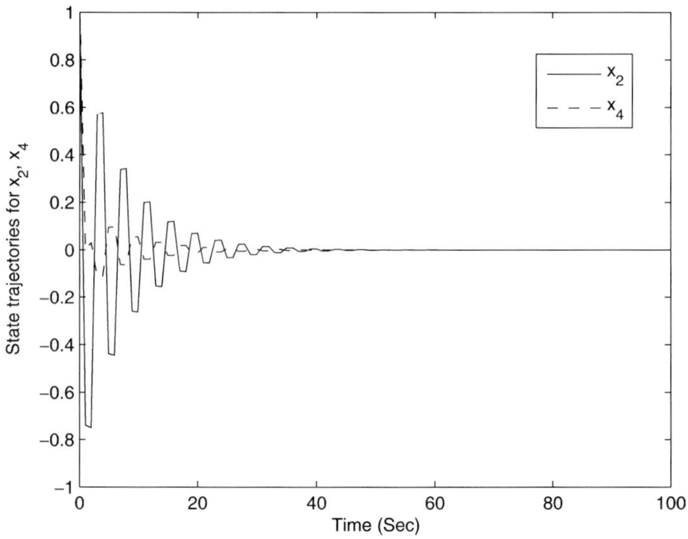

Fig. 9.7 Trajectories of state x_2 and x_4

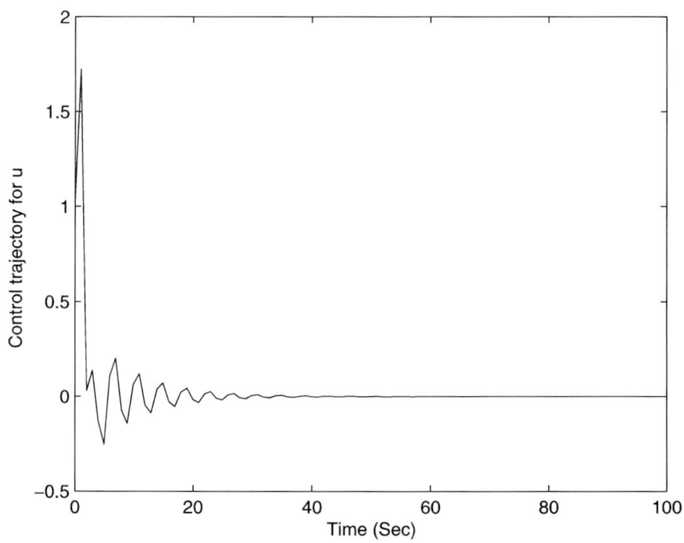

Fig. 9.8 Trajectory of control u

9.3.1 Problem Formulation

Consider a continuous-time nonaffine nonlinear zero-sum game described by the state equation

$$\dot{x}(t) = f(x(t), u(t), w(t)), \;\; x(t_0) = x_0 \tag{9.29}$$

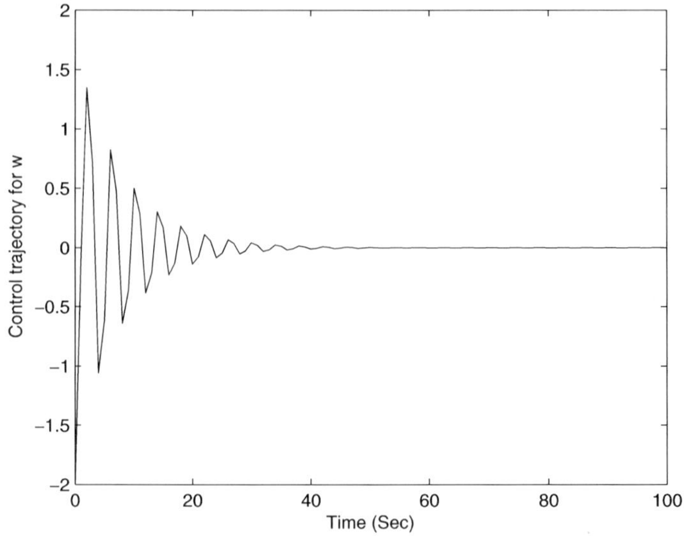

Fig. 9.9 Trajectory of control w

with the finite horizon cost functional

$$\begin{aligned} J(x_0,u,w) = {} & \frac{1}{2}x^{\mathrm{T}}(t_f)F(x(t_f))x(t_f) \\ & + \frac{1}{2}\int_{t_0}^{t_f}\big[x^{\mathrm{T}}(t)Q(x(t))x(t)+u^{\mathrm{T}}(t)R(x(t))u(t) \\ & - w^{\mathrm{T}}(t)S(x(t))w(t)\big]\,\mathrm{d}t, \end{aligned} \tag{9.30}$$

where $x(t)\in\mathbb{R}^n$ is the state, $x(t_0)\in\mathbb{R}^n$ is the initial state, t_f is the terminal time, the control input $u(t)$ takes values in a convex and compact set $U\subset\mathbb{R}^{m_1}$, and $w(t)$ takes values in a convex and compact set $W\subset\mathbb{R}^{m_2}$. $u(t)$ seeks to minimize the cost functional $J(x_0,u,w)$, while $w(t)$ seeks to maximize it. The state-dependent weight matrices $F(x(t))$, $Q(x(t))$, $R(x(t))$, $S(x(t))$ are with suitable dimensions and $F(x(t))\geq 0$, $Q(x(t))\geq 0$, $R(x(t))>0$, $S(x(t))>0$. In this section, $x(t)$, $u(t)$, and $w(t)$ sometimes are described by x, u, and w for brevity. Our objective is to find the optimal policies for the above nonaffine nonlinear zero-sum games.

In the nonaffine nonlinear zero-sum game problem, nonlinear functions are implicit function with respect to controller input. It is very hard to obtain the optimal policies satisfying (9.29) and (9.30). For practical purposes one may just as well be interested in finding a near-optimal or an approximate optimal policy. Therefore, we present an iterative algorithm to deal with this problem. Nonaffine nonlinear zero-sum games are transformed into an equivalent sequence of linear quadratic zero-sum games which can use the linear quadratic zero-sum game theory directly.

9.3.2 Finite Horizon Optimal Control of Nonaffine Nonlinear Zero-Sum Games

Using a factored form to represent the system (9.29), we get

$$\begin{aligned} \dot{x}(t) &= f(x(t))x(t) + g(x(t), u(t))u(t) + k(x(t), w(t))w(t), \\ x(t_0) &= x_0, \end{aligned} \tag{9.31}$$

where $f\colon \mathbb{R}^n \to \mathbb{R}^{n\times n}$ is a nonlinear matrix-valued function of x, $g\colon \mathbb{R}^n \times \mathbb{R}^{m_1} \to \mathbb{R}^{n\times m_1}$ is a nonlinear matrix-valued function of both the state x and control input u, and $k\colon \mathbb{R}^n \times \mathbb{R}^{m_2} \to \mathbb{R}^{n\times m_2}$ is a nonlinear matrix-valued function of both the state x and control input w.

We use the following sequence of linear time-varying differential equations to approximate the state equation (9.31):

$$\begin{aligned} \dot{x}_i(t) &= f(x_{i-1}(t))x_i(t) + g(x_{i-1}(t), u_{i-1}(t))u_i(t) + k(x_{i-1}(t), w_{i-1}(t))w_i(t), \\ x_i(t_0) &= x_0,\ i \ge 0, \end{aligned} \tag{9.32}$$

with the corresponding cost functional

$$\begin{aligned} V_i(x_0, u, w) = {} & \frac{1}{2} x_i^T(t_f) F(x_{i-1}(t_f)) x_i(t_f) \\ & + \frac{1}{2} \int_{t_0}^{t_f} \Big[x_i^T(t) Q(x_{i-1}(t)) x_i(t) + u_i^T(t) R(x_{i-1}(t)) u_i(t) \\ & - w_i^T(t) S(x_{i-1}(t)) w_i(t) \Big] \mathrm{d}t,\ i \ge 0, \end{aligned} \tag{9.33}$$

where the superscript i represents the iteration index. For the first approximation, $i = 0$, we assume that the initial values $x_{i-1}(t) = x_0$, $u_{i-1}(t) = 0$, and $w_{i-1}(t) = 0$. Obviously, for the ith iteration, $f(x_{i-1}(t))$, $g(x_{i-1}(t), u_{i-1}(t))$, $k(x_{i-1}(t), w_{i-1}(t))$, $F(x_{i-1}(t_f))$, $Q(x_{i-1}(t))$, $R(x_{i-1}(t))$, and $S(x_{i-1}(t))$ are time-varying functions which do not depend on $x_i(t)$, $u_i(t)$, and $w_i(t)$. Hence, each approximation problem in (9.32) and (9.33) is a linear quadratic zero-sum game problem which can be solved by the existing classical linear quadratic zero-sum game theory.

The corresponding Riccati differential equation of each linear quadratic zero-sum game can be expressed as

$$\begin{aligned} \dot{P}_i(t) = {} & - Q(x_{i-1}(t)) - P_i(t) f(x_{i-1}(t)) - f^{\mathrm{T}}(x_{i-1}(t)) P_i(t) \\ & + P_i(t) \big[g(x_{i-1}(t), u_{i-1}(t)) R^{-1}(x_{i-1}(t)) \\ & \times g^{\mathrm{T}}(x_{i-1}(t), u_{i-1}(t)) - k(x_{i-1}(t), w_{i-1}(t)) \\ & \times S^{-1}(x_{i-1}(t)) k^{\mathrm{T}}(x^{[i-1]}(t), w_{i-1}(t)) \big] P_i(t), \\ P_i(t_f) = {} & F(x_{i-1}(t_f)),\ i \ge 0, \end{aligned} \tag{9.34}$$

where $P_i \in \mathbb{R}^{n\times n}$ is a real, symmetric and nonnegative definite matrix.

Assumption 9.23 It is assumed that $S(x_{i-1}(t)) > \hat{S}_i$, where the threshold value $\hat{S}_i$ is defined as $\hat{S}_i = \inf\{S_i(t) > 0,$ and (9.34) does not have a conjugate point on $[0, t_f]\}$.

If Assumption 9.23 is satisfied, the game admits the optimal policies given by

$$\begin{aligned} u_i(t) &= -R^{-1}(x_{i-1}(t))g^{\mathrm{T}}(x_{i-1}(t), u_{i-1}(t))P_i(t)x_i(t), \\ w_i(t) &= S^{-1}(x_{i-1}(t))k^{\mathrm{T}}(x_{i-1}(t), w_{i-1}(t))P_i(t)x_i(t), \ i \geq 0, \end{aligned} \tag{9.35}$$

where $x_i(t)$ is the corresponding optimal state trajectory, generated by

$$\begin{aligned} \dot{x}_i(t) =& \big[f(x_{i-1}(t)) - g(x_{i-1}(t), u_{i-1}(t))R^{-1}(x_{i-1}(t)) \\ & \times g^{\mathrm{T}}(x_{i-1}(t), u_{i-1}(t))P_i(t) + k(x_{i-1}(t), w_{i-1}(t)) \\ & \times S^{-1}(x_{i-1}(t))k^{\mathrm{T}}(x_{i-1}(t), w_{i-1}(t))P_i(t)\big]x_i(t), \\ x_i(t_0) =& x_0. \end{aligned} \tag{9.36}$$

By using the iteration between sequences (9.34) and (9.36) sequently, the limit of the solution of the approximating sequence (9.32) will converge to the unique solution of system (9.29), and the sequences of optimal policies (9.35) will converge, too. The convergence of iterative algorithm will be analyzed in the next section. Notice that the factored form in (9.31) does not need to be unique. The approximating linear time-varying sequences will converge whatever the representation of $f(x(t))$, $g(x(t), u(t))$, and $k(x(t), w(t))$.

Remark 9.24 For the fixed finite interval $[t_0, t_f]$, if $S(x_{i-1}(t)) > \hat{S}_i$, the Riccati differential equation (9.34) has a conjugate point on $[t_0, t_f]$. It means that $V_i(x_0, u, w)$ is strictly concave in w. Otherwise, since $V_i(x_0, u, w)$ is quadratic and $R(t) > 0$, $F(t) \geq 0$, $Q(t) \geq 0$, it follows that $V_i(x_0, u, w)$ is strictly convex in u. Hence, for linear quadratic zero-sum games (9.32) with the performance index function (9.34) there exists a unique saddle point; they are the optimal policies.

The convergence of the algorithm described above requires the following:

1. The sequence $\{x_i(t)\}$ converges on $C([t_0, t_f]; \mathbb{R}^n)$, which means that the limit of the solution of approximating sequence (9.32) converges to the unique solution of system (9.29).
2. The sequences of optimal policies $\{u_i(t)\}$ and $\{w_i(t)\}$ converge on $C([t_0, t_f]; \mathbb{R}^{m1})$ and $C([t_0, t_f]; \mathbb{R}^{m2})$, respectively.

For simplicity, the approximating sequence (9.32) is rewritten as

$$\begin{aligned} &\dot{x}_i(t) = f(x_{i-1}(t))x_i(t) + G(x_{i-1}(t), u_{i-1}(t))x_i(t) + K(x_{i-1}(t), w_{i-1}(t))x_i(t), \\ &x_i(t_0) = x_0, \ i \geq 0, \end{aligned} \tag{9.37}$$

where

$$G(x_{i-1},u_{i-1}) \stackrel{\Delta}{=} -g(x_{i-1}(t),u_{i-1}(t))R^{-1}(x_{i-1}(t))^{\mathrm{T}}(x_{i-1}(t),u_{i-1}(t))P_i(t),$$

$$K(x_{i-1},w_{i-1}) \stackrel{\Delta}{=} k(x_{i-1}(t),w_{i-1}(t))S^{-1}(x_{i-1}(t))k^{\mathrm{T}}(x_{i-1}(t),w_{i-1}(t))P_i(t).$$

The optimal policies for zero-sum games are rewritten as

$$\begin{aligned} u_i(t) &= M(x_{i-1}(t),u_{i-1}(t))x_i(t), \\ w_i(t) &= N(x_{i-1}(t),w_{i-1}(t))x_i(t), \end{aligned} \tag{9.38}$$

where

$$M(x_{i-1},u_{i-1}) \stackrel{\Delta}{=} -R^{-1}(x_{i-1}(t))g^{\mathrm{T}}(x_{i-1}(t),u_{i-1}(t))P_i(t),$$

$$N(x_{i-1},w_{i-1}) \stackrel{\Delta}{=} S^{-1}(x_{i-1}(t))k^{\mathrm{T}}(x_{i-1}(t),w_{i-1}(t))P_i(t).$$

Assumption 9.25 $g(x,u)$, $k(x,w)$, $R^{-1}(x)$, $S^{-1}(x)$, $F(x)$ and $Q(x)$ are bounded and Lipschitz continuous in their arguments x, u, and w, thus satisfying:

(C1) $\|g(x,u)\| \le b$, $\|k(x,u)\| \le e$
(C2) $\|R^{-1}(x)\| \le r$, $\|S^{-1}(x)\| \le s$
(C3) $\|F(x)\| \le f$, $\|Q(x)\| \le q$

for $\forall x \in \mathbb{R}^n$, $\forall u \in \mathbb{R}^{m_1}$, $\forall w \in \mathbb{R}^{m_2}$, and for finite positive numbers b, e, r, s, f, and q.

Define $\Phi_{i-1}(t,t_0)$ as the transition matrix generated by $f_{i-1}(t)$. It is well known that

$$\|\Phi_{i-1}(t,t_0)\| \le \exp\left[\int_{t_0}^{t} \mu(f(x_{i-1}(\tau)))\mathrm{d}\tau\right], \quad t \ge t_0, \tag{9.39}$$

where $\mu(f)$ is the measure of matrix f, $\mu(f) = \lim_{h\to 0+} \frac{\|I+hf\|-1}{h}$. We use the following lemma to get an estimate for $\Phi_{i-1}(t,t_0) - \Phi_{i-2}(t,t_0)$.

The following lemma is relevant for the solution of the Riccati differential equation (9.34), which is the basis for proving the convergence.

Lemma 9.26 *Let Assumption* 9.25 *hold*; *the solution of the Riccati differential equation* (9.34) *satisfies*:

1. $P_i(t)$ *is Lipschitz continuous.*
2. $P_i(t)$ *is bounded, if the linear time-varying system* (9.32) *is controllable.*

Proof First, let us prove that $P_i(t)$ is Lipschitz continuous. We transform (9.34) into the form of a matrix differential equation:

$$\begin{bmatrix} \dot{\lambda}_i(t) \\ \dot{X}_i(t) \end{bmatrix} = \begin{bmatrix} -f(x_{i-1}(t)) & -Q(x_{i-1}(t)) \\ \Xi & f(x_{i-1}(t)) \end{bmatrix} \begin{bmatrix} \lambda_i(t) \\ X_i(t) \end{bmatrix}, \quad \begin{bmatrix} \lambda_i(t_f) \\ X_i(t_f) \end{bmatrix} = \begin{bmatrix} F(t_f) \\ I \end{bmatrix},$$

where

$$\begin{aligned}\Xi =& g(x_{i-1}(t), u_{i-1}(t))R^{-1}(x_{i-1}(t))g^{\mathrm{T}}(x_{i-1}(t), u_{i-1}(t))\\ &- k(x_{i-1}(t), w_{i-1}(t))S^{-1}(x_{i-1}(t))k^{\mathrm{T}}(x_{i-1}(t), w_{i-1}(t)).\end{aligned}$$

Thus, the solution $P_i(t)$ of the Riccati differential equations (9.34) becomes

$$P_i(t) = \lambda_i(t)\,(X_i(t))^{-1}. \tag{9.40}$$

If Assumption 9.25 is satisfied, such that $f(x)$, $g(x,u)$, $k(x,w)$, $R^{-1}(x)$, $S^{-1}(x)$, $F(x)$, and $Q(x)$ are Lipschitz continuous, then $X_i(t)$ and $\lambda_i(t)$ are Lipschitz continuous. Furthermore, it is easy to verify that $(X_i(t))^{-1}$ also satisfies the Lipschitz condition. Hence, $P_i(t)$ is Lipschitz continuous.

Next, we prove that $P_i(t)$ is bounded.

If the linear time varying system (9.32) is controllable, there must exist $\hat{u}_i(t)$, $\hat{w}_i(t)$ such that $x(t_1) = 0$ at $t = t_1$. We define $\bar{u}_i(t)$, $\bar{w}_i(t)$ as

$$\bar{u}_i(t) = \begin{cases} \hat{u}_i(t), & t \in [0, t_1)\\ 0, & t \in [t_1, \infty)\end{cases}$$

$$\bar{w}_i(t) = \begin{cases} \hat{w}_i(t) = S^{-1}(x_{i-1}(t))k^{\mathrm{T}}(x_{i-1}(t), w_{i-1}(t))P_i(t)x_i(t), & t \in [0, t_1)\\ 0, & t \in [t_1, \infty)\end{cases}$$

where $\hat{u}_i(t)$ is any control policy making $x(t_1) = 0$, $\hat{w}_i(t)$ is defined as the optimal policy. We have $t \geq t_1$, and we let $\bar{u}_i(t)$ and $\bar{w}_i(t)$ be 0, the state $x(t)$ will still hold at 0.

The optimal cost functional $V_i^*(x_0, u, w)$ described as

$$\begin{aligned}V_i^*(x_0, u, w) =& \frac{1}{2}x_i^{\mathrm{T}}(t_f)F(x_{i-1}(t_f))x_i(t_f) + \frac{1}{2}\int_{t_0}^{t_f}\big[x_i^{\mathrm{T}}(t)Q(x_{i-1}(t))x_i(t)\\ &+ u_i^{*\mathrm{T}}(t)R(x_{i-1}(t))u_i^*(t) - w_i^{*\mathrm{T}}(t)S(x_{i-1}(t))w_i^*(t)\big]\mathrm{d}t,\end{aligned} \tag{9.41}$$

where $u_i^*(t)$ and $w_i^*(t)$ are the optimal policies. $V_i^*(x_0, u, w)$ is minimized by $u^*(t)$ and maximized by $w_i^*(t)$.

For the linear system, $V_i^*(x_0, u, w)$ can be expressed as $V_i^*(x_0, u, w) = 1/(2x_i^{\mathrm{T}}(t)P_i(t)x_i(t))$. Since $x_i(t)$ is arbitrary, if $V_i^*(x_0, u, w)$ is bounded, then $P_i(t)$ is bounded. Next, we discuss the boundedness of $V_i^*(x_0, u, w)$ in two cases:

Case 1: $t_1 < t_f$; we have

$$\begin{aligned}V_i^*(x_0, u, w) \leq& \frac{1}{2}x_i^{\mathrm{T}}(t_f)F(x_{i-1}(t_f))x_i(t_f) + \frac{1}{2}\int_{t_0}^{t_f}\big[x_i^{\mathrm{T}}(t)Q(x_{i-1}(t))x_i(t)\\ &+ \hat{u}_i^{\mathrm{T}}(t)R(x_{i-1}(t))\hat{u}_i(t) - w_i^{*\mathrm{T}}(t)S(x_{i-1}(t))w_i^*(t)\big]\mathrm{d}t\end{aligned}$$

$$
\begin{aligned}
&= \frac{1}{2}\int_{t_0}^{t_1} \big[x_i^{\mathrm{T}}(t)Q(x_{i-1}(t))x_i(t) \\
&\quad + \hat{u}_i^{\mathrm{T}}(t)R(x_{i-1}(t))\hat{u}_i(t) - w_i^{*\mathrm{T}}(t)S(x_{i-1}(t))w_i^*(t)\big]\mathrm{d}t \\
&= V_{t_1 i}(x) \\
&< \infty. \qquad (9.42)
\end{aligned}
$$

Case 2: $t_1 \geq t_f$; we have

$$
\begin{aligned}
V_i^*(x_0, u, w) &\leq \frac{1}{2}x_i^{\mathrm{T}}(t_f)F(x_{i-1}(t_f))x_i(t_f) + \frac{1}{2}\int_{t_0}^{t_f} \big[x_i^{\mathrm{T}}(t)Q(x_{i-1}(t))x_i(t) \\
&\quad + \hat{u}_i^{\mathrm{T}}(t)R(x_{i-1}(t))\hat{u}_i(t) - w_i^{*\mathrm{T}}(t)S(x_{i-1}(t))w_i^*(t)\big]\mathrm{d}t \\
&\leq \frac{1}{2}\int_{t_0}^{\infty} \big[x_i^{\mathrm{T}}(t)Q(x_{i-1}(t))x_i(t) \\
&\quad + \bar{u}_i^{\mathrm{T}}(t)R(x_{i-1}(t))\bar{u}_i(t) - \bar{w}_i^{\mathrm{T}}(t)S(x_{i-1}(t))\bar{w}_i(t)\big]\mathrm{d}t \\
&= \frac{1}{2}\int_{t_0}^{t_1} \big[x_i^{\mathrm{T}}(t)Q(x_{i-1}(t))x_i(t) \\
&\quad + \hat{u}_i^{\mathrm{T}}(t)R(x_{i-1}(t))\hat{u}_i(t) - w_i^{*\mathrm{T}}(t)S(x_{i-1}(t))w_i^*(t)\big]\mathrm{d}t \\
&= V_{t_1 i}(x) \\
&< \infty. \qquad (9.43)
\end{aligned}
$$

From (9.42) and (9.43), we know that $V_i^*(x)$ has an upper bound, independent of t_f. Hence, $P_i(t)$ is bounded. □

According to Lemma 9.26, $P_i(t)$ is bounded and Lipschitz continuous. If Assumption 9.25 is satisfied, then $M(x,u)$, $N(x,w)$, $G(x,w)$, and $K(x,w)$ are bounded and Lipschitz continuous in their arguments, thus satisfying:

(C4) $\|M(x,u)\| \leq \delta_1$, $\|N(x,w)\| \leq \sigma_1$,
(C5) $\|M(x_1,u_1) - M(x_2,u_2)\| \leq \delta_2\|x_1 - x_2\| + \delta_3\|u_1 - u_2\|$, $\|N(x_1,w_1) - N(x_2,w_2)\| \leq \sigma_2\|x_1 - x_2\| + \sigma_3\|w_1 - w_2\|$,
(C6) $\|G(x,u)\| \leq \zeta_1$, $\|K(x,w)\| \leq \xi_1$,
(C7) $\|G(x_1,u_1) - G(x_2,u_2)\| \leq \zeta_2\|x_1 - x_2\| + \zeta_3\|u_1 - u_2\|$, $\|K(x_1,w_1) - K(x_2,w_2)\| \leq \xi_2\|x_1 - x_2\| + \xi_3\|w_1 - w_2\|$,

$\forall x \in \mathbb{R}^n$, $\forall u \in \mathbb{R}^{m_1}$, $\forall w \in \mathbb{R}^{m_2}$, and for finite positive numbers δ_j, σ_j, ζ_j, ξ_j, $j = 1, 2, 3$.

Theorem 9.27 (cf. [16]) *Consider the system* (9.29) *of nonaffine nonlinear zero-sum games with the cost functional* (9.30), *the approximating sequences* (9.32) *and* (9.33) *can be introduced. We have* $F(x(t)) \geq 0$, $Q(x(t)) \geq 0$, $R(x(t)) > 0$, *and the terminal time* t_f *is specified. Let Assumption* 9.25, *and Assumptions* (A1) *and* (A2) *hold and* $S(x(t)) > \tilde{S}$, *for small enough* t_f *or* x_0; *then the limit of the solution of the approximating sequence* (9.32) *converges to the unique solution of system* (9.29) *on* $C([t_0,t_f];\mathbb{R}^n)$. *Meanwhile, the approximating sequences of optimal policies given by* (9.35) *also converge on* $C([t_0,t_f];\mathbb{R}^{m_1})$ *and* $C([t_0,t_f];\mathbb{R}^{m_2})$, *if*

$$\|\Psi(t)\| < 1, \tag{9.44}$$

where

$$\begin{aligned}
\Psi(t) &= \begin{bmatrix} \psi_1 & \psi_2 & \psi_3 \\ \psi_4 & \psi_5 & \psi_6 \\ \psi_7 & \psi_8 & \psi_9 \end{bmatrix}, \\
\psi_1(t) &= \frac{\left\{\left[\frac{\zeta_2+\xi_2}{\zeta_1+\xi_1} + \alpha(t-t_0)\right]e^{(\zeta_1+\xi_1)(t-t_0)} - \frac{\zeta_2+\xi_2}{\zeta_1+\xi_1}\right\}}{\left(1+\frac{\zeta_1+\xi_1}{\mu_0}\left(1-e^{\mu_0(t-t_0)}\right)\right)}\|x_0\| e^{\mu_0(t-t_0)}, \\
\psi_2(t) &= \frac{\frac{\zeta_3}{\zeta_1+\xi_1}\|x_0\| e^{\mu_0(t-t_0)}(e^{(\zeta_1+\xi_1)(t-t_0)}-1)}{\left(1+\frac{\zeta_1+\xi_1}{\mu_0}\left(1-e^{\mu_0(t-t_0)}\right)\right)}, \\
\psi_3(t) &= \frac{\frac{\xi_3}{\zeta_1+\xi_1}\|x_0\| e^{\mu_0(t-t_0)}(e^{(\zeta_1+\xi_1)(t-t_0)}-1)}{\left(1+\frac{\zeta_1+\xi_1}{\mu_0}\left(1-e^{\mu_0(t-t_0)}\right)\right)}, \\
\psi_4(t) &= \delta_1\psi_1(t) + \delta_2\|x_0\| e^{(\mu_0+\zeta_1+\xi_1)(t-t_0)}, \\
\psi_5(t) &= \delta_1\psi_2(t) + \delta_3\|x_0\| e^{(\mu_0+\zeta_1+\xi_1)(t-t_0)}, \\
\psi_6(t) &= \delta_1\psi_3(t), \\
\psi_7(t) &= \sigma_1\psi_1(t) + \sigma_2\|x_0\| e^{(\mu_0+\zeta_1+\xi_1)(t-t_0)}, \\
\psi_8(t) &= \sigma_1\psi_2(t), \\
\psi_9(t) &= \sigma_1\psi_3(t) + \sigma_3\|x_0\| e^{(\mu_0+\zeta_1+\xi_1)(t-t_0)}, \\
\tilde{S} &= \max\{\hat{S}_i\}.
\end{aligned}$$

Proof The approximating sequence (9.37) is an nonhomogeneous differential equation, whose solution can be given by

$$\begin{aligned}
x_i(t) = \Phi_{i-1}(t,t_0)x_i(t_0) + \int_{t_0}^{t} \Phi_{i-1}(t,s)\big[G(x_{i-1}(s),u_{i-1}(s)) \\
+K(x_{i-1}(s),w_{i-1}(s))\big]x_i(s)\mathrm{d}s.
\end{aligned} \tag{9.45}$$

Then,

$$\|x_i(t)\| \le \|\Phi_{i-1}(t,t_0)\|\,\|x_i(t_0)\| + \int_{t_0}^{t}\|\Phi_{i-1}(t,s)\|\big[\,\|G(x_{i-1},u_{i-1})\|$$

$$+\left\|K(x_{i-1},w_{i-1})\right\|\Big]\left\|x_i(s)\right\|\mathrm{d}s. \tag{9.46}$$

According to inequality (9.39) and assuming (C6) to hold, we obtain

$$e^{-\mu_0 t}\left\|x_i(t)\right\| \le e^{-\mu_0 t_0}\left\|x_0\right\| + \int_{t_0}^{t}(\zeta_1+\xi_1)e^{-\mu_0 s}\left\|x_i(s)\right\|\mathrm{d}s. \tag{9.47}$$

On the basis of Gronwall–Bellman's inequality

$$\left\|x_i(t)\right\| \le \left\|x_0\right\| e^{(\mu_0+\zeta_1+\xi_1)(t-t_0)}, \tag{9.48}$$

which is bounded by a small time interval $t\in[t_0,t_f]$ or small x_0.

From (9.45) we have

$$\begin{aligned}
x_i(t)-x_{i-1}(t) = {} & \big[\Phi_{i-1}(t,t_0)-\Phi_{i-2}(t,t_0)\big]x_0 \\
& + \int_{t_0}^{t}\Phi_{i-1}(t,s)G(x_{i-1},u_{i-1})\big[x_i(s)-x_{i-1}(s)\big]\mathrm{d}s \\
& + \int_{t_0}^{t}\Phi_{i-1}(t,s)K(x_{i-1},w_{i-1})\big[x_i(s)-x_{i-1}(s)\big]\mathrm{d}s \\
& + \int_{t_0}^{t}\Phi_{i-1}(t,s)\big[G(x_{i-1},u_{i-1})-G(x_{i-2},u_{i-2})\big]x_{i-1}(s)\mathrm{d}s \\
& + \int_{t_0}^{t}\Phi_{i-1}(t,s)\big[K(x_{i-1},w_{i-1})-K(x_{i-2},w_{i-2})\big]x_{i-1}(s)\mathrm{d}s \\
& + \int_{t_0}^{t}\big[\Phi_{i-1}(t,s)-\Phi_{i-2}(t,s)\big]G(x_{i-2},u_{i-2})x_{i-1}(s)\mathrm{d}s \\
& + \int_{t_0}^{t}\big[\Phi_{i-1}(t,s)-\Phi_{i-2}(t,s)\big]K(x_{i-2},w_{i-2})x_{i-1}(s)\mathrm{d}s.
\end{aligned} \tag{9.49}$$

Consider the supremum to both sides of (9.49) and let

$$\begin{aligned}
\beta_i(t) &= \sup_{s\in[t_0,t]}\left\|x_i(s)-x_{i-1}(s)\right\|, \\
\gamma_i(t) &= \sup_{s\in[t_0,t]}\left\|u_i(s)-u_{i-1}(s)\right\|, \\
\eta_i(t) &= \sup_{s\in[t_0,t]}\left\|w_i(s)-w_{i-1}(s)\right\|.
\end{aligned}$$

By using (9.39), (C6), and (C7), we get

$$\beta_i(t) \le \alpha\left\|x_0\right\| e^{\mu_0(t-t_0)}(t-t_0)\beta_{i-1}(t) + (\zeta_1+\xi_1)\int_{t_0}^{t} e^{\mu_0(t-s)}\beta_i(s)\mathrm{d}s$$

$$+ \|x_0\| e^{\mu_0(t-t_0)} \int_{t_0}^{t} e^{(\zeta_1+\xi_1)(s-t_0)} \big[\zeta_2\beta_{i-1}(s) + \zeta_3\gamma_{i-1}(s)\big] \mathrm{d}s$$

$$+ \|x_0\| e^{\mu_0(t-t_0)} \int_{t_0}^{t} e^{(\zeta_1+\xi_1)(s-t_0)} \big[\xi_2\beta_{i-1}(s) + \xi_3\eta_{i-1}(s)\big] \mathrm{d}s$$

$$+ \alpha\zeta_1 \|x_0\| e^{\mu_0(t-t_0)} \int_{t_0}^{t} e^{(\zeta_1+\xi_1)(s-t_0)} (t-s)\beta_{i-1}(s)\mathrm{d}s$$

$$+ \alpha\xi_1 \|x_0\| e^{\mu_0(t-t_0)} \int_{t_0}^{t} e^{(\zeta_1+\xi_1)(s-t_0)} (t-s)\beta_{i-1}(s)\mathrm{d}s. \tag{9.50}$$

Combining similar terms, we have

$$\beta_i(t) \le \psi_1(t)\beta_{i-1}(t) + \psi_2(t)\gamma_{i-1}(t) + \psi_3(t)\eta_{i-1}(t), \tag{9.51}$$

where $\psi_1(t)$ through $\psi_3(t)$ are described in (9.44).

Similarly, from (9.38), we get

$$\begin{aligned} u_i(t) - u_{i-1}(t) &= M(x_{i-1}, u_{i-1})\big[x_i(t) - x_{i-1}(t)\big] \\ &\quad + \big[M(x_{i-1}, u_{i-1}) - M(x_{i-2}, u_{i-2})\big]x_{i-1}(t) \\ w_i(t) - w_{i-1}(t) &= N(x_{i-1}, w_{i-1})\big[x_i(t) - x_{i-1}(t)\big] \\ &\quad + \big[N(x_{i-1}, w_{i-1}) - N(x_{i-2}, w_{i-2})\big]x_{i-1}(t). \end{aligned} \tag{9.52}$$

According to (C4), (C5), and (9.48), we have

$$\begin{aligned} \gamma_i(t) &\le \psi_4(t)\beta_{i-1}(t) + \psi_5(t)\gamma_{i-1}(t) + \psi_6(t)\eta_{i-1}(t) \\ \eta_i(t) &\le \psi_7(t)\beta_{i-1}(t) + \psi_8(t)\gamma_{i-1}(t) + \psi_9(t)\eta_{i-1}(t), \end{aligned} \tag{9.53}$$

where $\psi_4(t)$ through $\psi_9(t)$ are shown in (9.44).

Then, combining (9.51) and (9.53), we have

$$\Theta_i(t) \le \Psi(t)\Theta_{i-1}(t), \tag{9.54}$$

where $\Theta_i(t) = \begin{bmatrix} \beta_i(t) \\ \gamma_i(t) \\ \eta_i(t) \end{bmatrix}$ and $\Psi(t) = \begin{bmatrix} \psi_1 & \psi_2 & \psi_3 \\ \psi_4 & \psi_5 & \psi_6 \\ \psi_7 & \psi_8 & \psi_9 \end{bmatrix}$.

By induction, Θ_i satisfies

$$\Theta_i(t) \le \Psi^{i-1}(t)\Theta^{[1]}(t), \tag{9.55}$$

which implies that we have $x_i(t)$, $u_i(t)$ and Cauchy sequences in Banach spaces $C([t_0, t_f]; \mathbb{R}^n)$, $C([t_0, t_f]; \mathbb{R}^n)$, $C([t_0, t_f]; \mathbb{R}^{m_1})$, and $C([t_0, t_f]; \mathbb{R}^{m_2})$, respectively. If $\{x_i(t)\}$ converges on $C([t_0, t_f]; \mathbb{R}^n)$, and the sequences of optimal policies $\{u_i\}$ and $\{w_i\}$ also converge on $C([t_0, t_f]; \mathbb{R}^{m_1})$ and $C([t_0, t_f]; \mathbb{R}^{m_2})$ on $[t_0, t_f]$.

It means that $x_{i-1}(t) = x_i(t)$, $u_{i-1}(t) = u_i(t)$, $w_{i-1}(t) = w_i(t)$ when $i \to \infty$. Hence, the system (9.29) has a unique solution on $[t_0, t_f]$, which is given by the limit of the solution of approximating sequence (9.32). □

Based on the iterative algorithm described in Theorem 9.27, the design procedure of optimal policies for nonlinear nonaffine zero-sum games is summarized as follows:

1. Give x_0, maximum iteration times $i_{\max}$ and approximation accuracy ε.
2. Use a factored form to represent the system as (9.31).
3. Set $i=0$. Let $x_{i-1}(t)=x_0$, $u_{i-1}(t)=0$ and $w_{i-1}(t)=0$. Compute the corresponding matrix-valued functions $f(x_0)$, $g(x_0,0)$, $k(x_0,0)$, $F(x_0)$, $Q(x_0)$, $R(x_0)$, and $S(x_0)$.
4. Compute $x^{[0]}(t)$ and $P^{[0]}(t)$ according to differential equations (9.34) and (9.36) with $x(t_0)=x_0$, $P(t_f)=F(x_f)$.
5. Set $i=i+1$. Compute the corresponding matrix-valued functions $f(x_{i-1}(t))$, $g(x_{i-1}(t),u_{i-1}(t))$, $k(x_{i-1}(t),w_{i-1}(t))$, $Q(x_{i-1}(t))$, $R(x_{i-1}(t))$, $F(x_{i-1}(t_f))$, and $S(x_{i-1}(t))$.
6. Compute $x_i(t)$ and $P_i(t)$ by (9.34) and (9.36) with $x(t_0)=x_0$, $P(t_f)=F(x_{tf})$.
7. If $\|x_i(t)-x_{i-1}(t)\|<\varepsilon$, go to Step 9); otherwise, go to Step 8.
8. If $i>i_{\max}$, then go to Step 9; else, go to Step 5.
9. Stop.

9.3.3 Simulations

Example 9.28 We now show the power of our iterative algorithm for finding optimal policies for nonaffine nonlinear zero-sum games.

In the following, we introduce an example of a control system that has the form (9.29) with control input $u(t)$, subject to a disturbance $w(t)$ and a cost functional $V(x_0,u,w)$. The control input $u(t)$ is required to minimize the cost functional $V(x_0,u,w)$. If the disturbance has a great effect on the system, the single disturbance $w(t)$ has to maximize the cost functional $V(x_0,u,w)$. The conflicting design can guarantee the optimality and strong robustness of the system at the same time. This is a zero-sum game problem, which can be described by the state equations

$$\begin{aligned}\dot{x}_1(t)=&-2x_1(t)+x_2^2(t)-x_1(t)u(t)+u^2(t)-3x(t)w(t)+5w^2(t),\\ \dot{x}_2(t)=&5x_1^2(t)-2x_2(t)+x_2^2(t)+u^2(t)+w^2(t).\end{aligned}\tag{9.56}$$

Define the finite horizon cost functional to be of the form (9.30), where $F=0.01\,I_{2\times2}$, $Q=0.01\,I_{2\times2}$, $R=1$ and $S=1$, where I is an identity matrix. Clearly, (9.56) is not affine in $u(t)$ and $w(t)$, it has the control nonaffine nonlinear structure. Therefore, we represent the system (9.56) in the factored form $f(x(t))x(t)$, $g(x(t),u(t))u(t)$ and $k(x(t),w(t))w(t)$, which, given the wide selection of possible representations, have been chosen as

$$f(x(t))=\begin{bmatrix}2 & x_2(t)\\ 5x_1(t) & -2+x_2(t)\end{bmatrix},\quad g(x(t),u(t))=\begin{bmatrix}x_1(t)+u(t)\\ u(t)\end{bmatrix},$$

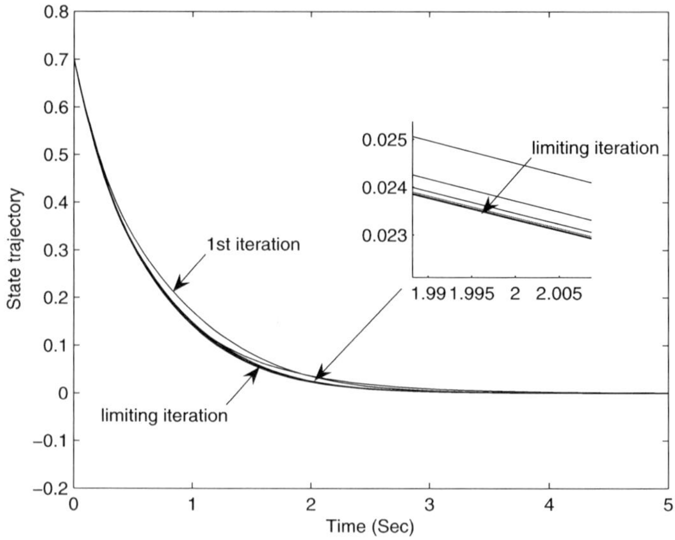

Fig. 9.10 The state trajectory $x_1(t)$ of each iteration

$$k(x(t), w(t)) = \begin{bmatrix} -3x_1(t) + 5w(t) \\ w(t) \end{bmatrix}. \tag{9.57}$$

The optimal policies designs given by Theorem 9.27 can now be applied to (9.31) with the dynamics (9.57).

The initial state vectors are chosen as $x_0 = [0.6, 0]^{\mathrm{T}}$ and the terminal time is set to $t_f = 5$. Let us define the required error norm between the solutions of the linear time-vary differential equations by $\|x_i(t) - x_{i-1}(t)\| < \varepsilon = 0.005$, which needs to be satisfied if convergence is to be achieved. The factorization is given by (9.57). Implementing the present iterative algorithm, it just needs six sequences to satisfy the required bound, $\|x^{[6]}(t) - x^{[5]}(t)\| = 0.0032$. With increasing of number of times of iterations, the approximation error will reduce obviously. When the iteration number $i = 25$, the approximation error is just 5.1205×10^{-10}.

Define the maximum iteration times $i_{\max} = 25$. Figure 9.10 represents the convergence trajectories of the state trajectory of each linear quadratic zero-sum game. It can be seen that the sequence is obviously convergent. The magnifications of the state trajectories are given in the figure, which shows that the error will be smaller as the number of times of iteration becomes bigger. The trajectories of control input $u(t)$ and disturbance input $w(t)$ of each iteration are also convergent, which is shown in Figs. 9.11 and 9.12. The approximate optimal policies $u^*(t)$ and $w^*(t)$ are obtained by the last iteration. Substituting the approximate optimal policies $u^*(t)$ and $w^*(t)$ into the system of zero-sum games (9.56), we get the state trajectory. The norm of the error between this state trajectory and the state trajectory of the last iteration is just 0.0019, which proves that the approximating iterative approach developed in this section is highly effective.

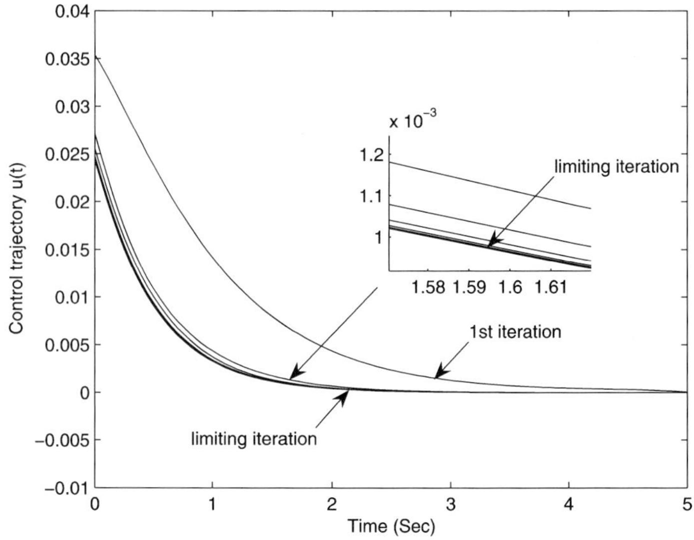

Fig. 9.11 The trajectory $u(t)$ of each iteration

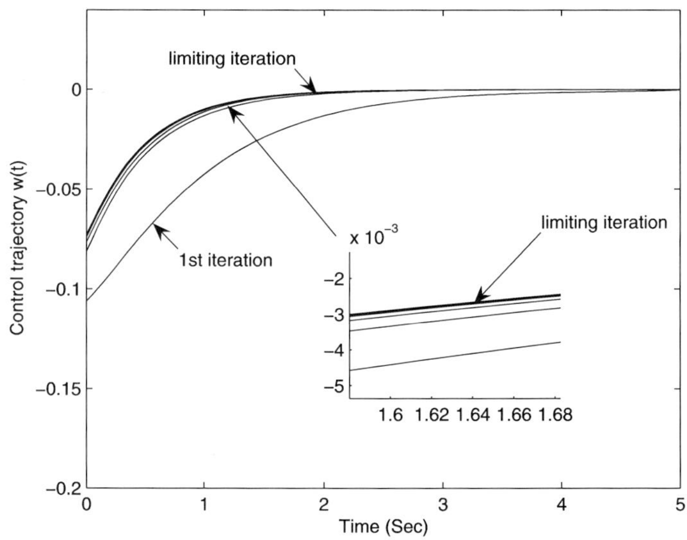

Fig. 9.12 The trajectory $w(t)$ of each iteration

9.4 Non-Zero-Sum Games for a Class of Nonlinear Systems Based on ADP

In this section, a near-optimal control scheme is developed for the non-zero-sum differential games of continuous-time nonlinear systems. The single network ADP is utilized to obtain the optimal control policies which make the cost functions reach the Nash equilibrium of non-zero-sum differential games, where only one critic network is used for each player, instead of the action-critic dual network used in a typical ADP architecture. Furthermore, novel weight tuning laws for critic neural networks are developed, which not only ensure the Nash equilibrium to be reached, but also guarantee the stability of the system. No initial stabilizing control policy is required for each player. Moreover, Lyapunov theory is utilized to demonstrate the uniform ultimate boundedness of the closed-loop system.

9.4.1 Problem Formulation of Non-Zero-Sum Games

Consider the following continuous-time nonlinear systems:

$$\dot{x}(t) = f(x(t)) + g(x(t))u(t) + k(x(t))w(t), \tag{9.58}$$

where $x(t) \in \mathbb{R}^n$ is the state vector, $u(t) \in \mathbb{R}^m$ and $d(t) \in \mathbb{R}^q$ are the control input vectors. Assume that $f(0) = 0$ and that $f(x)$, $g(x)$, $k(x)$ are locally Lipschitz.

The cost functional associated with u is defined as

$$J_1(x, u, w) = \int_t^{\infty} r_1(x(\tau), u(\tau), w(\tau))\mathrm{d}\tau, \tag{9.59}$$

where $r_1(x, u, w) = Q_1(x) + u^{\mathrm{T}} R_{11} u + w^{\mathrm{T}} R_{12} w$, $Q_1(x) \geq 0$ is the penalty on the states, $R_{11} \in \mathbb{R}^{m \times m}$ is a positive definite matrix, and $R_{12} \in \mathbb{R}^{q \times q}$ is a positive semidefinite matrix.

The cost functional associated with w is defined as

$$J_2(x, u, w) = \int_t^{\infty} r_2(x(\tau), u(\tau), w(\tau))\mathrm{d}\tau, \tag{9.60}$$

where $r_2(x, u, w) = Q_2(x) + u^{\mathrm{T}} R_{21} u + w^{\mathrm{T}} R_{22} w$, $Q_2(x) \geq 0$ is the penalty on the states, $R_{21} \in \mathbb{R}^{m \times m}$ is a positive semidefinite matrix, and $R_{22} \in \mathbb{R}^{q \times q}$ is a positive definite matrix.

For the above non-zero-sum differential games, the two feedback control policies u and w are chosen by player 1 and player 2, respectively, where player 1 tries to minimize the cost functional (9.59), while player 2 attempts to minimize the cost functional (9.60).

Definition 9.29 $u = \mu_1(x)$ and $w = \mu_2(x)$ are defined as admissible with respect to (9.59) and (9.60) on $\Omega \in \mathbb{R}^n$, denoted by $\mu_1 \in \psi(\Omega)$ and $\mu_2 \in \psi(\Omega)$, respectively, if $\mu_1(x)$ and $\mu_2(x)$ are continuous on Ω, $\mu_1(0) = 0$ and $\mu_2(0) = 0$, $\mu_1(x)$ and $\mu_2(x)$ stabilize (9.58) on Ω, and (9.59) and (9.60) are finite, $\forall x_0 \in \Omega$.

Definition 9.30 The policy set (u^*, w^*) is a Nash equilibrium policy set if the inequalities

$$\begin{aligned} J_1(u^*, w^*) &\le J_1(u, w^*), \\ J_2(u^*, w^*) &\le J_2(u^*, w) \end{aligned} \tag{9.61}$$

hold for any admissible control policies u and w.

Next, define the Hamilton functions for the cost functionals (9.59) and (9.60) with associated admissible control input u and w, respectively, as follows:

$$\begin{aligned} H_1(x, u, w) = {} & Q_1(x) + u^{\mathrm{T}} R_{11} u + w^{\mathrm{T}} R_{12} w \\ & + \nabla J_1^{\mathrm{T}}(f(x) + g(x)u + k(x)w), \end{aligned} \tag{9.62}$$

$$\begin{aligned} H_2(x, u, w) = {} & Q_2(x) + u^{\mathrm{T}} R_{21} u + w^{\mathrm{T}} R_{22} w \\ & + \nabla J_2^{\mathrm{T}}(f(x) + g(x)u + k(x)w), \end{aligned} \tag{9.63}$$

where ∇J_i is the partial derivative of the cost function $J_i(x, u, w)$ with respect to x, $i = 1, 2$.

According to the stationarity conditions of optimization, we have

$$\begin{aligned} \partial H_1(x, u, w)/\partial u = 0, \\ \partial H_2(x, u, w)/\partial w = 0. \end{aligned}$$

Therefore, the associated optimal feedback control policies u^* and w^* are found and revealed to be

$$u^* = -\frac{1}{2} R_{11}^{-1} g^{\mathrm{T}}(x) \nabla J_1, \tag{9.64}$$

$$w^* = -\frac{1}{2} R_{22}^{-1} k^{\mathrm{T}}(x) \nabla J_2. \tag{9.65}$$

The optimal feedback control policies u^* and w^* provide a Nash equilibrium for the non-zero-sum differential games among all the feedback control policies.

Considering $H_1(x, u^*, w^*) = 0$ and $H_2(x, u^*, w^*) = 0$, and substituting the optimal feedback control policy (9.64) and (9.65) into the Hamilton functions (9.62) and (9.63), we have

$$\begin{aligned} Q_1(x) - {} & \frac{1}{4} \nabla J_1^{\mathrm{T}} g(x) R_{11}^{-1} g^{\mathrm{T}}(x) \nabla J_1 + \nabla J_1^{\mathrm{T}} f(x) \\ & + \frac{1}{4} \nabla J_2^{\mathrm{T}} k(x) R_{22}^{-1} R_{12} R_{22}^{-1} k^{\mathrm{T}}(x) \nabla J_2 \\ & - \frac{1}{2} \nabla J_1^{\mathrm{T}} k(x) R_{22}^{-1} k^{\mathrm{T}}(x) \nabla J_2 = 0, \end{aligned} \tag{9.66}$$

$$\begin{aligned} Q_2(x) &- \frac{1}{4}\nabla J_2^{\mathrm{T}} k(x) R_{22}^{-1} k^{\mathrm{T}}(x) \nabla J_2 + \nabla J_2^{\mathrm{T}} f(x) \\ &+ \frac{1}{4}\nabla J_1^{\mathrm{T}} g(x) R_{11}^{-1} R_{21} R_{11}^{-1} g^{\mathrm{T}}(x) \nabla J_1 \\ &- \frac{1}{2}\nabla J_2^{\mathrm{T}} g(x) R_{11}^{-1} g^{\mathrm{T}}(x) \nabla J_1 = 0. \end{aligned} \tag{9.67}$$

If the coupled HJ equations (9.66) and (9.67) can be solved for the optimal value functions $J_1(x, u^*, w^*)$ and $J_2(x, u^*, w^*)$, the optimal control can then be implemented by using (9.64) and (9.65). However, these equations are generally difficult or impossible to solve due to their inherently nonlinear nature. To overcome this difficulty, a near-optimal control scheme is developed to learn the solution of coupled HJ equations online using a single network ADP in order to obtain the optimal control policies.

Before presenting the near-optimal control scheme, the following lemma is required.

Lemma 9.31 *Given the system* (9.58) *with associated cost functionals* (9.59) *and* (9.60) *and the optimal feedback control policies* (9.64) *and* (9.65). *For player* i, $i = 1, 2$, *let* $L_i(x)$ *be a continuously differentiable, radially unbounded Lyapunov candidate such that* $\dot{L}_i = \nabla L_i^{\mathrm{T}} \dot{x} = \nabla L_i^{\mathrm{T}}(f(x) + g(x)u^* + k(x)w^*) < 0$, *with* ∇L_i *being the partial derivative of* $L_i(x)$ *with respect to* x. *Moreover, let* $\bar{Q}_i(x) \in \mathbb{R}^{n\times n}$ *be a positive definite matrix satisfying* $\|\bar{Q}_i(x)\| = 0$ *if and only if* $\|x\| = 0$ *and* $\bar{Q}_{i\min} \le \|\bar{Q}_i(x)\| \le \bar{Q}_{i\max}$ *for* $\|\chi_{\min}\| \le \|x\| \le \chi_{\max}$ *with positive constants* $\bar{Q}_{i\min}$, $\bar{Q}_{i\max}$, $\chi_{\min}$, $\chi_{\max}$. *In addition, let* $\bar{Q}_i(x)$ *satisfy* $\lim_{x\to\infty} \bar{Q}_i(x) = \infty$ *as well as*

$$\nabla J_i^{*\mathrm{T}} \bar{Q}_i(x) \nabla L_i = r_i(x, u^*, w^*). \tag{9.68}$$

Then the following relation holds:

$$\nabla L_i^{\mathrm{T}}(f(x) + g(x)u^* + k(x)w^*) = -\nabla L_i^{\mathrm{T}} \bar{Q}_i(x) \nabla L_i. \tag{9.69}$$

Proof When the optimal control u^* and w^* in (9.64) and (9.65) are applied to the nonlinear system (9.58), the value function $J_i(x, u^*, w^*)$ becomes a Lyapunov function, $i = 1, 2$. Then, for $i = 1, 2$, differentiating the value function $J_i(x, u^*, w^*)$ with respect to t, we have

$$\begin{aligned} \dot{J}_i^* &= \nabla J_i^{*\mathrm{T}}(f(x) + g(x)u^* + k(x)w^*) \\ &= -r_i(x, u^*, w^*). \end{aligned} \tag{9.70}$$

Using (9.68), (9.70) can be rewritten as

$$\begin{aligned} (f(x) + g(x)u^* + k(x)w^*) &= -(\nabla J_i^* \nabla J_i^{*\mathrm{T}})^{-1} \nabla J_i^* r_i(x, u^*, w^*) \\ &= -(\nabla J_i^* \nabla J_i^{*\mathrm{T}})^{-1} \nabla J_i^* \nabla J_i^{*\mathrm{T}} \bar{Q}_i(x) \nabla L_i \\ &= -\bar{Q}_i(x) \nabla L_i. \end{aligned} \tag{9.71}$$

Next, multiplying both sides of (9.71) by ∇L_i^{T}, (9.69) can be obtained. This completes the proof. □

9.4.2 *Optimal Control of Nonlinear Non-Zero-Sum Games Based on ADP*

To begin the development, we rewrite the cost functions (9.59) and (9.60) by NNs as

$$J_1(x) = W_{c1}^{\mathrm{T}}\phi_1(x) + \varepsilon_1, \tag{9.72}$$

$$J_2(x) = W_{c2}^{\mathrm{T}}\phi_2(x) + \varepsilon_2, \tag{9.73}$$

where W_i, $\phi_i(x)$, and ε_i are the critic NN ideal constant weights, the critic NN activation function vector and the NN approximation error for player i, $i = 1, 2$, respectively.

The derivative of the cost functions with respect to x can be derived as

$$\nabla J_1 = \nabla\phi_1^{\mathrm{T}} W_{c1} + \nabla\varepsilon_1, \tag{9.74}$$

$$\nabla J_2 = \nabla\phi_2^{\mathrm{T}} W_{c2} + \nabla\varepsilon_2, \tag{9.75}$$

where $\nabla\phi_i \triangleq \partial\phi_i(x)/\partial x$, $\nabla\varepsilon_i \triangleq \partial\varepsilon_i/\partial x$, $i = 1, 2$.

Using (9.74) and (9.75), the optimal feedback control policies (9.64) and (9.65) can be rewritten as

$$u^* = -\frac{1}{2}R_{11}^{-1}g^{\mathrm{T}}(x)\nabla\phi_1^{\mathrm{T}} W_{c1} - \frac{1}{2}R_{11}^{-1}g^{\mathrm{T}}(x)\nabla\varepsilon_1, \tag{9.76}$$

$$w^* = -\frac{1}{2}R_{22}^{-1}k^{\mathrm{T}}(x)\nabla\phi_2^{\mathrm{T}} W_{c2} - \frac{1}{2}R_{22}^{-1}k^{\mathrm{T}}(x)\nabla\varepsilon_2, \tag{9.77}$$

and the coupled HJ equations (9.66) and (9.67) can be rewritten as

$$\begin{aligned} Q_1(x) &- \frac{1}{4}W_{c1}^{\mathrm{T}}\nabla\phi_1 D_1 \nabla\phi_1^{\mathrm{T}} W_{c1} + W_{c1}^{\mathrm{T}}\nabla\phi_1 f(x) \\ &+ \frac{1}{4}W_{c2}^{\mathrm{T}}\nabla\phi_2 S_2 \nabla\phi_2{}^{\mathrm{T}} W_{c2} - \frac{1}{2}W_{c1}^{\mathrm{T}}\nabla\phi_1 D_2 \nabla\phi_2^{\mathrm{T}} W_{c2} - \varepsilon_{\mathrm{HJ}1} = 0, \end{aligned} \tag{9.78}$$

$$\begin{aligned} Q_2(x) &- \frac{1}{4}W_{c2}^{\mathrm{T}}\nabla\phi_2 D_2 \nabla\phi_2^{\mathrm{T}} W_{c2} + W_{c2}^{\mathrm{T}}\nabla\phi_2 f(x) \\ &+ \frac{1}{4}W_{c1}^{\mathrm{T}}\nabla\phi_1 S_1 \nabla\phi_1{}^{\mathrm{T}} W_{c1} - \frac{1}{2}W_{c2}^{\mathrm{T}}\nabla\phi_2 D_1 \nabla\phi_1^{\mathrm{T}} W_{c1} - \varepsilon_{\mathrm{HJ}2} = 0, \end{aligned} \tag{9.79}$$

where

$$
\begin{aligned}
D_1 &= g(x)R_{11}^{-1}g^{\mathrm{T}}(x),\\
D_2 &= k(x)R_{22}^{-1}k^{\mathrm{T}}(x),\\
S_1 &= g(x)R_{11}^{-1}R_{21}R_{11}^{-1}g^{\mathrm{T}}(x),\\
S_2 &= k(x)R_{22}^{-1}R_{12}R_{22}^{-1}k^{\mathrm{T}}(x).
\end{aligned}
\tag{9.80}
$$

The residual error due to the NN approximation for player 1 is

$$
\begin{aligned}
\varepsilon_{\mathrm{HJ1}} = &- \nabla\varepsilon_1^{\mathrm{T}}\bigg(f(x) - \frac{1}{2}D_1(\nabla\phi_1^{\mathrm{T}}W_{c1} + \nabla\varepsilon_1)\\
&- \frac{1}{2}D_2(\nabla\phi_2^{\mathrm{T}}W_{c2} + \nabla\varepsilon_2)\bigg) - \frac{1}{4}\nabla\varepsilon_1^{\mathrm{T}}D_1\nabla\varepsilon_1 + \frac{1}{2}W_{c1}^{\mathrm{T}}\nabla\phi_1 D_2\nabla\varepsilon_2\\
&- \frac{1}{2}\nabla\varepsilon_2^{\mathrm{T}}S_2\nabla\phi_2^{\mathrm{T}}W_{c2} - \frac{1}{4}\nabla\varepsilon_2^{\mathrm{T}}S_2\nabla\varepsilon_2.
\end{aligned}
\tag{9.81}
$$

The residual error due to the NN approximation for player 2 is

$$
\begin{aligned}
\varepsilon_{\mathrm{HJ2}} = &- \nabla\varepsilon_2^{\mathrm{T}}\bigg(f(x) - \frac{1}{2}D_1(\nabla\phi_1^{\mathrm{T}}W_{c1} + \nabla\varepsilon_1)\\
&- \frac{1}{2}D_2(\nabla\phi_2^{\mathrm{T}}W_{c2} + \nabla\varepsilon_2)\bigg) - \frac{1}{4}\nabla\varepsilon_2^{\mathrm{T}}D_2\nabla\varepsilon_2 + \frac{1}{2}W_{c2}^{\mathrm{T}}\nabla\phi_2 D_1\nabla\varepsilon_1\\
&- \frac{1}{2}\nabla\varepsilon_2^{\mathrm{T}}S_1\nabla\phi_1^{\mathrm{T}}W_{c2} - \frac{1}{4}\nabla\varepsilon_1^{\mathrm{T}}S_1\nabla\varepsilon_1.
\end{aligned}
\tag{9.82}
$$

Let $\hat{W}_{c1}$ and $\hat{W}_{c2}$ be the estimates of W_{c1} and W_{c2}, respectively. Then we have the estimates of $V_1(x)$ and $V_2(x)$ as follows:

$$
\hat{J}_1(x) = \hat{W}_{c1}^{\mathrm{T}}\phi_1(x), \tag{9.83}
$$

$$
\hat{J}_2(x) = \hat{W}_{c2}^{\mathrm{T}}\phi_2(x). \tag{9.84}
$$

Substituting (9.83) and (9.84) into (9.64) and (9.65), respectively, the estimates of optimal control policies can be written as

$$
\hat{u} = -\frac{1}{2}R_{11}^{-1}g^{\mathrm{T}}(x)\nabla\phi_1^{\mathrm{T}}\hat{W}_{c1}, \tag{9.85}
$$

$$
\hat{w} = -\frac{1}{2}R_{22}^{-1}k^{\mathrm{T}}(x)\nabla\phi_2^{\mathrm{T}}\hat{W}_{c2}. \tag{9.86}
$$

Applying (9.85) and (9.86) to the system (9.58), we have the closed-loop system dynamics as follows:

$$
\dot{x} = f(x) - \frac{D_1\nabla\phi_1^{\mathrm{T}}\hat{W}_{c1}}{2} - \frac{D_2\nabla\phi_2^{\mathrm{T}}\hat{W}_{c2}}{2}. \tag{9.87}
$$

Substituting (9.83) and (9.84) into (9.62) and (9.63), respectively, the approximate Hamilton functions can be derived as follows:

$$\begin{aligned} H_1(x, \hat{W}_{c1}, \hat{W}_{c2}) &= Q_1(x) - \frac{1}{4}\hat{W}_{c1}^{\mathrm{T}} \nabla\phi_1 D_1 \nabla\phi_1^{\mathrm{T}} \hat{W}_{c1} + \hat{W}_{c1}^{\mathrm{T}} \nabla\phi_1 f(x) \\ &\quad + \frac{1}{4}\hat{W}_{c2}^{\mathrm{T}} \nabla\phi_2 S_2 \nabla\phi_2^{\mathrm{T}} \hat{W}_{c2} - \frac{1}{2}\hat{W}_{c1}^{\mathrm{T}} \nabla\phi_1 D_2 \nabla\phi_2^{\mathrm{T}} \hat{W}_{c2} \\ &= e_1, \end{aligned} \tag{9.88}$$

$$\begin{aligned} H_2(x, \hat{W}_{c1}, \hat{W}_{c2}) &= Q_2(x) - \frac{1}{4}\hat{W}_{c2}^{\mathrm{T}} \nabla\phi_2 D_2 \nabla\phi_2^{\mathrm{T}} \hat{W}_{c2} + \hat{W}_{c2}^{\mathrm{T}} \nabla\phi_2 f(x) \\ &\quad + \frac{1}{4}\hat{W}_{c1}^{\mathrm{T}} \nabla\phi_1 S_1 \nabla\phi_1^{\mathrm{T}} \hat{W}_{c1} - \frac{1}{2}\hat{W}_{c2}^{\mathrm{T}} \nabla\phi_2 D_1 \nabla\phi_1^{\mathrm{T}} \hat{W}_{c1} \\ &= e_2. \end{aligned} \tag{9.89}$$

It is desired to select $\hat{W}_{c1}$ and $\hat{W}_{c2}$ to minimize the squared residual error $E = e_1^{\mathrm{T}} e_1/2 + e_2^{\mathrm{T}} e_2/2$. Then we have $\hat{W}_{c1} \to W_{c1}$, $\hat{W}_{c2} \to W_{c2}$, and $e_1 \to \varepsilon_{\mathrm{HJ}1}$, $e_2 \to \varepsilon_{\mathrm{HJ}2}$. In other words, the Nash equilibrium of the non-zero-sum differential games of continuous-time nonlinear system (9.58) can be obtained. However, tuning the critic NN weights to minimize the squared residual error E alone does not ensure the stability of the nonlinear system (9.58) during the learning process of critic NNs. Therefore, we propose the novel weight tuning laws of critic NNs for two players, which cannot only minimize the squared residual error E but also guarantee the stability of the system as follows:

$$\begin{aligned} \dot{\hat{W}}_1 = &- \alpha_1 \frac{\bar{\sigma}_1}{m_{s_1}} \bigg(Q_1(x) - \frac{1}{4}\hat{W}_{c1}^{\mathrm{T}} \nabla\phi_1 D_1 \nabla\phi_1^{\mathrm{T}} \hat{W}_{c1} \\ &+ \hat{W}_{c1}^{\mathrm{T}} \nabla\phi_1 f(x) + \frac{1}{4}\hat{W}_{c2}^{\mathrm{T}} \nabla\phi_2 S_2 \nabla\phi_2^{\mathrm{T}} \hat{W}_{c2} - \frac{1}{2}\hat{W}_{c1}^{\mathrm{T}} \nabla\phi_1 D_2 \nabla\phi_2^{\mathrm{T}} \hat{W}_{c2} \bigg) \\ &+ \frac{\alpha_1}{4} \nabla\phi_1 D_1 \nabla\phi_1^{\mathrm{T}} \hat{W}_{c1} \frac{\bar{\sigma}_1^{\mathrm{T}}}{m_{s_1}} \hat{W}_{c1} + \frac{\alpha_2}{4} \nabla\phi_1 S_1 \nabla\phi_1^{\mathrm{T}} \hat{W}_{c1} \frac{\bar{\sigma}_2^{\mathrm{T}}}{m_{s_2}} \hat{W}_{c2} \\ &+ \Sigma(x, \hat{u}, \hat{w}) \left(\frac{\alpha_1 \nabla\phi_1 D_1 \nabla L_1}{2} + \frac{\alpha_1 \nabla\phi_1 D_1 \nabla L_2}{2} \right) \\ &- \alpha_1 (F_1 \hat{W}_{c1} - F_2 \bar{\sigma}_1^{\mathrm{T}} \hat{W}_{c1}), \end{aligned} \tag{9.90}$$

$$\begin{aligned} \dot{\hat{W}}_2 = &- \alpha_2 \frac{\bar{\sigma}_2}{m_{s_2}} \bigg(Q_2(x) - \frac{1}{4}\hat{W}_{c2}^{\mathrm{T}} \nabla\phi_2 D_2 \nabla\phi_2^{\mathrm{T}} \hat{W}_{c2} \\ &+ \hat{W}_{c2}^{\mathrm{T}} \nabla\phi_2 f(x) + \frac{1}{4}\hat{W}_{c1}^{\mathrm{T}} \nabla\phi_1 S_1 \nabla\phi_1^{\mathrm{T}} \hat{W}_{c1} - \frac{1}{2}\hat{W}_{c2}^{\mathrm{T}} \nabla\phi_2 D_1 \nabla\phi_1^{\mathrm{T}} \hat{W}_{c1} \bigg) \end{aligned}$$

$$
\begin{aligned}
&+\frac{\alpha_2}{4}\nabla\phi_2 D_2 \nabla\phi_2^{\mathrm{T}}\hat{W}_{c2}\frac{\bar{\sigma}_2^{\mathrm{T}}}{m_{s_2}}\hat{W}_{c2}+\frac{\alpha_2}{4}\nabla\phi_2 S_2 \nabla\phi_2^{\mathrm{T}}\hat{W}_{c2}\frac{\bar{\sigma}_1^{\mathrm{T}}}{m_{s_1}}\hat{W}_{c1}\\
&+\Sigma(x,\hat{u},\hat{w})\left(\frac{\alpha_2\nabla\phi_2 D_2\nabla L_2}{2}+\frac{\alpha_2\nabla\phi_2 D_2\nabla L_1}{2}\right)\\
&-\alpha_2(F_3\hat{W}_{c2}-F_4\bar{\sigma}_2^{\mathrm{T}}\hat{W}_{c2}),
\end{aligned}
\tag{9.91}
$$

where $\bar{\sigma}_i=\hat{\sigma}_i/(\hat{\sigma}_i^{\mathrm{T}}\hat{\sigma}_i+1)$, $\hat{\sigma}_i=\nabla\phi_i(f(x)-D_1\nabla\phi_1^{\mathrm{T}}\hat{W}_{c1}/2-D_2\nabla\phi_2^{\mathrm{T}}\hat{W}_{c2}/2)$, $m_{s_i}=\hat{\sigma}_i^{\mathrm{T}}\hat{\sigma}_i+1$, $\alpha_i>0$ is the adaptive gain, ∇L_i is described in Lemma 9.31, $i=1,2$. F_1, F_2, F_3, and F_4 are design parameters. The operator $\Sigma(x,\hat{u},\hat{w})$ is given by

$$
\Sigma(x,\hat{u},\hat{w})=\begin{cases}0 \text{ if } \ \nabla L_1\dot{x}\le 0 \ \text{ and } \nabla L_2\dot{x}\le 0,\\ 1 \text{ else,}\end{cases}
\tag{9.92}
$$

where $\dot{x}$ is given as (9.87).

Remark 9.32 The first terms in (9.90) and (9.91) are utilized to minimize the squared residual error E and derived by using a normalized gradient descent algorithm. The other terms are utilized to guarantee the stability of the closed-loop system while the critic NNs learn the optimal cost functions and are derived by following Lyapunov stability analysis. The operator $\Sigma(x,\hat{u},\hat{w})$ is selected based on the Lyapunov's sufficient condition for stability, which means that the state x is stable if $L_i(x)>0$ and $\nabla L_i\dot{x}<0$ for player i, $i=1,2$. When the system (9.58) is stable, the operator $\Sigma(x,\hat{u},\hat{w})=0$ and it will not take effect. When the system (9.58) is unstable, the operator $\Sigma(x,\hat{u},\hat{w})=1$ and it will be activated. Therefore, no initial stabilizing control policies are needed due to the introduction of the operator $\Sigma(x,\hat{u},\hat{w})$.

Remark 9.33 From (9.88) and (9.89), it can be seen that the approximate Hamilton functions $H_1(x,\hat{W}_{c1},\hat{W}_{c2})=e_1=0$ and $H_2(x,\hat{W}_{c1},\hat{W}_{c2})=e_2=0$ when $x=0$. For this case, the tuning laws of critic NN weights for two players (9.90) and (9.91) cannot achieve the purpose of optimization anymore. This can be considered as a persistency of the requirement of excitation for the system states. Therefore, the system states must be persistently excited enough for minimizing the squared residual errors E to drive the critic NN weights toward their ideal values. In order to satisfy the persistent excitation condition, probing noise is added to the control input.

Define the weight estimation errors of critic NNs for two players to be $\tilde{W}_{c1}=W_{c1}-\hat{W}_{c1}$ and $\tilde{W}_{c2}=W_{c2}-\hat{W}_{c2}$, respectively. From (9.78) and (9.79), we observe that

$$
\begin{aligned}
Q_1(x)=&\frac{1}{4}W_{c1}^{\mathrm{T}}\nabla\phi_1 D_1\nabla\phi_1^{\mathrm{T}}W_{c1}-W_{c1}^{\mathrm{T}}\nabla\phi_1 f(x)-\frac{1}{4}W_{c2}^{\mathrm{T}}\nabla\phi_2 S_2\nabla{\phi_2}^{\mathrm{T}}W_{c2}\\
&+\frac{1}{2}W_{c1}^{\mathrm{T}}\nabla\phi_1 D_2\nabla\phi_2^{\mathrm{T}}W_{c2}+\varepsilon_{\mathrm{HJ1}},
\end{aligned}
\tag{9.93}
$$

$$Q_2(x) = \frac{1}{4} W_{c2}^{\mathrm{T}} \nabla\phi_2 D_2 \nabla\phi_2^{\mathrm{T}} W_{c2} - W_{c2}^{\mathrm{T}} \nabla\phi_2 f(x) - \frac{1}{4} W_{c1}^{\mathrm{T}} \nabla\phi_1 S_1 \nabla\phi_1{}^{\mathrm{T}} W_{c1}$$
$$+ \frac{1}{2} W_{c2}^{\mathrm{T}} \nabla\phi_2 D_1 \nabla\phi_1^{\mathrm{T}} W_{c1} + \varepsilon_{\mathrm{HJ2}}. \tag{9.94}$$

Combining (9.90) with (9.93), we have

$$\begin{aligned}\dot{\tilde{W}}_1 =& \alpha_1 \frac{\bar{\sigma}_1}{m_{s_1}} \Bigg[-\tilde{W}_{c1}^{\mathrm{T}} \hat{\sigma}_1 + \frac{1}{4} \tilde{W}_{c1}^{\mathrm{T}} \nabla\phi_1 D_1 \nabla\phi_1^{\mathrm{T}} \tilde{W}_{c1} + \frac{1}{2} W_{c1}^{\mathrm{T}} \nabla\phi_1 D_2 \nabla\phi_2^{\mathrm{T}} \tilde{W}_{c2} \\
&- \frac{1}{2} \tilde{W}_{c2}^{\mathrm{T}} \nabla\phi_2 S_2 \nabla\phi_2^{\mathrm{T}} W_{c2} + \frac{1}{4} \tilde{W}_{c2}^{\mathrm{T}} \nabla\phi_2 S_2 \nabla\phi_2^{\mathrm{T}} \tilde{W}_{c2} + \varepsilon_{\mathrm{HJ1}} \Bigg] \\
&- \frac{\alpha_1}{4} \nabla\phi_1 D_1 \nabla\phi_1^{\mathrm{T}} \hat{W}_{c1} \frac{\bar{\sigma}_1^{\mathrm{T}}}{m_{s_1}} \hat{W}_{c1} - \frac{\alpha_2}{4} \nabla\phi_1 S_1 \nabla\phi_1^{\mathrm{T}} \hat{W}_{c1} \frac{\bar{\sigma}_2^{\mathrm{T}}}{m_{s_2}} \hat{W}_{c2} \\
&- \Sigma(x, \hat{u}, \hat{w}) \left(\frac{\alpha_1 \nabla\phi_1 D_1 \nabla L_1}{2} + \frac{\alpha_1 \nabla\phi_1 D_1 \nabla L_2}{2} \right) \\
&+ \alpha_1 (F_1 \hat{W}_{c1} - F_2 \bar{\sigma}_1^{\mathrm{T}} \hat{W}_{c1}).\end{aligned} \tag{9.95}$$

Similarly, combining (9.91) with (9.94), we have

$$\begin{aligned}\dot{\tilde{W}}_2 =& \alpha_2 \frac{\bar{\sigma}_2}{m_{s_2}} \Bigg[-\tilde{W}_{c2}^{\mathrm{T}} \hat{\sigma}_2 + \frac{1}{4} \tilde{W}_{c2}^{\mathrm{T}} \nabla\phi_2 D_2 \nabla\phi_2^{\mathrm{T}} \tilde{W}_{c2} + \frac{1}{2} W_{c2}^{\mathrm{T}} \nabla\phi_2 D_1 \nabla\phi_1^{\mathrm{T}} \tilde{W}_{c1} \\
&- \frac{1}{2} \tilde{W}_{c1}^{\mathrm{T}} \nabla\phi_1 S_1 \nabla\phi_1^{\mathrm{T}} W_{c1} + \frac{1}{4} \tilde{W}_{c1}^{\mathrm{T}} \nabla\phi_1 S_1 \nabla\phi_1^{\mathrm{T}} \tilde{W}_{c1} + \varepsilon_{\mathrm{HJ2}} \Bigg] \\
&- \frac{\alpha_2}{4} \nabla\phi_2 D_2 \nabla\phi_2^{\mathrm{T}} \hat{W}_{c2} \frac{\bar{\sigma}_2^{\mathrm{T}}}{m_{s_2}} \hat{W}_{c2} - \frac{\alpha_2}{4} \nabla\phi_2 S_2 \nabla\phi_2^{\mathrm{T}} \hat{W}_{c2} \frac{\bar{\sigma}_1^{\mathrm{T}}}{m_{s_1}} \hat{W}_{c1} \\
&- \Sigma(x, \hat{u}, \hat{w}) \left(\frac{\alpha_2 \nabla\phi_2 D_2 \nabla L_2}{2} + \frac{\alpha_2 \nabla\phi_2 D_2 \nabla L_1}{2} \right) \\
&+ \alpha_2 (F_3 \hat{W}_{c2} - F_4 \bar{\sigma}_2^{\mathrm{T}} \hat{W}_{c2}).\end{aligned} \tag{9.96}$$

In the following, the stability analysis will be performed. First, the following assumption is made, which can reasonably be satisfied under the current problem settings.

Assumption 9.34

(a) $g(\cdot)$ and $k(\cdot)$ are upper bounded, i.e., $\|g(\cdot)\| \le g_M$ and $\|k(\cdot)\| \le k_M$ with g_M and k_M being positive constants.
(b) The critic NN approximation errors and their gradients are upper bounded so that $\|\varepsilon_i\| \le \varepsilon_{iM}$ and $\|\nabla\varepsilon_i\| \le \varepsilon_{idM}$ with ε_{iM} and ε_{idM} being positive constants, $i = 1, 2$.
(c) The critic NN activation function vectors are upper bounded, so that $\|\phi_i\| \le \phi_{iM}$ and $\|\nabla\phi_i\| \le \phi_{idM}$, with ϕ_{iM} and ϕ_{idM} being positive constants, $i = 1, 2$.

(d) The critic NN weights are upper bounded so that $\|W_i\| \le W_{iM}$ with W_{iM} being positive constant, $i = 1, 2$. The residual errors ε_{HJi} are upper bounded, so that $\|\varepsilon_{\text{HJi}}\| \le \varepsilon_{\text{HJ}iM}$ with $\varepsilon_{\text{HJ}iM}$ being positive constant, $i = 1, 2$.

Now we are ready to prove the following theorem.

Theorem 9.35 (cf. [17]) *Consider the system given by* (9.58). *Let the control input be provided by* (9.85) *and* (9.86), *and the critic NN weight tuning laws be given by* (9.90) *and* (9.91). *Then, the system state x and the weight estimation errors of critic NNs $\tilde{W}_{c1}$ and $\tilde{W}_{c2}$ are uniformly ultimately bounded (UUB). Furthermore, the obtained control input $\hat{u}$ and $\hat{w}$ in* (9.85) *and* (9.86) *are proved to converge to the Nash equilibrium policy of the non-zero-sum differential games approximately, i.e., $\hat{u}$ and $\hat{w}$ are closed for the optimal control input u^* and w^* with bounds ϵ_u and ϵ_w, respectively.*

Proof Choose the following Lyapunov function candidate:

$$L = L_1(x) + L_2(x) + \frac{1}{2}\tilde{W}_{c1}^{\mathrm{T}}\alpha_1^{-1}\tilde{W}_{c1} + \frac{1}{2}\tilde{W}_{c2}^{\mathrm{T}}\alpha_2^{-1}\tilde{W}_{c2}, \tag{9.97}$$

where $L_1(x)$ and $L_2(x)$ are given by Lemma 9.31.

The derivative of the Lyapunov function candidate (9.97) along the system (9.87) is computed as

$$\begin{aligned}\dot{L} &= \nabla L_1^{\mathrm{T}}\left(f(x) - \frac{D_1\nabla\phi_1^{\mathrm{T}}\hat{W}_{c1}}{2} - \frac{D_2\nabla\phi_2^{\mathrm{T}}\hat{W}_{c2}}{2}\right)\\ &\quad + \nabla L_2^{\mathrm{T}}\left(f(x) - \frac{D_1\nabla\phi_1^{\mathrm{T}}\hat{W}_{c1}}{2} - \frac{D_2\nabla\phi_2^{\mathrm{T}}\hat{W}_{c2}}{2}\right)\\ &\quad + \tilde{W}_{c1}^{\mathrm{T}}\alpha_1^{-1}\dot{\tilde{W}}_1 + \tilde{W}_{c2}^{\mathrm{T}}\alpha_2^{-1}\dot{\tilde{W}}_2.\end{aligned} \tag{9.98}$$

Then, substituting (9.95) and (9.96) into (9.98), we have

$$\begin{aligned}\dot{L} &= \nabla L_1^{\mathrm{T}}\left(f(x) - \frac{D_1\nabla\phi_1^{\mathrm{T}}\hat{W}_{c1}}{2} - \frac{D_2\nabla\phi_2^{\mathrm{T}}\hat{W}_{c2}}{2}\right)\\ &\quad + \nabla L_2^{\mathrm{T}}\left(f(x) - \frac{D_1\nabla\phi_1^{\mathrm{T}}\hat{W}_{c1}}{2} - \frac{D_2\nabla\phi_2^{\mathrm{T}}\hat{W}_{c2}}{2}\right)\\ &\quad + \tilde{W}_{c1}^{\mathrm{T}}\bar{\sigma}_1\left(-\bar{\sigma}_1^{\mathrm{T}}\tilde{W}_{c1} + \frac{\varepsilon_{\text{HJ1}}}{m_{s_1}}\right) + \tilde{W}_{c2}^{\mathrm{T}}\bar{\sigma}_2\left(-\bar{\sigma}_2^{\mathrm{T}}\tilde{W}_{c2} + \frac{\varepsilon_{\text{HJ2}}}{m_{s_2}}\right)\\ &\quad + \frac{1}{4}\tilde{W}_{c1}^{\mathrm{T}}\nabla\phi_1 D_1\nabla\phi_1^{\mathrm{T}}W_{c1}\frac{\bar{\sigma}_1^{\mathrm{T}}}{m_{s_1}}\tilde{W}_{c1} - \frac{1}{4}\tilde{W}_{c1}^{\mathrm{T}}\nabla\phi_1 D_1\nabla\phi_1^{\mathrm{T}}W_{c1}\frac{\bar{\sigma}_1^{\mathrm{T}}}{m_{s_1}}W_{c1}\end{aligned}$$

$$
\begin{aligned}
&+\frac{1}{4}\tilde{W}_{c1}^{\mathrm{T}}\nabla\phi_1 D_1\nabla\phi_1^{\mathrm{T}}\tilde{W}_{c1}\frac{\bar{\sigma}_1^{\mathrm{T}}}{m_{s_1}}W_{c1}\\
&+\frac{1}{4}\tilde{W}_{c2}^{\mathrm{T}}\nabla\phi_2 D_2\nabla\phi_2^{\mathrm{T}}W_{c2}\frac{\bar{\sigma}_2^{\mathrm{T}}}{m_{s_2}}\tilde{W}_{c2}-\frac{1}{4}\tilde{W}_{c2}^{\mathrm{T}}\nabla\phi_2 D_2\nabla\phi_2^{\mathrm{T}}W_{c2}\frac{\bar{\sigma}_2^{\mathrm{T}}}{m_{s_2}}W_{c2}\\
&+\frac{1}{4}\tilde{W}_{c2}^{\mathrm{T}}\nabla\phi_2 D_2\nabla\phi_2^{\mathrm{T}}\tilde{W}_{c2}\frac{\bar{\sigma}_2^{\mathrm{T}}}{m_{s_2}}W_{c2}\\
&+\frac{1}{2}\tilde{W}_{c1}^{\mathrm{T}}\frac{\bar{\sigma}_1}{m_{s_1}}W_{c1}^{\mathrm{T}}\nabla\phi_1 D_2\nabla\phi_2^{\mathrm{T}}\tilde{W}_{c2}-\frac{1}{2}\tilde{W}_{c1}^{\mathrm{T}}\frac{\bar{\sigma}_1}{m_{s_1}}\tilde{W}_{c2}^{\mathrm{T}}\nabla\phi_2 S_2\nabla\phi_2^{\mathrm{T}}W_{c2}\\
&+\frac{1}{2}\tilde{W}_{c2}^{\mathrm{T}}\frac{\bar{\sigma}_2}{m_{s_2}}W_{c2}^{\mathrm{T}}\nabla\phi_2 D_1\nabla\phi_1^{\mathrm{T}}\tilde{W}_{c1}-\frac{1}{2}\tilde{W}_{c2}^{\mathrm{T}}\frac{\bar{\sigma}_2}{m_{s_2}}\tilde{W}_{c1}^{\mathrm{T}}\nabla\phi_1 S_1\nabla\phi_1^{\mathrm{T}}W_{c1}\\
&+\frac{1}{4}\tilde{W}_{c1}^{\mathrm{T}}\nabla\phi_1 S_1\nabla\phi_1^{\mathrm{T}}W_{c1}\frac{\bar{\sigma}_2^{\mathrm{T}}}{m_{s_2}}\tilde{W}_{c2}-\frac{1}{4}\tilde{W}_{c1}^{\mathrm{T}}\nabla\phi_1 S_1\nabla\phi_1^{\mathrm{T}}W_{c1}\frac{\bar{\sigma}_2^{\mathrm{T}}}{m_{s_2}}W_{c2}\\
&+\frac{1}{4}\tilde{W}_{c1}^{\mathrm{T}}\nabla\phi_1 S_1\nabla\phi_1^{\mathrm{T}}\tilde{W}_{c1}\frac{\bar{\sigma}_2^{\mathrm{T}}}{m_{s_2}}W_{c2}\\
&+\frac{1}{4}\tilde{W}_{c2}^{\mathrm{T}}\nabla\phi_2 S_2\nabla\phi_2^{\mathrm{T}}W_{c2}\frac{\bar{\sigma}_1^{\mathrm{T}}}{m_{s_1}}\tilde{W}_{c1}-\frac{1}{4}\tilde{W}_{c2}^{\mathrm{T}}\nabla\phi_2 S_2\nabla\phi_2^{\mathrm{T}}W_{c2}\frac{\bar{\sigma}_1^{\mathrm{T}}}{m_{s_1}}W_{c1}\\
&+\frac{1}{4}\tilde{W}_{c2}^{\mathrm{T}}\nabla\phi_2 S_2\nabla\phi_2^{\mathrm{T}}\tilde{W}_{c2}\frac{\bar{\sigma}_1^{\mathrm{T}}}{m_{s_1}}W_{c1}\\
&-\Sigma(x,\hat{u},\hat{w})\left(\frac{\tilde{W}_{c1}^{\mathrm{T}}\nabla\phi_1 D_1\nabla L_1}{2}+\frac{\tilde{W}_{c1}^{\mathrm{T}}\nabla\phi_1 D_1\nabla L_2}{2}\right)\\
&-\Sigma(x,\hat{u},\hat{w})\left(\frac{\tilde{W}_{c2}^{\mathrm{T}}\nabla\phi_2 D_2\nabla L_2}{2}+\frac{\tilde{W}_{c2}^{\mathrm{T}}\nabla\phi_2 D_2\nabla L_1}{2}\right)\\
&+\tilde{W}_{c1}^{\mathrm{T}}F_1\hat{W}_{c1}-\tilde{W}_{c1}^{\mathrm{T}}F_2\bar{\sigma}_1^{\mathrm{T}}\hat{W}_{c1}\\
&+\tilde{W}_{c2}^{\mathrm{T}}F_3\hat{W}_{c2}-\tilde{W}_{c2}^{\mathrm{T}}F_4\bar{\sigma}_2^{\mathrm{T}}\hat{W}_{c2}.
\end{aligned}
\tag{9.99}
$$

In (9.99), the last two terms can be rewritten as

$$
\begin{aligned}
&\tilde{W}_{c1}^{\mathrm{T}}F_1\hat{W}_{c1}-\tilde{W}_{c1}^{\mathrm{T}}F_2\bar{\sigma}_1^{\mathrm{T}}\hat{W}_{c1}\\
&\quad=\tilde{W}_{c1}^{\mathrm{T}}F_1W_{c1}-\tilde{W}_{c1}^{\mathrm{T}}F_2\bar{\sigma}_1^{\mathrm{T}}\tilde{W}_{c1}-\tilde{W}_{c1}^{\mathrm{T}}F_2\bar{\sigma}_1^{\mathrm{T}}W_{c1}-\tilde{W}_{c1}^{\mathrm{T}}F_2\bar{\sigma}_1^{\mathrm{T}}\tilde{W}_{c1}\\
&\qquad+\tilde{W}_{c2}^{\mathrm{T}}F_3\hat{W}_{c2}-\tilde{W}_{c2}^{\mathrm{T}}F_4\bar{\sigma}_2^{\mathrm{T}}\hat{W}_{c2}\\
&\quad=\tilde{W}_{c2}^{\mathrm{T}}F_3W_{c2}-\tilde{W}_{c2}^{\mathrm{T}}F_3\tilde{W}_{c2}-\tilde{W}_{c2}^{\mathrm{T}}F_4\bar{\sigma}_2^{\mathrm{T}}W_{c2}-\tilde{W}_{c2}^{\mathrm{T}}F_4\bar{\sigma}_2^{\mathrm{T}}\tilde{W}_{c2}.
\end{aligned}
\tag{9.100}
$$

Define $z = [\bar{\sigma}_1^{\mathrm{T}} \tilde{W}_{c1}, \bar{\sigma}_2^{\mathrm{T}} \tilde{W}_{c2}, \tilde{W}_{c1}, \tilde{W}_{c2}]^{\mathrm{T}}$; then (9.99) can be rewritten as

$$
\begin{aligned}
\dot{L} = & -z^{\mathrm{T}} \begin{bmatrix} M_{11} & M_{12} & M_{13} & M_{14} \\ M_{21} & M_{22} & M_{23} & M_{24} \\ M_{31} & M_{32} & M_{33} & M_{34} \\ M_{41} & M_{42} & M_{43} & M_{44} \end{bmatrix} z + z^{\mathrm{T}} \delta \\
& + \nabla L_1^{\mathrm{T}} \left(f(x) - \frac{D_1 \nabla \phi_1^{\mathrm{T}} \hat{W}_{c1}}{2} - \frac{D_2 \nabla \phi_2^{\mathrm{T}} \hat{W}_{c2}}{2} \right) \\
& + \nabla L_2^{\mathrm{T}} \left(f(x) - \frac{D_1 \nabla \phi_1^{\mathrm{T}} \hat{W}_{c1}}{2} - \frac{D_2 \nabla \phi_2^{\mathrm{T}} \hat{W}_{c2})}{2} \right) \\
& - \Sigma(x, \hat{u}, \hat{w}) \left(\frac{\tilde{W}_{c1}^{\mathrm{T}} \nabla \phi_1 D_1 \nabla L_1}{2} + \frac{\tilde{W}_{c1}^{\mathrm{T}} \nabla \phi_1 D_1 \nabla L_2}{2} \right) \\
& - \Sigma(x, \hat{u}, \hat{w}) \left(\frac{\tilde{W}_{c2}^{\mathrm{T}} \nabla \phi_2 D_2 \nabla L_2}{2} + \frac{\tilde{W}_{c2}^{\mathrm{T}} \nabla \phi_2 D_2 \nabla L_1}{2} \right),
\end{aligned}
\tag{9.101}
$$

where the components of the matrix M are given by

$$
\begin{aligned}
M_{11} &= M_{22} = I, \\
M_{12} &= M_{21}^{\mathrm{T}} = 0, \\
M_{13} &= M_{31}^{\mathrm{T}} = -\frac{1}{4m_{s_1}} \nabla \phi_1 D_1 \nabla \phi_1^{\mathrm{T}} W_{c1} - \frac{F_2}{2}, \\
M_{14} &= M_{41}^{\mathrm{T}} = -\frac{1}{4m_{s_1}} \nabla \phi_2 D_2 \nabla \phi_1^{\mathrm{T}} W_{c1} + \frac{1}{8} \nabla \phi_2 S_2 \nabla \phi_2^{\mathrm{T}} W_{c2}, \\
M_{23} &= M_{32}^{\mathrm{T}} = -\frac{1}{4m_{s_2}} \nabla \phi_1 D_1 \nabla \phi_2^{\mathrm{T}} W_{c2} + \frac{1}{8} \nabla \phi_1 S_1 \nabla \phi_1^{\mathrm{T}} W_{c1}, \\
M_{24} &= M_{42}^{\mathrm{T}} = -\frac{1}{4m_{s_2}} \nabla \phi_2 D_2 \nabla \phi_2^{\mathrm{T}} W_{c2} - \frac{F_4}{2}, \\
M_{33} &= -\frac{1}{4m_{s_2}} \nabla \phi_1 S_1 \nabla \phi_1^{\mathrm{T}} W_{c2} \bar{\sigma}_2^{\mathrm{T}} + F_1, \\
M_{34} &= M_{43}^{\mathrm{T}} = 0, \\
M_{44} &= -\frac{1}{4m_{s_1}} \nabla \phi_2 S_2 \nabla \phi_2^{\mathrm{T}} W_{c1} \bar{\sigma}_1^{\mathrm{T}} + F_3,
\end{aligned}
$$

and the components of the vector $\delta = [d_1 \; d_2 \; d_3 \; d_4]^{\mathrm{T}}$ are given as

$$
d_1 = \frac{\varepsilon_{\mathrm{HJ1}}}{m_{s_1}},
$$

$$d_2 = \frac{\varepsilon_{\mathrm{HJ2}}}{m_{s_2}},$$

$$\begin{aligned} d_3 = & -\frac{1}{4m_{s_1}} \nabla\phi_1 D_1 \nabla\phi_1^{\mathrm{T}} W_{c1} \bar{\sigma}_1^{\mathrm{T}} W_{c1} \\ & -\frac{1}{4m_{s_2}} \nabla\phi_1 S_1 \nabla\phi_1^{\mathrm{T}} W_{c1} \bar{\sigma}_2^{\mathrm{T}} W_{c2} + F_1 W_{c1} - F_2 \bar{\sigma}_1^{\mathrm{T}} W_{c1}, \\ d_4 = & -\frac{1}{4m_{s_2}} \nabla\phi_2 D_2 \nabla\phi_2^{\mathrm{T}} W_{c2} \bar{\sigma}_2^{\mathrm{T}} W_{c2} \\ & -\frac{1}{4m_{s_1}} \nabla\phi_2 S_2 \nabla\phi_2^{\mathrm{T}} W_{c2} \bar{\sigma}_1^{\mathrm{T}} W_{c1} + F_3 W_{c2} - F_4 \bar{\sigma}_2^{\mathrm{T}} W_{c2}. \end{aligned}$$

According to Assumption 9.34 and observing the facts that $\bar{\sigma}_1 < 1$ and $\bar{\sigma}_2 < 1$, it can be concluded that δ is bounded by δ_M. Let the parameters F_1, F_2, F_3, and F_4 be chosen such that $M > 0$. Then, taking the upper bounds of (9.101) reveals

$$\begin{aligned} \dot{L} \le & \nabla L_1 \left(f(x) - \frac{D_1 \nabla\phi_1^{\mathrm{T}} \hat{W}_{c1}}{2} - \frac{D_2 \nabla\phi_2^{\mathrm{T}} \hat{W}_{c2}}{2} \right) \\ & + \nabla L_2 \left(f(x) - \frac{D_1 \nabla\phi_1^{\mathrm{T}} \hat{W}_{c1}}{2} - \frac{D_2 \nabla\phi_2^{\mathrm{T}} \hat{W}_{c2}}{2} \right) \\ & - \|z\|^2 \sigma_{\min}(M) + \|z\| \delta_M \\ & - \Sigma(x, \hat{u}, \hat{w}) \left(\frac{\tilde{W}_{c1}^{\mathrm{T}} \nabla\phi_1 D_1 \nabla L_1}{2} + \frac{\tilde{W}_{c1}^{\mathrm{T}} \nabla\phi_1 D_1 \nabla L_2}{2} \right) \\ & - \Sigma(x, \hat{u}, \hat{w}) \left(\frac{\tilde{W}_{c2}^{\mathrm{T}} \nabla\phi_2 D_2 \nabla L_2}{2} + \frac{\tilde{W}_{c2}^{\mathrm{T}} \nabla\phi_2 D_2 \nabla L_1}{2} \right). \end{aligned} \tag{9.102}$$

Now, the cases of $\Sigma(x, \hat{u}, \hat{w}) = 0$ and $\Sigma(x, \hat{u}, \hat{w}) = 1$ will be considered.

(1) When $\Sigma(x, \hat{u}, \hat{w}) = 0$, the first two terms are less than zero. Noting that $\|x\| > 0$ as guaranteed by the persistent excitation condition and using the operator defined in (9.92), it can be ensured that there exists a constant $\dot{x}_{\min}$ satisfying $0 < \dot{x}_{\min} < \|\dot{x}\|$. Then (9.102) becomes

$$\begin{aligned} \dot{L} \le & -\dot{x}_{\min}(\|\nabla L_1\| + \|\nabla L_2\|) - \|z\|^2 \sigma_{\min}(M) + \|z\| \delta_M \\ = & -\dot{x}_{\min}(\|\nabla L_1\| + \|\nabla L_2\|) - \sigma_{\min}(M) \left(\|z\| - \frac{\delta_M}{2\sigma_{\min}(M)} \right)^2 \\ & + \frac{\delta_M^2}{4\sigma_{\min}(M)}. \end{aligned} \tag{9.103}$$

Given that the following inequalities:

$$\|\nabla L_1\| \geq \frac{\delta_M^2}{4\sigma_{\min}(M)\dot{x}_{\min}} \triangleq B_{\nabla L_1}, \tag{9.104}$$

or

$$\|\nabla L_2\| \geq \frac{\delta_M^2}{4\sigma_{\min}(M)\dot{x}_{\min}} \triangleq B_{\nabla L_2}, \tag{9.105}$$

or

$$\|z\| \geq \frac{\delta_M}{\sigma_{\min}(M)} \triangleq B_z \tag{9.106}$$

hold, then $\dot{L} < 0$. Therefore, using Lyapunov theory, it can be concluded that $\|\nabla L_1\|$, $\|\nabla L_2\|$ and $\|z\|$ are UUB.

(2) When $\Sigma(x, \hat{u}, \hat{w}) = 1$, it implies that the feedback control input (9.85) and (9.86) may not stabilize the system (9.58). Adding and subtracting $\nabla L_1^{\mathrm{T}} D_1 \varepsilon_1/2 + \nabla L_2^{\mathrm{T}} D_2 \varepsilon_2/2$ to the right hand side of (9.102), and using (9.64), (9.65), and (9.80), we have

$$\begin{aligned}
\dot{L} \leq{}& \nabla L_1^{\mathrm{T}}(f(x) + g(x)u^* + k(x)w^*) + \nabla L_2^{\mathrm{T}}(f(x) + g(x)u^* + k(x)w^*) \\
&+ \frac{1}{2}\nabla L_1 D_1 \nabla\varepsilon_1 + \frac{1}{2}\nabla L_2 D_2 \nabla\varepsilon_2 + \frac{1}{2}\nabla L_1 D_2 \nabla\varepsilon_2 + \frac{1}{2}\nabla L_2 D_1 \nabla\varepsilon_1 \\
&- \sigma_{\min}(M)\left(\|z\| - \frac{\delta_M}{2\sigma_{\min}(M)}\right)^2 + \frac{\delta_M^2}{4\sigma_{\min}(M)}.
\end{aligned} \tag{9.107}$$

According to Assumption 9.34, D_i is bounded by D_{iM}, where D_{iM} is a known constant, $i = 1, 2$. Using Lemma 9.31 and recalling the boundedness of $\nabla\varepsilon_1$, $\nabla\varepsilon_2$, and δ, (9.107) can be rewritten as

$$\begin{aligned}
\dot{L} \leq{}& -\bar{Q}_{1\min}\|\nabla L_1\|^2 - \bar{Q}_{2\min}\|\nabla L_2\|^2 + \frac{1}{2}\|\nabla L_1\| D_{1M}\varepsilon_{1dM} \\
&+ \frac{1}{2}\|\nabla L_2\| D_{2M}\varepsilon_{2dM} + \frac{1}{2}\|\nabla L_1\| D_{2M}\varepsilon_{2dM} \\
&+ \frac{1}{2}\|\nabla L_2\| D_{1M}\varepsilon_{1dM} - \sigma_{\min}(M)\left(\|z\| - \frac{\delta_M}{2\sigma_{\min}(M)}\right)^2 + \frac{\delta_M^2}{4\sigma_{\min}(M)} \\
\leq{}& -\frac{1}{2}\bar{Q}_{1\min}\|\nabla L_1\|^2 - \frac{1}{2}\bar{Q}_{2\min}\|\nabla L_2\|^2 - \sigma_{\min}(M)\left(\|z\| - \frac{\delta_M}{2\sigma_{\min}(M)}\right)^2 + \eta,
\end{aligned} \tag{9.108}$$

where

$$\eta = \frac{D_{1M}^2\varepsilon_{1dM}^2}{4\bar{Q}_{1\min}} + \frac{D_{2M}^2\varepsilon_{2dM}^2}{4\bar{Q}_{2\min}} + \frac{D_{2M}^2\varepsilon_{2dM}^2}{4\bar{Q}_{1\min}} + \frac{D_{1M}^2\varepsilon_{1dM}^2}{4\bar{Q}_{2\min}} + \frac{\delta_M^2}{4\sigma_{\min}(M)}.$$

Given that the following inequalities:

$$\|\nabla L_1\| > \sqrt{\frac{2\eta}{\bar{Q}_{1\min}}} \triangleq B'_{\nabla L_1}, \tag{9.109}$$

or

$$\|\nabla L_2\| > \sqrt{\frac{2\eta}{\bar{Q}_{2\min}}} \triangleq B'_{\nabla L_2}, \tag{9.110}$$

or

$$\|z\| > \sqrt{\frac{\eta}{\sigma_{\min}(M)} + \frac{\delta_M}{2\sigma_{\min}(M)}} \triangleq B'_z \tag{9.111}$$

hold, then $\dot{L} < 0$. Therefore, using Lyapunov theory, it can be concluded that $\|\nabla L_1\|$, $\|\nabla L_2\|$, and $\|z\|$ are UUB.

In summary, for the cases $\Sigma(x, \hat{u}, \hat{w}) = 0$ and $\Sigma(x, \hat{u}, \hat{w}) = 1$, if inequalities $\|\nabla L_1\| > \max(B_{\nabla L_1}, B'_{\nabla L_1}) \triangleq \bar{B}_{\nabla L_1}$, or $\|\nabla L_2\| > \max(B_{\nabla L_2}, B'_{\nabla L_2}) \triangleq \bar{B}_{\nabla L_2}$ or $\|z\| > \max(B_z, B'_z) \triangleq \bar{B}_z$ hold, then $\dot{L} < 0$. Therefore, we can conclude that $\|\nabla L_1\|$, $\|\nabla L_2\|$ and $\|z\|$ are bounded by $\bar{B}_{\nabla L_1}$, $\bar{B}_{\nabla L_2}$, and $\bar{B}_z$, respectively. According to Lemma 9.31, the Lyapunov candidates ∇L_1 and ∇L_2 are radially unbounded and continuously differentiable. Therefore, the boundedness of $\|\nabla L_1\|$ and $\|\nabla L_2\|$ implies the boundedness of $\|x\|$. Specifically, $\|x\|$ is bounded by $\bar{B}_x = \max(B_{1x}, B_{2x})$, where B_{1x} and B_{2x} are determined by $\bar{B}_{\nabla L_1}$ and $\bar{B}_{\nabla L_2}$, respectively. Besides, note that if any component of z exceeds the bound, i.e., $\|\tilde{W}_{c1}\| > \bar{B}_z$ or $\|\tilde{W}_{c2}\| > \bar{B}_z$ or $\|\bar{\sigma}_1^{\mathrm{T}}\tilde{W}_{c1}\| > \bar{B}_z$ or $\|\bar{\sigma}_2^{\mathrm{T}}\tilde{W}_{c2}\| > \bar{B}_z$, the $\|z\|$ are bounded by $\bar{B}_z$, which implies that the critic NN weight estimation errors $\|\tilde{W}_{c1}\|$ and $\|\tilde{W}_{c2}\|$ are also bounded by B_z.

Next, we will prove $\|\hat{u} - u^*\| \le \epsilon_u$ and $\|\hat{w} - w^*\| \le \epsilon_w$. From (9.64) and (9.85) and recalling the boundedness of $\|\nabla\phi_1\|$ and $\|\tilde{W}_{c1}\|$, we have

$$\begin{aligned} \|\hat{u} - u^*\| &\le \left\| -\frac{1}{2} R_{11}^{-1} g^{\mathrm{T}} \nabla\phi_1^{\mathrm{T}} \tilde{W}_{c1} \right\| \\ &\le \lambda_{\max}(R_{11}^{-1}) \nabla\phi_{1M} \bar{B}_z \\ &\triangleq \epsilon_u. \end{aligned} \tag{9.112}$$

Similarly, from (9.65) and (9.86) and recalling the boundedness of $\|\nabla\phi_2\|$ and $\|\tilde{W}_{c2}\|$, we obtain $\|\hat{w} - w^*\| \le \epsilon_w$.

This completes the proof. □

Remark 9.36 In [10], each player needs two NNs consisting of a critic NN and an action NN to implement the online learning algorithm. By contrast with [10], only one critic NN is required for each player, the action NN is eliminated, resulting in a simpler architecture, and less computational burden.

Remark 9.37 In Remark 3 of [10] one pointed out that the NN weights can be initialized randomly but non-zero. That is because the method proposed in [10] requires initial stabilizing control policies for guaranteeing the stability of the system. By contrast, no initial stabilizing control policies are needed by adding an operator, which is selected by the Lyapunov's sufficiency condition for stability, on the critic NN weight tuning law for each player in this subsection.

9.4.3 Simulations

Example 9.38 An example is provided to demonstrate the effectiveness of the present control scheme.

Consider the affine nonlinear system as follows:

$$\dot{x} = f(x) + g(x)u + k(x)w, \tag{9.113}$$

where

$$f(x) = \begin{bmatrix} x_2 - 2x_1 \\ -x_2 - 0.5x_1 + 0.25x_2(\cos(2x_1 + 2))^2 + 0.25x_2(\sin(4x_1^2) + 2)^2 \end{bmatrix}, \tag{9.114}$$

$$g(x) = \begin{bmatrix} 0 \\ \cos(2x_1 + 2) \end{bmatrix}, \quad k(x) = \begin{bmatrix} 0 \\ \sin(4x_1^2) + 2) \end{bmatrix}. \tag{9.115}$$

The cost functionals for player 1 and player 2 are defined by (9.59) and (9.60), respectively, where $Q_1(x) = 2x^{\mathrm{T}}x$, $R_{11} = R_{12} = 2I$, $Q_2(x) = x^{\mathrm{T}}x$, $R_{21} = R_{22} = 2I$, and I denotes an identity matrix of appropriate dimensions.

For player 1, the optimal cost function is $V_1^*(x) = 0.25x_1^2 + x_2^2$. For player 2, the optimal cost function is $V_2^*(x) = 0.25x_1^2 + 0.5x_2^2$. The activation functions of critic NNs of two players are selected as $\phi_1 = \phi_2 = [x_1^2, x_1x_2, x_2^2]^{\mathrm{T}}$. Then, the optimal values of the critic NN weights for player 1 are $W_{c1} = [0.5, 0, 1]^{\mathrm{T}}$. The optimal values of the critic NN weights for player 2 are $W_{c2} = [0.25, 0, 0.5]^{\mathrm{T}}$. The estimates of the critic NN weights for two players are denoted $\hat{W}_{c1} = [W_{11}, W_{12}, W_{13}]^{\mathrm{T}}$ and $\hat{W}_{c2} = [W_{21}, W_{22}, W_{23}]^{\mathrm{T}}$, respectively. The adaptive gains for the critic NNs are selected as $a_1 = 1$ and $a_2 = 1$, and the design parameters are selected as $F_1 = F_2 = F_3 = F_4 = 10I$. All NN weights are initialized to zero, which means that no initial stabilizing control policies are needed for implementing the present control scheme. The system state is initialized as $[0.5, 0.2]^{\mathrm{T}}$. To maintain the excitation condition, probing noise is added to the control input for the first 250 s.

After simulation, the trajectories of the system states are shown in Fig. 9.13. The convergence trajectories of the critic NN weights for player 1 are shown in Fig. 9.14, from which we see that the critic NN weights for player 1 finally converge to $[0.4490, 0.0280, 0.9777]^{\mathrm{T}}$. The convergence trajectories of the critic NN weights for player 2 are shown in Fig. 9.15, from which we see that the critic NN weights for

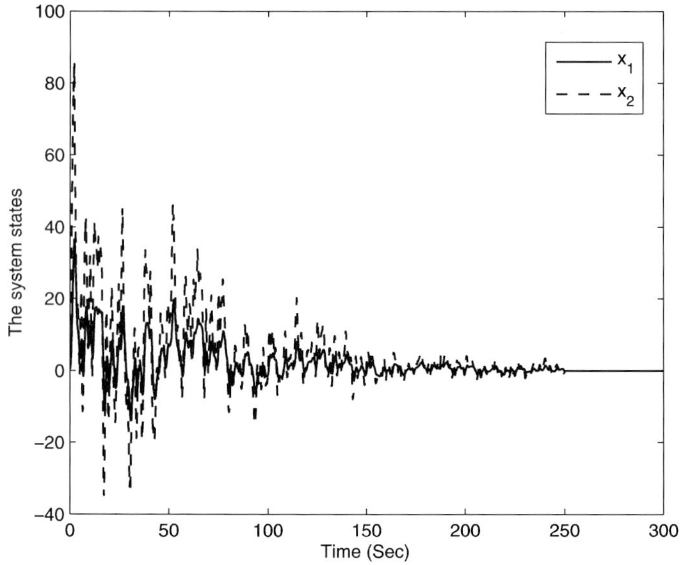

Fig. 9.13 The trajectories of system states

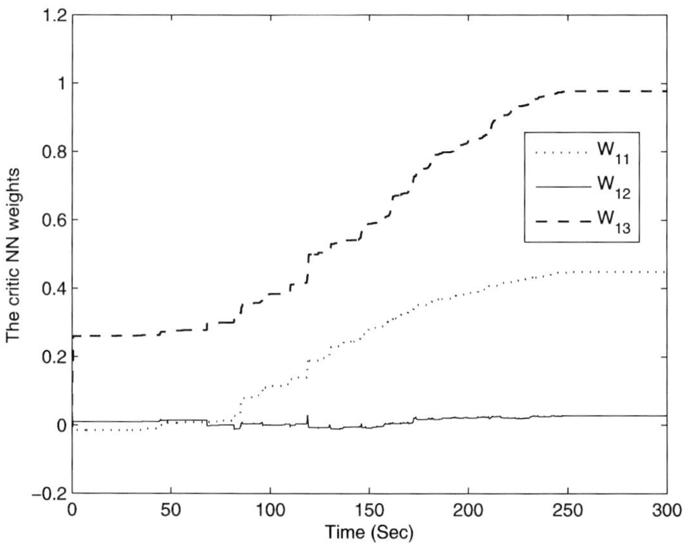

Fig. 9.14 The convergence trajectories of critic NN weights for player 1

player 2 finally converge to $[0.1974, 0.0403, 0.4945]^{\mathrm{T}}$. The convergence trajectory of $e_u = \hat{u} - u^*$ is shown in Fig. 9.16. The convergence trajectory of $e_w = \hat{w} - w^*$ is shown in Fig. 9.17. From Fig. 9.16, we see that the error between the estimated control $\hat{u}$ and the optimal control u^* for player 1 is close to zero when $t = 230$ s.

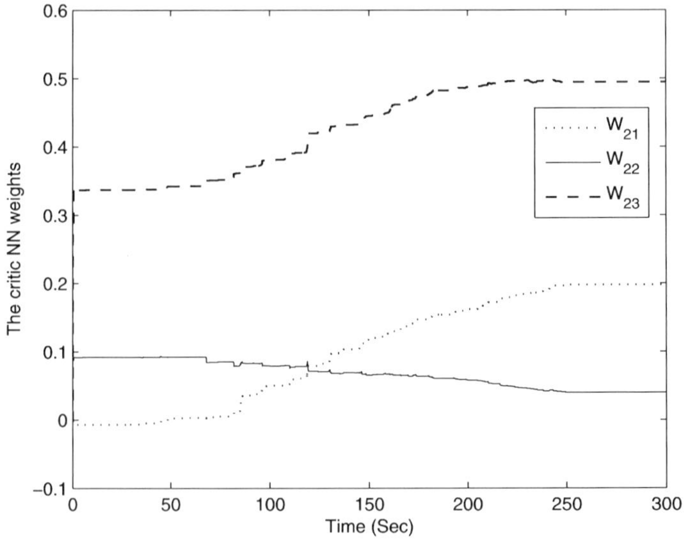

Fig. 9.15 The convergence trajectories of critic NN weights for player 2

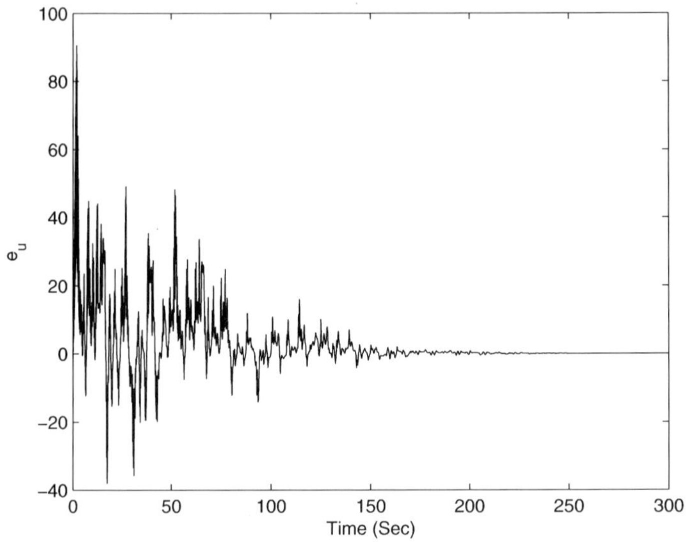

Fig. 9.16 The convergence trajectory of e_u

Similarly, it can been seen from Fig. 9.17 that the estimated control $\hat{w}$ and the optimal control w^* for player 2 are also close to zero when $t = 180$ s. Simulation results reveal that the present control scheme can make the critic NN learn the optimal cost function for each player and meanwhile guarantees stability of the closed-loop system.

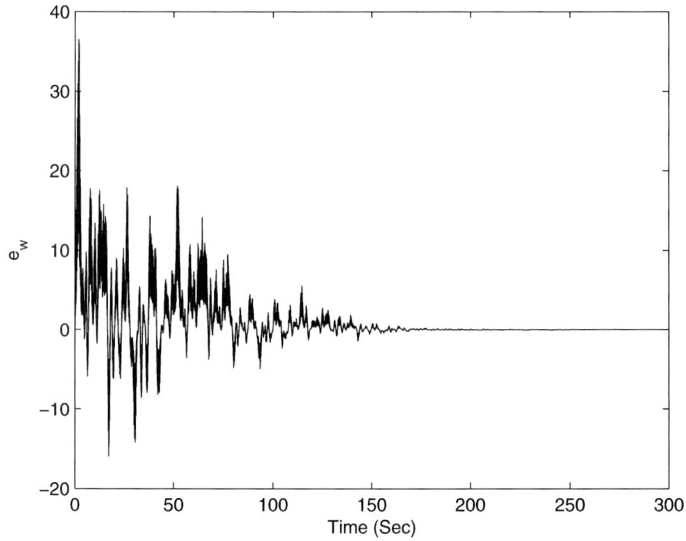

Fig. 9.17 The convergence trajectory of e_w

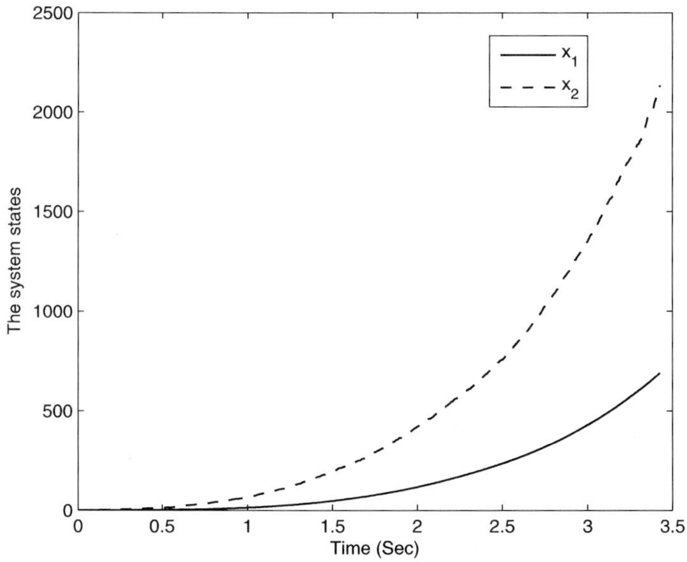

Fig. 9.18 The trajectories of system states obtained by the method in [10] with initial NN weights selected being zero

In order to compare with [10], we use the method proposed in [10] to solve the non-zero-sum games of system (9.113) where all NN weights are initialized to be zero, then obtain the trajectories of system states as shown in Fig. 9.18. It is shown

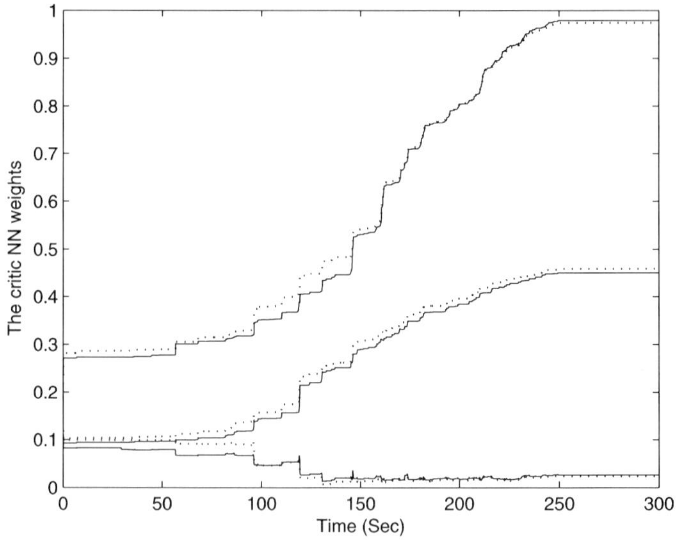

Fig. 9.19 The convergence trajectories of critic NN weights for player 1 (*solid line*: the method in [10]), *dashed line*: our method)

that the system is unstable, which implies that the method in [10] requires initial stabilizing control policies for guaranteeing the stability of the system. By contrast, the present method does not need the initial stabilizing control policies.

As pointed out earlier, one of the main advantages of the single ADP approach is that it results in less computational burden and eliminates the approximation error resulting from the action NNs. To demonstrate this quantitatively, we apply the method in [10] and our method to the system (9.113) with the same initial condition. Figures 9.19 and 9.20 show the convergence trajectories of the critic NN weights for player 1 and player 2, where the solid line and the dashed line represent the results from the method in [10] and our method, respectively. For the convenience of comparison, we define an evaluation function by $\mathrm{PER(i)} = \sum_{k=1}^{N} \|\tilde{W}_i(k)\|$, $i = 1, 2$, which means that the sum of the norm of the critic NN weights error during running time, where N is the number of sample points. The evaluation functions of the critic NN estimation errors as well as the time taken by the method in [10] and our method are calculated and shown in Table 9.1. It clearly indicates that the present method takes less time and obtains a smaller approximation error than [10].

9.5 Summary

In this chapter, we investigated the problem of continuous-time differential games based on ADP. In Sect. 9.2, we developed a new iterative ADP method to obtain

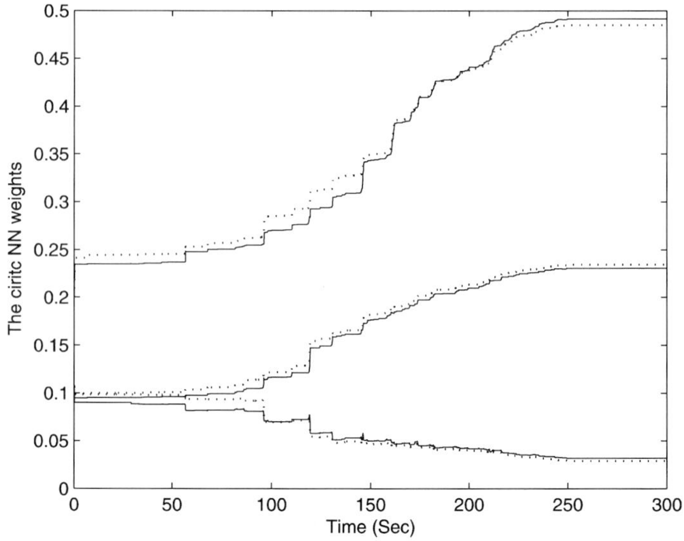

Fig. 9.20 The convergence trajectories of critic NN weights for player 2 (*solid line*: the method in [10]), *dashed line*: our method)

Table 9.1 Critic NN estimation errors and calculation time

Methods	PER(1)	PER(2)	Time
[10]	78.2312	29.4590	184.2024 s
Our method	70.2541	26.4152	111.1236 s

the optimal control pair or the mixed optimal control pair for a class of affine nonlinear zero-sum differential games. In Sect. 9.3, finite horizon zero-sum games for nonaffine nonlinear systems were studied. Then, in Sect. 9.4, the case of non-zero-sum differential games was studied using a single network ADP. Several numerical simulations showed that the present methods are effective.

References

1. Abu-Khalaf M, Lewis FL, Huang J (2006) Policy iterations on the Hamilton–Jacobi–Isaacs equation for H-infinity state feedback control with input saturation. IEEE Trans Autom Control 51:1989–1995
2. Abu-Khalaf M, Lewis FL, Huang J (2008) Neurodynamic programming and zero-sum games for constrained control systems. IEEE Trans Neural Netw 19:1243–1252
3. Al-Tamimi A, Lewis FL, Abu-Khalaf M (2007) Model-free Q-learning designs for linear discrete-time zero-sum games with application to H-infinity control. Automatica 43:473–481
4. Bardi M, Capuzzo-Dolcetta I (1997) Optimal control and viscosity solutions of Hamilton–Jacobi–Bellman equations. Birkhäuser, Germany

5. Birkhäuser (1995) H-infinity optimal control and related minimax design problems: a dynamical game approach. Birkhäuser, Berlin
6. Chang HS, Hu J, Fu MC (2010) Adaptive adversial multi-armed bandit approach to two-person zero-sum Markov games. IEEE Trans Autom Control 55:463–468
7. Chen BS, Tseng CS, Uang HJ (2002) Fuzzy differential games for nonlinear stochastic systems: suboptimal approach. IEEE Trans Fuzzy Syst 10:222–233
8. Laraki R, Solan E (2005) The value of zero-sum stopping games in continuous time. SIAM J Control Optim 43:1913–1922
9. Starr AW, Ho YC (1967) Nonzero-sum differential games. J Optim Theory Appl 3:184–206
10. Vamvoudakisand KG, Lewis FL (2011) Multi-player non-zero-sum games: online adaptive learning solution of coupled Hamilton–Jacobi equations. Automatica. doi:10.1016/j.automatica.2011.03.005
11. Wang X (2008) Numerical solution of optimal control for scaled systems by hybrid functions. Int J Innov Comput Inf Control 4:849–856
12. Wei QL, Zhang HG, Liu DR (2008) A new approach to solve a class of continuous-time nonlinear quadratic zero-sum game using ADP. In: Proceedings of IEEE international conference on networking, sensing and control, Sanya, China, pp 507–512
13. Wei QL, Zhang HG, Cui LL (2009) Data-based optimal control for discrete-time zero-sum games of 2-D systems using adaptive critic designs. Acta Autom Sin 35:682–692
14. Wei QL, Zhang HG, Dai J (2009) Model-free multiobjective approximate dynamic programming for discrete-time nonlinear systems with general performance index functions. Neurocomputing 7–9:1839–1848
15. Zhang HG, Wei QL, Liu DR (2011) An iterative adaptive dynamic programming method for solving a class of nonlinear zero-sum differential games. Automatica 47:207–214
16. Zhang X, Zhang HG, Wang XY (2011) A new iteration approach to solve a class of finite-horizon continuous-time nonaffine nonlinear zero-sum game. Int J Innov Comput Inf Control 7:597–608
17. Zhang HG, Cui LL, Luo YH (2012) Near-optimal control for non-zero-sum differential games of continuous-time nonlinear systems using single network ADP. IEEE Trans Syst Man Cybern, Part B, Cybern. doi:10.1109/TSMCB.2012.2203336

Chapter 10
Other Applications of ADP

10.1 Introduction

As is known, both modern wireless networks and automotive engines are complex non-linear dynamical systems. The corresponding optimal control problems have been investigated for many years by many researchers. However, there are still not results referring to the optimal control based on the ADP method. Due to the advantages of ADP in solving the optimal feedback control in a forward-time fashion, the ADP method is introduced in this chapter to solve the optimal control problem of modern wireless networks and automotive engines.

In the first part, a self-learning call admission control algorithm is developed for signal-to-interference ratio (SIR)-based power-controlled direct-sequence (DS)-code division multiple access (CDMA) cellular networks that provide both voice and data services [21]. The idea is built upon a novel learning control architecture with only a single module instead of two or three modules in adaptive critic designs. The call admission controller can perform learning in real time as well as in off-line environments and the controller improves its performance as it gains more experience. Another important contribution in the present work is the choice of utility function for the present self-learning control approach which makes the present learning process much more efficient than existing learning control methods.

In the second part, we apply an existing control algorithm to a production vehicle. The algorithm is presented according to certain criteria and calibrated for vehicle operation over the entire operating regime [24]. Actually, the algorithm has been optimized for the engine in terms of its performance, fuel economy, and tailpipe emissions through a significant effort in the research and development of calibration process. To further improve the engine performance through controller design, one may go through the traditional calibration and control procedures today. But an alternative to this traditional approach is to use the neural network-based learning control approach. The final result of our neural network learning process is a controller that has learned to provide optimal control signals under various operating conditions. We emphasize that such a neural network controller will be obtained

H. Zhang et al., *Adaptive Dynamic Programming for Control*,
Communications and Control Engineering, DOI 10.1007/978-1-4471-4757-2_10,

after a special learning process that performs adaptive dynamic programming algorithm. Once a controller is learned and obtained (off-line or on-line), it will be applied to perform the task of engine control. The performance of the controller can be further refined and improved through continuous learning in real-time vehicle operations.

10.2 Self-Learning Call Admission Control for CDMA Cellular Networks Using ADP

10.2.1 Problem Formulation

In the DS-CDMA cellular network model used in this section, we assume that separate frequency bands are used for the reverse link and the forward link, so that the mobiles only experience interference from the base stations and the base stations only experience interference from the mobiles. We consider cellular networks that support both voice and data services. Assume that there are K classes of services provided by the wireless networks under consideration, where $K \geq 1$ is an integer. We define a mapping $\sigma: Z^+ \to \{1, \dots, K\}$ to indicate the fact that the nth connection is from the service class $\sigma(n)$, where Z^+ denotes the set of non-negative integers. We assume that each connection in our network may be from a different service class, which requires a different quality of service target (e.g., in terms of different bit error rate for each service class). This includes the case when we allow each call to specify its own quality of service requirements. We assume that traffic from the same service class has the same data rate, the same activity factor, the same desired SIR (signal-to-interference ratio) value, and the same maximum power limit that can be received at the base station.

Consider a base station currently with N active connections. The power received at the base station from the user (mobile station) of the nth connection is denoted by S_n, $n = 1, \dots, N$. In an SIR-based power-controlled DS-CDMA network [2, 7, 13, 29], the desired value of S_n is a function of the number of active home connections and total other cell interference. If we assume that the maximum received power at a base station is limited to H_k for connections from service class $k = \sigma(n)$, then S_n is a random variable in the range of $(0, H_k]$. The maximum power limits H_k, $k = 1, \dots, K$, are determined by the power limit of mobile transmitters, the cell size, the path loss information, and the user's service class. They have been used in several previous works on call admission control [9, 17, 29].

In CDMA networks, the instantaneous bit SIR (or the bit energy-to-interference ratio) for the nth connection at the base station (in a cell) can be expressed in terms of the received powers of the various connections as [9]

$$(E_b/N_0)_n = \frac{S_n W}{I_n R_{\sigma(n)}}, \tag{10.1}$$

where S_n is the instantaneous power level of the nth connection received at the base station, W is the total spread bandwidth (or the chip rate), and $R_{\sigma(n)}$ is the data rate of service class $\sigma(n)$. I_n in (10.1) indicates the instantaneous total interference to the nth connection received at the base station and it is given by

$$I_n = (1+f)\sum_{i=1, i\neq n}^{N} \nu_{\sigma(i)} S_i + \eta_n,$$

where $\nu_{\sigma(i)}$ is the traffic (e.g., voice) activity factor of the ith connection, which is from the service class $\sigma(i)$, η_n is the background (or thermal) noise, N is the number of active connections in the cell, and f is called the intercell interference factor [32], having a typical value of 0.55. In the above, the value of f may not always be constant in a system. Its value can be calculated using existing measurements and can be updated periodically to reflect changes in traffic conditions and traffic distributions.

Assume that after the admission of a new call or a handoff call, the power control algorithm starts to evolve until convergence. Assume that the power control algorithm converges and it requires the power received at a base station from each connection in the system given by $S_n^*, n = 0, 1, \ldots, N$, where the total number of connections in the system is $N + 1$ and connection 0 is the newly admitted caller. Obviously, if $S_n^* > H_{\sigma(n)}$ or $S_n^* \leq 0$, for some n, $0 \leq n \leq N$, the admission should not be granted since it leads to an outage. Only when $0 < S_n^* \leq H_{\sigma(n)}$ for all n, $n = 0, 1, \ldots, N$, the admission decision can be considered as a correct decision. The goal of the present study is to develop a self-learning control algorithm that learns to achieve the correct admission decisions under various, possibly changing, environment conditions and user behavior and to optimize the grade of service (GoS) measure.

The GoS in cellular networks is mainly determined by the new call blocking probability and the handoff blocking probability. The former determines the fraction of new calls that are blocked, while the latter is closely related to the fraction of already admitted calls that cannot maintain their required quality of service (bit error rate) and are dropped. For example, many works have chosen to use the following definition for the GoS [1]:

$$\text{GoS} = P(\text{call blocking}) + w \times P(\text{handoff failure}), \tag{10.2}$$

where $P(a)$ is the probability of event a and w is typically chosen as, e.g., 10. In our simulation studies, we fix $w = 10$. The GoS defined in (10.2) provides a trade-off between the new call blocking rate and the handoff call blocking rate. The parameter w is a weighting factor that decides how much emphasis is placed on handoff calls. Keeping the GoS defined in (10.2) under a desired target level would require one to give much higher priority to handoff calls than to new calls when $w = 10$. On the other hand, quality of service (QoS) is usually defined according to the bit error rate in digital transmission. For example, the quality of service requirement for voice users is usually expressed as a bit error rate less than 10^{-3} in order to guarantee the quality of communication which can be satisfied by the power control mechanism keeping E_b/N_0 at a required value of 7 dB or higher [9, 17, 29].

For a given set of parameters, including traffic statistics and mobility characteristics, fixed call admission control schemes can sometimes yield optimal solutions [26] in terms of GoS. All such schemes (see [12, 23, 26, 27, 29]), however, by reserving a fixed part of capacity, cannot adapt to changes in the network conditions due to their static nature. Therefore, we develop in the present work a self-learning call admission control algorithm for CDMA wireless networks. The present call admission control algorithm based on adaptive critic designs has the capability to learn from the environment and the user behavior so that the performance of the algorithm will be improved through further learning.

10.2.2 A Self-Learning Call Admission Control Scheme for CDMA Cellular Networks

10.2.2.1 Adaptive Critic Designs for Problems with Finite Action Space

In the following, we provide a brief introduction to adaptive critic designs [18]. Suppose that one is given a discrete-time non-linear dynamical system

$$x(k+1) = F[x(k), u(k), k],$$

where $x \in \mathbb{R}^n$ represents the state vector of the system and $u \in \mathbb{R}^m$ denotes the control action. In the present section, the function F denotes a stochastic transition from the state $x(k)$ to the next state $x(k+1)$ under the given control action $u(k)$ at time k. Suppose that one associates with this system the cost functional

$$J[x(i), i] = \sum_{k=i}^{\infty} \gamma^{k-i} L[x(k), u(k), k], \tag{10.3}$$

where L is called the utility function and γ is the discount factor with $0 < \gamma \leq 1$. Note that J is dependent on the initial time i and the initial state $x(i)$, and it is referred to as the cost-to-go of state $x(i)$. The objective is to choose the control sequence $u(k)$, $k = i, i+1, \dots$, so that the cost functional J in (10.3) is minimized.

The class of ACDs considered in the present section is called action-dependent heuristic dynamic programming, or ADHDP, which is shown in Fig. 10.1 (see [19]). The critic network in this case will be trained by minimizing the following error measure over time:

$$\begin{aligned} \|E_q\| &= \sum_k E_q(k) \\ &= \sum_k \left[Q(k-1) - L(k) - \gamma Q(k)\right]^2, \end{aligned} \tag{10.4}$$

Fig. 10.1 A typical scheme of an action-dependent heuristic dynamic programming [19]

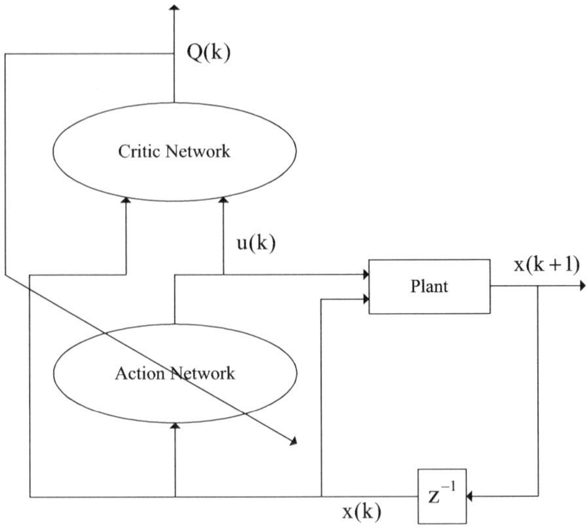

where $Q(k)$ is the critic network output at time k. When $E_q(k) = 0$ for all k, (10.4) implies that

$$\begin{aligned} Q(k-1) &= L(k) + \gamma Q(k) \\ &= L(k) + \gamma[L(k+1) + \gamma Q(k+1)] \\ &= \cdots \\ &= \sum_{i=k}^{\infty} \gamma^{i-k} L(i). \end{aligned} \tag{10.5}$$

Comparing (10.5) to (10.3), we see that when minimizing the error function in (10.4), we have a neural network trained so that its output becomes an estimate of the cost functional defined in dynamic programming for $i = k+1$, i.e., the value of the cost functional in the immediate future [19].

The input–output relationship of the critic network in Fig. 10.1 is given by

$$Q(k) = Q\left[x(k), u(k), k, W_C^{(p)}\right],$$

where $W_C^{(p)}$ represents the weights of the critic network after the pth weight update. There are two methods to train the critic network according to the error function (10.4) in the present case, which are described in [19]. We will use the so-called forward-in-time method.

We can train the critic network at time $k-1$, with the output target given by $L(k) + \gamma Q(k)$. The training of the critic network is to realize the mapping given by

$$C_f : \begin{Bmatrix} x(k-1) \\ u(k-1) \end{Bmatrix} \to \{L(k) + \gamma Q(k)\}. \tag{10.6}$$

In this case, the output from the network to be trained is $Q(k-1)$ and the input to the network to be trained is composed of $x(k-1)$ and $u(k-1)$. The target output value for the critic network training is calculated using its output at time k as indicated in (10.6). The goal of learning the function given by (10.6) is to have the critic network output satisfy

$$Q(k-1) \approx L(k) + \gamma Q(k) \quad \text{for all } k,$$

which is required by (10.5) for adaptive dynamic programming solutions.

The critic network training procedure is described as follows using the strategy of [16]:

1. Initialize two critic networks: cnet1 = cnet2;
2. Use cnet2 to get $Q(k)$, and then train cnet1 for 50 epochs using the Levenberg–Marquardt algorithm [11];
3. Copy cnet1 to cnet2, i.e., let cnet2 = cnet1;
4. Repeat Steps 2 and 3, e.g., four times;
5. Repeat Steps 1–4, e.g., ten times (start from different initial weights);
6. Pick the best cnet1 obtained as the trained critic network.

After the critic network's training is finished, the action network's training starts with the objective of minimizing the critic network output $Q(k)$. In this case, we can choose the target of the action network training as zero, i.e., we will update the action network's weights so that the output of the critic network becomes as small as possible. In general, a good critic network should not output negative values if $L(k)$ is non-negative. This is particularly true when $L(k)$ is chosen as the square error function in tracking control problems [31]. The desired mapping which will be used for the training of the action network in Fig. 10.1 is given by

$$A\colon \{x(k)\} \to \{0(k)\}, \tag{10.7}$$

where $0(k)$ indicates the target values of zero. We note that during the training of the action network, it will be connected to the critic network as shown in Fig. 10.1. The target in (10.7) is for the output of the whole network, i.e., the output of the critic network after it is connected to the action network as shown in Fig. 10.1.

There are many problems in practice that have a control action space that is finite. Typical examples include bang-bang control applications where the control signal only takes a few (finite) extreme values or vectors. When the application has only a finite action space, the decisions that can be made are constrained to a limited number of choices, e.g., a binary choice in the case of call admission control problem. When a new call or a handoff call arrives at a base station requesting admission, the decisions that a base station can make are constrained to two choices, i.e., to accept the call or to reject the call. Let us denote the two options by using $u(k) = +1$ for "accept" and $u(k) = -1$ for "reject". It is important to realize that in the present case the control actions are limited to a binary choice, or to only two possible options. Because of this, the ACDs introduced in Fig. 10.1 can be further simplified, so that only the critic network is needed. Our self-learning call admission control

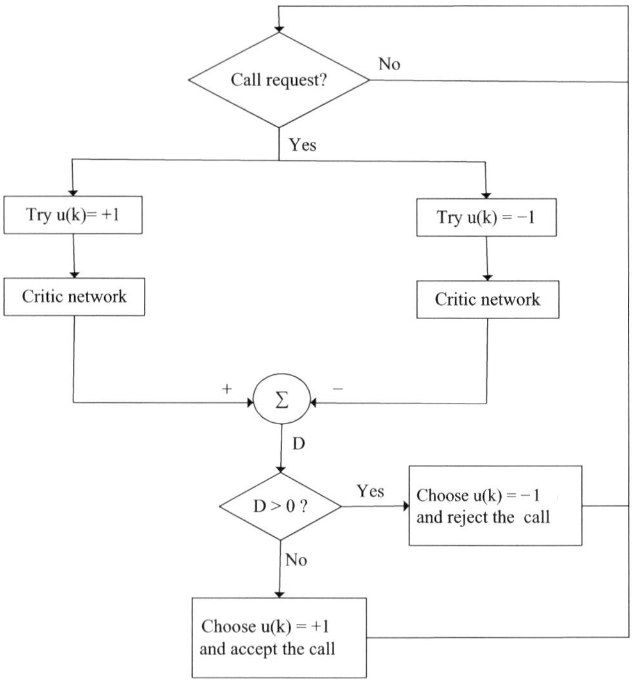

Fig. 10.2 The block diagram of the present adaptive critic approach

scheme for wireless cellular networks using adaptive critic designs is illustrated in Fig. 10.2. When a new call or a handoff call arrives at a base station requesting admission, we can first ask the critic network to see whether $u(k) = +1$ (accept) or $u(k) = -1$ (reject) will give a smaller output value. We will then choose the control action, choosing from $u(k) = +1$ and $u(k) = -1$, that gives a smaller critic network output. As in the case of Fig. 10.1, the critic network would also take the states of the system as input. We note that Fig. 10.2 is only a schematic diagram that shows how the computation takes place while making a call admission control decision. The two blocks for the critic network in Fig. 10.2 represent the same network or computer code in software. The block diagram in Fig. 10.2 indicates that the critic network will be used twice in calculations (with different values of $u(k)$) to make a decision on whether or not to accept a call.

The above description assumes that the critic network has been trained successfully. Once the critic network training is done, it can be applied as in Fig. 10.2. To guarantee that the overall system will achieve optimal performance now and in future environments which may be significantly different from what they are now, we will allow the critic network to perform further learning when needed in the future. In the next section, we describe how the critic network learning is performed. In particular, we describe how the training data are collected at each time step and how the utility function $L(k)$ is defined. We note that once the training data are collected, the training of the critic network can use, e.g., the forward-in-time method, described in this section. We also note that the description here for the critic network training applies to both the initial training of the critic network and further training of the critic network when needed in the future.

10.2.2.2 Self-learning Call Admission Control for CDMA Cellular Networks

We use a utility function as reward or penalty to the action made by the call admission control scheme. When the call admission control scheme makes a decision about accepting a call, it will lead to two distinct results. The first is that the decision of accepting a call is indeed the right decision due to the guarantee of quality of service during the entire call duration. In this case we should give a reward to the decision of accepting a call. Otherwise, a penalty is assigned to this decision. On the other hand, if rejecting a call would have been the right decision due to call dropping or system outage after the call acceptance, we will also give a reward to the decision of rejecting a call. Otherwise, a penalty is assigned to this decision.

It is generally accepted in practice that handoff calls will be given higher priority than new calls [12, 26, 27, 29]. This is accomplished in our call admission control scheme by using different thresholds for new calls and handoff calls. A handoff call can be admitted if $0 < S_n^* \leq H_{\sigma(n)}$ for all $n = 0, 1, \ldots, N$. A new call can only be admitted if $0 < S_0^* \leq T_{\sigma(0)}$ and $0 < S_n^* \leq H_{\sigma(n)}$ for $n = 1, 2, \ldots, N$, where $T_{\sigma(0)} < H_{\sigma(0)}$ is the threshold for new calls (where connection 0 is the new caller).

For handoff calls, when $0 < S_n^* \leq H_{\sigma(n)}$ for $n = 0, 1, \ldots, N$, accepting a handoff call will be given a reward and rejecting a handoff call will be given a penalty. On the other hand, when $S_n^* > H_{\sigma(n)}$ for some $n,\ 0 \leq n \leq N$, accepting a handoff call (i.e., the zeroth caller) will be given a penalty and rejecting a handoff call will be given a reward. Obviously, when $S_n^* \leq 0$ for some $n,\ 0 \leq n \leq N$, the call should be rejected. We note that if the power control algorithm leads to either $S_n^* > H_{\sigma(n)}$ or $S_n^* \leq 0$ for some n, the network will enter an outage, i.e., some calls will have to be terminated prematurely since they cannot maintain required quality of service (bit error rate). In this case, the action of rejection is given a reward and the action of acceptance is given a penalty. Note that S_n^*, $n = 0, 1, \ldots, N$, are power levels for all connections after the call admission decision and the power control algorithm convergence.

We first define the cost functional for handoff calls as follows ($S_n^* \geq 0$ for $n = 0, 1, \ldots, N$):

$$E_n = \xi \max \left\{ u(k) \left[\frac{S_n^*}{H_{\sigma(n)}} - 1 \right], 0 \right\}, \tag{10.8}$$

where $\xi > 0$ is a coefficient, and $u(k) = 1$ represents accepting a call and $u(k) = -1$ represents rejecting a call. We emphasize that the conditions, $0 \leq S_n^* \leq H_{\sigma(n)}$ for $n = 0, 1, \ldots, N$, must hold for the entire duration of all calls in order for the system to give reward to the action of accepting a handoff call.

For new calls, when $0 < S_0^* \le T_{\sigma(0)}$ and $0 < S_n^* \le H_{\sigma(n)}$ for $n = 1, 2, \ldots, N$, we give a reward to the action of accepting a new call, and we give a penalty to the action of rejecting a new call. When $S_0^* > T_{\sigma(0)}$, or $S_n^* > H_{\sigma(n)}$ for some n, $n = 1, 2, \ldots, N$, or $S_n^* \le 0$ for some n, $n = 0, 1, \ldots, N$, we give penalty for accepting a new call and we give a reward for rejecting a new call. The cost functional for new calls is defined as ($S_n^* \ge 0$ for $n = 0, 1, \ldots, N$):

$$E_n = \begin{cases} \xi \max\left\{\left[\dfrac{S_n^*}{T_{\sigma(n)}} - 1\right], 0\right\}, & \text{when } u(k) = 1, \\ \xi \max\left\{\left[1 - \dfrac{S_n^*}{H_{\sigma(n)}}\right], 0\right\}, & \text{when } u(k) = -1, \end{cases} \tag{10.9}$$

where $T_{\sigma(n)} < H_{\sigma(n)}, n = 0, 1, \ldots, N$. We note again that the conditions, $0 < S_n^* \le H_{\sigma(n)}$ for $n = 0, 1, \ldots, N$, must hold for the entire duration of all calls in order for the system to give reward to the action of accepting a new call, even though the admission decision is according to the condition $0 < S_0^* \le T_{\sigma(0)}$ for the new call.

The functions defined above satisfy the condition that $E_n \ge 0$ for all n, $n = 0, 1, \ldots, N$. From (10.8) and (10.9), we see that when the action is "accept", if the value of the utility function of any connection is larger than 0, this action should be penalized. Also, when the action is "reject", if the value of the utility function of any connection is zero, this action should be rewarded. Therefore, from the system's point of view, the cost function should be chosen as

$$E = \begin{cases} \max\limits_{0 \le n \le N} (E_n), & \text{if } u(k) = 1, \\ \min\limits_{0 \le n \le N} (E_n), & \text{if } u(k) = -1. \end{cases} \tag{10.10}$$

The cost function defined in (10.10) indicates that the goal of our call admission control algorithm is to minimize the value of function E, i.e., to reach its minimum value of zero and to avoid its positive values.

The utility function L (used in (10.3)) in our present work is chosen as

$$L(u) = \frac{E}{1 + E}. \tag{10.11}$$

Figures 10.3 and 10.4 show plots of this utility function for handoff calls and new calls, respectively, when $\xi = 10$. The parameter ξ is used to obtain a desired shape of the utility function. Since our algorithm will search for the optimal performance that corresponds to small values of the utility function, the shape of the utility function will have some effects on the optimization process. When $\xi = 10$, we see that the utility function becomes more selective than $\xi = 1$ for any condition indicated by signal power. From Figs. 10.3 and 10.4 we see that the choice of the present utility functions in (10.11) clearly shows minimum points (the flat areas) that our

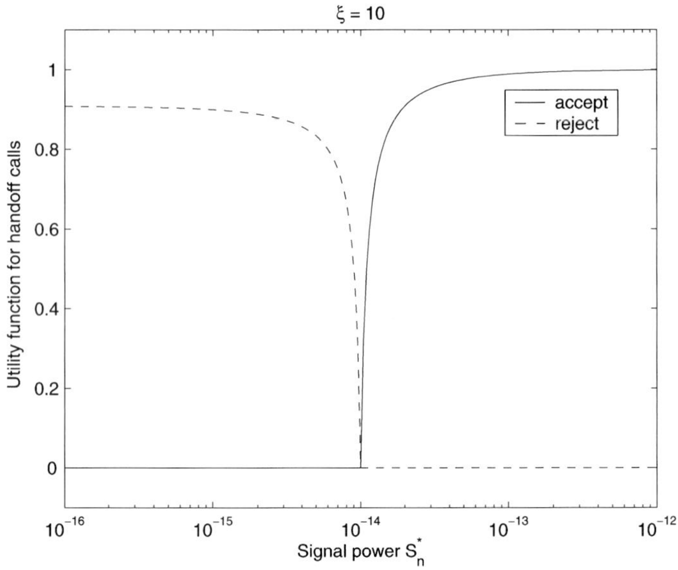

Fig. 10.3 Utility function for handoff calls

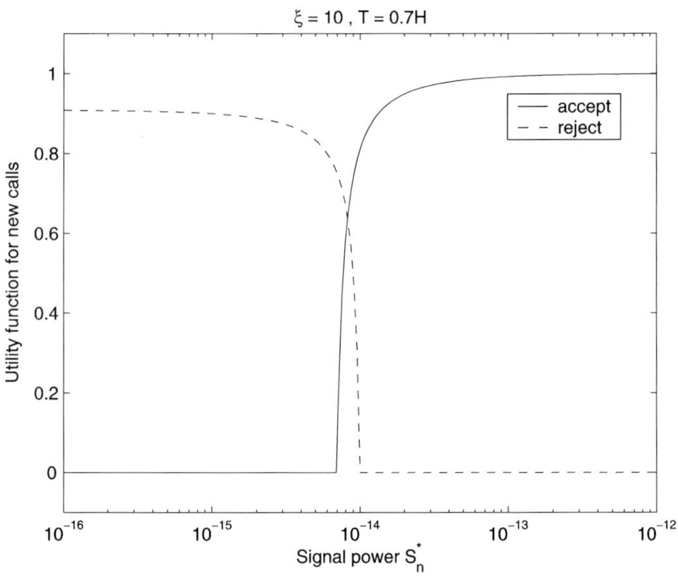

Fig. 10.4 Utility function for new calls

call admission control scheme tries to reach and the points with high penalty that our scheme should avoid. In addition, the conversion from E to U guarantees the convergence of the performance index of dynamic programming, which is defined as in (10.3). With the present utility function given by (10.10) and (10.11), we have

$$0 < J(k) < \frac{1}{1-\gamma},$$

since L in (10.11) satisfies $0 \le L < 1$. Without the conversions in (10.11), there is no guarantee that the infinite summation in (10.3) will be bounded. We note that the present critic network produces an output that approximates the cost functional $J(k)$ in (10.3), and the admission action chosen each time will minimize the critic network output, to achieve approximate optimal control.

In stationary environment, where user traffic statistics (patterns) remain unchanged, a simple static call admission control algorithm [20] will be able to achieve the admission objective described above. However, traffic patterns including user arrival rate, call holding times, user mobility patterns, etc., may show significant changes from time to time. To deal with changing environments, static call admission control algorithm would not be appropriate. The present call admission control algorithm based on adaptive critic designs will be able to deal with environment changes through further learning in the future. Another benefit of the present self-learning call admission control algorithm is its ability to improve performance through further learning as the controller gains more and more experience.

The development of the present self-learning call admission control scheme involves the following four steps:

1. Collecting data: During this phase, when a call comes, we can accept or reject the call with any scheme and calculate the utility function for the system as presented above. In the present section, we simply accept and reject calls randomly with the same probability of 0.5. At the same time, we collect the states corresponding to each action. The states (environment) collected for each action include total interference, call type (new call or handoff call), call class (voice or data), etc.
2. Training critic network: Using the data collected to train the critic network as mentioned in the previous section. Examples of input variables chosen for the critic network will be given in our simulation examples.
3. Applying critic network: The trained critic network is then applied as shown in Fig. 10.2.
4. Further updating critic network: The critic network will be updated as needed while it is used in application to accommodate environment changes, for example, user pattern and behavior changes or new requirements for the system. Data collection has to be performed again and the training of the critic network as well. In this case, the above three steps will be repeated.

The critic network will be updated when there are changes in call admission requirements or if the already trained ACD scheme does not satisfy new requirements. In fact, ACD is going to learn the rules imposed by the utility function of the system. Therefore, rule changes can be accommodated by modifying the utility function to accommodate new requirements. For example, to satisfy certain requirements, we modify the cost function in (10.9) to become

$$E_n = \begin{cases} \xi \max\left\{\left[\dfrac{S_n^*}{H_{\sigma(n)}} - 1\right], 0\right\}, & \text{when } u(k) = 1 \\ & \text{and } n_a \le N_h, \\ \xi \max\left\{\left[\dfrac{S_n^*}{T_{\sigma(n)}} - 1\right], 0\right\}, & \text{when } u(k) = 1 \\ & \text{and } n_a > N_h, \\ \xi \max\left\{\left[1 - \dfrac{S_n^*}{H_{\sigma(n)}}\right], 0\right\}, & \text{when } u(k) = -1, \end{cases} \tag{10.12}$$

where n_a is the number of active handoff calls in the cell and N_h is a fixed parameter indicating the threshold for low traffic load.

When a call arrives at the base station, the admission decision would be either "accept" or "reject". If the decision is taken to accept a call, there will be two kinds of data collected. One is that the decision is incorrect due to call dropping, and the other one is that the decision is correct since $0 < S_n^* \le H_{\sigma(n)}$, $n = 0, 1, \dots, N$, is maintained for the entire duration of all calls. In the former case, a penalty is recorded and in the latter case, a reward is recorded. If the decision is to reject a call, there will be also be two kinds of data collected that correspond to a reward and a penalty. Note that in the case of a "reject" decision, the value of the utility function is determined as in (10.8)–(10.11) where the values of S_n^*, $n = 0, 1, \dots, N$, are calculated according to [3, 20, 30].

10.2.3 Simulations

We first conduct simulation studies for a network with a single class of service (e.g., voice). The network parameters used in the present simulation are taken similarly as the parameters used in [13, 29] (see Table 10.1).

The arrival rate consists of the new call attempt rate λ_c and the handoff call attempt rate λ_h. The new call attempt rate λ_c depends on the expected number of subscribers per cell. The handoff call attempt rate λ_h depends on such network parameters as traffic load, user velocity, and cell coverage areas [10, 12]. In our simulation, we assume that $\lambda_c : \lambda_h = 5 : 1$ [12]. A channel is released by call completion or handoff to a neighboring cell. The channel occupancy time is assumed to be exponentially distributed [10, 12] with the same mean value of $1/\mu = 3$ minutes. For each neural network training in our simulation studies, we generated 35000 data points according to the data collection procedure in the previous section.

In the following, we conduct comparison studies between the present self-learning call admission control algorithm and the algorithm developed in [20] with fixed thresholds for new calls given by $T = H$, $T = 0.8H$, and $T = 0.5H$, respectively. The arrival rate in all neighboring cells is fixed at 18 calls/minutes.

Table 10.1 Network parameters

Parameters	Values	Parameters	Values
W	1.2288 Mcps	R	9.6 kbps
η	1×10^{-14} W	H	1×10^{-14} W
E_b/N_0	7 dB	υ	3/8

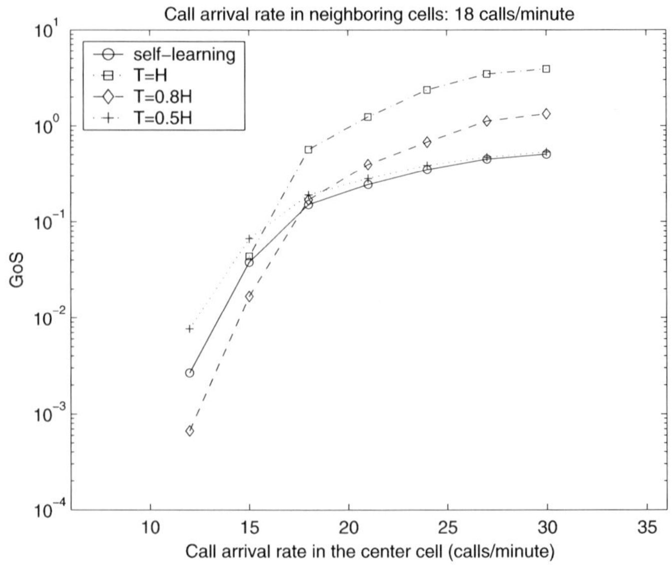

Fig. 10.5 Comparison study using utility function defined in (10.9)

The training data are collected as mentioned in the previous section. We choose $T_{\sigma(n)} = 0.5H_{\sigma(n)}$ and $\xi = 10$ in (10.9). The critic network has three inputs. The first is the total interference received at the base station, the second is the action (1 for accepting, -1 for rejecting), and the third is the call type (1 for new calls, -1 for handoff calls). The critic network is a multilayer feedforward neural network with 3–6–1 structure, i.e., three neurons at the input layer, six neurons at the hidden layer, and one neuron at the output layer. Both the hidden and the output layers use the hyperbolic tangent function as the activation function. Figure 10.5 shows the simulation results. We see from the figure that the performance of the self-learning algorithm is similar to the case of static algorithm with $T = 0.5H$, because we choose $T_{\sigma(n)} = 0.5H_{\sigma(n)}$ in (10.9) for our learning control algorithm. When the call arrival rate is low, the self-learning algorithm is not so good because it reserves too much for handoff calls and as a result it rejects too many new calls. That is why the GoS is worse than the other two cases of static algorithms ($T = 1.0H$ and $T = 0.8H$). In this case, the self-learning algorithm is trained to learn a call admission control scheme that gives higher priority to handoff calls.

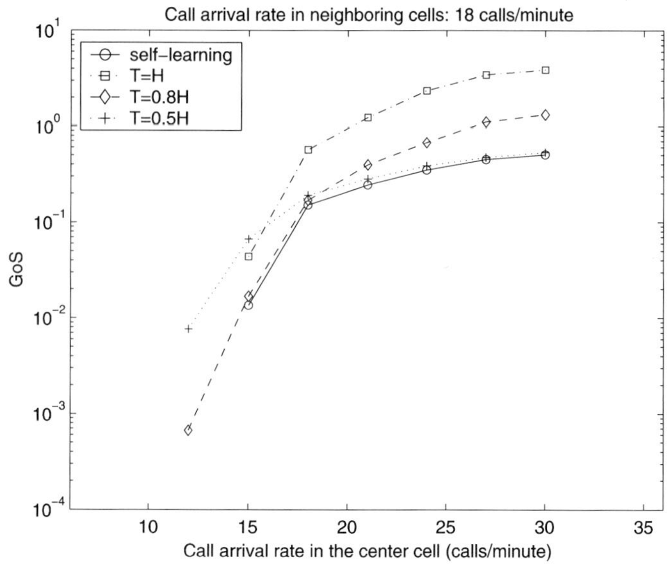

Fig. 10.6 Comparison study using utility function defined in (10.12)

In order to improve the GoS when the call arrival rate is low, we use the modified cost function for new calls as in (10.12), where we choose $N_h = 15$ in our simulation. Using this new utility function we collect the training data using one of the static algorithms with fixed threshold or the previous critic network. Then we train a new critic network with the newly collected data. In this time the critic network has four inputs. Three of them are the same as in the previous critic network. The new input is equal to 1 when $n_a \leq N_h$ and otherwise it is equal to -1. The critic network in this case has a structure given by 4–8–1. Figure 10.6 shows the result of applying the new critic network to the same traffic pattern as in Fig. 10.5. From the figure we see that the self-learning algorithm using the new critic network has the best GoS. We see that by simply changing the cost function from (10.11) to (10.12), the self-learning algorithm can significantly improve its performance to outperform static admission control algorithms. One of the benefits of self-learning call admission control algorithm is that we can easily and efficiently design call admission control algorithms by modifying the cost function (equivalently, the utility function) to satisfy the requirements or to accommodate new environment changes.

The traffic load in telephony systems is typically time varying. Figure 10.7 shows a pattern concerning call arrivals during a typical 24 hour business day, beginning at midnight [8]. It can be seen that the peak hours occur around 11:00 am and 4:00 pm. Next, we use our newly trained critic network above for this traffic pattern. Figure 10.8 gives the simulation results under the assumption that the traffic load was spatially uniformly distributed among cells, but followed the time-varying pattern given in Fig. 10.7. Figure 10.8 compares the four call admission control algorithms and shows that the self-learning algorithm has the best GoS among all the algorithms tested. We note that the self-learning call admission control algorithm was

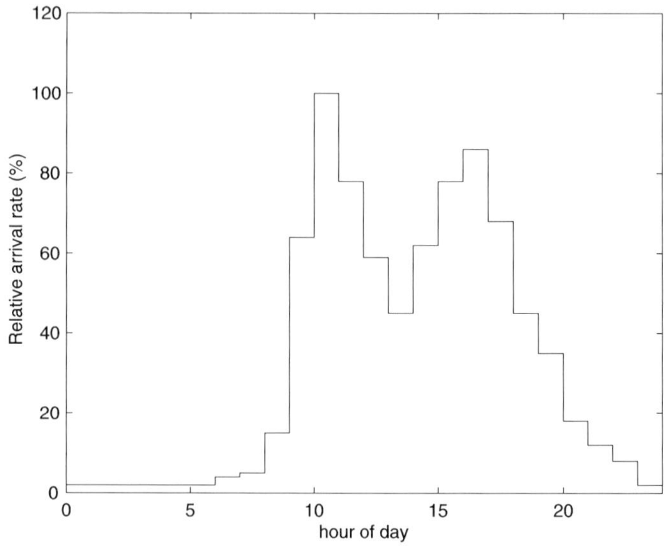

Fig. 10.7 A traffic pattern of a typical business day

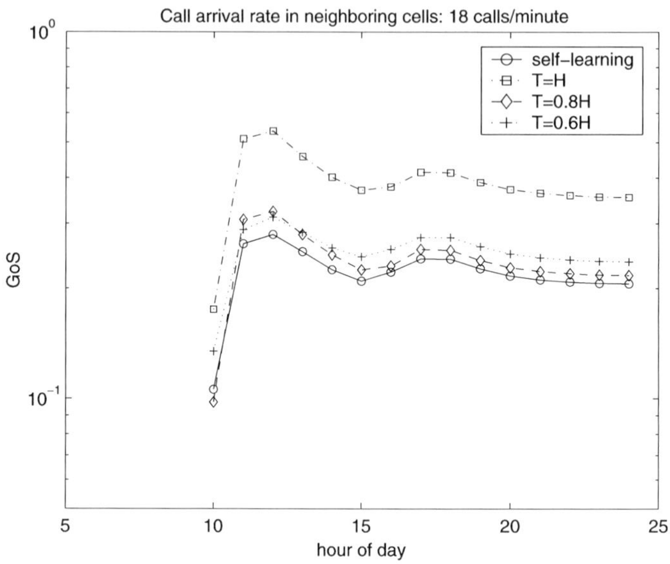

Fig. 10.8 Comparison study for varying traffic

not retrained from the previous case, i.e., we used the same critic network in the simulation results of Fig. 10.8 as in Fig. 10.6.

In the following, we conduct comparison studies between the present self-learning call admission control algorithm and that of [29]. Using the algorithm

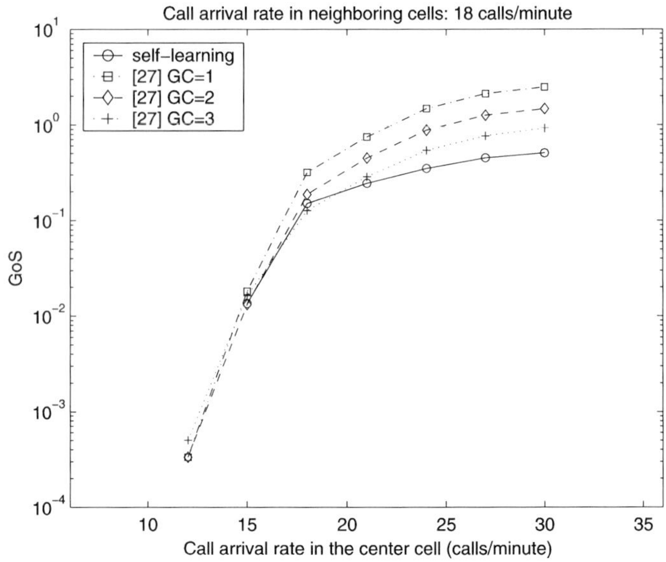

Fig. 10.9 Comparison studies with the algorithm in [29]

in [29], the base station controller reads the current interference from the power strength measurer. It then estimates the current interference margin (CIM) and handoff interference margin (HIM), where CIM $<$ HIM. A total interference margin (TIM) is set according to the quality of service target. If CIM $>$ TIM, reject the call admission request. If HIM $<$ TIM, accept the call request. If CIM $<$ TIM $<$ HIM, then only handoff calls will be accepted. Figure 10.9 compares the present self-learning call admission control algorithm with the algorithm in [29] that reserves one, two, and three channels for handoff calls, respectively. The arrival rate in all neighboring cells is fixed at 18 calls/minute. We assume the use of a hexagonal cell structure. From Fig. 10.9, we see that the present algorithm has the best GoS. That is because the algorithm in [29] is a kind of guard channel algorithm used in CDMA systems. Therefore, when the load is low, GC $= 1$ performs the best, and when the load is high, GC $= 3$ performs the best. However, our algorithm can adapt to varying traffic load conditions. It has the best overall performance under various traffic loads. Again, we used the same critic network in the simulation results of Fig. 10.9 as in Fig. 10.6.

Finally, we conduct simulation studies for cellular networks with two classes of services. One class is voice service and the other is data service. Network parameters in our simulations are chosen in reference to the parameters used in [14, 28] (see Table 10.2). In our simulation, the data traffic is similar to that in [14], i.e., low resolution video or interactive data. In this case, the data traffic can be specified by a constant transmission rate. The background noise in this case is chosen the same as in Table 10.1. The utility function is defined for voice and data calls as in (10.9) and (10.11). In (10.12), we choose $T_{\sigma(n)} = 0.6H_{\sigma(n)}$ and $\xi = 10$ for both voice calls and

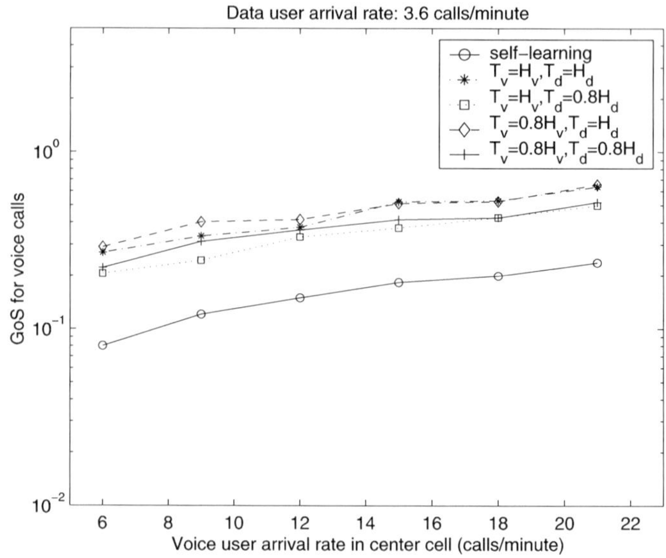

Fig. 10.10 GoS for voice calls

Table 10.2 Network parameters

Voice users		Data users	
Parameters	Values	Parameters	Values
W_v	4.9152 Mcps	W_d	4.9152 Mcps
R_v	9.6 kbps	R_d	38.4 kbps
H_v	1×10^{-14} W	H_d	1×10^{-13} W
$(E_b/N_0)_v$	7 dB	$(E_b/N_0)_d$	9 dB
ν_v	3/8	ν_d	1

data calls. N_h is chosen as 20 and 4 for voice calls and data calls, respectively. The critic network now has five inputs. The newly added input is the call class which is 1 for voice calls and -1 for data calls. The critic network structure is chosen as 5–10–1. Figures 10.10 and 10.11 compare our self-learning call admission control algorithm and the static algorithm [20] with fixed thresholds given by $T = H$ and $T = 0.8H$. The arrival rates of voice users and data users in all neighboring cells are fixed at 20 calls/minute and 3 calls/minute, respectively. From Figs. 10.10 and 10.11, we see that the present self-learning algorithm has the best GoS for almost all call arrival rates tested. We conclude that the present self-learning algorithm performs better than the fixed algorithms due to the fact that the self-learning algorithm can adapt to varying traffic conditions and environment changes.

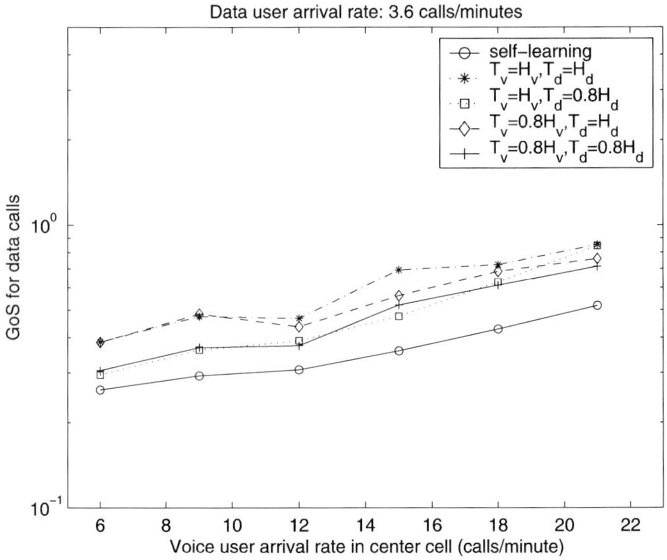

Fig. 10.11 GoS for data calls

10.3 Engine Torque and Air–Fuel Ratio Control Based on ADP

10.3.1 Problem Formulation

A test vehicle with a V8 engine and 4-speed automatic transmission is instrumented with engine and transmission torque sensors, wide-range air–fuel ratio sensors in the exhaust pipe located before and after the catalyst on each bank, as well as exhaust gas pressure and temperature sensors. The vehicle is also equipped with a dSPACE rapid prototyping controller for data collection and controller implementation. Data are collected at each engine event under various driving conditions, such as the Federal Test Procedure (FTP cycles), as well as more aggressive driving patterns, for a length of about 95,000 samples during each test. The engine is run under closed-loop fuel control using switching-type oxygen sensors. The dSPACE is interfaced with the power-train control module (PCM) in a by-pass mode.

We build a neural network model for the test engine with a structure compatible with the mathematical engine model developed by Dobner [5, 6, 22] and others. Due to the complexity of modern automotive engines, in the present work, we use the time-lagged recurrent neural networks (TLRNs) for engine modeling. In practice, TLRNs have been used often for function approximation and it is believed that they are more powerful than the networks with only feedforward structures (cf. [25, 33]).

For the neural network engine model, we choose air–fuel ratio (AFR) and engine torque (TRQ) as the two outputs. We choose throttle position (TPS), electrical fuel pulse width (FPW), and spark advance (SPA) as the three control inputs. These are input signals to be generated using our new adaptive critic learning control algorithm. We choose intake manifold pressure (MAP), mass air flow rate (MAF), and engine speed (RPM) as reference input. The time-lagged recurrent neural network used for the engine combustion module has six input neurons, a single hidden layer with eight neurons, and two output neurons.

Validation results for the output TRQ and AFR of the neural network engine model indicate a very good match between the real vehicle data and the neural network model output during the validation phase [15].

10.3.2 *Self-learning Neural Network Control for Both Engine Torque and Exhaust Air–Fuel Ratio*

Suppose that one is given a discrete-time non-linear dynamical system

$$x(k+1) = F[x(k), u(k), k], \tag{10.13}$$

where $x \in \mathbb{R}^n$ represents the state vector of the system and $u \in \mathbb{R}^m$ denotes the control action. Suppose that one associates with this system the cost functional (or cost)

$$J[x(i), i] = \sum_{k=i}^{\infty} \gamma^{k-i} L[x(k), u(k), k], \tag{10.14}$$

where L is called the utility function or local cost function and γ is the discount factor with $0 < \gamma \le 1$. Note that J is dependent on the initial time i and the initial state $x(i)$, and it is referred to as the cost-to-go of state $x(i)$. The objective is to choose the control sequence $u(k)$, $k = i, i+1, \dots$, so that the cost functional J (i.e., the cost) in (10.14) is minimized.

Adaptive critic designs (ACDs) are defined as designs that approximate dynamic programming in the general case, i.e., approximate optimal control over time in noisy, non-linear environments. A typical design of ACDs consists of three modules—Critic (for evaluation), Model (for prediction), and Action (for decision). When in ACDs the critic network (i.e., the evaluation module) takes the action/control signal as part of its input, the designs are referred to as action-dependent ACDs (ADACDs). We use an action-dependent version of ACDs that does not require the explicit use of the model network in the design. The critic network in this case will be trained by minimizing the following error measure over time:

$$\begin{aligned} \|E_q\| &= \sum_k E_q(k) \\ &= \sum_k \left[Q(k-1) - L(k) - \gamma Q(k)\right]^2, \end{aligned} \tag{10.15}$$

where $Q(k) = Q[x(k), u(k), k, W_C]$. When $E_q(k) = 0$ for all k, (10.15) implies that

$$\begin{aligned} Q(k-1) &= L(k) + \gamma Q(k) \\ &= L(k) + \gamma[L(k+1) + \gamma Q(k+1)] \\ &= \cdots \\ &= \sum_{i=k}^{\infty} \gamma^{i-k} L(k). \end{aligned} \tag{10.16}$$

We see that when minimizing the error function in (10.15), we have a neural network trained so that its output at time k becomes an estimate of the cost functional defined in dynamic programming for $i = k + 1$, i.e., the value of the cost functional in the immediate future [19].

The input–output relationship of the critic network is given by

$$Q(k) = Q\left[x(k), u(k), k, W_C\right],$$

where W_C represents the weight vector of the critic network.

We can train the critic network at time $k - 1$, with the desired output target given by $L(k) + \gamma Q(k)$. The training of the critic network is to realize the mapping given by

$$C_f\colon \begin{Bmatrix} x(k-1) \\ u(k-1) \end{Bmatrix} \to \{L(k) + \gamma Q(k)\}. \tag{10.17}$$

We consider $Q(k - 1)$ in (10.15) as the output from the network to be trained and the target output value for the critic network is calculated using its output at time k.

After the critic network's training is finished, the action network's training starts with the objective of minimizing $Q(k)$. The goal of the action network training is to minimize the critic network output $Q(k)$. In this case, we can choose the target of the action network training as zero, i.e., we will train the action network so that the output of the critic network becomes as small as possible. The desired mapping which will be used for the training of the action network in the present ADHDP is given by

$$A\colon \{x(k)\} \to \{0(k)\}, \tag{10.18}$$

where $0(k)$ indicates the target values of zero. We note that during the training of the action network, it will be connected to the critic network to form a larger neural network. The target in (10.18) is for the output of the whole network, i.e., the output of the critic network after it is connected to the action network.

After the action network's training cycle is completed, one may check the system's performance, then stop or continue the training procedure by going back to the critic network's training cycle again, if the performance is not acceptable yet.

Assume that the control objective is to have $x(k)$ in (10.13) track another signal given by $x^*(k)$. We define in this case the local cost function $L(k)$ as

$$L(k) = \frac{1}{2}e^{\mathrm{T}}(k)e(k) = \frac{1}{2}[x(k) - x^*(k)]^{\mathrm{T}}[x(k) - x^*(k)].$$

Using the ADHDP introduced earlier in this section, we can design a controller to minimize

$$J(k) = \sum_{i=k}^{\infty} \gamma^{i-k} L(i),$$

where $0 < \gamma < 1$. We note that in this case our control objective is to minimize an infinite summation of $L(k)$ from the current time to the infinite future, while in conventional tracking control designs, the objective is often to minimize $L(k)$ itself.

10.3.3 Simulations

The objective of the present engine controller design is to provide control signals so that the torque generated by the engine will track the torque measurement as in the data and the air–fuel ratio will track the required values also as in the data. The measured torque values in the data are generated by the engine using the existing controller. Our learning controller will assume no knowledge about the control signals provided by the existing controller. It will generate a set of control signals that are independent of the control signals in the measured data. Based on the data collected, we use our learning controller to generate control signals TPS, FPW, and SPA, with the goal of producing exactly the same torque and air–fuel ratio as in the data set. That is to say, we keep our system under the same requirements as the data collected, and we build a controller that provides control signals which achieve the torque control and air–fuel ratio control performance of the engine.

As described in the previous section, the development of an adaptive critic learning controller involves two stages: the training of a critic network and the development of a controller/action network. We describe in the rest of the present section the learning control design for tracking the TRQ and AFR measurements in the data set. This is effectively a torque-based controller, i.e., a controller that can generate control signals given the torque demand. The block diagram of the present adaptive critic engine control (including air–fuel ratio control) is shown in Fig. 10.12. The diagram shows how adaptive critic designs can be applied to engine control through adaptive dynamic programming.

10.3.3.1 Critic Network

The critic network is chosen as a 8–15–1 structure with eight input neurons and 15 hidden layer neurons:

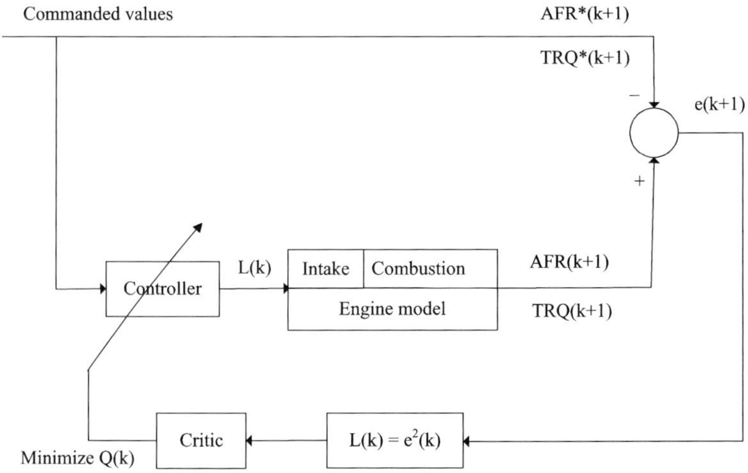

Fig. 10.12 Structure of adaptive critic learning engine controller

- The eight inputs to the critic network are TRQ, TRQ*, MAP, MAF, RPM, TPS, FPW, and SPA, where TRQ* is read from the data set, indicating the desired torque values for the present learning control algorithm to track.
- The hidden layer of the critic network uses a sigmoidal function, i.e., the `tansig` function in MATLAB [4], and the output layer uses the linear function purelin.
- The critic network outputs the function Q, which is an approximation to the function $J(k)$ defined as in (10.14).
- The local cost functional L defined in (10.14) in this case is chosen as

$$L(k) = \frac{1}{2}[\mathrm{TRQ}(k) - \mathrm{TRQ}^*(k)]^2 + \frac{1}{2}[\mathrm{AFR}(k) - \mathrm{AFR}^*(k)]^2,$$

  where TRQ and AFR are the engine torque and air–fuel ratio generated using the proposed controller, respectively, and TRQ* and AFR* are the demanded TRQ value and the desired AFR value, respectively. Both TRQ* and AFR* are taken from the actual measured data in the present case. The utility function chosen in this way will lead to a control objective of TRQ following TRQ* and AFR following AFR*.
- Utilizing the MATLAB Neural Network Toolbox, we apply `traingdx` (gradient descent algorithm) for the training of the critic network. We note that other algorithms implemented in MATLAB, such as `traingd`, `traingda`, `traingdm`, `trainlm` are also applicable. We employ batch training for the critic network, i.e., the training is performed after each trial of a certain number of steps (e.g., 10000 steps). We choose $\gamma = 0.9$ in the present experiments.

10.3.3.2 Controller/Action Network

The structure of the action network is chosen as 6–12–3 with six input neurons, 12 hidden layer neurons, and three output neurons:

- The six inputs to the action network are TRQ, TRQ*, MAP, MAF, THR, and RPM, where THR indicates the driver's throttle command.
- Both the hidden layer and the output layer use the sigmoidal function `tansig`.
- The outputs of the action network are TPS, FPW, and SPA, which are the three control input signals used in the engine model.
- The training algorithm we choose to use is `traingdx`. We employ batch training for the action network as well.

10.3.3.3 Simulation Results

In the present simulation studies, we first train the critic network for many cycles with 500 training epochs in each cycle. At the end of each training cycle, we check the performance of the critic network. Once the performance is found to be satisfactory, we stop critic network training. This process usually takes about 6–7 hours.

After the critic network training is finished, we start the action network training. We train the controller network for 200 epochs after each trial. We check to see the performance of the neural network controller at the end of each trial.

We choose to use 4000 data points from the data (16000–20000 in the data set) for the present critic and action network training.

We first show the TRQ and AFR output due to the initial training of our neural network controller when TRQ* and AFR* are chosen as random signals during training. Figures 10.13 and 10.14 show the controller performance when it is applied with TRQ* and AFR* chosen as the measured values in the data set. The neural network controller in this case is trained for 15 cycles using randomly generated target signal TRQ* and AFR*. Figures 10.13 and 10.14 show that very good tracking control of the commanded torque signal (TRQ) and the exhaust AFR are achieved. We note that at the present stage of the research we have not attempted to regulate the AFR at the stoichiometric value but to track a given command. In these experiments we simply try to track the measured engine-out AFR values so that the control signal obtained can directly be validated against the measured control signals in the vehicle. In Fig. 10.16, it appears that better tracking of AFR was achieved on the rich side of stoichiometric value, possibly due to more frequent rich excursions encountered during model training. This could also have been caused by intentional fuel enrichments (i.e., wall-wetting compensation) during vehicle accelerations.

Figures 10.15 and 10.16 show the TRQ and AFR output after refined training when TRQ* and AFR* are chosen as the measured values in the data. The neural network controller in this case is trained for 15 cycles using target signal TRQ* and AFR* as in the data. Figures 10.15 and 10.16 show that excellent tracking control results for the commanded TRQ and AFR are achieved.

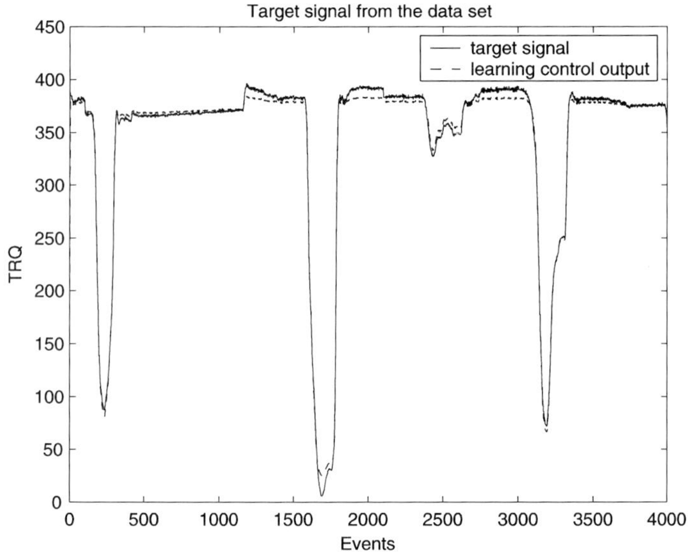

Fig. 10.13 Torque output generated by the neural network controller

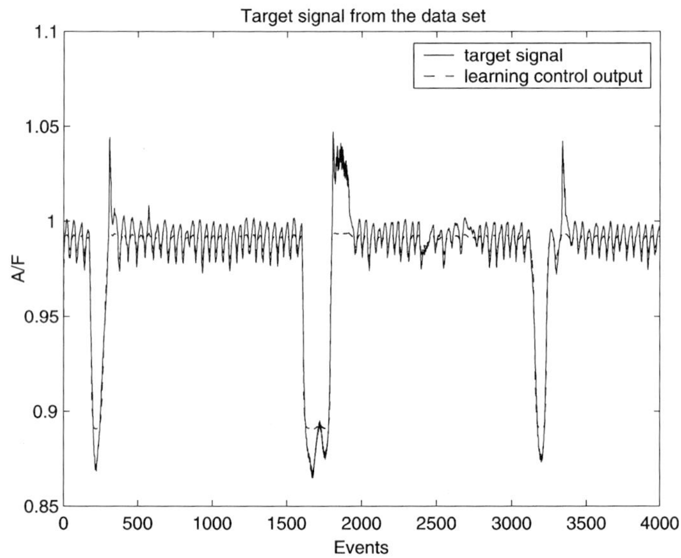

Fig. 10.14 Air–fuel ratio output generated by the neural network controller

The simulation results indicate that the present learning controller design based on adaptive dynamic programming (adaptive critic designs) is effective in training a neural network controller to track the desired TRQ and AFR sequences through proper control actions.

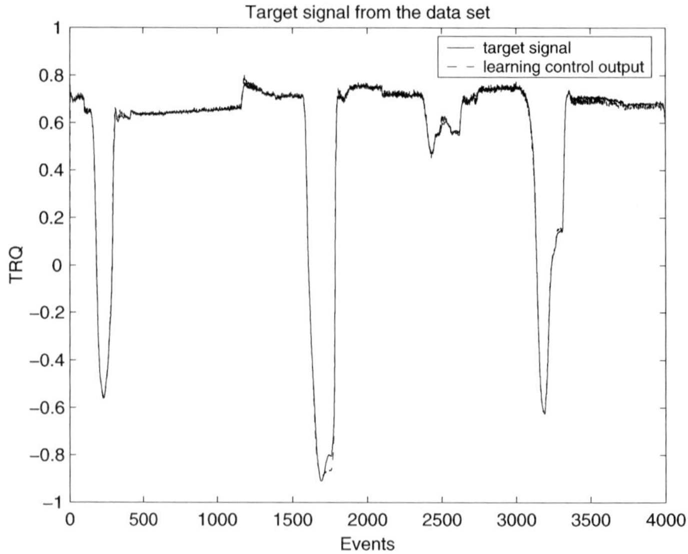

Fig. 10.15 Torque output generated by the refined neural network controller

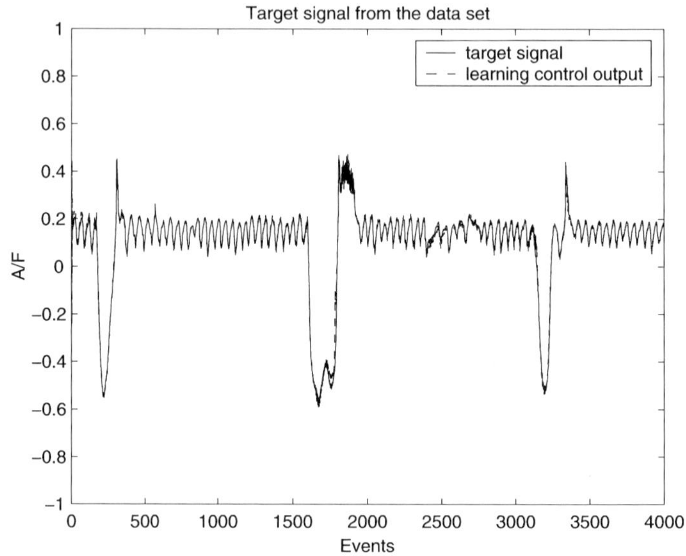

Fig. 10.16 Air–fuel ratio output generated by the refined neural network controller

10.4 Summary

In this chapter, we have investigated the optimal control problem of modern wireless networks and automotive engines by using ADP methods. First, we developed a self-learning call admission control algorithm based on ADP for multiclass traffic in SIR-based power-controlled DS-CDMA cellular networks. The most important benefit of our self-learning call admission control algorithm is that we can easily and efficiently design call admission control algorithms to satisfy the system requirement or to accommodate new environments. We note that changes in traffic conditions are inevitable in reality. Thus, fixed call admission control policies are less preferable in applications. Simulation results showed that when the traffic condition changes, the self-learning call admission control algorithm can adapt to changes in the environment, while the fixed admission policy will suffer either from a higher new call blocking rate, higher handoff call blocking rate, or interference being higher than the tolerance.

Next a neural network learning control using adaptive dynamic programming was developed for engine calibration and control. After the network was fully trained, the present controller may have the potential to outperform existing controllers with regard to the following three aspects. First, the technique presented will automatically learn the inherent dynamics and non-linearities of the engine from real vehicle data and, therefore, do not require a mathematical model of the system to be developed. Second, the developed methods will further advance the development of a virtual power train for performance evaluation of various control strategies through the development of neural network models of the engine and transmission in a prototype vehicle. Third, the present controllers can learn to improve their performance during the actual vehicle operations, and will adapt to uncertain changes in the environment and vehicle conditions. This is an inherent feature of the present neural network learning controller. As such, these techniques may offer promise for use as real-time engine calibration tools. Simulation results showed that the present self-learning control approach was effective in achieving tracking control of the engine torque and air–fuel ratio control through neural network learning.

References

1. A guide to DECT features that influence the traffic capacity and the maintenance of high radio link transmission quality, including the results of simulations. ETSI technical report: ETR 042, July 1992. Available on-line at http://www.etsi.org
2. Ariyavisitakul S (1994) Signal and interference statistics of a CDMA system with feedback power control—Part II. IEEE Trans Commun 42:597–605
3. Bambos N, Chen SC, Pottie GJ (2000) Channel access algorithms with active link protection for wireless communication networks with power control. IEEE/ACM Trans Netw 8:583–597
4. Demuth H, Beale M (1998) Neural network toolbox user's guide. MathWorks, Natick
5. Dobner DJ (1980) A mathematical engine model for development of dynamic engine control. SAE paper no 800054
6. Dobner DJ (1983) Dynamic engine models for control development—Part I: non-linear and linear model formation. Int J Veh Des Spec Publ SP4:54–74
7. Dziong Z, Jia M, Mermelstein P (1996) Adaptive traffic admission for integrated services in CDMA wireless-access networks. IEEE J Sel Areas Commun 14:1737–1747

8. Freeman RL (1996) Telecommunication system engineering. Wiley, New York
9. Gilhousen KS, Jacobs IM, Padovani R, Viterbi AJ, Weaver LA, Wheatley CE III (1991) On the capacity of a cellular CDMA system. IEEE Trans Veh Technol 40:303–312
10. Guerin RA (1987) Channel occupancy time distribution in a cellular radio system. IEEE Trans Veh Technol 35:89–99
11. Hagan MT, Menhaj MB (1994) Training feedforward networks with the Marquardt algorithm. IEEE Trans Neural Netw 5:989–993
12. Hong D, Rappaport SS (1986) Traffic model and performance analysis for cellular mobile radio telephone systems with prioritized and nonprioritized handoff procedures. IEEE Trans Veh Technol 35:77–92
13. Kim DK, Sung DK (2000) Capacity estimation for an SIR-based power-controlled CDMA system supporting ON–OFF traffic. IEEE Trans Veh Technol 49:1094–1100
14. Kim YW, Kim DK, Kim JH, Shin SM, Sung DK (2001) Radio resource management in multiple-chip-rate DS/CDMA systems supporting multiclass services. IEEE Trans Veh Technol 50:723–736
15. Kovalenko O, Liu D, Javaherian H (2001) Neural network modeling and adaptive critic control of automotive fuel-injection systems. In: Proceedings of IEEE international symposium on intelligent control, Taipei, Taiwan, pp 368–373
16. Lendaris GG, Paintz C (1997) Training strategies for critic and action neural networks in dual heuristic programming method. In: Proceedings of international conference on neural networks, Houston, TX, pp 712–717
17. Liu Z, Zarki ME (1994) SIR-based call admission control for DS-CDMA cellular systems. IEEE J Sel Areas Commun 12:638–644
18. Liu D, Zhang Y (2002) A new learning control approach suitable for problems with finite action space. In: Proceedings of international conference on control and automation, Xiamen, China, pp 1669–1673
19. Liu D, Xiong X, Zhang Y (2001) Action-dependent adaptive critic designs. In: Proceedings of INNS-IEEE international joint conference on neural networks, Washington, DC, pp 990–995
20. Liu D, Zhang Y, Hu S (2004) Call admission policies based on calculated power control setpoints in SIR-based power-controlled DS-CDMA cellular networks. Wirel Netw 10:473–483
21. Liu D, Zhang Y, Zhang H (2005) A self-learning call admission control scheme for CDMA cellular networks. IEEE Transactions on Neural Networks 16:1219–1228
22. Liu D, Hu S, Zhang HG (2006) Simultaneous blind separation of instantaneous mixtures with arbitrary rank. IEEE Trans Circuits Syst I, Regul Pap 53:2287–2298
23. Liu D, Xiong X, DasGupta B, Zhang HG (2006) Motif discoveries in unaligned molecular sequences using self-organizing neural networks. IEEE Trans Neural Netw 17:919–928
24. Liu D, Javaherian H, Kovalenko O (2008) Adaptive critic learning techniques for engine torque and air–fuel ratio control. IEEE Trans Syst Man Cybern, Part B, Cybern 38:988–993
25. Puskorius GV, Feldkamp LA, Davis LL (1996) Dynamic neural network methods applied to on-vehicle idle speed control. Proc IEEE 84:1407–1420
26. Ramjee R, Towsley D, Nagarajan R (1997) On optimal call admission control in cellular networks. Wirel Netw 3:29–41
27. Rappaport SS, Purzynski C (1996) Prioritized resource assignment for mobile cellular communication systems with mixed services and platform types. IEEE Trans Veh Technol 45:443–458
28. Sampath A, Holtzman JM (1997) Access control of data in integrated voice/data CDMA systems: benefits and tradeoffs. IEEE J Sel Areas Commun 15:1511–1526
29. Shin SM, Cho CH, Sung DK (1999) Interference-based channel assignment for DS-CDMA cellular systems. IEEE Trans Veh Technol 48:233–239
30. Veeravalli VV, Sendonaris A (1999) The coverage-capacity tradeoff in cellular CDMA systems. IEEE Trans Veh Technol 48:1443–1450

31. Visnevski NA, Prokhorov DV (1996) Control of a nonlinear multivariable system with adaptive critic designs. In: Proceedings of conference on artificial neural networks in engineering, St Louis, MO, pp 559–565
32. Viterbi AJ, Viterbi AM, Zehavi E (1994) Other-cell interference in cellular power-controlled CDMA. IEEE Trans Commun 42:1501–1504
33. Werbos PJ, McAvoy T, Su T (1992) Neural networks, system identification, and control in the chemical process industries. In: White DA, Sofge DA (eds) Handbook of intelligent control: neural, fuzzy, and adaptive approaches, Van Nostrand Reinhold, New York, NY

Index

H. Zhang et al., *Adaptive Dynamic Programming for Control*,
Communications and Control Engineering, DOI 10.1007/978-1-4471-4757-2,